Wolfgang Hackbusch

Multi-Grid Methods and Applications

With 43 Figures and 48 Tables

Springer-Verlag
Berlin Heidelberg New York Tokyo

Wolfgang Hackbusch
Institut für Informatik und Praktische Mathematik,
Christian-Albrechts-Universität Kiel
Olshausenstraße 40–60, D-2300 Kiel 1
Federal Republic of Germany

Mathematics Subject Classification (1980):
65F10, 65F15, 65N20, 65N25, 65N30, 65R05, 35J99, 35P99, 45-00

ISBN 3-540-12761-5 Springer-Verlag Berlin Heidelberg New York Tokyo
ISBN 0-387-12761-5 Springer-Verlag New York Heidelberg Berlin Tokyo

Printed in Germany

Typesetting: Daten- und Lichtsatz-Service Würzburg
Offsetprinting: Ruksaldruck, Berlin. Bookbinding: B. Helm, Berlin
2141/3020-543210

Preface

Although multi-grid methods had already been described in the early 1960's, it was not until the mid-seventies that they were realised to be very efficient methods of solution with a broad area of application. Over the past ten years, the number of publications on this topic has grown rapidly with the result that in searching for information about multi-grid methods at the present time, one is confronted with an abundance of articles scattered in a wide range of periodicals and proceedings.

This monograph is an attempt to describe the basic concepts of multi-grid methods. Different groups of readers may be interested in different parts of the book. The first part, which concentrates on the algorithmic details, is intended for readers interested in the theory and practice of multi-grid methods. The second part is devoted to the mathematical analysis, and is especially intended for mathematicians. Readers interested in engineering and technical applications will find numerous chapters on specific multi-grid applications and additional techniques: the primary emphasis is on applications in the field of fluid dynamics. A special chapter is dedicated to the multi-grid algorithms for integral equations, a topic which up until now has found little recognition in the literature.

I would like to thank numerous colleagues here and abroad for the many fruitful discussions and ideas they have shared with me. My special thanks to my colleagues S. McKee, G. Shaw, and S. Trickett at Oxford for anglicising my manuscript. I would also like to thank Mrs. B. Koberling for carefully typing the major portion of the manuscript. I am very grateful to my colleagues D. Braess and R. Verfürth and my associates G. Hofmann and G. Wittum for their help in proofreading. Last but not least, I would like to thank Springer-Verlag for their friendly cooperation.

Kiel, June 1985 W. Hackbusch

Contents

1. Preliminaries

1.1 Introduction

The numerical solution of boundary value problems is indispensable in almost all fields of physics and engineering sciences. The recent development, e.g. the study of three-dimensional problems, leads to systems of a larger and larger number of equations. Although the computers have become faster and vector computers are available, new numerical methods are required. A step in this direction was the development of fast Poisson solvers in the late sixties. At that time it seemed that there exist faster numerical methods the simpler the discrete elliptic problem. The first multi-grid methods have also been applied to Poisson's equation and show an efficiency similar to that of the direct solvers. But differently from other numerical methods, the efficiency is not lost when more involved problems are to be solved.

The characteristic feature of the multi-grid iteration is its fast convergence. The convergence speed does not deteriorate when the discretisation is refined, whereas classical iterative methods slow down for decreasing grid size. As a consequence one obtains an acceptable approximation of the discrete problem at the expense of computational work proportional to the number of unknowns, which is also the number of the equations in the system. It is not only the complexity which is optimal, also the constant of proportionality is so small that other methods can hardly surpass the multi-grid efficiency.

The previous characterisation does not mean that there is a fixed multi-grid algorithm applying to all boundary value problems. There is rather a multi-grid technique fixing only the framework of the algorithm. The efficiency of the multi-grid algorithm depends on the adjustment of its components to the problem in question. Therefore it is one task of this book to describe the basic framework and the scope of several of the different multi-grid components.

The book is divided into five parts:

Part I	(§§ 2–5):	linear multi-grid algorithms
Part II	(§§ 6–8):	convergence analysis
Part III	(§§ 9–12):	special multi-grid applications
Part IV	(§§ 13–15):	additional techniques
Part V	(§§ 16):	application to integral equations

Part I presents the basic algorithmic multi-grid concept almost without theoretical considerations. The exception is § 2, where we analyse in some detail a one-dimensional problem in order to understand the characteristic behaviour of multi-grid iterations.

To contradict the impression the multi-grid algorithms are based only on heuristical considerations, Part II is devoted to a comprehensive convergence analysis.

The adaptation of multi-grid algorithms to more general elliptic problems is described in Part III. Here the reader finds applications to nonlinear problems, to convection diffusion equations, and to the important Stokes and Navier-Stokes equations.

In Part IV we suggest numerical techniques which can easily be combined with the multi-grid solution process to yield an extra efficiency.

Less attention has been paid to the application of multi-grid algorithms to integral equations. As described in Part V the extremely fast multi-grid iteration of the second kind applies not only to common integral equations but also to more general equations of the second kind, which leads to several interesting applications.

1.2 Notation

The formulae are numbered separately in each subsection. For instance, the tenth formula of Subsection 3.1 is denoted by (3.1.10). Within the same subsection we refer to this formula as

(10),

whereas we write

(1.10)

in other parts of Sect. 3. The full notation

(3.1.10)

is used outside of Sect. 3.

Below we list those symbols which have a special meaning throughout the book.

a, $a(\cdot,\cdot)$	sesqui-linear form; see § 1.4.5
$\mathbb{C}$	set of complex numbers
C, C_0, C_1, etc.	generic constants
$C^s(\Omega)$	Hölder spaces; see § 1.4.1
d	dimension of Ω; see § 3.1.1
f, f_l	right-hand sides; see § 1.3.1, § 3.1.1

$\mathscr{F}_l$	linear space of right-hand functions; see § 6.1
$\mathscr{F}_l(\varrho)$	see § 9.5.1
h_l	discretisation parameter (grid size) at level l; see § 2.1
$\mathscr{H}$, $\mathscr{H}^1$	Hilbert space, on which the variational formulation is based; see § 3.1.3, § 6.3.1.1
$\mathscr{H}^s$	see § 6.3.1.1
$\mathscr{H}_l$	finite element subspace of $\mathscr{H}$; see § 3.1.3
$H^k(\Omega)$, $H_0^k(\Omega)$, $H^s(\Omega)$, etc.	Sobolev spaces; see § 1.4.2
I	identity mapping, identity matrix
l	level number; see § 2.1
L, L_Ω, L_Γ	operators associated with the boundary value problem; see § 3.1.1, § 3.1.3
L_l	matrix of the linear system; see § 1.3.1
L_l, $L_l(v_l)$	Jacobian of $\mathscr{L}_l$ (at v_l); see § 9.1
$\mathscr{L}_l$, $\mathscr{L}_l(u_l)$	nonlinear operator at level l; see § 9.1
m	$2m$ is the order of the differential operator; see § 3.1.1
M_l, M_l', $M_l(v_1, v_2)$, etc.	iteration matrix (in particular of the two- and multi-grid iteration); see § 1.3.1, § 6.1.1
$\mathbb{N}$	set of natural numbers
N_l	see § 1.3.1
$\mathscr{N}_l$, $\mathscr{N}_l(\lambda)$, $\mathscr{N}_l^*$, etc.	eigenspaces; see § 12.1.1
n_l	number of variables at level l; see § 4.3
0	zero matrix
$O(\cdot)$, $o(\cdot)$	Landau symbols
p	prolongation from level $l-1$ to l; see § 2.3
$\tilde{p}$	see § 5.1
P_l	prolongation P_l: $\mathscr{U}_l \to \mathscr{H}_l$ in the finite element case; see § 3.1.3, § 3.6
$\hat{P}_l$	see § 3.6
r	restriction r: $\mathscr{F}_l \to \mathscr{F}_{l-1}$; see § 2.3
$\tilde{r}$	restriction $\tilde{r}$: $\mathscr{U}_l \to \mathscr{U}_{l-1}$; see § 9.3.2
r_{inj}	trivial injection; see § 3.5
$\mathbb{R}$	set of real numbers
R_l	restriction R_l: $\mathscr{H}' \to \mathscr{F}_l$; see § 3.6
$\hat{R}_l$	see § 3.6
$\tilde{R}_l$	restriction $\tilde{R}_l$: $\mathscr{U} \to \mathscr{U}_l$; see § 14.1
$\mathscr{S}_l$, $\mathscr{S}_l^{(\nu)}$, $\mathscr{S}_l^\nu$	linear or nonlinear smoothing method; see § 2.2, § 3.2

S_l	iteration matrix of $\mathscr{S}_l$; see § 2.2
$s_{\mathscr{F}}, s_{\mathscr{U}}$	see (6.2.7)
T_l	see § 2.2
$\mathscr{T}_l$	finite element triangulation; see § 3.8
u_l	discrete solution; see § 1.3.1, § 9.1
$\tilde{u}_l$	approximation to u_l
u_l^i	i^{th} iterate; see § 1.3.1
$\mathscr{U}_l(\varrho)$	see § 9.5.1
$\mathbf{x}, \mathbf{x}', \mathbf{y}$ etc.	spatial vectors in $\mathbb{R}^d$
(x, y)	coordinates in $\mathbb{R}^2$
$\mathbb{Z}$	set of integers
α	exponent involved in the smoothing and approximation properties; see § 6.1.3
γ	number of iterations on the coarser grid; see § 4.1
Γ	boundary of Ω
ζ, ζ_ν	contraction number (i.e. bound on the norm of the iteration matrix); see § 1.3.2, § 2.4
$\eta(\nu)$	function involved in the smoothing property; see § 6.1.3
$\eta_0(s)$	function defined in (6.2.1)
κ	often: consistency order; see § 5.2
ν	number of smoothing iterations; see § 3.2
ν_1	number of pre-smoothing iterations
ν_2	number of post-smoothing iterations
$\varrho(A)$	spectral radius of A; see § 1.3.1
ϱ, ϱ_ν	bound on the convergence rate; see § 2.4
ϱ_B	smoothing rate; see § 2.6.3, § 8.2.3
ϱ_L	smoothing number; see § 2.6.3
$\sigma(A)$	spectrum of A; see § 1.3.1
ω, ω_l	relaxation factor; see § 3.3
$\boldsymbol{\omega}$	vector of relaxation factors for semi-iterative methods; see § 3.3.5
Ω	domain of the boundary value problem
Ω_l	grid at level l
$\Omega_l^i, \Omega_l^r, \Omega_l^b$	subsets of Ω_l; see § 3.3.1
$(\cdot,\cdot)_{\mathscr{U}}, \langle\cdot,\cdot\rangle_{\mathscr{U}}$	scalar product in a Hilbert space $\mathscr{U}$; see § 1.4.5
$(\cdot,\cdot)$	scalar product, in particular, of $L^2(\Omega)$
$\langle\cdot,\cdot\rangle$	scalar product in $\mathscr{U}_l$; see § 3.5

$\|\cdot\|_{\mathscr{U}}, \|\cdot\|_{\mathscr{F}}$	norms of $\mathscr{U}_l$ and $\mathscr{F}_l$, respectively; see § 6.1.1
$\vert\cdot\vert_0$	Euclidean norm on $\mathscr{U}_l$ and $\mathscr{F}_l$; see § 1.3.2, § 6.2.1
$\vert\cdot\vert_s$	discrete Sobolev norm of order sm; see § 2.6.2, § 6.2.1
$\|\cdot\|_{\mathscr{F}\leftarrow\mathscr{U}}, \|\cdot\|_{\mathscr{U}\leftarrow\mathscr{U}}$, etc.	matrix norms associated with $\|\cdot\|_{\mathscr{U}}$ and $\|\cdot\|_{\mathscr{F}}$; see § 1.3.2
$\vert\cdot\vert_{s\leftarrow t}$	matrix norm associated with $\vert\cdot\vert_s$ and $\vert\cdot\vert_t$; see § 6.2.1
$\|\cdot\|$	spectral norm ($\|\cdot\| = \vert\cdot\vert_{0\leftarrow 0}$); see § 1.3.2

1.3 Some Elements from Linear Algebra

For the sake of completeness some notation, definitions, and statements from the field of linear algebra are given. In the first subsection we recall some results concerning linear iterations. Thereafter we remind the reader of the definition of norms, matrix norms, and scalar products.

1.3.1 Analysis of Iterative Processes

Let $\mathscr{U}_l$ be an n_l-dimensional vector space [1] and consider a system

$$L_l u_l = f_l \qquad (u_l, f_l \in \mathscr{U}_l) \tag{1.3.1}$$

of n_l linear equations. An iteration for solving Eq. (1) is a process $u_l^0 \mapsto u_l^1 \mapsto \ldots \mapsto u_l^{j-1} \mapsto u_l^j \mapsto \ldots$ generated by some linear mapping φ_l:

$$u_l^{j+1} = \varphi_l(u_l^j, f_l).$$

The multi-grid iteration and also the well-known classical iterations are of this form. As φ_l is linear it can be represented by $\varphi_l(u_l, f_l) = M_l u_l + N_l f_l$. The iteration becomes

$$u_l^{j+1} = M_l u_l^j + N_l f_l. \tag{1.3.2}$$

M_l is called the iteration matrix. The explicit representation of the jth iterate u_l^j is

$$u_l^j = M_l^j u_l^0 + N_l^{(j)} f_l, \qquad N_l^{(j)} = \sum_{k=0}^{j-1} M_l^k N_l. \tag{1.3.3}$$

An obvious condition on the iteration (2) is that the solution of (1) is a fixed point of (2):

$$u_l = M_l u_l + N_l f_l \qquad (u_l \text{ is the solution of (1)}). \tag{1.3.4}$$

1 Usually, we shall consider real vector spaces. Complex spaces are needed only in connection with Fourier transforms

Assuming (4) for all $f_l \in \mathscr{U}_l$ one obtains

$$I = M_l + N_l L_l \tag{1.3.5}$$

(I always denotes the identity matrix). When L_l is non-singular, Eq. (5) yields the representation

$$N_l = (I - M_l)\, L_l^{-1} \tag{1.3.6}$$

of N_l. Hence, the iteration is completely described by means of its iteration matrix M_l.

The error of the jth iterate u_l^j is $u_l^j - u_l$. From (2) and (3) one concludes

$$u_l^{j+1} - u_l = M_l(u_l^j - u_l), \tag{1.3.7 a}$$

(where u_l is the solution of (1))

$$u_l^j - u_l = M_l^j(u_l^0 - u_l). \tag{1.3.7 b}$$

Thus, the iteration matrix is the amplification matrix of the error.

The *spectral radius* of a matrix A is defined by

$$\varrho(A) = \max\{|\lambda|: \lambda \in \sigma(A)\}, \tag{1.3.8 a}$$

where

$$\sigma(A) = \{\lambda \in \mathbb{C}: \lambda \text{ is an eigenvalue of } A\} \tag{1.3.8 b}$$

is the spectrum of A. The spectral radius is of essential importance because of

Lemma 1.3.1. The iteration (2) converges for all initial guesses $u_l^0 \in \mathscr{U}_l$ if and only if $\varrho(M_l) < 1$.

The spectral radius of the iteration matrix M_l is called the *rate of convergence*. The convergence rate governs the asymptotic behaviour of the errors $u_l^j - u_l$. Since fast iterative processes should terminate after a very small number of iterations, the spectral radius is not suited for describing such iterations. To estimate the error after a few iterations, we need norm estimates.

1.3.2 Norms

$\|\cdot\|_{\mathscr{U}}$ is a *norm* on a vector space $\mathscr{U}$, if

$$\|u+v\|_{\mathscr{U}} \leqq \|u\|_{\mathscr{U}} + \|v\|_{\mathscr{U}} \qquad \text{for all } u, v \in \mathscr{U},$$
$$\|\lambda u\|_{\mathscr{U}} = |\lambda|\ \|u\|_{\mathscr{U}} \qquad \text{for all } u \in \mathscr{U},\ \lambda \in \mathbb{R}\ (\mathbb{C}, \text{resp.}),$$
$$\|u\|_{\mathscr{U}} > 0 \qquad \text{for all } 0 \neq u \in \mathscr{U}.$$

A *scalar product* on $\mathscr{U} \times \mathscr{U}$ satisfies

$$\langle u + \lambda v, w\rangle_{\mathscr{U}} = \langle u, w\rangle_{\mathscr{U}} + \lambda\,\langle v, w\rangle_{\mathscr{U}} \qquad \text{for all } u, v, w \in \mathscr{U},\ \lambda \in \mathbb{R}\ (\mathbb{C}),$$
$$\langle u, v\rangle_{\mathscr{U}} = \overline{\langle v, u\rangle_{\mathscr{U}}} \qquad \text{for all } u, v \in \mathscr{U},$$
$$\langle u, u\rangle_{\mathscr{U}} > 0 \qquad \text{for all } 0 \neq u \in \mathscr{U}.$$

$(\mathscr{U}, \|\cdot\|_{\mathscr{U}})$ is a Hilbert space [2], if $\|\cdot\|_{\mathscr{U}}$ is generated by a scalar product:

$$\|u\|_{\mathscr{U}} := \sqrt{\langle u, u\rangle_{\mathscr{U}}}. \tag{1.3.9}$$

Example 1.3.2. Let u_i, $1 \leqq i \leqq n$, be the components of the vector $u \in \mathscr{U}$. The *Euclidean scalar product* and the *Euclidean norm* are defined by

$$\langle u, v\rangle_{\mathscr{U}} = c \sum_{i=1}^{n} u_i \bar{v}_i, \qquad \|u\|_{\mathscr{U}} = \sqrt{c \sum_{i=1}^{n} |u_i|^2}, \tag{1.3.10}$$

where c is a suitable scaling factor.

Let $\mathscr{F}$ be a second vector space with norm $\|\cdot\|_{\mathscr{F}}$. We include the case $\mathscr{U} = \mathscr{F}$ with $\|\cdot\|_{\mathscr{U}} = \|\cdot\|_{\mathscr{F}}$ or $\|\cdot\|_{\mathscr{U}} \neq \|\cdot\|_{\mathscr{F}}$. The associated matrix norm of a linear mapping $A: \mathscr{U} \to \mathscr{F}$ is

$$\|A\|_{\mathscr{F}\leftarrow\mathscr{U}} := \sup_{0 \neq u \in \mathscr{U}} \frac{\|Au\|_{\mathscr{F}}}{\|u\|_{\mathscr{U}}}. \tag{1.3.11}$$

Analogously, $\|\cdot\|_{\mathscr{U}\leftarrow\mathscr{F}}$, $\|\cdot\|_{\mathscr{U}\leftarrow\mathscr{U}}$, $\|\cdot\|_{\mathscr{F}\leftarrow\mathscr{F}}$ are defined for mappings $\mathscr{F} \to \mathscr{U}$, $\mathscr{U} \to \mathscr{U}$, $\mathscr{F} \to \mathscr{F}$, respectively.

Definition (11) implies the inequality

$$\|Au\|_{\mathscr{F}} \leqq \|A\|_{\mathscr{F}\leftarrow\mathscr{U}} \|u\|_{\mathscr{U}} \qquad (A: \mathscr{U} \to \mathscr{F},\ u \in \mathscr{U}).$$

Moreover, the associated matrix norms are submultiplicative. For three normed spaces $(\mathscr{U}, \|\cdot\|_{\mathscr{U}})$, $(\mathscr{V}, \|\cdot\|_{\mathscr{V}})$, $(\mathscr{W}, \|\cdot\|_{\mathscr{W}})$ and mappings $A: \mathscr{V} \to \mathscr{W}$, $B: \mathscr{U} \to \mathscr{V}$ the inequality

$$\|AB\|_{\mathscr{W}\leftarrow\mathscr{U}} \leqq \|A\|_{\mathscr{W}\leftarrow\mathscr{V}} \|B\|_{\mathscr{V}\leftarrow\mathscr{U}} \tag{1.3.12a}$$

is valid. In particular,

$$\|A^n\|_{\mathscr{U}\leftarrow\mathscr{U}} \leqq \|A\|^n_{\mathscr{U}\leftarrow\mathscr{U}} \qquad (A: \mathscr{U} \to \mathscr{U},\ n \geqq 0, \text{ an integer}) \tag{1.3.12b}$$

holds. The norm bounds the spectral radius:

$$\varrho(A) \leqq \|A\|_{\mathscr{U}\leftarrow\mathscr{U}}. \tag{1.3.13}$$

Let $(\mathscr{U}, \langle\cdot,\cdot\rangle_{\mathscr{U}})$ and $(\mathscr{V}, \langle\cdot,\cdot\rangle_{\mathscr{V}})$ be Hilbert spaces. $A: \mathscr{U} \to \mathscr{V}$ has the adjoint mapping $A^*: \mathscr{V} \to \mathscr{U}$ defined by

$$\langle Au, v\rangle_{\mathscr{V}} = \langle u, A^*v\rangle_{\mathscr{U}} \qquad \text{for all } u \in \mathscr{U},\ v \in \mathscr{V}. \tag{1.3.14}$$

Note that A^* depends on the choice of the scalar products.

Example 1.3.3. Let $\langle\cdot,\cdot\rangle_{\mathscr{U}}$, $\langle\cdot,\cdot\rangle_{\mathscr{V}}$ be as in Example 1.3.2 with $c = c_{\mathscr{U}}$ and $c = c_{\mathscr{V}}$, respectively. Then $A^* = (c_{\mathscr{V}}/c_{\mathscr{U}})\bar{A}^T$, where A^T is the transposed matrix.

The *spectral norm* $\|\cdot\|_{\mathscr{U}\leftarrow\mathscr{U}}$ is the matrix norm associated with the Euclidean norm $\|\cdot\|_{\mathscr{U}}$ (cf. (10)). It is independent of the scaling factor c in $\|\cdot\|_{\mathscr{U}}$. The name 'spectral norm' originates from

2 Here, $\dim(\mathscr{U}) < \infty$ is assumed. Otherwise, $(\mathscr{U}, \|\cdot\|_{\mathscr{U}})$ must be complete

Lemma 1.3.4. Let $A: \mathcal{U} \to \mathcal{U}$, $(\mathcal{U}, \|\cdot\|_{\mathcal{U}})$ a Hilbert space. Then

$$\|A\|_{\mathcal{U}\leftarrow\mathcal{U}} = \sqrt{\varrho(A^*A)}, \tag{1.3.15a}$$

$$\|A\|_{\mathcal{U}\leftarrow\mathcal{U}} = \varrho(A) \quad \text{if } A \text{ is normal (i.e. } AA^* = A^*A). \tag{1.3.15b}$$

In particular, (15 b) holds for symmetric operators $A = A^*$.

Proof. cf. Varga [1, p. 11]. □

Taking norms of both sides of Eqs. (7 a, b) we get

$$\|u_l^{j+1} - u_l\|_{\mathcal{U}} \leqq \|M_l\|_{\mathcal{U}\leftarrow\mathcal{U}} \, \|u_l^j - u_l\|_{\mathcal{U}}, \tag{1.3.16a}$$

$$\|u_l^j - u_l\|_{\mathcal{U}} \leqq \|M_l\|_{\mathcal{U}\leftarrow\mathcal{U}}^j \, \|u_l^0 - u_l\|_{\mathcal{U}} \tag{1.3.16b}$$

(cf. (12 b)). Obviously $\|M_l\|_{\mathcal{U}\leftarrow\mathcal{U}} < 1$ implies convergence. The norm $\|M_l\|_{\mathcal{U}\leftarrow\mathcal{U}}$ is called the *contraction number* (with respect to $\|\cdot\|_{\mathcal{U}}$).

1.3.3 Symmetric Matrices

Let $(\mathcal{U}, \|\cdot\|_{\mathcal{U}})$ be a Hilbert space over the field $\mathbb{R}$ (or $\mathbb{C}$). A mapping $A = A^*: \mathcal{U} \to \mathcal{U}$ is called *symmetric* (or *Hermitian*, resp.). Since real spaces are the standard cases in our applications we shall always use the term 'symmetric'.

A matrix A is called *positive definite* (*positive semi-definite*/*negative definite*/*negative semi-definite*) if A is symmetric with eigenvalues being positive ($\geqq 0$, < 0, $\leqq 0$, respectively). The notation

$$A > B \tag{1.3.17}$$

signifies that A and B are symmetric and that $A - B$ is positive definite. $A \geqq B$, $A < B$, $A \leqq B$ are defined analogously. $A > 0$ (0: zero matrix) is an abbreviation for "A is positive definite".

A symmetric matrix admits a factorisation $A = QDQ^*$, where Q is unitary (i.e. $Q^{-1} = Q^*$) and $D = \operatorname{diag}\{\lambda_1, \dots, \lambda_n\}$ is diagonal and real. Let $f: \sigma(A) \to \mathbb{C}$ be a mapping defined on the spectrum $\sigma(A)$ of A (cf. (8 b)). The matrix $f(A)$ is defined by

$$f(A) = Q \operatorname{diag}\{f(\lambda_1), \dots, f(\lambda_n)\} \, Q^*. \tag{1.3.18}$$

The choices $f(x) = \sqrt{x}$ and $f(x) = x^s$ yield the first parts of

Lemma 1.3.5. Assume $A = A^*$.

(i) $A = A^* \geqq 0$ admits a unique symmetric square root $A^{1/2} \geqq 0$.

(ii) General powers $A^s \geqq 0$ ($s \geqq 0$ or $s \in \mathbb{R}$) are well-defined if $A \geqq 0$ or $A > 0$, respectively. The norms are

$$\|A^s\|_{\mathcal{U}\leftarrow\mathcal{U}} = \|A\|_{\mathcal{U}\leftarrow\mathcal{U}}^s \qquad (A \geqq 0,\ s \geqq 0). \tag{1.3.19a}$$

(iii) Let f be defined on the spectrum of the normal matrix A. Then

$$\|f(A)\|_{\mathcal{U}\leftarrow\mathcal{U}} = \max\{|f(\lambda)|: \lambda \text{ is an eigenvalue of } A\}. \tag{1.3.19b}$$

1.4 Some Elements from Functional Analysis

In this section notation describing the space of continuous, k-times continuously differentiable and Hölder continuous functions is introduced (§ 1.4.1). A precise description of the properties of elliptic boundary value problems requires the use of Sobolev spaces (§ 1.4.2); furthermore, the formulation of regularity results needs spaces of negative and non-integer order (§ 1.4.3, § 1.4.4). The multi-grid convergence proof will use interpolation techniques that are provided in § 1.4.4. The variational formulation of a boundary value problem as well as the finite element method are based on bilinear or sesqui-linear forms introduced in § 1.4.5.

1.4.1 Continuous Functions and Hölder Spaces

Let $\Omega \subset \mathbb{R}^d$ be an open subset with closure $\bar{\Omega}$. The linear space of continuous functions $\Omega \to \mathbb{R}^m$ (or $\mathbb{C}^m$) is denoted by $C(\Omega, \mathbb{R}^m)$ or simply by $C(\Omega)$. Partial derivatives of order k can be written as $D^{\mathbf{v}} u(x)$, where $\mathbf{v} = (v_1, \ldots, v_d)$, $v_j \geqq 0$, is a multi-index of length

$$|\mathbf{v}| = v_1 + \ldots + v_d \tag{1.4.1}$$

being equal to k and where

$$D^{\mathbf{v}} = (\partial/\partial x_1)^{v_1} (\partial/\partial x_2)^{v_2} \ldots (\partial/\partial x_d)^{v_d}. \tag{1.4.2}$$

The space of k-times continuously differentiable functions is

$$C^k(\Omega) = \{u\colon D^{\mathbf{v}} u \text{ exists and belongs to } C(\Omega) \text{ for all } |\mathbf{v}| \leqq k\}$$

where by $C^0(\Omega)$ we mean $C(\Omega)$. Functions with finite norm

$$\|u\|_{C^k(\Omega)} = \|u\|_{C^k} := \sup\{|D^{\mathbf{v}} u(x)|_\infty\colon |\mathbf{v}| \leqq k,\ x \in \Omega\}$$

($|\cdot|_\infty$ maximum norm on $\mathbb{R}^m$) form the subset

$$\bar{C}^k(\Omega) = \{u \in C^k(\Omega)\colon \|u\|_{C^k} < \infty\}$$

equipped with the norm $\|\cdot\|_{C^k}$. $(\bar{C}^k(\Omega), \|\cdot\|_{C^k})$ is a Banach space (i.e. a normed and complete vector space).

Functions $u \in C(\Omega)$ are (globally) *Hölder continuous* with exponent $\lambda \in (0, 1)$ if $\|u\|_{C^\lambda} < \infty$, where

$$\|u\|_{C^\lambda} = \max\{\|u\|_{C^0}, \sup\{|u(x) - u(y)|_\infty / |x - y|_\infty^\lambda\colon x, y \in \Omega,\ x \neq y\}\}.$$

The *Hölder space* of order k (and exponent $\lambda \in (0, 1)$) is

$$\bar{C}^{k+\lambda}(\Omega) = \{u \in C^k\colon \|u\|_{C^{k+\lambda}} < \infty\} \qquad (0 < \lambda < 1,\ 0 \leqq k, \text{ an integer})$$

with norm

$$\|u\|_{C^{k+\lambda}(\Omega)} = \|u\|_{C^{k+\lambda}} = \max\{\|D^{\mathbf{v}} u\|_{C^\lambda}\colon |\mathbf{v}| \leqq k\} \qquad (0 < \lambda < 1,\ 0 \leqq k, \text{ an integer}).$$

Hence, $\bar{C}^s(\Omega)$ is well-defined for all $s \geqq 0$.

For the set of linear and continuous mappings from one Banach space $\mathscr{U}$ into another space $\mathscr{V}$ we use the notation $L(\mathscr{U}, \mathscr{V})$.

1.4.2 Sobolev Spaces

$L^p(\Omega)$ or more precisely $L^p(\Omega, \mathbb{R}^m)$ or $L^p(\Omega, \mathbb{C}^m)$, $1 \leqq p \leqq \infty$, is the space of measurable functions with finite norm

$$\| u \|_{L^p(\Omega)} = \| u \|_{L^p} = \begin{cases} \{\int_\Omega |u(x)|_p^p \, dx\}^{1/p} & (1 \leqq p < \infty), \\ \operatorname{vrai\,max}_{x \in \Omega} |u(x)|_p & (p = \infty), \end{cases}$$

where $\operatorname{vrai\,max} |u(x)| = \inf\{\| v \|_{C^0} \colon v = u \text{ a.e.}\}$. $|\cdot|_p$ is the p-norm on $\mathbb{R}^m$ (or $\mathbb{C}^m$): $|v|_p = \left\{ \sum_{i=1}^m |v_i|^p \right\}^{1/p}$ if $1 \leqq p < \infty$, $|v|_\infty = \max_{i=1}^n |v_i|$ if $p = \infty$. $L^p(\Omega)$ is a Banach space for $p \in [1, \infty]$ and a Hilbert space for $p = 2$. In the latter case the scalar product is denoted by round brackets:

$$(u, v) = \int_\Omega \sum_{i=1}^m u_i(x) \overline{v_i(x)} \, dx \qquad (u, v \in L^2(\Omega, \mathbb{R}^m) \text{ or } L^2(\Omega, \mathbb{C}^m)). \tag{1.4.3}$$

If necessary we add the subscript $L^2(\Omega)$: $(\cdot, \cdot)_{L^2(\Omega)}$.

For the definition of derivatives $D^\nu u$ in the sense of distributions we refer e.g. to Adams [1]. If $D^\nu u \in L^p(\Omega)$ for all $|\nu| \leqq k$, u belongs to the Sobolev space $W^{p,k}(\Omega)$ of order k with norm

$$\| u \|_{W^{p,k}} = [\sum_{|\nu| \leqq k} \| D^\nu u \|_{L^p}^p]^{1/p}.$$

For $p = 2$ we write $H^k(\Omega)$ instead of $W^{2,k}(\Omega)$:

$$H^k(\Omega) = \{u \colon D^\nu u \in L^2(\Omega) \text{ for } |\nu| \leqq k\},$$

$$\| u \|_{H^k(\Omega)} = \| u \|_{H^k} = [\sum_{|\nu| \leqq k} \| D^\nu u \|_{L^2(\Omega)}^2]^{1/2}. \tag{1.4.4}$$

$H^0(\Omega)$ coincides with $L^2(\Omega)$. There are various other ways to define the Sobolev space $H^s(\Omega)$ for non-integer order $s \geqq 0$ (cf. § 1.4.4, Adams [1], Lions-Magenes [1]).

An important subspace of $H^s(\Omega)$ is $H_0^s(\Omega)$, which is the completion of $C_0^\infty(\Omega)$ with respect to $\| \cdot \|_{H^s}$ (cf. Lions-Magenes [1, p. 54]).

1.4.3 Dual Spaces [3]

Let $\mathscr{U}$ be a Hilbert space, and $\mathscr{V} \subset \mathscr{U}$ a dense continuously embedded Hilbert subspace (i.e. $\| v \|_{\mathscr{U}} \leqq C \| v \|_{\mathscr{V}}$ for all $v \in \mathscr{V}$ or equivalently $\| I \|_{\mathscr{U} \leftarrow \mathscr{V}} < \infty$).

---

3 This subsection can be skipped by readers not interested in the mathematical details

The *(anti-)dual space*[4] $\mathscr{V}' = L'(\mathscr{V}, \mathbb{C})$ endowed with the *dual norm* $\| v' \|_{\mathscr{V}'} = \sup\{|v'(v)|: v \in \mathscr{V}, \| v \|_{\mathscr{V}} = 1\}$ can be regarded as an extension of $\mathscr{U}$:

$$\mathscr{V}' \supset \mathscr{U} \supset \mathscr{V} \quad \text{with continuous and dense embedding,} \tag{1.4.5}$$

if we identify $u \in \mathscr{U}$ with $v' \in \mathscr{V}'$ defined by $v'(v) := (u, v)_{\mathscr{U}}$. $\mathscr{U}$ is called the *pivot space*

Note 1.4.1. There is a unique extension of the scalar product $(\cdot,\cdot)_{\mathscr{U}}$: $\mathscr{V} \times \mathscr{U} \subset \mathscr{U} \times \mathscr{U} \to \mathbb{C}$ to the dual form[4] on $\mathscr{V} \times \mathscr{V}'$. Therefore, we shall use the notation $(u, v)_{\mathscr{U}}$ for $u, v \in \mathscr{U}$ as well as for $u \in \mathscr{V}$, $v \in \mathscr{V}'$ or $u \in \mathscr{V}'$, $v \in \mathscr{V}$.

$H^s(\Omega)$ and $H_0^s(\Omega)$ are special dense and continuously embedded Hilbert subspaces of $L^2(\Omega)$. The dual spaces of $H^s(\Omega)$ and $H^s(\Omega) \cap H_0^t(\Omega)$ $(0 \leqq t \leqq s)$ are written as $(H^s(\Omega))'$ and $(H^s(\Omega) \cap H_0^t(\Omega))'$, whereas the dual space of $H_0^s(\Omega)$ has the symbol

$$H^{-s}(\Omega) = (H_0^s(\Omega))' \qquad (s \geqq 0).$$

As mentioned in Note 1.4.1, the product $(u, v) = (u, v)_{L^2(\Omega)}$ is also used for e.g., $u \in H^{-s}(\Omega)$ and $v \in H_0^s(\Omega)$.

1.4.4 Hilbert Scales[5]

The results of this subsection are not only needed for the definition of $H^s(\Omega)$, $s \neq$ integer, but also form the background of some technical arguments in § 6.

Given the situation (5) with separable Hilbert spaces $\mathscr{U}$ and $\mathscr{V}$, there is a selfadjoint and positive operator Λ with the following properties:

$$\Lambda: \mathscr{V} \xrightarrow{\text{onto}} \mathscr{U} \quad \text{bounded,} \tag{1.4.6a}$$

$$\Lambda: \mathscr{U} \xrightarrow{\text{onto}} \mathscr{V}' \quad \text{bounded,} \tag{1.4.6b}$$

$$(v, w)_{\mathscr{V}} = (\Lambda v, \Lambda w)_{\mathscr{U}} \qquad \text{for } v, w \in \mathscr{V}, \tag{1.4.6c}$$

$$\| v \|_{\mathscr{V}} = \| \Lambda v \|_{\mathscr{U}} \quad (v \in \mathscr{V}), \qquad \| v' \|_{\mathscr{V}'} = \| \Lambda^{-1} v' \|_{\mathscr{U}} \quad (v' \in \mathscr{V}') \tag{1.4.6d}$$

(cf. Lions-Magenes [1, p. 9][6]). The powers Λ^s, $s \in \mathbb{R}$, are well-defined. The *intermediate space*

$$[\mathscr{V}, \mathscr{U}]_\Theta = \text{domain of } \Lambda^{1-\Theta} \qquad (0 \leqq \Theta \leqq 1) \tag{1.4.7a}$$

endowed with the norm

$$\| u \|_{[\mathscr{V}, \mathscr{U}]_\Theta} = \| \Lambda^{1-\Theta} u \|_{\mathscr{U}} \tag{1.4.7b}$$

satisfies $[\mathscr{V}, \mathscr{U}]_0 = \mathscr{V}$, $[\mathscr{V}, \mathscr{U}]_1 = \mathscr{U}$, and $[\mathscr{V}, \mathscr{U}]_\eta \subset [\mathscr{V}, \mathscr{U}]_\Theta$ for $0 \leqq \eta \leqq \Theta \leqq 1$.

---

4 $\varphi: \mathscr{V} \to C$ is anti-linear if $\bar{\varphi}$ defined by $\bar{\varphi}(v) = \overline{\varphi(v)}$ is linear. The anti-dual space is $\mathscr{V}' = L'(\mathscr{V}, \mathbb{C}) := \{\varphi: \mathscr{V} \to \mathbb{C}$ anti-linear and continuous$\}$. The (anti-)dual form $(\cdot,\cdot)$: $\mathscr{V} \times \mathscr{V}' \to \mathbb{C}$ is defined by $(v, v') = v'(v)$. For simplicity we shall omit the prefix "anti" in the following

5 This subsection can be skipped by readers not interested in the mathematical details

6 The actual operator Λ equals $(I + \Lambda^2)^{1/2}$ with Λ from Lions-Magenes [1]

A family $\{\mathscr{H}^s\colon -t \leqq s \leqq t\}$ of Hilbert spaces $\mathscr{H}^s$ is called a *Hilbert scale*, if

$$\mathscr{H}^{-t} = (\mathscr{H}^t)' \supset \mathscr{H}^0 \supset \mathscr{H}^t \qquad \text{dense and continuously embedded,} \tag{1.4.8a}$$

$$\mathscr{H}^s = [\mathscr{H}^t, \mathscr{H}^{-t}]_{(t-s)/(2t)} \qquad \text{for all} \qquad -t \leqq s \leqq t, \tag{1.4.8b}$$

$$|u|_s := \|u\|_{\mathscr{H}^s} = \|\Lambda^s u\|_{\mathscr{H}^0} \qquad \text{for all } u \in \mathscr{H}^s, \quad -t \leqq s \leqq t, \tag{1.4.8c}$$

for some selfadjoint and positive definite operator Λ. The spaces $\mathscr{H}^s$ are uniquely determined by $\mathscr{H}^0$ and $\mathscr{H}^t$. Λ is uniquely defined by $\|u\|_{\mathscr{H}^t} = \|\Lambda^t u\|_{\mathscr{H}^0}$.

For a proof of the next lemma we refer to Kreĭn-Petunin-Semenov [1] and Lions-Magenes [1]:

Lemma 1.4.2. Let $\{\mathscr{H}^s\colon -t \leqq s \leqq t\}$ be a Hilbert scale. Then

$$[\mathscr{H}^r, \mathscr{H}^s]_\Theta = \mathscr{H}^{r+\Theta(s-r)} \qquad (-t \leqq s \leqq r \leqq t, \quad 0 \leqq \Theta \leqq 1), \tag{1.4.9a}$$

$$(\mathscr{H}^s)' = \mathscr{H}^{-s} \qquad (-t \leqq s \leqq t), \tag{1.4.9b}$$

$$(u, v)_{\mathscr{H}^0} \leqq \|u\|_s \|v\|_{-s} \qquad (-t \leqq s \leqq t, \quad u, v \in \mathscr{H}^{|s|}), \tag{1.4.9c}$$

$$\|u\|_s \leqq \|u\|_q^{(r-s)/(r-q)} \|u\|_r^{(s-q)/(r-q)} \qquad (-t \leqq q \leqq s \leqq r \leqq t, \quad u \in \mathscr{H}^r), \tag{1.4.9d}$$

$$\|u\|_s \leqq \varepsilon \|u\|_r + C(\varepsilon) \|u\|_q \qquad \text{for all } \varepsilon > 0, \quad -t \leqq q \leqq s < r \leqq t, \quad u \in \mathscr{H}^r, \tag{1.4.9e}$$

with $C(\varepsilon) = \left[\dfrac{s-q}{\varepsilon(r-q)}\right]^{(s-q)/(r-s)} \dfrac{r-s}{r-q}$.

The operator norm $\|A\|_{\mathscr{K} \leftarrow \mathscr{H}}$ for $A \in L(\mathscr{H}, \mathscr{K})$ is defined by (3.11).

Lemma 1.4.3. Let $\{\mathscr{H}^s\colon -t \leqq s \leqq t\}$ and $\{\mathscr{K}^s\colon -t \leqq s \leqq t\}$ be two Hilbert scales. $A \in L(\mathscr{H}^q, \mathscr{K}^q)$ and $A \in L(\mathscr{H}^r, \mathscr{K}^r)$ for some $-t \leqq q \leqq r \leqq t$ imply $A \in L(\mathscr{H}^s, \mathscr{K}^s)$ for all $s \in [q, r]$ and

$$\|A\|_{\mathscr{K}^s \leftarrow \mathscr{H}^s} \leqq \|A\|_{\mathscr{K}^q \leftarrow \mathscr{H}^q}^{\frac{r-s}{r-q}} \|A\|_{\mathscr{K}^r \leftarrow \mathscr{H}^r}^{\frac{s-q}{r-q}} \qquad (-t \leqq q \leqq s \leqq r \leqq t). \tag{1.4.10a}$$

If $\{\mathscr{H}^s\}$ is a Hilbert scale, then the spaces $\mathscr{K}^s := \mathscr{H}^{(q'(r-s)+r'(s-q))/(r-q)}$ form also a Hilbert scale, provided that $r' \geqq q'$ and $r > q$. Hence (10a) yields

$$\|A\|_{\mathscr{H}^{s'} \leftarrow \mathscr{H}^s} \leqq \|A\|_{\mathscr{H}^{q'} \leftarrow \mathscr{H}^q}^{\alpha} \|A\|_{\mathscr{H}^{r'} \leftarrow \mathscr{H}^r}^{1-\alpha} \tag{1.4.10b}$$

for $r' \geqq q'$, $r \geqq q$, $0 \leqq \alpha \leqq 1$, $s = \alpha q + (1-\alpha) r$, $s' = \alpha q' + (1-\alpha) r'$.

If a family $\{\mathscr{H}^s\colon -t \leqq s \leqq t\}$ satisfies (8b) but not necessarily (8a, c), then Eq. (9a) is still valid, whereas (9d, e) and (10a, b) hold with extra factors C (depending on s, q, r) on the right sides. If (8a) is satisfied as well, then (9b) and (9c) with a possible factor $C(s)$ on the right side are also valid.

The Sobolev spaces of order s almost form a Hilbert scale.

Proposition 1.4.4. Let the boundary of Ω be smooth enough.

(i) The scale

$$\mathscr{H}^s = H^s(\Omega) \quad (s \geqq 0), \qquad \mathscr{H}^s = (H^{-s}(\Omega))' \quad (s < 0) \tag{1.4.11 a}$$

satisfies (8 a, b) for arbitrary t. The original norms of $H^s(\Omega)$, $s =$ integer, differ from (8 c) but both norms are equivalent.

(ii) The scale

$$\mathscr{H}^s = H_0^s(\Omega) \quad (s \geqq 0), \qquad \mathscr{H}^s = H^s(\Omega) \quad (s \leqq 0) \tag{1.4.11 b}$$

has the same property except when $s = \frac{1}{2} +$ an integer.

(iii) The scale

$$\mathscr{H}^s = \begin{cases} H_0^s(\Omega) & (0 \leqq s \leqq m) \\ H^s(\Omega) \cap H_0^m(\Omega) & (m \leqq s) \end{cases}, \qquad \mathscr{H}^s = H^s(\Omega) \quad (s \leqq 0) \tag{1.4.11 c}$$

has the same property except when $s = k + \frac{1}{2}$ for integral $k < m$.

For a proof we refer to Lions-Magenes [1, pp. 43, 64].

The following result about orthogonal projections will be used in § 6.3.1.

Proposition 1.4.5. Let $\{\mathscr{H}^s\colon -\infty < s \leqq t\}$ be a Hilbert scale and $\mathscr{H} \subset \mathscr{H}^t$ a finite dimensional subspace. By Q_s we denote the projection onto $\mathscr{H}$, which is orthogonal with respect to the norm of $\mathscr{H}^s$, i.e.

$$v^* = Q_s u \in \mathscr{H} \ \text{ satisfies } \ \| v^* - u \|_{\mathscr{H}^s} \leqq \| v - u \|_{\mathscr{H}^s} \quad \text{for all } v \in \mathscr{H}. \tag{1.4.12}$$

Suppose the existence of some numbers $r < t$ and $\eta \geqq 0$ such that

$$\| I - Q_r \|_{\mathscr{H}^r \leftarrow \mathscr{H}^t} \leqq \eta^{t-r}. \tag{1.4.13 a}$$

Then

$$\| I - Q_s \|_{\mathscr{H}^\sigma \leftarrow \mathscr{H}^\tau} \leqq \eta^{\tau-\sigma} \quad \text{for all } 2s - t \leqq \sigma \leqq s \leqq \tau \leqq t,\ s \leqq r. \tag{1.4.13 b}$$

Proof (cf. Babuška-Aziz [1]). We use the following abbreviations:

$$P_s = I - Q_s, \quad |\cdot|_p = \|\cdot\|_{\mathscr{H}^p}, \quad |\cdot|_{p \leftarrow q} = \|\cdot\|_{\mathscr{H}^p \leftarrow \mathscr{H}^q}.$$

Step 1. Interpolation of $|P_r|_{r \leftarrow t} \leqq \eta^{t-r}$ and $|P_r|_{r \leftarrow r} = 1$ yields $|P_r|_{r \leftarrow \tau} \leqq \eta^{\tau - r}$ for all $r \leqq \tau \leqq t$ (cf. (10 b)). $\tau' \in [r, t]$ implies $\sigma := 2r - \tau' \in [2r - t, r]$. $\mathscr{H}^\sigma$ is the dual space of $\mathscr{H}^{\tau'}$ with respect to the pivot space $\mathscr{H}^r$, and P_r is selfadjoint in $\mathscr{H}^r$. Hence, $|P_r|_{\sigma \leftarrow r} = |P_r|_{r \leftarrow \tau'} \leqq \eta^{\tau' - r} = \eta^{r - \sigma}$ proves that

$$|P_r|_{\sigma \leftarrow \tau} = |P_r^2|_{\sigma \leftarrow \tau} = |P_r|_{\sigma \leftarrow r}\, |P_r|_{r \leftarrow \tau} \leqq \eta^{r-\sigma} \eta^{\tau - r} = \eta^{\tau - \sigma}$$

for all $2r - t \leqq \sigma \leqq r \leqq \tau \leqq t$, i.e. (13 b) holds for $s = r$.

Step 2. We show that (13 b) for $s = r$ implies (13 b) for all $s \in [2r - t, r]$. Let $s \in [2r - t, r]$ and $u \in \mathscr{H}^t$. By

$$|P_s u|_s = \inf\{|u - v|_s\colon v \in \mathscr{H}\} \leqq |u - Q_r u|_s = |P_r u|_s \leqq |P_r|_{s \leftarrow t}\, |u|_t \leqq \eta^{t-s}\, |u|_t$$

one proves $|P_s|_{s\leftarrow t} \leqq \eta^{t-s}$, i.e. (13a) holds for all $s \in [2r-t, r]$ instead of r. Therefore, by Step 1 inequality (13b) can be shown for all $s \in [2r-t, r]$.

Step 3. Set $r' := 2r - t$. Applying Step 2 to r' instead of r, we prove (13b) for all $s \in [2r'-t, r'] = [r-3(t-r), 2r-t]$. Repeating Step 2 again, we obtain (13b) also for $s \in [r-7(t-r), r-3(t-r)]$, etc. □

1.4.5 Sesqui-Linear Forms

Sesqui-linear forms are needed for the formulation of boundary value problems as variational problems. In particular, the finite element method makes extensive use of sesqui-linear forms. For real spaces sesqui-linear forms are also called *bi-linear* forms.

Let $\mathscr{V}$ be a Hilbert space with scalar product $(\cdot,\cdot)_{\mathscr{V}}$ and norm $\|u\|_{\mathscr{V}} = (u,u)_{\mathscr{V}}^{1/2}$. The mapping $a: \mathscr{V} \times \mathscr{V} \to \mathbb{C}$ is said to be

(i) *sesqui-linear* if

$$\begin{aligned} a(u+\alpha v, w) &= a(u,w) + \alpha a(v,w) \quad (u,v,w \in \mathscr{V}, \quad \alpha \in \mathbb{C}), \\ a(u, v+\alpha w) &= a(u,v) + \bar{\alpha} a(u,w) \quad (u,v,w \in \mathscr{V}, \quad \alpha \in \mathbb{C}); \end{aligned} \tag{1.4.14}$$

(ii) *bounded* (or *continuous*) if

$$|a(u,v)| \leqq C \|u\|_{\mathscr{V}} \|v\|_{\mathscr{V}} \quad (u,v \in \mathscr{V}) \quad \text{for some } C; \tag{1.4.15}$$

(iii) *symmetric* if $a = a^*$, where the *adjoint form* is

$$a^*(u,v) = \overline{a(v,u)} \quad (u,v \in \mathscr{V}); \tag{1.4.16}$$

(iv) *$\mathscr{V}$-elliptic* if there is an $\alpha > 0$ such that

$$\operatorname{Re} a(v,v) \geqq \alpha \|v\|_{\mathscr{V}}^2 \quad (v \in \mathscr{V}). \tag{1.4.17}$$

The real part of a has the form $\frac{1}{2}(a + a^*)$.

A *variational problem* involves finding $u \in \mathscr{V}$ satisfying

$$a(u,v) = f(v) \quad \text{for all } v \in \mathscr{V} \tag{1.4.18}$$

with given $f \in \mathscr{V}'$. $f(v)$ denotes the value of f at v.

Example 1.4.6. The Poisson equation $-\Delta u = f_\Omega$ in $\Omega \subset \mathbb{R}^d$, $u = 0$ on the boundary, can be expressed by Eq. (18) with $\mathscr{V} = H_0^1(\Omega)$ and

$$a(u,v) = \sum_{i=1}^{d} \int_\Omega \frac{\partial u}{\partial x_i} \frac{\partial \bar{v}}{\partial x_i} dx, \qquad f(v) = \int_\Omega f_\Omega(x)\, \overline{v(x)}\, dx.$$

a is sesqui-linear, bounded and symmetric. For bounded Ω it is also $H_0^1(\Omega)$-elliptic.

Proposition 1.4.7. There is a one-to-one correspondence between bounded sesqui-linear forms a and bounded mappings $A: \mathscr{V} \to \mathscr{V}'$ by means of

$$a(u,v) = (Au)(v) \quad (u,v \in \mathscr{V}).$$

A is called the *associated operator*. Equation (18) is equivalent to the linear equation $Au = f$ in $\mathscr{V}'$. The operator A^* associated with the adjoint form a^* is the adjoint of A. Under hypothesis (17), Eq. (18) has a unique solution $u = A^{-1}f \in \mathscr{V}$, where $A^{-1}, A^{*-1} \in L(\mathscr{V}', \mathscr{V})$ satisfy

$$\| A^{-1} \|_{\mathscr{V} \leftarrow \mathscr{V}'} = \| A^{*-1} \|_{\mathscr{V} \leftarrow \mathscr{V}'} \leqq 1/\alpha .$$

Consider the situation (5), $\mathscr{V}' \supset \mathscr{U} \supset \mathscr{V}$. A sesqui-linear form a is called *$\mathscr{V}$-coercive (relative to $\mathscr{U}$)*, if there are $\lambda_0 \in \mathbb{R}$ and $\alpha > 0$ such that

$$\operatorname{Re} a(v, v) + \lambda_0 \| v \|_{\mathscr{U}}^2 \geqq \alpha \| v \|_{\mathscr{V}}^2 \qquad \text{for all } v \in \mathscr{V}. \tag{1.4.19}$$

The left-hand side defines a symmetric and $\mathscr{V}$-elliptic form $b = (a + a^*)/2 + \lambda_0(\cdot, \cdot)_{\mathscr{U}}$. The properties of the associated operator $B = \frac{1}{2}(A + A^*) + \lambda_0 I$ are stated in

Proposition 1.4.8. Let a be sesqui-linear, bounded, symmetric and $\mathscr{V}$-elliptic. The associated operator A can be regarded as an (unbounded) selfadjoint and positive operator on $\mathscr{U}$ with domain $D(A)$ dense in $\mathscr{V}$. $A^{1/2}$ is well-defined with $D(A^{1/2}) = [D(A), \mathscr{U}]_{1/2} = \mathscr{V}$. $a(u, v)$ equals $(A^{1/2}u, A^{1/2}v)_{\mathscr{U}}$ for $u, v \in \mathscr{V}$.

For a proof compare Kato [1, p. 331].

Corollary 1.4.9. The Proposition 1.4.8 admits the introduction of the Hilbert scale

$$\mathscr{H}^s = D(A^{s/2}), \quad \mathscr{H}^{-s} = (\mathscr{H}^s)' \quad (0 \leqq s \leqq 2) \tag{1.4.20}$$

with norms $\| u \|_{\mathscr{H}^s} = \| A^{s/2} u \|_{\mathscr{U}}$ $(-2 \leqq s \leqq 2)$. The following spaces coincide: $\mathscr{H}^0 = \mathscr{U}$, $\mathscr{H}^1 = \mathscr{V}$, $\mathscr{H}^{-1} = \mathscr{V}'$, $\| \cdot \|_{\mathscr{H}^0}$ equals $\| \cdot \|_{\mathscr{U}}$, whereas $\| u \|_{\mathscr{H}^1} = \sqrt{a(u, u)}$ is equivalent to $\| \cdot \|_{\mathscr{V}}$.

1.5 Exercises

The exercises 1.5.1 to 4 concern § 1.3. Hilbert spaces that appear are of finite dimension.

Exercise 1.5.1. Consider two linear iterations

$$u_l^{j+1} = \varphi_l'(u_l^j, f_l) := M_l' u_l^j + N_l' f_l, \quad u_l^{j+1} = \varphi_l''(u_l^j, f_l) := M_l'' u_l^j + N_l'' f_l.$$

Define the composition by $\varphi_l(u_l, f_l) := \varphi_l''(\varphi_l'(u_l, f_l), f_l)$. Prove:

(i) φ has the representation $\varphi_l(u_l, f_l) = M_l u_l + N_l f_l$ with

$$M_l = M_l'' M_l', \quad N_l = M_l'' N_l' + N_l''. \tag{1.5.1}$$

(ii) A fixed point of φ_l' and φ_l'' is a fixed point of φ_l.
(iii) If $\{M_l', N_l'\}$ and $\{M_l'', N_l''\}$ satisfy (3.5), then so does $\{M_l, N_l\}$.

Exercise 1.5.2. Let $\| \cdot \|_{\mathscr{U}}$ and $\| \cdot \|_{\mathscr{V}}$ be vector norms. Show:

(i) Every regular matrix $A\colon \mathscr{U} \to \mathscr{U}$ generates a new norm on $\mathscr{W} = \mathscr{U}$:

$$\| u \|_{\mathscr{W}} := \| Au \|_{\mathscr{U}}.$$

(ii) The matrix norms involving $\| \cdot \|_{\mathscr{W}}$ satisfy

$$\| B \|_{\mathscr{W} \leftarrow \mathscr{V}} = \| AB \|_{\mathscr{U} \leftarrow \mathscr{V}}, \qquad \| B \|_{\mathscr{V} \leftarrow \mathscr{W}} = \| BA^{-1} \|_{\mathscr{V} \leftarrow \mathscr{U}},$$
$$\| B \|_{\mathscr{W} \leftarrow \mathscr{W}} = \| ABA^{-1} \|_{\mathscr{U} \leftarrow \mathscr{U}}.$$

Exercise 1.5.3. Let $\mathscr{U} = \mathscr{V} = \mathscr{W}$ be Hilbert spaces, possibly with different scalar products. Denote the corresponding norms and dual norms by $\| \cdot \|_{\mathscr{V}}$, $\| \cdot \|_{\mathscr{W}}$, $\| \cdot \|_{\mathscr{V}*}$, $\| \cdot \|_{\mathscr{W}*}$. Define the adjoint mapping A^* with respect to $\mathscr{U}$. Show that

$$\| A \|_{\mathscr{W} \leftarrow \mathscr{V}} = \| A^* \|_{\mathscr{V}* \leftarrow \mathscr{W}*}. \tag{1.5.2}$$

Exercise 1.5.4. Let $A = A^*$ and $B = B^*$.

(i) Prove that

$$\| A \|_{\mathscr{U} \leftarrow \mathscr{U}} \leqq c \quad \text{is equivalent to} \quad -cI \leqq A \leqq cI, \tag{1.5.3a}$$
$$A > 0 \quad \text{if and only if} \quad A^{-1} > 0, \tag{1.5.3b}$$
$$A \leqq B \quad \text{implies} \quad C^*AC \leqq C^*BC \quad \text{for arbitrary} \quad C. \tag{1.5.3c}$$

(ii) Let $A = A^*$ be positive definite, while $C\colon \mathscr{U} \to \mathscr{V}$ may map into another space $\mathscr{V}$. Prove:

$$C^*C \leqq A \quad \text{if and only if} \quad CA^{-1}C^* \leqq I. \tag{1.5.3d}$$

Exercise 1.5.5. Let $\{\mathscr{H}^q\colon 0 \leqq q \leqq t\}$ be a Hilbert scale. Let $r, s \in [0, t]$ be such that $r < t$ or $s < t$. Prove: For all $\varepsilon > 0$ there exists $C(\varepsilon)$ such that

$$\| u \|_{\mathscr{H}^r} \| u \|_{\mathscr{H}^s} \leqq \varepsilon \| u \|^2_{\mathscr{H}^t} + C(\varepsilon) \| u \|^2_{\mathscr{H}^0} \qquad \text{for all } u \in H^t. \tag{1.5.4}$$

Exercise 1.5.6. Let $\{\mathscr{H}^s\colon 0 \leqq s \leqq t\}$ be a Hilbert scale and set $\mathscr{U} = \mathscr{H}^0$, $\mathscr{V} = \mathscr{H}^t$. Assume that the sesqui-linear form a can be split into $a = a_0 + b$, where a_0 is $\mathscr{V}$-elliptic, or at least $\mathscr{V}$-coercive, and that b satisfies

$$|b(u, v)| \leqq C \| u \|_{\mathscr{H}^r} \| v \|_{\mathscr{H}^s} \qquad \text{for all } u \in \mathscr{H}^r,\ v \in \mathscr{H}^s \tag{1.5.5}$$

with $0 \leqq r, s \leqq t$, $r < t$ or $s < t$. Prove: a is also $\mathscr{V}$-coercive.

Exercise 1.5.7. Let $A \in L(\mathscr{V}, \mathscr{V}')$ be the operator associated with the sesqui-linear form a and assume $\dim(\mathscr{V}) < \infty$. Show:

(i) $A^{-1} \in L(\mathscr{V}', \mathscr{V})$ exists if there is $\varepsilon > 0$ with

$$\sup \{|a(v, w)|\colon v \in \mathscr{V}, \| v \|_{\mathscr{V}} = 1\} \geqq \varepsilon > 0 \qquad \text{for all } w \in \mathscr{V}. \tag{1.5.6}$$

(ii) The largest ε satisfying (6) equals the inverse of

$$\| A^{-1} \|_{\mathscr{V} \leftarrow \mathscr{V}'} = \sup_{\substack{w \in \mathscr{V} \\ \|w\|_{\mathscr{V}} = 1}} \ \inf_{\substack{v \in \mathscr{V} \\ \|v\|_{\mathscr{V}} = 1}} 1/|a(v, w)| = 1/\varepsilon. \tag{1.5.7}$$

2. Introductory Model Problem

2.1 One-Dimensional Model Problem

A natural model of an elliptic equation is the two-dimensional Poisson equation in a square. Unfortunately, the analysis of the multi-grid algorithm for this problem is still an involved exercise. Therefore, we start with the simplest *one*-dimensional Dirichlet boundary value problem

$$-u''(x) = f(x) \quad \text{in} \quad \Omega = (0,1), \qquad u(x) = 0 \quad \text{at} \quad x \in \Gamma = \{0,1\}. \tag{2.1.1}$$

A finite difference discretisation is characterised by the choice of the difference scheme and the *grid size*, which will always be denoted by h. The grid corresponding to $h = 1/(n+1)$ consists of the points $x_\nu = \nu h (\nu = 1, \ldots, n)$. From the beginning we introduce not only one grid but a sequence of grids. The largest possible grid size is $h_0 = \frac{1}{2}$. The corresponding grid Ω_0 consists of the single grid point $x_1 = \frac{1}{2}$ (cf. Fig. 2.1.1). For reasons that will become clear in § 2.3, the next finer grid should be defined by halving the step size: $h_1 = \frac{1}{4}$. Thus, the sequence

$$h_0 > h_1 > h_2 > \ldots > h_l > \ldots \quad \text{with} \quad h_l = 2^{-1-l} \tag{2.1.2}$$

results. The subscript l is called the *level number*. The grid of level l (cf. Fig. 2.1.1) is

$$\Omega_l = \{\nu h_l \colon \nu = 1, 2, \ldots, 2^{1+l} - 1\}. \tag{2.1.3}$$

For every grid point $x_\nu \in \Omega_l$ the differential Eq. (1) has to be replaced by a difference formula. The simplest one approximating the negative second derivative is

$$h_l^{-2}[-u(x_{\nu-1}) + 2u(x_\nu) - u(x_{\nu+1})] = -u''(x_\nu) + O(h_l^2). \tag{2.1.4}$$

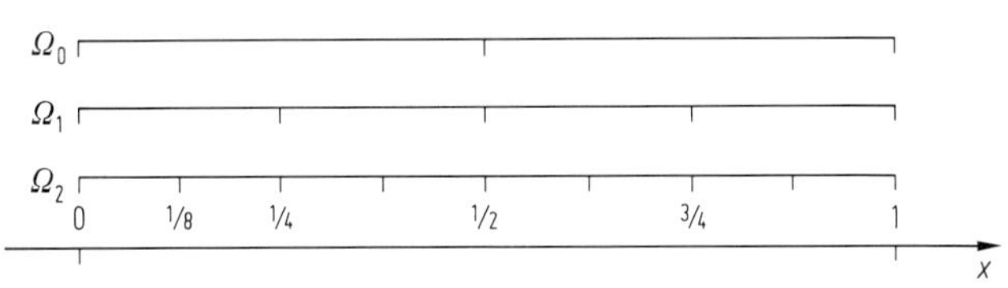

Fig. 2.1.1. Different grids

Setting $f_l(\nu h_l) := f(\nu h_l)$ and defining the vectors

$$u_l = (u_l(\nu h_l))_{\nu=1}^{n_l}, \quad f_l = (f_l(\nu h_l))_{\nu=1}^{n_l} \qquad \left(n_l = \frac{1}{h_l} - 1\right),$$

we obtain the system (5) of n_l difference equations for the n_l unknown elements of the vector u_l:

$$h_l^{-2} \begin{bmatrix} 2 & -1 & & & & \\ -1 & 2 & -1 & & & \\ & -1 & 2 & -1 & & \\ & & \ddots & \ddots & \ddots & \\ & & & -1 & 2 & -1 \\ & & & & -1 & 2 \end{bmatrix} u_l = f_l. \tag{2.1.5}$$

The boundary conditions $u_l(0) = u_l(1) = 0$ are incorporated into the system by eliminating $u(0)$ and $u(1)$ from Eq. (4) for $\nu = 1$ and $\nu = n_l$, respectively.

Occasionally, the difference Eq. (5) are abbreviated by the 'difference stencil'

$$h_l^{-2}[-1 \;\; 2 \;\; -1]\, u_l = f_l. \tag{2.1.5'}$$

This notation does not indicate the treatment of the boundary condition.

From now on, the matrix of system (5) will be called L_l; Eq. (5) is rewritten as

$$L_l u_l = f_l. \tag{2.1.5''}$$

Because L_l is tridiagonal, it is very easy to solve system (5″) directly. Nevertheless, one-dimensional boundary value problems are suitable for the analysis of iterative methods. Multi-grid iterations as well as other iterations have many properties that do not depend on the dimension. Moreover, the multi-grid method can be described very easily for the present case, whereas in a more general situation (cf. § 3) there are many details which distract attention from the basic principles.

2.2 Smoothing Effect of Classical Iterations

Among all the classical iterations we choose the *Jacobi iteration*, since its analysis is the most transparent one.

The system $L_l u_l = f_l$ can be written as $D_l u_l = B_l u_l + f_l$ with $B_l := D_l - L_l$, where D_l is the diagonal $2h_l^{-2} I$ (I: identity matrix) of L_l. This representation leads us to the iteration

$$u_l^{j+1} = D_l^{-1}(B_l u_l^j + f_l) \tag{2.2.1}$$

due to Jacobi [1]. The jth iterate is always denoted by the superscript j. An

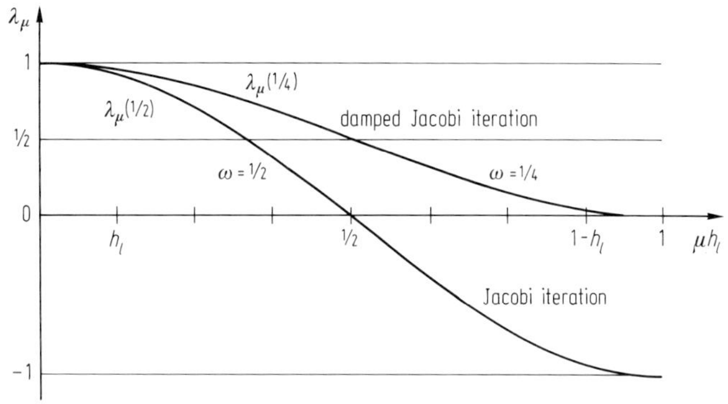

Fig. 2.2.1. Eigenvalues of the iteration matrix M_l as function of μh_l

equivalent formulation of (1) is

$$u_l^{j+1} = u_l^j - D_l^{-1}(L_l u_l^j - f_l), \tag{2.2.1'}$$

showing that the correction $u_l^{j+1} - u_l$ is obtained from the defect[1] $L_l u_l^j - f_l$.

It will turn out that the Jacobi iteration (1) is less interesting for our purpose than the *damped Jacobi iteration*

$$u_l^{j+1} = u_l^j - \Theta D_l^{-1}(L_l u_l^j - f_l) \quad \text{with} \quad \Theta \in (0,1). \tag{2.2.2}$$

In our situation D_l^{-1} equals $\frac{1}{2}h_l^2 I$. Hence, setting $\omega = \Theta/2$, we rewrite (2) as

$$u_l^{j+1} = u_l^j - \omega h_l^2 (L_l u_l^j - f_l) \quad \text{with} \quad \omega \in (0, \tfrac{1}{2}). \tag{2.2.3}$$

To analyse the convergence of iteration (3), note that the eigenvectors of L_l are

$$e_l^\mu = \sqrt{2h_l}(\sin(\nu\mu\pi h_l))_{\nu=1}^{n_l} \qquad \text{for } \mu = 1, \dots, n_l. \tag{2.2.4}$$

The corresponding eigenvalues of L_l are $4h_l^{-2}\sin^2(\mu\pi h_l/2)$:

$$L_l e_l^\mu = 4h_l^{-2}\sin^2(\mu\pi h_l/2)\, e_l^\mu. \tag{2.2.5}$$

Since

$$M_l = I - \omega h_l^2 L_l$$

is the iteration matrix (cf. § 1.3.1) of iteration (3), it has the same eigenvectors e_l^μ and the eigenvalues

$$\lambda_\mu(\omega) = 1 - 4\omega\sin^2(\mu\pi h_l/2) \qquad (1 \leqq \mu \leqq n_l = h_l^{-1} - 1).$$

For $\omega = \frac{1}{2}$ and $\omega = \frac{1}{4}$ these eigenvalues are shown in Fig. 2.2.1. As can be seen from the graph, the maximum absolute value $\varrho(M_l)$ is taken at $\mu = 1$. The

1 We use the name 'defect' instead of 'residual'. Usually, the residual is defined with opposite sign: $f_l - L_l u_l$

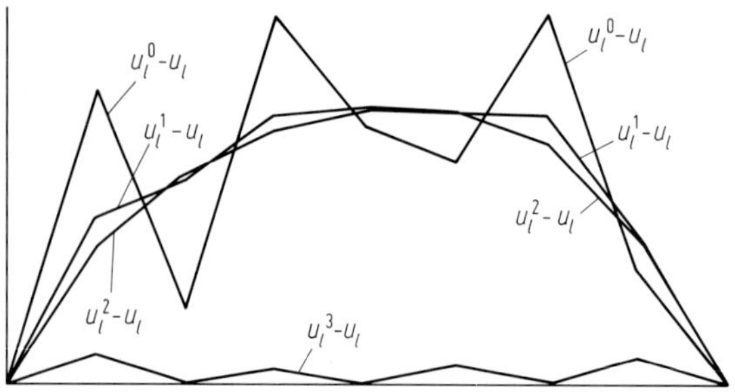

Fig. 2.2.2. Smoothing effect of the damped Jacobi iteration. u_l, exact discrete solution; $u_l^j (j = 0, 1, 2)$, iterates of the smoothing iteration; u_l^3, result of coarse-grid correction

choice $\omega = \frac{1}{2}$ yields the Jacobi iteration (1). Its rate of convergence is

$$\varrho(M_l) = \lambda_1(\tfrac{1}{2}) = 1 - 2\sin^2(\pi h_l/2) = 1 - \tfrac{1}{2}\pi^2 h_l^2 + O(h_l^4),$$

proving the very slow convergence of the Jacobi iteration. For any other choice $\omega \in (0, \frac{1}{2})$ the rates are even worse. E.g., $\omega = \frac{1}{4}$ yields $\varrho(M_l) = \lambda_1(\frac{1}{4}) = \cos^2(\pi h_l/2) = 1 - \frac{1}{4}\pi^2 h_l^2 + O(h_l^4)$. Thus, iteration (3) with $\omega = \frac{1}{4}$ requires twice as many iteration steps as with $\omega = \frac{1}{2}$.

In the following we fix ω to be $\frac{1}{4}$. Even though the iteration (3) proves very slow, Fig. 2.2.1 shows that components e_l^μ with frequency $\mu \geqq 1/(2h_l)$ are reduced at least by a factor $\frac{1}{2}$ per iteration. This means that the convergence rate of the damped Jacobi iteration restricted to the subspace span$\{e_l^\mu : \frac{1}{2} \leqq \mu h_l < 1\}$ of *high frequencies* is $\frac{1}{2}$. The iteration is rapidly convergent with respect to the high frequencies. The slow convergence is caused by the lower frequencies only. The construction of the multi-grid iteration is based on this observation.

The initial error $u_l^0 - u_l$ can be represented by $\sum \alpha_\mu e_l^\mu$. ν steps of the damped Jacobi iteration yield the error $u_l^\mu - u_l = \sum \beta_\mu e_l^\mu$ with $\beta_\mu = \alpha_\mu [\lambda_\mu(\frac{1}{4})]^\nu$ (cf. (1.3.7 b)). The preceding consideration shows that $\beta_\mu \approx \alpha_\mu$ for low frequencies, but $|\beta_\mu| \ll |\alpha_\mu|$ for high frequencies. This fact can be expressed by saying that the error $u_l^\nu - u_l$ is *smoother* than $u_l^0 - u_l$. The first three graphs of Fig. 2.2.2 illustrate the increasing smoothness of $u_l^\nu - u_l$ $(\nu = 0, 1, 2)$. Therefore, the iteration (3) serves as a 'smoothing iteration'.

In the following the matrices M_l and N_l (cf. (1.3.2)) will be reserved for the two- or multi-grid iterations. For the smoothing iteration we replace M_l, N_l by S_l, T_l:

$$u_l^{j+1} = S_l u_l^j + T_l f_l. \tag{2.2.6 a}$$

In the special case of the damped Jacobi iteration, the matrices are

$$S_l = I - \omega h_l^2 L_l, \qquad T_l = \omega h_l^2 I. \tag{2.2.7}$$

The iteration will also be denoted by $\mathscr{S}_l(u_l, f_l)$:

$$u_l^{j+1} = \mathscr{S}_l(u_l^j, f_l). \tag{2.2.6b}$$

The ν-fold application of $\mathscr{S}_l$ is symbolised by $\mathscr{S}_l^\nu$:

$$u_l^{j+\nu} = \mathscr{S}_l^\nu(u_l^j, f_l). \tag{2.2.6c}$$

2.3 Two-Grid Method

The foregoing section showed that the simple damped Jacobi iteration (2.3) with $\omega = \frac{1}{4}$,

$$u_l^{j+1} = \mathscr{S}_l(u_l^j, f_l) = S_l u_l^j + T_l f_l, \qquad S_l = I - \tfrac{1}{4} h_l^2 L_l, \tag{2.3.1}$$

is quite an efficient method for reducing the high-frequency components. Convergence is only lacking with respect to low frequencies (smooth components). Therefore, one should combine this iteration with a second one having complementary properties. In particular, the second iteration should reduce the smooth error very well. Such a complementary iteration can be constructed by means of the coarse grid with step size $h_{l-1} = 2h_l$ (cf. Fig. 2.1.1).

Let u_l^{old} be some given approximation to $u_l = L_l^{-1} f_l$. A small number of iterations of method (1) will result in an intermediate value $\bar{u}_l$. From the previous section we know that the error $v_l = \bar{u}_l - u_l$ is smooth (more precisely, smoother than $u_l^{\text{old}} - u_l$). v_l can also be regarded as the exact correction since $u_l = \bar{u}_l - v_l$.

Inserting $\bar{u}_l$ into the equation $L_l u_l - f_l = 0$ we obtain the *defect*

$$d_l = L_l \bar{u}_l - f_l \tag{2.3.2}$$

of $\bar{u}_l$, which vanishes if and only if $\bar{u}_l$ is the exact solution u_l. Because of $L_l v_l = L_l \bar{u}_l - L_l u_l = L_l \bar{u}_l - f_l = d_l$ the exact correction v_l is the solution of

$$L_l v_l = d_l. \tag{2.3.3}$$

Equation (3) is of the same form as the original equation $L_l u_l = f_l$. The *exact* solution of the former equation is as difficult as for the latter one. Nonetheless, v_l can be *approximated* better than u_l, since v_l is a smooth grid function. Only smooth functions can be represented well by means of coarser grids.

To approximate the problem $L_l v_l = d_l$ by a coarse-grid equation

$$L_{l-1} v_{l-1} = d_{l-1} \tag{2.3.4}$$

we have to choose d_{l-1} reasonably. Note that the matrix L_k is already defined by (1.5) for all levels $k \geqq 0$, especially for $k = l - 1$. d_{l-1} should depend linearly on the original right-hand side d_l. Hence, a linear mapping r, called a *restric-*

tion, defines d_{l-1} by

$$d_{l-1} = r d_l. \tag{2.3.5}$$

The simplest choice of a restriction is the *trivial injection* r_{inj} defined by

$$(r_{inj} d_l)(x) = d_l(x) \quad \text{for all } x \in \Omega_{l-1} \subset \Omega_l. \tag{2.3.6}$$

Although this restriction is the easiest to perform, it has some disadvantages (cf. § 3.5).

It is safer to take into account the components $d_l(x)$ at $x \in \Omega_l \backslash \Omega_{l-1}$, too. We shall use the restriction r given by

$$(r d_l)(x) = \tfrac{1}{4}[d_l(x - h_l) + 2 d_l(x) + d_l(x + h_l)] \quad \text{for } x \in \Omega_{l-1}. \tag{2.3.7 a}$$

The corresponding matrix is

$$r = \frac{1}{4} \begin{bmatrix} 1 & 2 & 1 & & & & & O \\ & & 1 & 2 & 1 & & & \\ & & & & \ddots & & & \\ & & & & 1 & 2 & 1 & \\ & O & & & & & 1 & 2 & 1 \end{bmatrix}. \tag{2.3.7 b}$$

It is a special *weighted* restriction. The justification of its weights $\frac{1}{4}, \frac{1}{2}, \frac{1}{4}$ will be given in Exercises 2.7.2–4.

Having defined d_{l-1} by (5) from the defect d_l, we obtain $v_{l-1} = L_{l-1}^{-1} d_{l-1}$ as the exact solution of Eq. (4). We expect v_{l-1} to be an approximation to the exact correction v_l. However, v_{l-1} is only defined on the coarse grid Ω_{l-1}. We have to interpolate this coarse-grid function by

$$\tilde{v}_l = p v_{l-1}, \tag{2.3.8}$$

where the *prolongation* p describes a coarse-to-fine interpolation. The simplest interpolation is the piecewise linear one:

$$(p v_{l-1})(x) = \begin{cases} v_{l-1}(x) & \text{if } x \in \Omega_{l-1}, \\ [v_{l-1}(x - h_l) + v_{l-1}(x + h_l)]/2 & \text{otherwise}. \end{cases} \tag{2.3.9 a}$$

In this definition we formally set $v_{l-1}(0) = v_{l-1}(1) = 0$. p can be represented by the rectangular $n_l \times n_{l-1}$ matrix

$$p = \frac{1}{2} \begin{bmatrix} 1 & & & \\ 2 & & & \\ 1 & 1 & & \\ & 2 & & \\ & & \ddots & \\ & & 2 & \\ & & 1 & 1 \\ & & & 2 \\ & & & 1 \end{bmatrix}. \tag{2.3.9 b}$$

Since $u_l = \bar{u}_l - v_l$ is the exact solution and $\tilde{v}_l = p v_{l-1}$ is supposed to approximate v_l, one tries to improve the value $\bar{u}_l$ by

$$u_l^{\text{new}} = \bar{u}_l - \tilde{v}_l. \tag{2.3.10}$$

The step from $\bar{u}_l$ to u_l^{new} by (2–10) is called the *coarse-grid correction*. Combining the separate parts (2–10), we obtain the compact formula

$$\bar{u}_l \mapsto \bar{u}_l - p L_{l-1}^{-1} r (L_l \bar{u}_l - f_l) \tag{2.3.11}$$

for the coarse-grid correction.

Figure 2.2.2 shows the errors $u_l^i - u_l$ after $i = 0, 1, 2$ damped Jacobi iterations. The coarse-grid correction applied to $\bar{u}_l = u_l^2$ yields u_l^3. The graph of the error $u_l^3 - u_l$ in Fig. 2.2.2 proves the success of the coarse-grid correction.

Although the coarse-grid correction seems to be efficient, it cannot be used as an iteration by itself, as remarked in

Note 2.3.1. The coarse-grid correction (11) is a non-convergent iteration.

Proof. The restriction r has a nontrivial kernel. Choose $0 \neq w_l \in \ker(r)$ and set $u_l^0 := L_l^{-1}(f_l + w_l)$. As $d_l = w_l$ and $d_{l-1} = r w_l = 0$, the resulting iterates are $u_l^j = u_l^0 \not\to u_l$. □

It is the combination of smoothing iteration and coarse-grid correction that is rapidly convergent, whereas both components by themselves converge slowly or not at all. The combination is called a *two-grid iteration* since two levels l and $l-1$ are involved. We summarise below:

two-grid iteration for solving $L_l u_l = f_l$ (2.3.12)

u_l^j given iterate,

smoothing step:

$$\bar{u}_l := \mathscr{S}_l^{\nu}(u_l^j, f_l) \qquad \text{(cf. (2.6c))} \tag{2.3.12a}$$

(application of ν damped Jacobi iterations to u_l^j),

coarse-grid correction:

$d_l := L_l \bar{u}_l - f_l$	(calculation of the defect)	(2.3.12 b$_1$)
$d_{l-1} := r d_l$	(restriction of the defect)	(2.3.12 b$_2$)
$v_{l-1} := L_{l-1}^{-1} d_{l-1}$	(solution of coarse-grid equation)	(2.3.12 b$_3$)
$u_l^{j+1} := \bar{u}_l - p v_{l-1}$	(correction of $\bar{u}_l$)	(2.3.12 b$_4$)

As stated in (11) the coarse-grid correction (12 b$_1$–12 b$_4$) can be condensed to

$$u_l^{j+1} = \bar{u}_l - p L_{l-1}^{-1} r (L_l \bar{u}_l - f_l). \tag{2.3.12b}$$

The number ν of smoothing iterations can be chosen independently of the grid size h_l. Its influence on convergence will be described in the following section.

The following algorithms are written in quasi-ALGOL. This presentation has the benefit of showing immediately what has to be programmed, and in the author's opinion it is more readable than flow-charts or FORTRAN texts.

The ALGOL description of the two-grid iteration (12) is given below:

two-grid iteration $TGM^{(\nu)}$ (2.3.12′)

procedure $TGM(l, u, f)$; **integer** l; **array** u, f;
if $l = 0$ **then** $u := L_0^{-1} * f$ **else**
begin array v, d;

$u := \mathscr{S}_l^{\nu}(u, f)$; (2.3.12 a′)

$d := r * (L_l * u - f)$; (2.3.12 $\mathrm{b}'_{1,2}$)

$v := L_{l-1}^{-1} * d$; (2.3.12 b'_3)

$u := u - p * v$ (2.3.12 b'_4)

end TGM;

The procedure performs *one* iteration step at level l. The third parameter f is the right-hand side f_l of the equation to be solved. The input value of the second parameter u is the given j^{th} iterate u_l^j that is overwritten by the output value $u = u_l^{j+1}$. The second line "**if** $l = 0$ **then**..." is added to have a well-defined algorithm for all levels $l \geqq 0$.

The program (12′) is denoted by $TGM^{(\nu)}$ since it depends on ν. Iteration (12) can be regarded as a prototype of the two-grid method. However, there are many modifications. Instead of applying the smoothing first and thereafter the coarse-grid correction, we can interchange the two parts. More generally, ν_1 smoothing iterations can be performed before, and ν_2 iterations after the coarse-grid correction. The resulting algorithm is

two-grid iteration $TGM^{(\nu_1, \nu_2)}$ (2.3.13)

procedure $TGM(l, u, f)$; **integer** l; **array** u, f;
if $l = 0$ **then** $u := L_0^{-1} * f$ **else**
begin array v, d;

$u := \mathscr{S}_l^{\nu_1}(u, f)$; (pre-smoothing) (2.3.13 a)

$u := u - p * L_{l-1}^{-1} * r * (L_l * u - f)$; (2.3.13 b)

$$u := \mathscr{S}_l^{\nu_2}(u, f); \qquad \text{(post-smoothing)} \tag{2.3.13c}$$

end *TGM*;

. . . .

2.4 Convergence of the Two-Grid Iteration

The one-dimensional model problem is chosen since the proof of convergence is relatively easy. A discrete Fourier analysis yields explicit results. It will turn out that most of the results can be extended to the multi-dimensional case.

The eigenvectors of L_l and S_l (from (3.1)) are the grid functions e_l^μ, $1 \leqq \mu \leqq n_l$ as defined in (2.4). These vectors form an orthonormal basis of $\mathbb{R}^{n_l}$, since

$$\sum_{\nu=1}^{n_l} \sin(\mu \pi \nu h_l) \sin(\chi \pi \nu h_l) = \begin{cases} 0 & \text{if } 1 \leqq \mu \neq \chi \leqq n_l, \\ n_l/2 & \text{if } 1 \leqq \mu = \chi \leqq n_l. \end{cases}$$

Therefore, the matrix

$$Q_l = [e_l^1, e_l^{n_l}, e_l^2, e_l^{n_l-1}, \ldots, e_l^{n_{l-1}}, e_l^{n_l+1-n_{l-1}}, e_l^{n_{l-1}+1}]$$

consisting of the vectors e_l^μ in the special ordering $1, n_l, \ldots, \mu, n_l + 1 - \mu, \ldots, n_{l-1}, n_l + 1 - n_{l-1}, n_{l-1} + 1$ (where $n_{l-1} = h_{l-1}^{-1} - 1$) as columns is unitary, i.e. $Q_l^{-1} = Q_l^*$. The two-grid iteration matrix M_l will be transformed into the block diagonal matrix

$$Q_l^{-1} M_l Q_l = \begin{bmatrix} M_l^{(1)} & & & & \\ & M_l^{(2)} & & & \\ & & \ddots & & \\ & & & M_l^{(n_{l-1})} & \\ & & & & M_l^{(n_{l-1}+1)} \end{bmatrix} \tag{2.4.1}$$

where $M_l^{(\mu)}$, $1 \leqq \mu \leqq n_{l-1} = h_{l-1}^{-1} - 1$, are 2×2 matrices and $M_l^{(n_{l-1}+1)}$ is 1×1. Since Q_l is unitary, M_l and $Q_l^{-1} M_l Q_l$ have the same spectral radius and spectral norm. By the following lemma, whose proof is left to the reader, it is sufficient to study the submatrices $M_l^{(\mu)}$:

Lemma 2.4.1. Let M_l be given by (1), where Q_l is unitary. Then

$$\varrho(M_l) = \max\{\varrho(M_l^{(\mu)}) : 1 \leqq \mu \leqq n_{l-1} + 1\}, \tag{2.4.2}$$

$$\| M_l \| = \max\{\| M_l^{(\mu)} \| : 1 \leqq \mu \leqq n_{l-1} + 1\}, \tag{2.4.3}$$

where $\| \cdot \|$ is the spectral norm (cf. (1.3.15a)).

The two-grid iteration (3.12) has a representation (1.3.2), $u_l^{j+1} = M_l u_l^j + N_l f_l$, where $M_l = M_l(\nu)$ depends on the number ν.

Lemma 2.4.2. The iteration matrix of the two-grid iteration (3.12) is

$$M_l = M_l(v) = (I - pL_{l-1}^{-1} r L_l) S_l^v, \qquad S_l = I - \tfrac{1}{4} h_l^2 L_l. \tag{2.4.4}$$

Proof. Apply the iteration to $f_l = 0$, u_l arbitrary. Use (1.3.3) with v and S_l instead of j and M_l for the smoothing step and (3.12b) for the coarse-grid correction. □

It will be shown that the subspace spanned by e_l^μ and $e_l^{n_l+1-\mu}$ forms an invariant subspace under multiplication by M_l, where $1 \leqq \mu \leqq n_{l-1}$. We introduce the index μ' as a function of μ by

$$\mu' = n_l + 1 - \mu \qquad \text{for } \mu = 1, \ldots, n_{l-1}.$$

Since $\mu < \mu'$, e_l^μ and $e_l^{\mu'}$ represent a low and a high frequency component, respectively, of this subspace.

e_{l-1}^μ is a grid function of level $l-1$ defined by (2.4) with h_l and n_l replaced by h_{l-1} and n_{l-1}. Note that e_l^μ and $\sqrt{2}\, e_{l-1}^\mu$ coincide at common grid points. Formally, $e_{l-1}^{\mu'}$ can be defined but it does not differ from $-e_{l-1}^\mu$. Moreover, $e_{l-1}^\mu = 0$ for $\mu = \mu' = n_{l-1} + 1$. Again the matrix

$$Q_{l-1} = [e_{l-1}^1, e_{l-1}^2, \ldots, e_{l-1}^{n_{l-1}-1}, e_{l-1}^{n_{l-1}}]$$

is unitary.

Define the transformed matrices

$$\hat{M}_l = Q_l^{-1} M_l Q_l, \qquad \hat{S}_l = Q_l^{-1} S_l Q_l, \qquad \hat{L}_l = Q_l^{-1} L_l Q_l,$$
$$\hat{L}_{l-1} = Q_{l-1}^{-1} L_{l-1} Q_{l-1}, \qquad \hat{p} = Q_l^{-1} p Q_{l-1}, \qquad \hat{r} = Q_{l-1}^{-1} r Q_l.$$

The proof of the following lemma is postponed to the end of this section.

Lemma 2.4.3. The matrices $\hat{S}_l$, $\hat{L}_l$, $\hat{p}$, $\hat{L}_{l-1}$, $\hat{r}$ have diagonal block structure:

$$\hat{S}_l = \begin{bmatrix} S_l^{(1)} & & & \\ & S_l^{(2)} & & \\ & & \ddots & \\ & & & S_l^{(n_{l-1}+1)} \end{bmatrix}, \qquad \hat{L}_l = \begin{bmatrix} L_l^{(1)} & & & \\ & L_l^{(2)} & & \\ & & \ddots & \\ & & & L_l^{(n_{l-1}+1)} \end{bmatrix},$$

$$\hat{p} = \begin{bmatrix} p^{(1)} & & & \\ & p^{(2)} & & \\ & & \ddots & \\ & & & p^{(n_{l-1})} \\ 0 & 0 \ldots & & 0 \end{bmatrix}, \qquad \hat{r} = \begin{bmatrix} r^{(1)} & & & & 0 \\ & r^{(2)} & & & 0 \\ & & \ddots & & \vdots \\ & & & r^{(n_{l-1})} & 0 \end{bmatrix},$$

$$\hat{L}_{l-1} = \operatorname{diag}\{L_{l-1}^{(1)}, \ldots, L_{l-1}^{(n_{l-1}+1)}\}.$$

Let $s_\mu^2 = \sin^2\left(\frac{\mu\pi h_l}{2}\right)$, $c_\mu^2 = \cos^2\left(\frac{\mu\pi h_l}{2}\right)$. The blocks are defined by

$$L_l^{(\mu)} = 4h_l^{-2}\begin{bmatrix} s_\mu^2 & 0 \\ 0 & c_\mu^2 \end{bmatrix} \quad (1 \leqq \mu \leqq n_{l-1}), \qquad L_l^{(n_{l-1}+1)} = 2h_l^{-2}, \tag{2.4.5}$$

$$S_l^{(\mu)} = I - \tfrac{1}{4}h_l^2 L_l^{(\mu)} = \begin{bmatrix} c_\mu^2 & 0 \\ 0 & s_\mu^2 \end{bmatrix} \quad (1 \leqq \mu \leqq n_{l-1}), \qquad S_l^{(n_{l-1}+1)} = \tfrac{1}{2} \tag{2.4.6}$$

$$r^{(\mu)} = \frac{1}{\sqrt{2}}[c_\mu^2 \quad - s_\mu^2], \qquad p^{(\mu)} = \sqrt{2}\begin{bmatrix} c_\mu^2 \\ - s_\mu^2 \end{bmatrix} \quad (1 \leqq \mu \leqq n_{l-1}) \tag{2.4.7}$$

$$L_{l-1}^{(\mu)} = 4h_l^{-2} s_\mu^2 c_\mu^2 \quad (1 \leqq \mu \leqq n_{l-1} + 1). \tag{2.4.8}$$

By Lemma 2.4.2, the matrix $\hat{M}_l = Q_l^{-1} M_l Q_l$ equals

$$\hat{M}_l = (I - \hat{p}\hat{L}_{l-1}^{-1}\hat{r}\hat{L}_l)\hat{S}_l^\nu .$$

From Lemma 2.4.3 one concludes that $\hat{M}_l$ has the diagonal block structure stated in (1). Its blocks are

$$M_l^{(\mu)} = [I - p^{(\mu)} L_{l-1}^{(\mu)} r^{(\mu)} L_l^{(\mu)}](S_l^{(\mu)})^\nu \quad (1 \leqq \mu \leqq n_{l-1}),$$
$$M_l^{(n_{l-1}+1)} = (S_l^{(n_{l-1}+1)})^\nu,$$

hence,

$$M_l^{(\mu)} = \begin{bmatrix} s_\mu^2 & c_\mu^2 \\ s_\mu^2 & c_\mu^2 \end{bmatrix}\begin{bmatrix} c_\mu^2 & 0 \\ 0 & s_\mu^2 \end{bmatrix}^\nu \quad (1 \leqq \mu \leqq n_{l-1}), \qquad M_l^{(n_{l-1}+1)} = 2^{-\nu}. \tag{2.4.9}$$

The matrix $\operatorname{diag}\{c_\mu^2, s_\mu^2\}^\nu$ corresponds to the damped Jacobi iteration and shows that the high frequency $e_l^{\mu'}$ is well reduced by $s_\mu^{2\nu}$, since $0 < s_\mu^2 < \frac{1}{2}$, whereas the low frequency e_l^μ is diminished by the factor $c_\mu^{2\nu}$ $(\frac{1}{2} < c_\mu^2 < 1)$ that approaches 1 for small μ.

The first factor $\begin{bmatrix} s_\mu^2 & c_\mu^2 \\ s_\mu^2 & c_\mu^2 \end{bmatrix}$ stems from the coarse-grid correction. It is plain that the coarse-grid correction has the complementary property. It is the low frequency e_l^μ which is reduced by $s_\mu^2 < \frac{1}{2}$ while the high frequency term $e_l^{\mu'}$ is decreased only by $c_\mu^2 > \frac{1}{2}$.

Convergence is proved in

Theorem 2.4.4. Let the two-grid iteration of level l be defined by (3.12) with $\nu \geqq 1$. The spectral radius of the iteration matrix $M_l = M_l(\nu)$ (cf. (4)) is bounded by

$$\varrho(M_l(\nu)) \leqq \max\{\xi(1-\xi)^\nu + (1-\xi)\xi^\nu : 0 \leqq \xi \leqq \tfrac{1}{2}\} =: \varrho_\nu < 1 \tag{2.4.10}$$

uniformly for all $l \geqq 0$; hence convergence follows. The spectral norm is bounded (uniformly in l) by

$$\| M_l(\nu) \| \leqq \max\{\sqrt{2[\xi^2(1-\xi)^{2\nu} + (1-\xi)^2\xi^{2\nu}]} : 0 \leqq \xi \leqq \tfrac{1}{2}\} =: \zeta_\nu < 1. \tag{2.4.11}$$

The asymptotic behaviour of ϱ_ν and ζ_ν as $\nu \to \infty$ is

$$\varrho_\nu \approx c_1/\nu, \qquad \zeta_\nu \approx c_2/\nu \tag{2.4.12}$$

with $c_1 = 1/e \approx 0.3679$, $c_2 = \sqrt{2}/e \approx 0.5203$. (Some values are listed in Table 2.4.1.)

Table 2.4.1. Uniform bounds for spectral radius and norm of the two-grid iteration matrix (4)

ν	1	2	3	4	5	10
ϱ_ν	$\frac{1}{2}$	$\frac{1}{4}$	$\frac{1}{8}$	0.0832	0.0671	0.0350
ζ_ν	$\frac{1}{2}$	$\frac{1}{4}$	0.150	0.1159	0.0947	0.0496

Proof. $M_l^{(\mu)} = \begin{bmatrix} \alpha & \beta \\ \alpha & \beta \end{bmatrix}$ with $\alpha = s_\mu^2 c_\mu^{2\nu}$, $\beta = c_\mu^2 s_\mu^{2\nu}$. Since

$$\varrho\left(\begin{bmatrix} \alpha & \beta \\ \alpha & \beta \end{bmatrix}\right) = \alpha + \beta \quad (\alpha, \beta \geqq 0),$$

$$\left\|\begin{bmatrix} \alpha & \beta \\ \alpha & \beta \end{bmatrix}\right\| = \left\|\begin{bmatrix} \alpha & \beta \\ \alpha & \beta \end{bmatrix}\begin{bmatrix} \alpha & \alpha \\ \beta & \beta \end{bmatrix}\right\|^{1/2} = \sqrt{\alpha^2 + \beta^2}\left\|\begin{bmatrix} 1 & 1 \\ 1 & 1 \end{bmatrix}\right\|^{1/2} = \sqrt{2(\alpha^2 + \beta^2)}$$

(cf. (1.3.15a)) one obtains

$$\varrho(M_l^{(\mu)}) = \varrho_\nu(s_\mu^2), \quad \|M_l^{(\mu)}\| = \zeta_\nu(s_\mu^2) \qquad \text{for } \mu = 1, \ldots, n_{l-1}, \tag{2.4.13}$$

where

$$\varrho_\nu(\xi) = \xi(1-\xi)^\nu + (1-\xi)\,\xi^\nu, \quad \zeta_\nu(\xi) = \sqrt{2[\xi^2(1-\xi)^{2\nu} + (1-\xi)^2\,\xi^{2\nu}]}. \tag{2.4.14}$$

As $\varrho_\nu(\frac{1}{2}) = \zeta_\nu(\frac{1}{2}) = 2^{-\nu}$, Eq. (13) holds also for $\mu = n_{l-1} + 1$ when $M_l^{(n_{l-1}+1)}$ is the 1×1 matrix $2^{-\nu}$ (cf. (9)). Lemma 2.4.1 implies

$$\begin{aligned} \varrho(M_l) &= \max\{\varrho_\nu(s_\mu^2)\colon 1 \leqq \mu \leqq n_{l-1} + 1\}, \\ \|M_l\| &= \max\{\zeta_\nu(s_\mu^2)\colon 1 \leqq \mu \leqq n_{l-1} + 1\}. \end{aligned} \tag{2.4.15}$$

The arguments $s_\mu^2 = \sin^2(\mu\pi h_l/2)$ lie in the interval $(0, \frac{1}{2})$; hence (10) and (11) follow.

The asymptotic formula (12) can be derived from the fact that the leading terms of $\varrho_\nu(\xi)$ and $\zeta_\nu(\xi)$ are $\xi(1-\xi)^\nu$ and $\sqrt{2}\,\xi(1-\xi)^\nu$. Their maxima are attained at $\xi_0 = 1/(\nu+1)$ and result in $\varrho_\nu(\xi_0) \approx \xi_0(1-\xi_0)^\nu = (1 - 1/(\nu+1))^\nu/(\nu+1) \approx 1/(e\nu)$ and $\zeta_0(\xi_0) \approx \sqrt{2}/(e\nu)$. $\square$

The bounds ϱ_ν and ζ_ν are uniform with respect to l, hence uniform with respect to the grid size h_l. The spectral radii of the classical iterative methods depend on h_l and tend to 1 for $h_l \to 0$. Examples are Jacobi's iteration and the Gauß-Seidel iteration with $\varrho(M_l) = 1 - O(h_l^2)$ and successive overrelaxation with $\varrho(M_l) = 1 - O(h_l)$ at best. In sharp contrast to this behaviour, the two-grid iteration as well as the multi-grid iteration defined below have spectral radius and contraction number uniformly bounded by some number smaller than 1. As a consequence, an accuracy ε can be obtained by $j = O(\log 1/\varepsilon)$ iterations, where j is independent of h_l. Of course the numbers $\varrho(M_l)$ and $\| M_l \|$ depend on l, but they approach their bounds ϱ_ν and ζ_ν very quickly (cf. Table 2.4.2).

Table 2.4.2. Values of $\varrho(M_l)$ and $\| M_l \|$ for $\nu = 3$

l	1	2	3	4	
h_l	$\frac{1}{4}$	$\frac{1}{8}$	$\frac{1}{16}$	$\frac{1}{32}$	bounds
$\varrho(M_l(3))$	0.0938	0.1123	0.1248	0.1250	$\varrho_3 = \frac{1}{8}$
$\| M_l(3)) \|$	0.1288	0.1471	0.1484	0.1501	$\zeta_3 = 0.1501$

The reason is stated in

Note 2.4.5. $\varrho(M_l(\nu)) = \varrho_\nu - O(h_l^2)$, $\| M_l(\nu) \| = \zeta_\nu - O(h_l^2)$.

Proof. $\hat{\varrho}_\nu(x) := \varrho_\nu(\sin^2(x\frac{\pi}{2}))$ (cf. (14)) takes its maximum ϱ_ν at some inner point $x_0 \in (0, \frac{1}{2})$. Choose $x' = \mu h_l (1 \leqq \mu \leqq n_{l-1})$ closest to x_0. Equation (15) implies $\varrho(M_l(\nu)) \geqq \hat{\varrho}_\nu(x')$. Because of $\frac{d}{dx} \hat{\varrho}_\nu(x_0) = 0$, $\hat{\varrho}_\nu(x') = \hat{\varrho}_\nu(x_0) + O(|x_0 - x'|^2) = \varrho_\nu + O(h_l^2)$ holds and completes the proof. □

The rate ϱ_ν and the contraction number ζ_ν improve with increasing ν. However, they do not decrease exponentially ($\approx C\varepsilon^\nu$) but as C/ν. This behaviour will be shown to hold for much more general boundary value problems also. Sometimes C/ν has to be changed to C/ν^α. For less regular problems $\alpha < 1$ has to be expected, while $\alpha > 1$ would require a prolongation p which is better than piecewise linear interpolation.

A consequence of $\zeta_\nu \approx C/\nu$ is the recommendation not to choose ν too large. Doubling of ν decreases ζ_ν to $\zeta_{2\nu} \approx \zeta_\nu/2$. But also the computational work is almost doubled (at least in the multi-grid version). Thus, the choice of ν is better than 2ν if $\zeta_\nu^2 \leqq \zeta_\nu/2$, i.e. if $\zeta_\nu \leqq \frac{1}{2}$. A more precise discussion is given in §4.3.

It remains to prove Lemma 2.4.3. Equation (5) can be deduced from Eq. (2.5) and $s_{\mu'}^2 = c_\mu^2$.

Equation (6) is an immediate consequence of $S_l = I - \frac{1}{4} h_l^2 L_l$.

To prove the first part of (7) one has to show

$$r e_l^\mu = \frac{1}{\sqrt{2}} c_\mu^2 e_{l-1}^\mu, \qquad r e_l^{\mu'} = -\frac{1}{\sqrt{2}} s_\mu^2 e_{l-1}^\mu \quad (\mu' = n_l + 1 - \mu). \tag{2.4.16}$$

The first equation follows from

$$\begin{aligned}(re_l^\mu)(x) &= \sqrt{2h_l}\,\tfrac{1}{4}[\sin(\mu\pi(x-h_l)) + 2\sin(\mu\pi x) + \sin(\mu\pi(x+h_l))]\\ &= \sqrt{2h_l}\,\tfrac{1}{2}[\cos(\mu\pi h_l)+1]\sin(\mu\pi x) = \sqrt{2h_l}\,c_\mu^2\sin(\mu\pi x)\\ &= \frac{1}{\sqrt{2}}c_\mu^2 e_{l-1}^\mu(x)\end{aligned}$$

for all $x\in\Omega_{l-1}$, since $\frac{1}{2}(\cos(\xi)+1)=\cos^2(\xi/2)$. The second equation of (16) follows formally from the first one since $c_{\mu'}^2 = s_\mu^2$ and $e_{l-1}^{\mu'} = -e_{l-1}^\mu$. The relation $p = 2r^T$ (cf. (3.7 b), (3.9 b)) yields $\hat{p} = 2\hat{r}^T$. Hence the first part of (7) implies the second one. The last row of $\hat{p}$ and the last column of $\hat{r}$ vanish since $re_l^\mu = 0$ for $\mu = n_{l-1}+1 = h_{l-1}^{-1}$. For the proof of $re_l^\mu = 0$ use (16) and note that $e_{l-1}^\mu = 0$ since $e_{l-1}^\mu(vh_{l-1}) = \sqrt{2h_{l-1}}\sin(\mu\pi v h_{l-1}) = \sqrt{2h_{l-1}}\sin(\pi v) = 0$.

The eigenvalue of L_{l-1} is $4h_{l-1}^{-2}\sin^2\frac{\mu\pi h_{l-1}}{2}$ (cf. (5)). Equation (8) follows from $4h_{l-1}^{-2} = h_l^{-2}$ and $\sin^2\frac{\mu\pi h_{l-1}}{2} = \sin^2\mu\pi h_l = 4c_\mu^2 s_\mu^2$. □

2.5 Multi-Grid Method

In the previous section the two-grid method proved to be a very fast iteration. However, in the more interesting case of multi-dimensional boundary value problems the two-grid iteration is impractical because the exact solution of $L_{l-1}v_{l-1} = d_{l-1}$ is required in (3.12 b$_3$). The solution v_{l-1} is needed to compute the correction $\tilde{v}_l = pv_{l-1}$. Since $\tilde{v}_l$ is only an approximation to $v_l = L_l^{-1}d_l$, there is no need for the exact computation of $\tilde{v}_l$ and then of v_{l-1}. It suffices to calculate an approximate solution $\tilde{v}_{l-1}$ of

$$L_{l-1}v_{l-1} = d_{l-1} \tag{2.5.1}$$

For example, the approximation $\tilde{v}_{l-1}$ can be obtained by some iterative process:

$$v_{l-1}^0 := 0 \mapsto v_{l-1}^1 \mapsto \ldots \mapsto v_{l-1}^\gamma =: \tilde{v}_{l-1} \quad (\gamma\text{: number of iterations}).$$

Note that the coarse-grid system (1) is of the same form as the original problem $L_l u_l = f_l$ (only l is replaced by $l-1$). Hence, the two-grid iteration of Sect. 2.3 (with levels $\{l-1, l-2\}$ instead of $\{l, l-1\}$) can be used as an iterative solver of Eq. (1) provided that $l-1>0$. This combination yields a three-grid method involving the levels $l, l-1, l-2$. The exact solution of Eq. (1) is replaced by γ steps of the two-grid iteration at the levels $l-1, l-2$ involving the solution of new auxiliary equations $L_{l-2}v_{l-2} = d_{l-2}$ for γ different right-hand sides d_{l-2}.

If $l-2>0$, the exact solution of $L_{l-2}v_{l-2} = d_{l-2}$ can again be approximated by a two-grid iteration at the levels $l-2$ and $l-3$. The resulting algorithm would be a four-grid method. This process can be repeated until all $l+1$ levels $\{l, l-1, l-2, \ldots, 1, 0\}$ are involved. Equations at level 0 must be solved exactly or approximated by some other iteration. Level 0 corresponds to the coarsest grid and therefore to the smallest system of equations. In our model example (1.5) there is only one ($n_0 = 2^{0+1} - 1 = 1$) unknown $u_0(\frac{1}{2})$ at level 0. The resulting $(l+1)$-level iteration, which is now called a multi-grid iteration, can easily be described by the following recursive program.

multi-grid iteration for solving $L_l u_l = f_l$ (2.5.2)

```
procedure MGM(l,u,f); integer l; array u,f;
if l = 0 then u := L_0^{-1} * f else                      (2.5.2a)
begin integer j; array v,d;
      u := S_l^ν(u,f);                                    (2.5.2b)
      d := r * (L_l * u - f);                             (2.5.2c1)
      v := 0;                                             (2.5.2c2)
      for j := 1 step 1 until γ do MGM(l - 1,v,d);        (2.5.2c3)
      u := u - p * v                                      (2.5.2c4)
end;
```

The meaning of the parameters l, u, f is the same as for the procedure *TGM* from (3.12′). One call of *MGM* performs one iteration of the multi-grid method (at level l).

Comparison of procedure *TGM* with *MGM* shows that the only modification is the replacement of (3.12 b′$_3$) by (2 c$_2$–c$_3$). Instead of solving Eq. (1) exactly, one applies γ iterations of the multi-grid iteration at level $l-1$ with right-hand side d. One call of *MGM* at level l gives rise to γ calls of *MGM* at level $l-1$ involving a further γ^2 calls at level $l-2$, etc. The recursive process ends at level 0, where the auxiliary problems $L_0 v_0 = d_0$ are solved exactly (cf. (2 a)).

A popular language permitting recursive procedure calls is PASCAL (cf. Jensen-Wirth [1]). A complete multi-grid solution of the model problem (1.5) with $f(x) = x$ in this language is given below.

PASCAL multi-grid program (2.5.3)

```
Program multi-grid (output);
const max=128; (* step size 1/128 corresponds to maximal level l = 6 *)
type grid-function=array [0..max] of real;
var i: integer; x: real; u, f, aux, uexact: grid-function;
```

```
function n(l: integer): integer; var k, i: integer; (* n:=2^(l+1) - 1 *)
begin k:= 2; for i:= 1 to l do k:= 2*k; n:= k - 1 end;

procedure compare (i: integer); var s: real; j: integer;
begin s:= 0.0; for j:= 1 to max do s:= s + sqr(u[j] - uexact[j]);
      writeln(i, sqrt(s/max))
end;

procedure mgm(l: integer; var u, f: grid-function);
const ny = 2; gamma = 2;
var i, nl, nl1: integer; v, d: grid-function; h2: real;
procedure restriction-of-defect (var d, u, f: grid-function);
var i: integer;
begin for i:= 1 to nl do aux[i]:= (2.0*u[i] - u[i - 1] - u[i + 1])/h2 - f[i];
      for i:= 1 to nl1 do d[i]:= aux[2*i]/2.0 + (aux[2*i - 1]
                                               + aux[2*i + 1])/4.0
end;
procedure subtract-prolongation (var u, v: grid-function); var i: integer;
begin for i:= 0 to nl1 do
      begin u[2*i]:= u[2*i] - v[i];
            u[2*i + 1]:= u[2*i + 1] - (v[i] + v[i + 1])/2.0
end   end;
procedure smoothing (var u, f: grid-function);
const omega = 0.25; var i: integer;
begin for i:= 1 to nl do aux[i]:= 2.0*u[i] - u[i + 1] - u[i - 1] - h2*f[i];
      for i:= 1 to nl do u[i]:= u[i]-omega*aux[i]
end;
begin if l = 0 then u[1]:= f[1]/8.0 else
      begin nl:= n(l); nl1:= n(l - 1); h2:= 1.0/sqr(nl + 1);
            for i:= 1 to ny do smoothing (u, f);
            restriction-of-defect (d, u, f);
            for i:= 0 to nl1 + 1 do v[i]:= 0.0;
            for i:= 1 to gamma do mgm(l - 1, v, d);
            subtract-prolongation(u, v)
end   end;

begin for i:= 0 to max do
      begin x:= i/max; f[i]:= x; uexact[i]:= x*(1.0 - x*x)/6.0; u[i]:= 0.0
      end;
      compare(0); for i:= 1 to 9 do begin mgm(6, u, f); compare(i) end
end.
```

. . . .

Program (3) produces the output

0	+4.6004370510357E-02	5	+2.3313836339897E-09
1	+1.2710489301310E-05	6	+4.9554907461052E-10
2	+3.4648355529475E-07	7	+1.0811551050020E-10
3	+5.7830989963782E-08	8	+2.4106600310668E-11
4	+1.1321089598505E-08	9	+5.4702747572260E-12

For readers not familiar with recursive algorithms we refer to the flow chart and the FORTRAN program in Sect. 4.1.

We shall see that $\gamma = 1$ or $\gamma = 2$ are appropriate values of the iteration number γ. In the special case of $\gamma = 1$ the multi-grid algorithm *MGM* can be written equivalently as

procedure $MGM(l, u_l, f_l)$;
begin for $k := l$ **step** -1 **until** 1 **do**
  begin $u_k := \mathscr{S}_k^\nu(u_k, f_k)$; $f_{k-1} := r * (L_k * u_k - f_k)$ (2.5.4)
  end;
  $u_0 := L_0^{-1} * f_0$;
  for $k := 1$ **step** 1 **until** l **do** $u_k := u_k - p * u_{k-1}$
end;

The sequence of operations during one step of the multi-grid iteration (4) (i.e. for $\gamma = 1$) is depicted in Fig. 2.5.1 for the example $l = 4$. The stave symbolises the scale of levels.

In the case when $\gamma = 2$ the multi-grid iteration cannot be represented in such a simple form as (4). Figure 2.5.2 shows the details of one multi-grid step at level 4 involving 2 (4, 8) iterations at level 3 (2, 1, resp.).

Due to the form of Figs. 2.5.1–2 the iteration with $\gamma = 1$ is also called a *V*-cycle, while the name *W*-cycle is used for $\gamma = 2$.

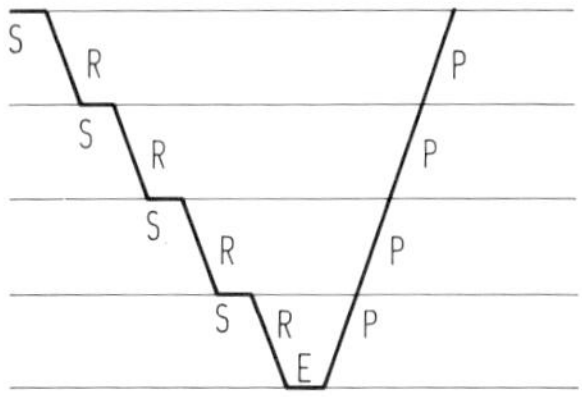

Fig. 2.5.1. One multi-grid iteration for $\gamma = 1$, $l = 4$. S, smoothing step; R, restriction of the defect; E, exact solving at level 0; P, correction $u_k \mapsto u_k - p u_{k-1}$

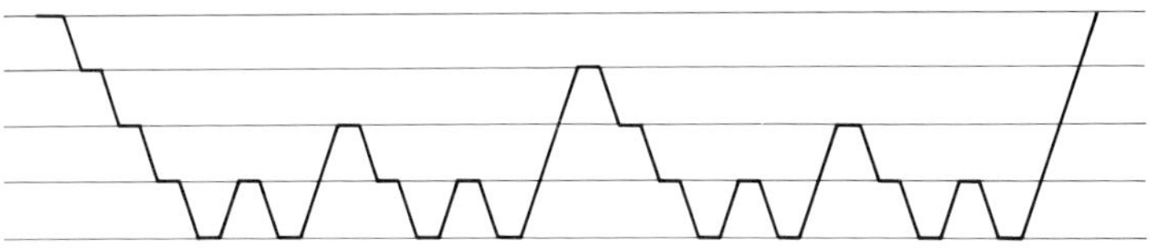

Fig. 2.5.2. One multi-grid iteration for $\gamma = 2$, $l = 4$

In Sect. 7 it will be shown that the properties of the multi-grid iteration are almost the same as those of the two-grid iteration. The contraction numbers of both methods are closely related. The bounds of the multi-grid contraction numbers (and of the convergence rates) are not only independent of the step sizes $\{h_l, h_{l-1}, \ldots, h_0\}$ but also independent of the total number of levels involved in the iteration.

2.6 Comments

2.6.1 Variants of the Two- and Multi-Grid Iteration

Other smoothing iterations. It has already been mentioned that the damped Jacobi iteration (3.1) is a very special example of a smoothing iteration. It was chosen since then the analysis of the two-grid convergence was relatively easy. The choice $\omega = \frac{1}{4}$ in (2.3) is by no means optimal (cf. § 2.6.3). Usually, the Gauß-Seidel iteration is preferred for smoothing. However, Exercise 2.7.6 shows why this iteration cannot serve as a model example. In § 3.3 numerous further smoothing iterations will be discussed.

Post-smoothing. The two-grid iteration (3.13) uses pre- and post-smoothing, i.e. smoothing before and after the coarse-grid correction. The analysis of § 2.4 can easily be extended to this modification. Let $M_l(\nu_1, \nu_2)$ be the two-grid iteration matrix, where ν_1 (ν_2) is the number of pre- (post-)smoothing iterations. The spectral radii of $M_l(\nu_1, \nu_2)$ depend only on the sum $\nu_1 + \nu_2$ (cf. Exercise 2.7.7). However, the spectral norms of $M_l(\nu_1, \nu_2)$ are different as can be seen from Table 2.6.1.

Table 2.6.1. Spectral norms $\| M_l(\nu_1, \nu_2) \|$

ν_2 \ ν_1	0	1	2	3	4
0	$\sqrt{2}$	$\frac{1}{2}$	$\frac{1}{4}$	0.150	0.115
1	1	$\frac{1}{4}$	0.150	0.849	
2	1	0.150	0.849		
3	1	0.115			
4	1				

Table 2.6.2. Norms $| M_l(\nu_1, \nu_2) |_{1 \leftarrow 1}$

ν_2 \ ν_1	0	1	2	3	4
0	1	$\frac{1}{2}$	0.289	0.238	0.222
1	$\frac{1}{2}$	$\frac{1}{4}$	0.128	0.092	
2	0.289	0.128	0.083		
3	0.238	0.092			
4	0.222				

The entries in Table 2.6.1 are the contraction numbers

$$\zeta_{v_1, v_2} = \sup\{\| M_l(v_1, v_2) \| : l = 0, 1, \ldots\}.$$

Obviously, the values $\zeta_{v_1,0}$ coincide with ζ_{v_1} from Table 2.4.1. The choice $v_1 = 0$, $v_2 > 0$ (only post-smoothing) seems to be very unfavourable. The reason for this fact is discussed in § 2.6.2.

Coarse-grid correction. The coarse-grid correction may be modified also. One can change the restriction r or the prolongation p. For example, the trivial injection (3.6) results in spectral radii $\varrho_v = 2^{-v}$ and spectral norms $\zeta_v \approx 1/(e\,v)$ which are unrepresentative for the multi-dimensional case.

We may also try to apply the idea of over- or under-relaxation to the coarse-grid correction. The step (3.12 b$_4$), $u_l^{j+1} = \bar{u}_l - p v_{l-1}$, can be replaced by $u_l^{j+1} = \bar{u}_l - \omega p v_{l-1}$ with suitable ω as proposed by Meis-Branca [1]. However, for the one-dimensional model problem one can show that $\zeta_v \geqq \varrho_v \geqq \max\{1, |1 - \omega|/2^v\}$. Thus the spectral radius (norm) cannot be improved by any choice of ω as long as $v \leqq 3$ ($v \leqq 2$) (cf. Table 2.4.1).

2.6.2 Contraction Numbers with Respect to Other Norms

The contraction number $\| M_l \|_{\mathscr{U} \leftarrow \mathscr{U}}$ depends on the choice of the norm $\| \cdot \|_{\mathscr{U}}$. The standard choice was the Euclidean norm for vectors and the associated spectral norm for matrices. For these norms we introduce new symbols:

$$|\cdot|_0 := \text{Euclidean norm}, \qquad |\cdot|_{0\leftarrow 0} := \text{spectral norm}. \tag{2.6.1 a}$$

From (1.3.16a), $|u_l^{j+1} - u_l|_0 \leqq |M_l|_{0\leftarrow 0}\, |u_l^j - u_l|_0$ ($u_l = L_l^{-1} f_l$), we see that during each iteration the Euclidean norm of the error $u_l^j - u_l$ decreases by the contraction number $|M_l|_{0\leftarrow 0}$.

Instead of observing the error $u_l^j - u_l$, one might be interested in the defect $d_l^j = L_l u_l^j - f_l$. The jth and $(j+1)^{\text{st}}$ iterates are connected by

$$d_l^{j+1} = L_l(u_l^{j+1} - u_l) = L_l M_l (u_l^j - u_l) = L_l M_l L_l^{-1} d_l^j.$$

As $L_l M_l L_l^{-1}$ and M_l are similar, the spectral radii coincide, but in general the new contraction number $|L_l M_l L_l^{-1}|_{0\leftarrow 0}$ governing the reduction of $|d_l^j|_0$,

$$|d_l^{j+1}|_0 \leqq |L_l M_l L_l^{-1}|_{0\leftarrow 0}\, |d_l^j|_0, \tag{2.6.2}$$

differs from $|M_l|_{0\leftarrow 0}$. For vectors v_l and matrices A_l set

$$|v_l|_2 := |L_l v_l|_0, \qquad |A_l|_{2\leftarrow 2} := |L_l A_l L_l^{-1}|_{0\leftarrow 0}. \tag{2.6.1 b}$$

By Exercise 1.5.2, $|\cdot|_2$ is a norm and $|\cdot|_{2\leftarrow 2}$ represents the associated matrix norm. As $|u_l^j - u_l|_2$ equals $|d_l^j|_0$, inequality (2) is equivalent to

$$|u_l^{j+1} - u_l|_2 \leqq |M_l|_{2\leftarrow 2}\, |u_l^j - u_l|_2. \tag{2.6.2'}$$

One learns from (2) and (2′) that there is no fundamental distinction between convergence estimates of the error and convergence estimates of the defect. One is equivalent to the other if the norms are chosen suitably.

In § 2.6.1 we discussed the two-grid iteration (3.13) with pre- and post-smoothing and obtained unfavourable contraction numbers $\zeta_{0,\nu} = 1$ when pre-smoothing was omitted ($\nu_1 = 0$). The situation reverses if we consider contraction numbers with respect to $|\cdot|_2$. Exercise 2.7.8 shows that

$$|M_l(\nu_1, \nu_2)|_{0\leftarrow 0} = |M_l(\nu_2, \nu_1)|_{2\leftarrow 2}. \tag{2.6.3}$$

The rôles of pre- and post-smoothing alternate when the reduction of the error $|u_l^j - u_l|_0$ is replaced by the reduction of the defect $|d_l^j|_0$.

A norm intermediate between $|\cdot|_0$ and $|\cdot|_2$ is $|\cdot|_1$ with associated matrix norm $|\cdot|_{1\leftarrow 1}$:

$$|v_l|_1 := |L_l^{1/2} v_l|_0, \qquad |A_l|_{1\leftarrow 1} := |L_l^{1/2} A_l L_l^{-1/2}|_{0\leftarrow 0}. \tag{2.6.4}$$

($L_l^{1/2}$ and $L_l^{-1/2}$ are well-defined positive definite matrices; cf. Lemma 1.3.5). Because of the representation $|v_l|_1^2 = \langle v_l, L_l v_l\rangle$ this norm is also called the (discrete) *energy norm*. The contraction numbers corresponding to $|\cdot|_1$ behave symmetrically with respect to pre- and post-smoothing:

$$|M_l(\nu_1, \nu_2)|_{1\leftarrow 1} = |M_l(\nu_2, \nu_1)|_{1\leftarrow 1}. \tag{2.6.5}$$

Their uniform bounds ζ_{ν_1,ν_2} are given in Table 2.6.2. The numbers ζ_{ν_1,ν_2}, $\nu_1 + \nu_2 = \nu$ even, are minimised when $\nu_1 = \nu_2 = \nu/2$. Moreover, the contraction number $|M_l(\nu/2, \nu/2)|_{1\leftarrow 1}$ equals the spectral radius $\varrho(M_l(\nu/2, \nu/2))$ (cf. Exercise 2.7.8 and Lemma 1.3.4).

The subscripts 0, 1, 2 of the norms $|\cdot|_0, |\cdot|_1, |\cdot|_2$ indicate the order of the differences involved. This is obvious for 0 and 2. The norm $|v_l|_1$ can be rewritten as $|\partial v_l|_0$ where $\partial v_l(x) = [v_l(x+h) - v_l(x)]/h$, $v_l(0) = v_l(1) = 0$. The norms $|\cdot|_k$ are discrete counterparts of the Sobolev norm $\|\cdot\|_{H^k(0,1)}$ ($k = 0, 1, 2$) of the spaces $H^0(0,1) = L^2(0,1)$, $H_0^1(0,1)$, $H^2(0,1) \cap H_0^1(0,1)$, respectively.

2.6.3 Measuring of the Smoothing Effect

The crucial condition of effective multi-grid convergence is the smoothing effect of the 'smoothing' iteration $\mathscr{S}_l$. Is it possible to measure the smoothing by $\mathscr{S}_l$? It must be recalled that it is not the result $\bar u_l = \mathscr{S}_l(u_l^j, f_l)$ but its error which has to become smooth. The error is

$$\bar v_l = \bar u_l - u_l = \mathscr{S}_l^{\nu}(u_l^j, f_l) - u_l = S_l^{\nu} v_l \quad \text{with} \quad v_l = u_l^j - u_l.$$

One can define the smoothness of $\bar v_l$ by the norm of differences of $\bar v_l$. A special difference operator of second order is the matrix L_l. Thus, $L_l \bar v_l = L_l S_l^{\nu} v_l$ should be relatively small, where 'relative' means that $\| L_l S_l^{\nu} v_l \|$ has to be compared with $\| L_l \| \, \| v_l \|$. We thus obtain the '*smoothing number*'

$$\varrho_L(v) = \sup_{l \geq 1} \| L_l S_l^v \| / \| L_l \| . \tag{2.6.6}$$

No smoothing (e.g. for $v = 0$) is indicated by $\varrho_L = 1$.

Another '*smoothing rate*' is introduced by Brandt [3]. In the model problem the eigenfunctions e_l^μ are divided into low frequencies $1 \leqq \mu \leqq n_{l-1}$ and high frequencies $n_{l-1} + 1 \leqq \mu \leqq n_l$. Let σ_μ be the μth eigenvalue of S_l, $S_l e_l^\mu = \sigma_\mu e_l^\mu$, and define

$$\varrho_B(v) = \sup_{l \geq 1} \max \{ |\sigma_\mu|^v : n_{l-1} + 1 \leqq \mu \leqq n_l \} . \tag{2.6.7}$$

Table 2.6.3 suggests a close relationship between ϱ_L and ϱ_B and the spectral radii ϱ_v from Table 2.4.1. $\varrho_B(v)$ cannot approximate ϱ_v for large v since it decreases exponentially in contrast to $\varrho_v \approx 1/(e\,v)$. The rate ϱ_B is discussed more precisely in § 8.2.

Table 2.6.3. Smoothing numbers $\varrho_L(v)$ and $\varrho_B(v)$ compared with ϱ_v

v	0	1	2	3	4	5
$\varrho_L(v)$	1	$\frac{1}{4}$	0.148	0.105	0.0819	0.06698
$\varrho_B(v)$	1	$\frac{1}{2}$	$\frac{1}{4}$	$\frac{1}{8}$	0.0625	0.03125
ϱ_v	1	$\frac{1}{2}$	$\frac{1}{4}$	$\frac{1}{8}$	0.0832	0.06709

The primary purpose of the smoothing numbers is not to approach the convergence rate of the two- or multi-grid process, but to describe the smoothing ability of the smoothing step $u_l^j \mapsto \mathscr{S}_l(u_l^j, f_l)$ alone. An application follows.

Let $\mathscr{S}_l$ be the damped Jacobi iteration (2.3) with $\omega \in (0, \frac{1}{2})$. One ought to find an optimal ω minimising ϱ_v or ζ_v. E.g., in the first case one has to minimise $\varrho_v = \varrho_v(\omega) = \max \{ P_{v+1}(\xi, \omega) : 0 \leqq \xi \leqq \frac{1}{2} \}$, where P_{v+1} is a certain polynomial of degree $v + 1$. It is easier to optimise the smoothing numbers ϱ_L or ϱ_B:

$$\varrho_L(v;\omega) = \max \{ |1 - 4\omega|^v, \; v^v / [4\omega(1+v)^{1+v}] \},$$

$$\varrho_B(v;\omega) = \max \{ |1 - 2\omega|^v, \; |1 - 4\omega|^v \}.$$

2.6.4 The Multi-Grid Iteration as a New Type of Iteration

The name 'relaxation method' is used for the Gauß-Seidel iteration as well as for a class of generalisations. Examples of accelerated versions are the successive overrelaxation method (SOR) and the symmetric SOR (cf. Young [1, 2, 3], Varga [1]). In § 2.2 we found that the drawback of Jacobi's iteration is the poor reduction of smooth (low frequency) error components. The same is true for the original relaxation method (Gauß-Seidel iteration). The accelerated versions mentioned above have improved convergence properties with respect to the smooth components, but at the same time the convergence with respect to high frequencies deteriorates.

The multi-grid iteration fully preserves the good convergence with respect to high frequencies. The underlying Jacobi or Gauß-Seidel iteration is not accelerated. It is the combination with the coarse-grid correction (having the complementary convergence property) that yields at the same time good convergence with respect to the smooth error components. The smoothing step (e.g. by relaxation) and the coarse-grid correction are of same importance, although relaxation seems to dominate the multi-grid process. A similar situation arises for preconditioned *cg* methods where the combination of both parts (preconditioning and conjugate gradient method) yields the characteristic effectiveness.

2.6.5 Historical Comments

For about one century the Jacobi and Gauß-Seidel method ('relaxation') were the only tools for solving small linear systems iteratively. A lot of interesting variations were proposed which were well-fitted for hand calculations. In particular, Southwell [1], [2], [3] proposed special block and group relaxation methods by which groups of unknowns instead of individual unknowns were changed. Although this approach came near to the use of a coarse grid, it could not develop into a two- or multi-grid iteration since Southwell's techniques were oriented to the particular physical problem and not suited for an abstract algorithm.

It was first the successive overrelaxation method (SOR; cf. Young [1] 1950) which achieved an improvement by a slight systematical modification of the algorithm. From the view point of multi-grid methods the paper of Stiefel [1] is very interesting. He reviewed the difficulties of the Gauß-Seidel iteration, the approaches of Southwell and others, and concluded that all these modifications were developed for hand calculations and did not allow a systematical application as it is needed for a computer program. As a systematical improvement of the Jacobi iteration he invented the conjugate gradient method (*cg*).

Occasionally, also after SOR and *cg* were known, the idea of correcting blocks of unknowns instead of single unknowns was repeated by other authors (e.g. by Wachspress [1]). These approaches can be regarded as pre-versions of two-grid iterations.

The first correct two-grid iteration was described by Fedorenko [1]. He emphasized the complementary roles of the Jacobi iteration and of the coarse-grid correction.

It was also Fedorenko [2] who formulated the first multi-grid algorithm in 1964 and proved the typical convergence behaviour. His proof was restricted to Poisson's equation in a square. The much more complex situation of a difference scheme for a second order elliptic equation with variable coefficients was considered by Bachvalov [1] in 1966. His main aspect was the optimal order of complexity achieved by the multi-grid solution process rather than its practical efficiency.

In 1972 Brandt [1] following the papers mentioned above discovered the efficiency of the multi-grid algorithms. He laid emphasis on the combination of the multi-grid process with additional techniques which should yield a "multi-level adaptive" method. A precise description of his multi-grid algorithms was given in 1975/76 (cf. Brandt [2], [3]). However, in these and later papers the considerations about convergence remained very vague.

An important step towards the convergence analysis was made by Nicolaides [1], [2]. While his first paper described a two-grid iteration, the second one from 1977 studied the convergence of a finite element discretisation.

Bank and Dupont gave two different convergence proofs in a report from 1977, which has since been divided. The second part of the report is published in a journal (cf. Bank-Dupont [2]), whereas the first one, containing a new approach is available only as a report (cf. Bank-Dupont [1]).

In 1976 the author of this book discovered the multi-grid algorithm independently of the papers mentioned above (cf. Hackbusch [1]). The first papers contained software and a convergence analysis by means of the Fourier transform (cf. Hackbusch [1], [2]). In 1977 the results of Nicolaides were generalised (cf. Hackbusch [5]) and a revised version (Hackbusch [21], written in 1978, published 1981) made use of the smoothing and approximation properties.

Reports of Frederickson [1], [2] from 1974/75 describing algorithms closely related to multi-grid iterations inspired Wesseling [1] in 1977 to use a multi-grid method for a special Navier-Stokes problem. The convergence proof of Wesseling [2] followed the considerations of Bachvalov.

In 1980 multi-grid publications appeared by Hemker [1] and the group of Trottenberg, Stüben, Witsch. After 1981 when most of the fundamental multi-grid papers had appeared, the number of multi-grid publications increased rapidly.

Papers concerning special topics will be mentioned in the respective chapters. It remains to draw attention to aggregation-disaggregation methods which have a formal relationship with two-grid algorithms. This approach is a very natural one for problems from economics (cf. Sekerka [1] 1978, Sekerka-Typolt [1] 1968, Mandel [1] 1981. More recently this approach has been attempted for general algebraic problems (cf. Miranker [1], Chatelin-Miranker [1]).

2.7 Exercises

Exercise 2.7.1. Let $\varphi_\nu(x)$ $(1 \leqq \nu \leqq n_l = h_l^{-1} - 1)$ be the 'hat-functions' $\max(0, 1 - |x - \nu h_l|/h_l)$. The finite element solution $u(x) = \sum_{\nu=1}^{n_l} u_l(\nu h_l)\, \varphi_\nu(x)$ of Eq. (1.1) is the solution of

$$\int_0^1 u'(x)\, \varphi_\nu'(x)\, dx = \int_0^1 f(x)\, \varphi_\nu(x)\, dx \qquad \text{for all } 1 \leqq \nu \leqq n_l.$$

Show that $u_l = (u_l(\nu h_l))_{\nu=1}^{n_l}$ is the solution of system (1.5) with slightly changed f_l.

Exercise 2.7.2. The matrices r and p are given in (3.7 b) and (3.9 b). The transposed r^T does not equal p. However, prove that

$$r^* = p \tag{2.7.1}$$

where r^* is the adjoint with respect to the scalar products

$$\langle v_l, w_l \rangle = h_l \sum_{x \in \Omega_l} v_l(x)\, \overline{w_l(x)} \tag{2.7.2}$$

($l = 0, 1, \ldots$) that are special cases of (1.3.10).

Exercise 2.7.3. For L_l and L_{l-1} from (1.5) and p, r from (3.7/9) prove that

$$L_{l-1} = r\, L_l\, p. \tag{2.7.3}$$

Exercise 2.7.4. The coarse-grid correction $\bar{u}_l \mapsto \mathscr{C}_l(\bar{u}_l, f_l)$ is defined in (3.11). Prove: a) $r(L_l \mathscr{C}_l(u_l, f_l) - f_l) = 0$ for all u_l, f_l; b) $\mathscr{C}_l$ is a projection, i.e. $\mathscr{C}_l(\mathscr{C}_l(u_l, f_l), f_l) = \mathscr{C}_l(u_l, f_l)$.

The statements of the next two exercises do not extend to the multi-dimensional case.

Exercise 2.7.5. Show that $\tilde{u}_l = \mathscr{C}_l(u_l, f_l$ satisfies $L_l\, \tilde{u}_l(x) = f_l(x)$ at all coarse-grid points $x \in \Omega_{l-1} \subset \Omega_l$.

Exercise 2.7.6. Define the smoothing procedure $\mathscr{S}_l$ by the Gauß-Seidel iteration corresponding to the ordering

$$(2h, 4h, \ldots, 1 - 4h, 1 - 2h, h, 3h, \ldots, 1 - 3h, 1 - h)$$

of the grid points. Prove: a) For arbitrary u_l^0 let u_l^j be the iterates of the two-grid iteration (3.12) with the new $\mathscr{S}_l$ and with $\nu \geqq 1$. Then $u_l^j = u_l$ (exact solution) for $j \geqq 1$. b) The same result holds for the multi-grid iteration with $\gamma \geqq 1$.

Hint. Prove $L_l\, u_l^j(x) = f_l(x)$ for $x \in \Omega_l \backslash \Omega_{l-1}$ and use Exercise 2.7.5.

Remark. Exercise 2.7.6 shows a close relationship between the one-dimensional multi-grid iteration and the direct reduction method of Schröder-Trottenberg [1] (cf. also Schröder-Trottenberg-Witsch [1], Trottenberg [1]).

Exercise 2.7.7. Let $M_l(\nu_1, \nu_2)$ be the iteration matrix of the two-grid iteration (3.13). Prove $\varrho(M_l(\nu_1, \nu_2)) = \varrho(M_l(\nu_1', \nu_2'))$ for $\nu_1 + \nu_2 = \nu_1' + \nu_2'$. Why does this statement not extend to the multi-grid version of (3.13)?

Exercise 2.7.8. Let $M_l(\nu_1, \nu_2)$ be as in Exercise 2.7.7. Prove that

a) $M_l(\nu_1, \nu_2)^* = L_l\, M_l(\nu_2, \nu_1)\, L_l^{-1}$,
b) $L_l^{1/2}\, M_l(\nu, \nu)\, L_l^{-1/2}$ is symmetric.

3. General Two-Grid Method

The complete multi-grid process will be developed in three stages (§ 3: two-grid iteration, § 4: multi-grid iteration, § 5: nested iteration). The two-grid algorithm serves only as an (impractical) illustration preparing for the multi-grid method which might seem intricate if introduced at once. Although the multi-grid iteration of § 4 is already effective, it can still be improved by the nested iteration (§ 5).

The term 'general' in the heading of this section indicates that we do not restrict our description to a few model problems. However, there is no two- or multi-grid algorithm working as a 'black box' for all boundary value problems. The various components of the algorithm have to be adapted to the special nature of the actual problem as discussed in §§ 10–11.

3.1 The Boundary Value Problem and Its Discretisation

The boundary value problem considered below is a general scalar equation of order $2m$ in a d-dimensional domain. However, we shall often assume a second order problem in a two-dimensional domain to avoid complicated notation. Elliptic systems, which are excluded here, will be studied in § 11.

We consider finite difference discretisations as well as (conforming) finite element discretisations. It is not intended to present an introduction to discretisation techniques. The only purpose of §§ 3.1.2–3 is to provide some examples, to introduce the notation, and to illustrate some difficulties.

3.1.1 The Boundary Value Problem

The simplest elliptic problem is Poisson's equation

$$-\Delta u = f_\Omega \quad \text{in} \quad \Omega \subset \mathbb{R}^d \tag{3.1.1}$$

with Dirichlet boundary conditions

$$u = f_\Gamma \quad \text{on} \quad \Gamma. \tag{3.1.2a}$$

Ω is an (open) domain of the d-dimensional space. Γ is its boundary. The Laplacian operator is defined by

$$\Delta = \sum_{i=1}^{d} \partial^2/\partial x_i^2,$$

where $\mathbf{x} = (x_1, x_2, \ldots, x_d) \in \mathbb{R}^d$ are the arguments of u and f. In the two-dimensional case we use (x, y) instead of (x_1, x_2) and write $\Delta = \partial^2/\partial x^2 + \partial^2/\partial y^2$.

Alternative boundary conditions are the Neumann condition

$$\partial u/\partial n = f_\Gamma \quad \text{on} \quad \Gamma, \tag{3.1.2b}$$

where $\partial/\partial n$ is the directional derivative with respect to the outer normal $\mathbf{n}$ at the respective point $\mathbf{x} \in \Gamma$, and the mixed (or third) boundary condition

$$\alpha u + \partial u/\partial n = f_\Gamma \quad \text{on} \quad \Gamma. \tag{3.1.2c}$$

For special geometries of Ω one can also pose periodic conditions. Further different types of conditions may be required on different parts of Γ. The general notation for the boundary condition is

$$L_\Gamma u = f_\Gamma \quad \text{on} \quad \Gamma. \tag{3.1.3}$$

The general linear equation of second order in two dimensions is

$$[a_{11}(x, y)\, u_x]_x + [a_{22}(x, y)\, u_y]_y + a_1(x, y)\, u_x + a_2(x, y)\, u_y + a(x, y)\, u = f_\Omega \quad \text{in} \quad \Omega \tag{3.1.4a}$$

or with the mixed term

$$a_{11} u_{xx} + 2 a_{12} u_{xy} + a_{22} u_{yy} + a_1 u_x + a_2 u_y + a u = f_\Omega \quad \text{in} \quad \Omega. \tag{3.1.4b}$$

These equations are elliptic if and only if

$$a_{11}(x, y)\, a_{22}(x, y) > a_{12}^2(x, y) \qquad \text{for all } x, y \in \Omega, \tag{3.1.5a}$$

where $a_{12} = 0$ in the case (4a). Thanks to (5a) the coefficients a_{11} and a_{22} have constant sign. Thus, we may assume

$$a_{11}(x, y) < 0, \qquad a_{22}(x, y) < 0 \quad \text{in} \quad \Omega. \tag{3.1.5b}$$

A general linear differential equation of order $2m$ in a d-dimensional domain can be written as

$$\sum_{|\alpha| + |\beta| \leq 2m} (-1)^{|\alpha|} D^\alpha a_{\alpha\beta}(\mathbf{x})\, D^\beta u = f_\Omega \quad \text{in} \quad \Omega, \tag{3.1.6}$$

where $\alpha = (\alpha_1, \ldots, \alpha_d)$ and β are multi-indices with

$$|\alpha| = \alpha_1 + \ldots + \alpha_d, \quad D^\alpha = \partial^{|\alpha|}/\partial x_1^{\alpha_1}\, \partial x_2^{\alpha_2} \ldots \partial x_d^{\alpha_d}, \quad \xi^\alpha = \xi_1^{\alpha_1} \ldots \xi_d^{\alpha_d}.$$

The ellipticity of Eq. (6) is expressed by

$$\sum_{|\alpha|+|\beta|=2m} a_{\alpha\beta}(\mathbf{x})\,\xi^{\alpha+\beta}>0 \qquad \text{for all } \mathbf{x}\in\Omega,\ 0\neq\xi\in\mathbb{R}^d. \tag{3.1.7}$$

An elliptic equation of order $2m$ requires m boundary conditions. For example, the Dirichlet conditions are

$$(\partial/\partial n)^k u = f_{k,\Gamma} \quad \text{on} \quad \Gamma \qquad \text{for } 0\leqq k\leqq m-1. \tag{3.1.8}$$

These m boundary conditions may also be denoted by (3), $L_\Gamma u = f_\Gamma := (f_{0,\Gamma},\dots,f_{m-1,\Gamma})$.

The general notation for all Eqs. (1), (4a), (4b), (6) is

$$L_\Omega u = f_\Omega \quad \text{in} \quad \Omega. \tag{3.1.9}$$

We combine the differential equation $L_\Omega u = f_\Omega$ and the boundary condition (3) and obtain the abstract equation

$$Lu = f. \tag{3.1.10}$$

Boundary value problems can be posed in a variational (or weak) formulation. Equation (6) with boundary condition (8) becomes: Seek $u\in H_0^m(\Omega)$ with

$$\sum_{|\alpha|,|\beta|\leqq m}\int_\Omega (D^\alpha v(\mathbf{x}))\,a_{\alpha\beta}(D^\beta u(\mathbf{x}))\,d\mathbf{x} = \int_\Omega v f_\Omega\,d\mathbf{x} \qquad \text{for all } v\in H_0^m(\Omega), \tag{3.1.11}$$

provided $a_{\alpha\beta}=0$ in (6) for $|\alpha|>m$ or $|\beta|>m$. Replacing $H_0^m(\Omega)$ by $H^m(\Omega)$, we obtain the 'natural boundary conditions'.

More details about the weak formulation can be found, e.g., in Hackbusch [39], Babuška-Aziz [1] or Lions-Magenes [1, pp. 200]. For the general theory of elliptic equations we refer to Lions-Magenes [1], Miranda [1], Mizohato [1], Nečas [2], Wloka [1].

3.1.2 Finite Difference Discretisation

The differential Eq. (9), $L_\Omega u = f_\Omega$, can be replaced by a difference scheme. Because of the multi-grid approach one needs a sequence of grids. The most convenient choice is as follows. Let

$$h_l = h_0/2^l \qquad (l=0,1,\dots) \tag{3.1.12}$$

and introduce the infinite grid

$$Q_l = \{\mathbf{x}\in\mathbb{R}^d\colon \mathbf{x}=(\alpha_1 h_l,\dots,\alpha_d h_l),\ \alpha_\mu\in\mathbb{Z}\} \tag{3.1.13}$$

and the subset

$$\Omega_l = Q_l\cap\Omega. \tag{3.1.14}$$

The coarsest step size h_0 must be reasonable depending on the diameter and on the shape of the domain (cf. Fig. 2.1.1). The choice (12)–(14) is a very special example. Alternative sequences $\{\Omega_l\}_l$ are mentioned in § 3.4.1.

The well-known difference formula for Poisson's Eq. (1) in the two-dimensional case is the *five-point formula*

$$h_l^{-2}\begin{bmatrix} & -1 & \\ -1 & 4 & -1 \\ & -1 & \end{bmatrix} u_l(x,y)=f_l(x,y):=f(x,y) \quad \text{for } (x,y)\in\Omega_l. \tag{3.1.15}$$

The general nine-point star

$$h_l^{-2}\begin{bmatrix} c_{-1,1} & c_{0,1} & c_{1,1} \\ c_{-1,0} & c_{00} & c_{1,0} \\ c_{-1,-1} & c_{0,-1} & c_{1,-1} \end{bmatrix} u_l(x,y)=f_l(x,y) \tag{3.1.16}$$

is an abbreviation of

$$h_l^{-2}\sum_{-1\leqq i,j\leqq 1} c_{i,j}u_l(x+ih_l,y+jh_l)=f_l(x,y). \tag{3.1.16$'$}$$

The *Mehrstellenverfahren* of Collatz [1] is a scheme of order 4 for Poisson's equation:

$$\frac{1}{6h_l^2}\begin{bmatrix} -1 & -4 & -1 \\ -4 & 20 & -4 \\ -1 & -4 & -1 \end{bmatrix} u_l(x,y)=f_l(x,y):=\frac{1}{12}\begin{bmatrix} & 1 & \\ 1 & 8 & 1 \\ & 1 & \end{bmatrix} f(x,y). \tag{3.1.17}$$

For the construction of schemes of high order we refer, e.g., to Yserentant [1] and Lynch-Rice [1]. A difference formula of second order for Eq. (4a) is the scheme (16) with

$$\begin{aligned} c_{00} &= -[a_{11}(x+h_l/2,y)+a_{11}(x-h_l/2,y)_x \\ &\quad + a_{22}(x,y+h_l/2)+a_{22}(x,y-h_l/2)]+h_l^2\,a(x,y), \\ c_{\pm1,0} &= a_{11}(x\pm h_l/2,y)\pm h_l a_1(x,y)/2, \\ c_{0,\pm1} &= a_{22}(x,y\pm h_l/2)\pm h_l a_2(x,y)/2, \qquad c_{\pm1,\pm1}=0. \end{aligned} \tag{3.1.18}$$

The discretisation of Eq. (4b) requires at least a seven point formula (cf. Mitchell [1, pp. 45]). Analogously, one can define difference formulae for dimensions $d\geqq 3$ or higher order ($2m\geqq 4$).

The boundary value problem consists of the differential equation and the boundary condition. The discrete problem can be partitioned into a system of difference equations, $L_{l,\Omega}u_l=f_{l,\Omega}$, and a further system of discrete boundary conditions $L_{l,\Gamma}u_l=f_{l,\Gamma}$. The complete system would be

$$L_l u_l = f_l \tag{3.1.19}$$

with $L_l=\begin{pmatrix} L_{l,\Omega} \\ L_{l,\Gamma}\end{pmatrix}$, $f_l=\begin{pmatrix} f_{l,\Omega} \\ f_{l,\Gamma}\end{pmatrix}$. However, the usual and more advantageous approach is based on elimination by means of $L_{l,\Gamma}u_l=f_{l,\Gamma}$ as explained below.

Let $L_\Omega u = f_\Omega$ be of second order and assume $\Omega = (0,1)^2 = \{(x,y)\colon 0 < x, y < 1\}$. The equations (16) are to be satisfied at all $(x,y) \in \Omega_l \subset \Omega$. For some $(x,y) \in \Omega_l$ Eq. (16) involves neighbouring grid points $(x', y') \in Q_l \backslash \Omega_l$ outside of Ω_l. If $1/h_0 \in \mathbb{Z}$, all these points (x', y') lie on Γ. Hence the unknown $u_l(x', y')$ can be eliminated by means of the Dirichlet condition (2a): $u_l(x', y') = f_\Gamma(x', y')$. If for instance $(0, y)$ and $(0, y \pm h_l)$ are the only neighbours of $(x,y) = (h_l, y)$ not belonging to Ω_l, Eq. (16) has to be changed into

$$h_l^{-2} \begin{bmatrix} 0 & c_{0,1} & c_{1,1} \\ 0 & c_{0,0} & c_{1,0} \\ 0 & c_{0,-1} & c_{1,-1} \end{bmatrix} u_l(h_l, y) = f_l(h_l, y)$$

$$:= f(h_l, y) - h_l^{-2} \sum_{\mu=-1}^{1} c_{-1,\mu} f_\Gamma(0, y + \mu h_l). \tag{3.1.20}$$

In the case of the Neumann condition (2b), the variables $u_l(0, y)$ can be eliminated by means of $u_l(0,y) = u_l(h_l, y) + h_l f_\Gamma(0, y)$. We obtain finally n_l equations for n_l unknowns $\{u_l(x,y)\colon (x,y) \in \Omega_l\}$, where

$$n_l = \text{number of grid points in } \Omega_l. \tag{3.1.21}$$

Fixing some numbering of the grid points, we obtain the system (19) with an $n_l \times n_l$ matrix L_l.

In the case of general domains Ω, different techniques can be applied. The prototype of a first discretisation technique is the *Shortley-Weller scheme* for Poisson's equation (1) (cf. Shortley-Weller [1], Meis-Marcowitz [1]). In the configuration of Fig. 3.1.1 a the second derivative $-u_{xx}(x,y)$, $(x,y) \in \Omega_l$, can be discretised by

$$h_l^{-2} \left\{ \frac{-2}{s(1+s)} u_l(x - s h_l, y) + \frac{2}{s} u_l(x,y) + \frac{-2}{1+s} u_l(x + h_l, y) \right\}, \tag{3.1.22}$$

where $(x - s h_l, y)$ lies on Γ. Therefore, the unknown $u_l(x - s h_l, y)$ can be eliminated by using the Dirichlet boundary value $f_\Gamma(x - s h_l, y)$.

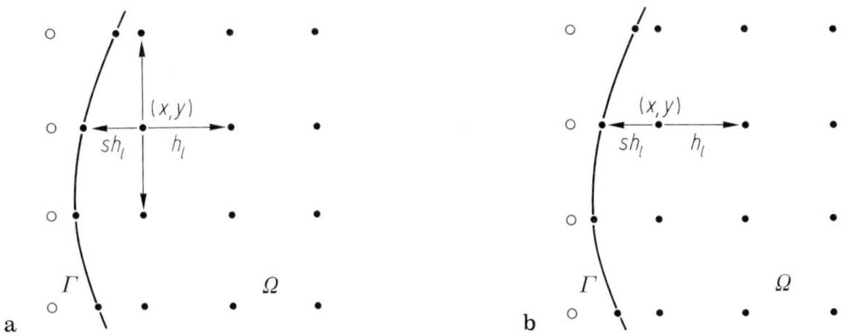

Fig. 3.1.1 a, b. Irregular discretisation near the boundary

The system (19) resulting from (22) is unsymmetric although the original boundary value problem is selfadjoint. A symmetric matrix L_l is obtained by a second discretisation method using the interpolation formula (cf. Fig. 3.1.1 b).

$$h_l^{-2}\left[\frac{1+s}{s}u_l(x,y)-u_l(x+h_l,y)\right]=h_l^{-2}s^{-1}f_\Gamma(x-sh_l,y) \tag{3.1.23}$$

as equation for the unknown $u_l(x, y)$ (cf. Collatz [1]). If $(x, y-h_l)$ is an interior point (i.e. with all neighbours in Ω_l) one must modify Eq. (23) by using in addition an analogous interpolation formula for the y-direction.

We recall that the discretisation near the boundary may be less accurate than the difference scheme used in the interior. E.g., the difference (22) is a first order accurate discretisation causing errors of third order $O(h_l^3)$ with respect to the maximum norm (cf. Bramble-Hubbard [1], Meis-Marcowitz [1]). High order discretisations near the boundary are discussed by Pereyra-Proskurowski-Widlund [1] and van Linde [1].

Difference schemes for Neumann and mixed boundary conditions are described, e.g., by Bramble-Hubbard [2, 3], Törnig [1, pp. 324]. Fourth order equations (e.g. the biharmonic problem) are studied by Bramble [1] and Zlámal [1]. For further monographs we refer to Babuška-Práger-Vitásek [1], Mitchell [1], Hackbusch [39]. A variational approach to difference schemes is described, e.g., by Marchuk [1], Oganesjan-Ruchovec [1].

3.1.3 Conforming Finite Element Discretisation

The variational formulation of the boundary value problem (e.g. (11)) will be abbreviated by

$$a(u,v)=(f,v), \quad u\in\mathscr{H}\subset H^m(\Omega), \qquad \text{for all } v\in\mathscr{H}. \tag{3.1.24}$$

In the case of (8) $\mathscr{H}$ equals $H_0^m(\Omega)$, whereas $\mathscr{H}=H^m(\Omega)$ holds for natural boundary conditions. f belongs to the dual space $\mathscr{H}'$. Due to $\mathscr{H}\subset L^2(\Omega)$, $f(v)$ may be written as $(f,v)=(f,v)_{L^2(\Omega)}$ (cf. Note 1.4.1). In the case of inhomogeneous boundary conditions (i.e. $f_\Gamma \neq 0$) the function $f\in\mathscr{H}'$ involves f_Ω and f_Γ.

Assume that a sequence of finitely dimensional subspaces

$$\mathscr{H}_l\subset\mathscr{H} \qquad (l=0,1,\ldots) \tag{3.1.25}$$

is given. The most convenient case for the multi-grid appraoch is a hierarchy of finite element subspaces

$$\mathscr{H}_0\subset\mathscr{H}_1\subset\ldots\subset\mathscr{H}_{l-1}\subset\mathscr{H}_l\subset\ldots\subset\mathscr{H}. \tag{3.1.26}$$

The discrete problem at level l consists in seeking $u^{\mathscr{H}}\in\mathscr{H}_l$ satisfying

$$a(u^{\mathscr{H}},v)=(f,v) \qquad \text{for all } v\in\mathscr{H}_l. \tag{3.1.27}$$

To obtain a system of linear equations one has to choose a basis for $\mathscr{H}_l$. This Rayleigh-Ritz-Galerkin method (cf. Ritz [1]) becomes a finite element method if the basis functions are 'finite elements' (i.e. if they have sufficiently small support). Let $\{\varphi_{l,\nu}: \nu = 1, \ldots, n_l\}$ be a basis, where

$$n_l = \dim \mathscr{H}_l. \tag{3.1.28}$$

Any $u \in \mathscr{H}_l$ can be written as

$$u(\mathbf{x}) = \sum_{\nu=1}^{n_l} \alpha_\nu \varphi_{l,\nu}(\mathbf{x}) \qquad (\mathbf{x} \in \Omega). \tag{3.1.29 a}$$

The 'coefficient vector' $(\alpha_1, \alpha_2, \ldots, \alpha_{n_l})$ is denoted by

$$u_l = (\alpha_1, \alpha_2, \ldots, \alpha_{n_l}). \tag{3.1.29 b}$$

We deliberately use the same symbol for grid functions as in the finite difference case. Usually, there are n_l 'nodal points' $\mathbf{x}^\nu$ $(1 \leqq \nu \leqq n_l)$ such that $\alpha_\nu = u(\mathbf{x}^\nu)$. The space of 'grid functions' or 'coefficient vectors' u_l is denoted by $\mathscr{U}_l$.

Note 3.1.1. Equation (29) defines a bijection

$$u = P_l u_l \in \mathscr{H}_l \tag{3.1.30}$$

from $\mathscr{U}_l$ onto $\mathscr{H}_l$ (and hence into $\mathscr{H}$).

Inserting the representation (29 a) (or equivalently (30)) into Eq. (27) with $v = \varphi_{l,\nu}$ $(1 \leqq \nu \leqq n_l)$, we obtain n_l linear equations for n_l unknowns α_ν. The system can again be rewritten as Eq. (19), $L_l u_l = f_l$. The solution is the coefficient vector u_l of the solution $u^{\mathscr{H}} = P_l u_l$ of problem (27). The stiffness matrix L_l and the right-hand side f_l can be described directly.

Lemma 3.1.2. Let $L: \mathscr{H} \to \mathscr{H}'$ be the operator associated with the sesquilinear form $a(u, v)$ according to Proposition 1.4.7: $a(u, v) = (Lu, v)$, where $(\cdot, \cdot) = (\cdot, \cdot)_{L^2(\Omega)}$. Let $\langle \cdot, \cdot \rangle$ be the Euclidean scalar product on $\mathscr{U}_l \times \mathscr{U}_l$ and define the mapping $R_l: \mathscr{H}' \to \mathscr{U}_l' = \mathscr{U}_l$ as adjoint of P_l:

$$R_l = P_l^* \qquad (P_l \text{ from } (30)). \tag{3.1.31}$$

Then L_l and f_l in Eq. (19), $L_l u_l = f_l$, are given by

$$L_l = R_l L P_l, \quad f_l = R_l f. \tag{3.1.32}$$

Proof. The solution u_l with $u^{\mathscr{H}} = P_l u_l$ has to satisfy $a(P_l u_l, v) = (f, v)$ for all $v \in \mathscr{H}_l$. By Note 3.1.1, this equation is equivalent to $a(P_l u_l, P_l v_l) = (f, P_l v_l)$ for all $v_l \in \mathscr{U}_l$. From

$$a(P_l u_l, P_l v_l) = (L P_l u_l, P_l v_l) = \langle P_l^* L P_l u_l, v_l \rangle, \qquad (f, P_l v_l) = \langle P_l^* f, v_l \rangle$$

we conclude that $\langle P_l^* L P_l u_l, v_l \rangle = \langle P_l^* f, v_l \rangle$ for all $v_l \in \mathscr{U}_l$ and therefore $P_l^* L P_l u_l = P_l^* f$. (31) implies (32). □

A perturbation of the stiffness matrix, e.g., arising from the numerical quadrature is discussed in Exercise 6.6.11.

The primal parameter in the hierarchy is the level number l. A *discretisation parameter* h_l is given indirectly by the supposition that an error estimate

$$\inf_{v \in \mathscr{H}_l} \| u - v \|_{H^m(\Omega)} \leqq C h_l \| u \|_{H^{m+1}(\Omega)} \quad \text{for all } u \in \mathscr{H} \cap H^{m+1}(\Omega) \qquad (3.1.33)$$

is valid. Usually, (33) holds with $h_l = \max\{\text{diameter}(\text{supp}(\varphi_{l,\nu})) \colon 1 \leqq \nu \leqq n_l\}$.

A special finite element hierarchy $\{\mathscr{H}_l\}_l$ will be constructed in § 3.8.2. For the general description of the finite element method we refer to Hinton-Owen [1], Mitchell-Wait [1], Oden-Reddy [1], Schwarz [1], Strang-Fix [1], Wachspress [2], Whiteman [1, 2], Zienckiewicz [1] and to the respective chapters in Marsal [1, pp. 409], Meis-Marcowitz [1, pp. 214], Törnig [1, pp. 330], Hackbusch [39]. The finite element theory is described by Babuška-Aziz [1] and Ciarlet [1]. Applications to Navier-Stokes equations are given by Girault-Raviart [1], Thomasset [1] and Temam [1].

3.2 Two-Grid Algorithm

The given system of linear equations is

$$L_l u_l = f_l \qquad (3.2.1)$$

(cf. (1.19)). The two-grid iteration is exactly the same as in the one-dimensional case given in § 2; we have only to explain the new meaning of the smoothing iteration $\mathscr{S}_l$, of the restriction r, of the prolongation p, and of the coarse-grid matrix L_{l-1}. In the multi-dimensional case there is more than one natural definition of $\mathscr{S}_l, r, p, L_{l-1}$. Various choices will be presented in subsequent sections §§ 3.3–7.

Assuming that $\mathscr{S}_l, r, p, L_{l-1}$ are defined suitably (according to §§ 3.3–7), we can repeat the formulation (2.3.13):[1]

. . . .

two-grid iteration $TGM^{(\nu_1,\nu_2)}$ for solving Eq. (1) (3.2.2)

. . . .

procedure $TGM(l,u,f)$; **integer** l; **array** u,f;
if $l = 0$ **then** $u := L_0^{-1} * f$ **else** (3.2.2a)
begin array d, v;
$u := \mathscr{S}_l^{(\nu_1)}(u,f)$; (pre-smoothing) (3.2.2b)

1 In (2.3.13) the symbol $\mathscr{S}_l^\nu$ denotes the ν-fold application of the smoothing iteration $\mathscr{S}_l$. Here we write $\mathscr{S}_l^{(\nu)}$ to include more general smoothing processes (cf. §§ 3.3.5–6)

$$\left.\begin{array}{l} d := r * (L_l * u - f); \\ v := L_{l-1}^{-1} * d; \\ u := u - p * v; \end{array}\right\} \text{(coarse-grid correction)} \qquad \begin{array}{r}(3.2.2\,\mathrm{c}_1)\\(3.2.2\,\mathrm{c}_2)\\(3.2.2\,\mathrm{c}_3)\end{array}$$

$$u := \mathscr{S}_l^{(\nu_2)}(u, f) \qquad \text{(post-smoothing)} \qquad (3.2.2\,\mathrm{d})$$

end;

. . . .

The procedure *TGM* performs one iteration step of the two-grid algorithm with ν_1 pre- and ν_2 post-smoothing iterations. Modifications have already been mentioned in § 2.6.1 and will be discussed in § 3.8.3.

Of course we cannot state any convergence result before we have fixed the components $\mathscr{S}_l, r, p, L_{l-1}$ of the algorithm. For a reasonable choice we shall prove that the contraction number (and hence also the convergence rate) is proportional to the smoothing number $\varrho_L(\nu)$, $\nu = \nu_1 + \nu_2$:

$$\left.\begin{array}{l}\text{contraction number}\\ \text{convergence rate}\end{array}\right\} \leqq C\varrho_L(\nu) \tag{3.2.3}$$

with $\varrho_L(\nu) = \sup\{\| L_l S_l^{(\nu)} \| / \| L_l \| : l \geqq 1\}$ introduced in § 2.6.3. C is always used as a generic constant independent of the level l (and therefore independent of the step size h_l). Since $\varrho_L(\nu) \to 0$ $(\nu \to \infty)$, inequality (3) implies convergence for a suitable choice of ν. Under further assumptions convergence (uniform with respect to l) follows for all $\nu > 0$ (cf. § 6).

3.3 Smoothing Iterations

In the following subsections several classic iterations are recalled which may serve as smoothing methods. For boundary value problems behaving like the Poisson equation all these smoothing iterations are effective. However, the special problems discussed later (§§ 10–11) require an adequate choice of the smoothing procedure.

3.3.1 Gauß-Seidel Iteration

The most convenient and often very effective smoothing process is the iteration due to Gauß and Seidel [1]. Let $u_{l,i}$ $(i = 1, \ldots, n_l)$ be the components of u_l and denote the entries of the matrix L_l by $L_{l,ij}$. One step of the Gauß-Seidel iteration is given by [2]

$$\textbf{for } i := 1\ (1)\ n_l \textbf{ do } u_{l,i} := \left[\sum_{\substack{j=1\\ j\neq i}}^{n_l} L_{l,ij} u_{l,j} - f_{l,i}\right] \Big/ L_{l,ii}. \tag{3.3.1}$$

2 The statement '**for** $a := b$ (c) d **do**' will be used as abbreviation of '**for** $a := b$ **step** c **until** d **do**'

For the matrix formulation we have to introduce the decomposition of L_l into

$$L_l = D_l - A_l - B_l, \tag{3.3.2}$$

D_l: diagonal matrix,

A_l (B_l): strictly lower (upper) triangular matrix.

Iteration (1) is equivalent to

$$u_l^{j+1} = (D_l - A_l)^{-1}(B_l u_l^j + f_l). \tag{3.3.3}$$

Therefore, $\mathscr{S}_l$ and its related matrices S_l, T_l (cf. (2.2.6 a)) are

$$\begin{aligned}\mathscr{S}_l(u_l, f_l) &= (D_l - A_l)^{-1}(B_l u_l + f_l) = S_l u_l + T_l f_l,\\ S_l &= (D_l - A_l)^{-1} B_l = I - (D_l - A_l)^{-1} L_l,\\ T_l &= (D_l - A_l)^{-1}.\end{aligned} \tag{3.3.4}$$

The splitting (2) is unique, but the representation of a discrete boundary value problem by an equation $L_l u_l = f_l$ is by no means unique. The construction of L_l depends on the numbering of the unknowns (i.e. of the grid points).

Note 3.3.1. The Gauß-Seidel iteration depends on the ordering of the unknowns.

For the two-dimensional case we list several orderings. Numbering the points $\mathbf{x}^i$, $1 \leqq i \leqq n_l$, of the grid Ω_l, we induce an ordering of the unknowns $u_{l,i} = u_l(\mathbf{x}^i)$.

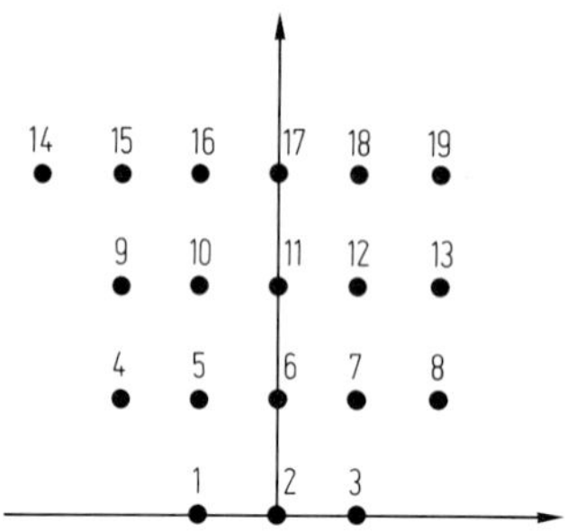

Fig. 3.3.1. Lexicographical ordering of Ω_l

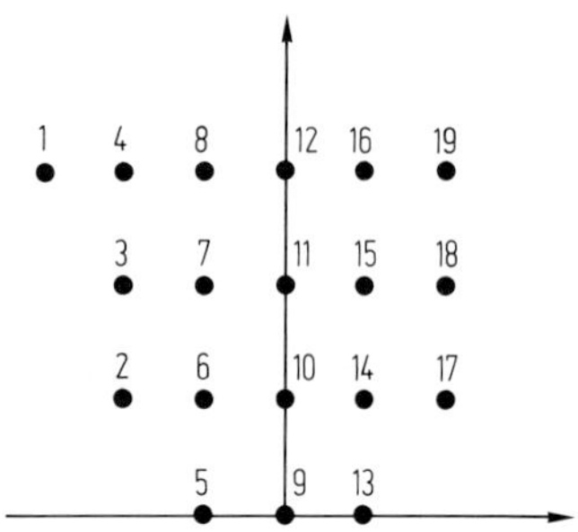

Fig. 3.3.2. Rotated lexicographical ordering

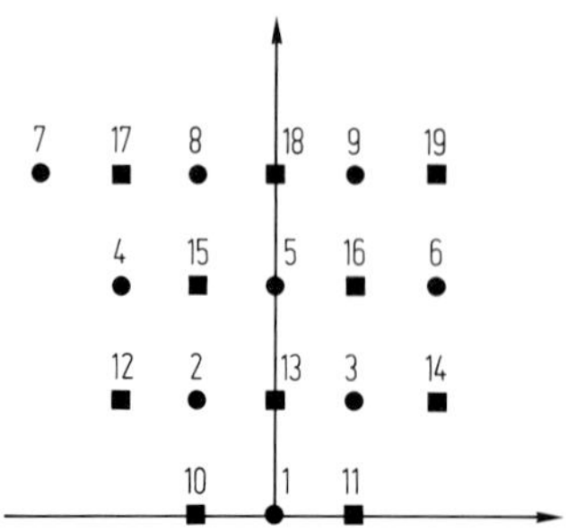

Fig. 3.3.3. Chequer-board ordering

Lexicographical ordering. A grid point $(x, y) \in \Omega_l$ precedes another point $(x', y') \in \Omega_l$ if and only if

$$y < y' \quad \text{or} \quad (y = y' \quad \text{and} \quad x < x')$$

(cf. Fig. 3.3.1).

Rotated lexicographical ordering. Interchanging the rôles of x and y we obtain the rotated lexicographical ordering (cf. Fig. 3.3.2).

Chequer-board ordering or *red-black ordering.* Assume

$$\Omega_l \subset Q_l = \{(\nu h_l, \mu h_l) \colon \nu, \mu \in \mathbb{Z}\}$$

and split Ω_l into 'red' points $\Omega_l^r = \{(\nu h_l, \mu h_l) \in \Omega_l \colon \nu + \mu \text{ even}\}$ and the remaining 'black' ones $\Omega_l^b = \Omega_l \backslash \Omega_l^r$. First, the (red) points of Ω_l^r are numbered in their lexicographical ordering; then, the black points are similarly lexicographically ordered (cf. Fig. 3.3.3).

Zebra-line ordering. Define 'red lines' $\Omega_l^r = \{(\nu h_l, \mu h_l) \in \Omega_l \colon \nu \text{ even}\}$ and 'black lines' $\Omega_l^b = \Omega_l \backslash \Omega_l^r$. Numbering first the points of Ω_l^r and then those of Ω_l^b, we obtain the 'zebra'-line ordering (cf. Fig. 3.3.4).

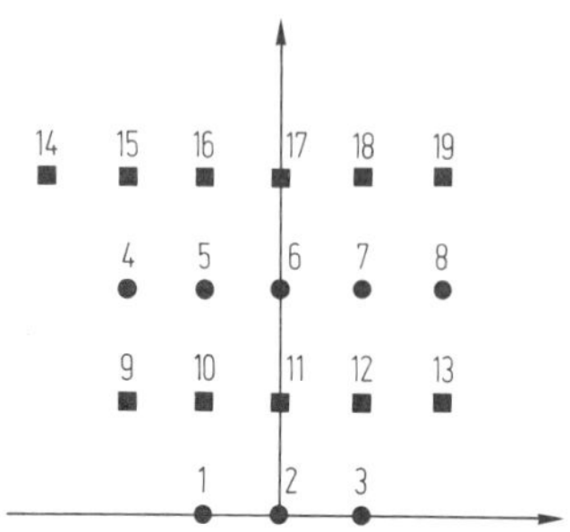

Fig. 3.3.4. Zebra-line ordering

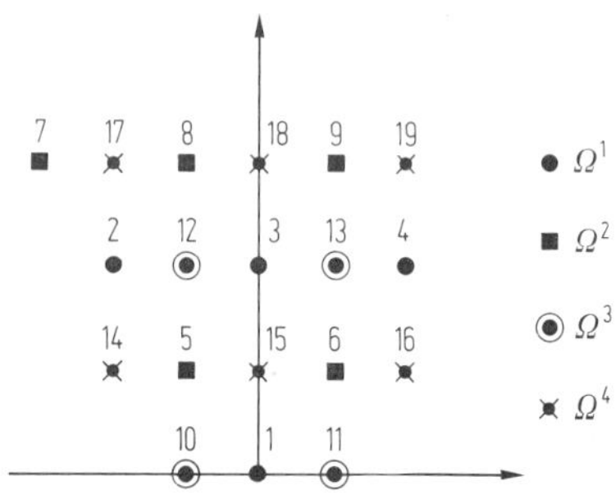

Fig. 3.3.5. Four-colour ordering

Four-colour ordering. Divide Ω_l into four subsets

$$\Omega_l^1 = \{(\nu h_l, \mu h_l) \colon \nu, \mu \text{ even}\},$$
$$\Omega_l^2 = \{(\nu h_l, \mu h_l) \colon \nu, \mu \text{ odd}\},$$
$$\Omega_l^3 = \{(\nu h_l, \mu h_l) \colon \nu \text{ even}, \mu \text{ odd}\},$$
$$\Omega_l^4 = \{(\nu h_l, \mu h_l) \colon \nu \text{ odd}, \mu \text{ even}\}.$$

Number the grid points in turn $\Omega_l^1, \Omega_l^2, \Omega_l^3, \Omega_l^4$. In each subset Ω_l^i, the points are ordered lexicographically (cf. Fig. 3.3.5).

The chequer-board ordering is well-suited for five-point formulae, while the four-colour ordering can be used for general nine-point schemes.

A variant of the Gauß-Seidel iteration is the *symmetric Gauß-Seidel iteration*. Its accelerated variant is known as SSOR (symmetric successive overrelaxation; cf. Young [3], Varga [1]). One iteration can be written as

for $i := 1, 2, \ldots, n_l - 1,\ n_l, n_l - 1, \ldots, 2, 1$ **do**

$$u_{l,i} = \left[\sum_{\substack{j=1 \\ j \neq i}}^{n_l} L_{l,ij} u_{l,j} - f_{l,i} \right] \Big/ L_{l,ii}. \tag{3.3.5}$$

We conclude with the remark that the accelerated Gauß-Seidel iteration (SOR; cf. Young [1, 2], Varga [1]) with optimal relaxation parameter cannot be used as a smoothing iteration. Although the convergence rate is better, the smoothing number is worse.

3.3.2 Modifications of Jacobi's Iterations

Let D_l be the diagonal of the matrix L_l. The original iteration of Jacobi [1] is

$$u_l^{j+1} = u_l^j - D_l^{-1}(L_l u_l^j - f_l)$$

or $u_l^{j+1} = D_l^{-1}[(A_l + B_l) u_l^j + f_l]$ with A_l, B_l from (2). Choosing $0 < \theta < 1$, e.g. $\theta = \frac{1}{2}$, we obtain the *damped Jacobi iteration*

$$u_l^{j+1} = u_l^j - \theta D_l^{-1}(L_l u_l^j - f_l), \qquad \theta \in (0, 1). \tag{3.3.6}$$

The corresponding smoothing iteration is

$$\mathscr{S}_l(u_l, f_l) = u_l - \theta D_l^{-1}(L_l u_l - f_l), \quad S_l = I - \theta D_l^{-1} L_l, \quad T_l = \theta D_l^{-1}. \tag{3.3.6'}$$

In contrast to Note 3.3.1 we have

Note 3.3.2. The Jacobi iteration (6) and the variants defined below are independent of the ordering of the unknowns.

Usually, the diagonal entries are of order $O(h_l^{2m})$, where $2m$ is the order of Eq. (1.9). Hence, in the case of constant diagonal elements, the factor θD_l^{-1} equals ωh_l^{2m} (cf. (2.2.2–3)). Iteration (6) becomes

$$u_l^{j+1} = u_l^j - \omega h_l^{2m}(L_l u_l^j - f_l) \tag{3.3.7}$$

and $\mathscr{S}_l$ is

$$\mathscr{S}_l(u_l, f_l) = u_l - \omega h_l^{2m}(L_l u_l - f_l), \quad S_l = I - \omega h_l^{2m} L_l, \quad T_l = \omega h_l^{2m} I. \tag{3.3.7'}$$

A possible choice of ω is

$$\omega = \omega_l = \| h_l^{2m} L_l \|^{-1} \qquad (\| \cdot \|\text{: spectral norm}) \tag{3.3.8 a}$$

or

$$\omega = 1/C_L, \quad C_L := \sup_{l \geq 1} \| h_l^{2m} L_l \|. \tag{3.3.8 b}$$

If $D_l \neq \text{const} \cdot I$, the analysis of (7) is simpler than the analysis of (6).

Replacing θD_l^{-1} in (6) or ωh_l^{2m} in (7) by a general matrix A_l, we are led to

$$u_l^{j+1} = u_l^j - A_l(L_l u_l^j - f_l) \tag{3.3.9}$$

which should not be called a modified or generalised Jacobi iteration, since any iteration can be presented in form (9) with $A_l = -T_l$ (cf. (2.2.6a)). However, the choice $A_l = \omega^2 h_l^{4m} L_l^*$ yields

$$u_l^{j+1} = u_l^j - \omega^2 h_l^{4m} L_l^*(L_l u_l^j - f_l) \tag{3.3.10}$$

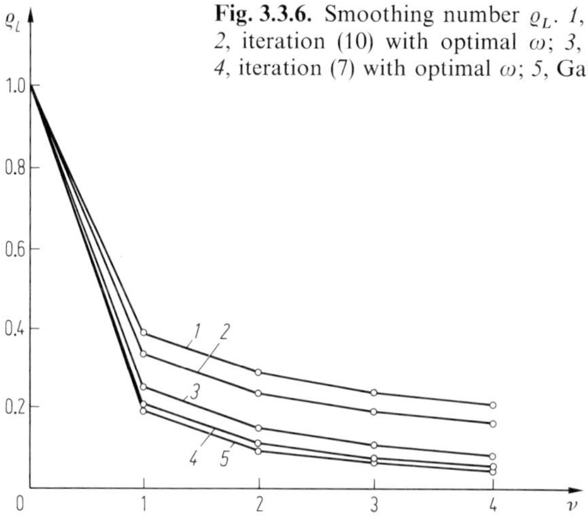

Fig. 3.3.6. Smoothing number ϱ_L. *1*, iteration (10) with ω from (8b); *2*, iteration (10) with optimal ω; *3*, iteration (7) with ω from (8b); *4*, iteration (7) with optimal ω; *5*, Gauß-Seidel iteration (3)

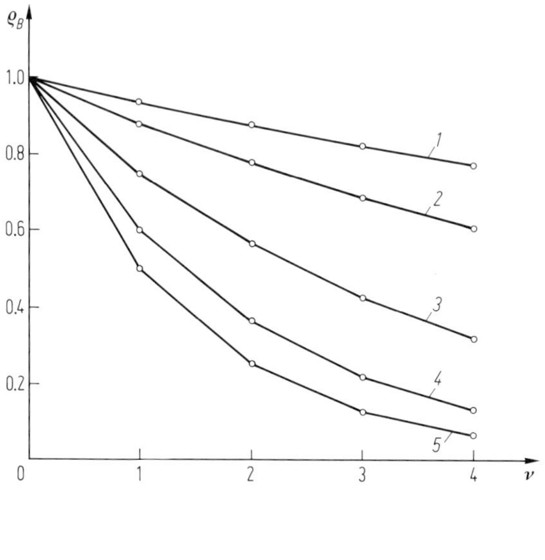

Fig. 3.3.7. Smoothing rate ϱ_B

which can be regarded as iteration (7) applied to the 'squared equation' $(L_l^* L_l) u_l = L_l^* f_l$. Using the choice (8 b) of the damping factor with the new matrix $L_l^* L_l$, we obtain $[\sup \| h_l^{4m} L_l^* L_l \|]^{-1} = [\sup \| h_l^{2m} L_l \|]^{-2} = \omega^2$ (cf. (1.3.15 a, b)) with ω from (8 b). The corresponding mapping $\mathscr{S}_l$ and its matrices S_l, T_l are

$$\mathscr{S}_l(u_l, f_l) = u_l - \omega^2 h_l^{4m} L_l^* (L_l u_l - f_l), \qquad S_l = I - \omega^2 h_l^{4m} L_l^* L_l, \\ T_l = \omega^2 h_l^{4m} L_l^* f_l. \tag{3.3.10'}$$

Usually the damped Jacobi iteration (7), (8 b) is better for smoothing than iteration (10), (8 b), but worse than the Gauß-Seidel iteration (3). We illustrate this statement by means of the smoothing numbers $\varrho_L(v)$ and $\varrho_B(v)$ for the five-point discretisation (1.15) of Poisson's equation (1.1) (ϱ_L and ϱ_B are defined in § 2.6.3; for the calculation of ϱ_L, ϱ_B in this particular case we refer to § 6.2.3 and Example 8.2.7). Equation (8 b) yields $\omega = 0.125$ as the relaxation parameter in the Jacobi iterations (7) and (10). This value is not optimal. In the case of (7) $\varrho_L(v)$ is minimised for $\omega_{\text{opt}} = \omega_{\text{opt}}(v)$ [values: 0.151, 0.167, 0.178, 0.186 for $v = 1, 2, 3, 4$], while ϱ_B is optimal at $\omega_{\text{opt}} = 0.2$. Even in the optimal case the smoothing numbers of the Jacobi iteration (7) are worse than for the Gauß-Seidel iteration. We remark that in the case of the Gauß-Seidel iteration ϱ_L refers to the chequer-board ordering, whereas ϱ_B is evaluated for the lexicographical ordering.

Further modifications of the Jacobi iteration are mentioned at the end of § 3.3.3 and in § 3.3.5.

3.3.3 Blockwise Iterations

The Gauß-Seidel iteration (1) is also called the pointwise Gauß-Seidel method. The original approach of Gauß is to solve the system iteratively with respect to blocks of unknowns. Such blocks arise naturally for discrete boundary value problems.

Let $(u_{l,1}, u_{l,2}, \ldots)$ be an ordering of the unknowns and divide the index set $I_l = \{1, \ldots, n_l\}$ into $k = k_l$ subsets

$$I_l^j = \{i_{j-1} + 1, i_{j-1} + 2, \ldots, i_j\} \quad \text{for } 1 \leqq j \leqq k \\ \text{with} \quad 0 = i_0 < i_1 < \ldots < i_k = n_l.$$

Then the unknowns $u_l^j := (u_{l,i})_{i \in I_l^j}$ form the jth block of u_l. Let L_l be decomposed into

$$L_l = D_l - A_l - B_l, \\ D_l\text{: block-diagonal matrix,} \\ A_l \ (B_l)\text{: strictly lower (upper) block-triangular matrix.} \tag{3.3.11}$$

The precise definition, e.g. of 'strictly lower block-triangular' is that $A_{l,\nu\mu} = 0$ whenever $\nu \in I_l^j$, $\mu \in I_l^{j'}$ with $j' \geqq j$.

The blockwise Gauß-Seidel iteration is again given by Eq. (3) and $\mathscr{S}_l$ is defined by (4). Loop (1) has to be replaced by

$$\textbf{for } j := 1 \ (1) \ k \ \textbf{do } u_l^j := (L_l^{jj})^{-1} \left[\sum_{\substack{\beta=1 \\ \beta \neq j}}^{k} L_l^{j\beta} u_l^\beta - f_l^j \right], \tag{3.3.12}$$

where the block $L_l^{\alpha\beta} (1 \leqq \alpha, \beta \leqq k)$ is the submatrix $(L_{l,\nu\mu})_{\nu \in I_l^\alpha, \mu \in I_l^\beta}$.

Note 3.3.3. The blockwise Gauß-Seidel iteration depends on the ordering of the blocks. The numbering within the blocks is irrelevant.

x-line Gauß-Seidel iteration. Assume $\Omega_l \subset Q_l := \{(\nu h_l, \mu h_l): \nu, \mu \in \mathbb{Z}\}$ and let $u_{l,i} (1 \leqq i \leqq n_l)$ be the value of u_l at the ith grid point $(x_i, y_i) \in \Omega_l$, which is numbered lexicographically. Let I_l^j be the maximal subset of $\{1, \dots, n_l\}$ such that the component y_i of $(x_i, y_i) \in \Omega_l$ coincides for all $i \in I_l^j$ and that i increases with increasing j. For example, in Fig. 3.3.1 we have $I_l^1 = \{1, 2, 3\}$, $I_l^2 = \{4, 5, 6, 7, 8\}$, $I_l^3 = \{9, \dots, 13\}$, $I_l^4 = \{14, \dots, 19\}$. The iteration (12) with respect to these blocks is called *lexicographical x-line Gauß-Seidel* method.

Let u_l be given in the zebra-line ordering and construct the blocks as above. Then resulting blocks are the x-lines of Fig. 3.3.4. The corresponding Gauß-Seidel iteration is named *x-zebra-line Gauß-Seidel* method.

y-line Gauß-Seidel iteration. Replace the lexicographical ordering by the rotated one (cf. Fig. 3.3.2). By analogous constructions we obtain the *lexicographical y-line Gauß-Seidel* iteration and the *y-zebra-line Gauß-Seidel* method.

For x- or y-line iterations (12) the solution of tridiagonal sub-problems $L_l^{jj} u_l^j = g_l^j$ is required because of

Note 3.3.4. Let the difference formula be some nine-point scheme (1.16). Then the blocks of D_l are tridiagonal. L_l is block-tridiagonal.

Usually, the linewise Gauß-Seidel methods have slightly better smoothing properties than the pointwise ones. For the five-point discretisation of Poisson's equation the smoothing numbers ϱ_B are $\varrho_B(\nu) = (0.25)^{\nu/2}$ for the (pointwise) lexicographical Gauß-Seidel method and $\varrho_B(\nu) = (0.2)^{\nu/2}$ for the lexicographical x-line version.

The *damped linewise Jacobi iteration* is given by Eq. (6) if D_l is not the diagonal but the block-diagonal of L_l (cf. (11)). As this iteration is independent of the ordering of the blocks (cf. Note 3.3.2), there is only one *damped x-line Jacobi iteration* or respectively one *damped y-line Jacobi iteration.*

3.3.4 ADI as a Smoothing Procedure

Split the difference operator L_l into x-differences L_l^x and y-differences L_l^y:

$$L_l = L_l^x + L_l^y. \tag{3.3.13}$$

The *alternating direction implicit iteration* (ADI) or *Peaceman-Rachford method* is defined by the double step

$$\begin{aligned} u_l^{j+1/2} &= [\omega h_l^{-2m} I + L_l^x]^{-1} [(\omega h_l^{-2m} I - L_l^y) u_l^j + f_l], \\ u_l^{j+1} &= [\omega h_l^{-2m} I + L_l^y]^{-1} [(\omega h_l^{-2m} I - L_l^x) u_l^{j+1/2} + f_l] \end{aligned} \tag{3.3.14}$$

(cf. Varga [1, pp. 209]). The matrices S_l and T_l of $\mathscr{S}_l$ are

$$\begin{aligned} S_l &= [\omega h_l^{-2m} I + L_l^y]^{-1} [\omega h_l^{-2m} I - L_l^x] [\omega h_l^{-2m} I + L_l^x]^{-1} [\omega h_l^{-2m} I - L_l^y], \\ T_l &= [\omega h_l^{-2m} I + L_l^y]^{-1} [I + (\omega h_l^{-2m} I - L_l^x)(\omega h_l^{-2m} I + L_l^x)^{-1}]. \end{aligned}$$

In the case of the discrete Poisson equation (five-point formula) in the unit square the optimal smoothing numbers are

$$\varrho_L(1) = 0.0685 \quad \text{for } \omega = 3.037, \qquad \varrho_B(1) = 0.1716 \quad \text{for } \omega = 2.8284.$$

Note that these optimal values of ω are quite different from the value $\omega_{\text{opt}} = 2 \sin(\pi h_l)$ minimising the spectral radius $\varrho(S_l)$.

3.3.5 Semi-Iterative Methods

The modified Jacobi iterations (6, 7, 10) as well as the ADI method involve a parameter ω. Instead of iterating with constant ω, one can choose different ω's. Denote a single iteration by $\mathscr{S}_l(u_l, f_l; \omega)$ to express the dependence on ω. ν steps by $\mathscr{S}_l$ with the parameters $(\omega_1, \omega_2, \ldots, \omega_\nu) =: \boldsymbol{\omega}$ are abbreviated by

$$\mathscr{S}_l^{(\nu)}(\cdot, \cdot; \boldsymbol{\omega}) = \mathscr{S}_l(\cdot, \cdot; \omega_\nu) \times \mathscr{S}_l(\cdot, \cdot; \omega_{\nu-1}) \times \ldots \times \mathscr{S}_l(\cdot, \cdot; \omega_1), \tag{3.3.15a}$$

where $(\mathscr{S}_l(\cdot, \cdot; \omega) \times \mathscr{S}_l(\cdot, \cdot; \omega'))(u_l, f_l) = \mathscr{S}_l(\mathscr{S}_l(u_l, f_l; \omega'), f_l; \omega)$. The iteration matrix of $\mathscr{S}_l^{(\nu)}$ is

$$S_l^{(\nu)} = S_l^{(\nu)}(\boldsymbol{\omega}) = S_l(\omega_\nu) S_l(\omega_{\nu-1}) \ldots S_l(\omega_2) S_l(\omega_1), \tag{3.3.15b}$$

where $S_l(\omega)$ is the iteration matrix corresponding to $\mathscr{S}_l(u_l, f_l; \omega)$.

The damped Jacobi iteration (7) with varying ω is usually called *Richardson's iteration* (cf. Varga [1, p. 141]). Consider a symmetric and positive definite matrix L_l. Let $\nu = 2$ and look for the optimal choice of $\boldsymbol{\omega} = (\omega_1, \omega_2)$ for the damped Jacobi iteration (7). The minimisation of $\varrho_L(2) = \| L_l S_l^{(2)}(\boldsymbol{\omega}) \| / \| L_l \|$ yields

$$\omega_1 = 2.1547 / \| h_l^{2m} L_l \|, \quad \omega_2 = \omega_1 / 2, \quad \varrho_L(2) = 0.0893.$$

For the discrete Poisson equation (1.15), the smoothing number $\varrho_B(2)$ is a minimum for

$$\omega_1 = 2.779/8, \quad \omega_2 = 1.123/8, \quad \varrho_B(2) = 0.2195.$$

$\mathscr{S}_l^{(2)}$ can also be regarded as one step of the iteration

$$u_l^{j+1} = u_l^j - [(\omega_1 + \omega_2) I - \omega_1 \omega_2 h_l^{2m} L_l] h_l^{2m} (L_l u_l^j - f_l)$$

which is called 'weighted Jacobi iteration' by Brandt [3, pp. 341/342]. It is a special case of iteration (9). In general, $\mathscr{S}_l^{(v)}$ can be written as one step of iteration (9) with $A_l = P_{v-1}(h_l^{2m} L_l)$, where $P_{v-1}(x) = \sum_{\mu=0}^{v-1} \alpha_\mu x^\mu$.

3.3.6 Conjugate Gradient Methods

A special class of semi-iterative methods are the conjugate gradient methods due to Stiefel and Hestenes (cf. Stiefel [1], Hestenes-Stiefel [1], Hestenes [1]). Assume $L_l = L_l^* > 0$, $W_l = W_l^* > 0$. The simplest choice $W_l = I$ in iteration (16) yields the standard conjugate gradient method.

$$\begin{aligned} &u_l^0\text{: starting value;}\quad r^0 := f_l - L_l u_l^0;\quad p^0 := W_l^{-1} r^0;\\ &q^k := L_l p^k;\quad a_k := \langle r^k, p^k\rangle / \langle p^k, q^k\rangle;\\ &u_l^{k+1} := u_l^k + a_k p^k;\quad r^{k+1} := r^k - a_k q^k;\quad v^{k+1} := W_l^{-1} r^{k+1};\\ &b_k := -\langle q^k, v^{k+1}\rangle / \langle p^k, q^k\rangle;\quad p^{k+1} := v^{k+1} + b_k p^k. \end{aligned} \tag{3.3.16}$$

Another variant is

$$\begin{aligned} &r^0 := f_l - L_l u_l^0;\quad p^0 := v^0 := W_l^{-1} r^0;\quad q^0 := L_l p^0;\\ &s^k := W_l^{-1} q^k;\quad a_k := \langle v^k, q^k\rangle / \langle q^k, s^k\rangle;\quad u_l^{k+1} := u_l^k + a_k p^k;\\ &r^{k+1} := r^k - a_k q^k;\quad v^{k+1} := v^k - a_k s^k;\quad w^{k+1} := L_l v^{k+1};\\ &b_k := -\langle s^k, w^{k+1}\rangle / \langle s^k, q^k\rangle;\quad p^{k+1} := v^{k+1} + b_k p^k;\quad q^{k+1} := w^{k+1} + b_k q^k. \end{aligned} \tag{3.3.17}$$

The vth iterate is denoted by $u_l^v = \mathscr{S}_l^{(v)}(u_l^0, f_l)$. Although $\mathscr{S}_l^{(v)}$ is nonlinear in u_l^0, there is a close relationship to semi-iterative methods.

Lemma 3.3.5.
(i) In the case of iteration (16), $\mathscr{S}_l^{(v)}(u_l^0, f_l) - L_l^{-1} f_l$ equals $\prod_{\mu=1}^{v} (I - \omega_\mu^* W_l^{-1} L_l)$ $(u_l^0 - L_l^{-1} f_l)$ for certain ω_μ^*, which are optimal in the sense of

$$|L_l^{1/2} [\mathscr{S}_l^{(v)}(u_l^0, f_l) - L_l^{-1} f_l]|_0 = \min_{\omega_\mu} \left| L_l^{1/2} \prod_{\mu=1}^{v} (I - \omega_\mu W_l^{-1} L_l)\, [u_l^0 - L_l^{-1} f_l] \right|_0. \tag{3.3.18 a}$$

(ii) In the case of iteration (17), a similar representation holds with ω_μ^* satisfying

$$\begin{aligned} &|W_l^{-1/2} L_l [\mathscr{S}_l^{(v)}(u_l^0, f_l) - L_l^{-1} f_l]|_0\\ &\quad = \min_{\omega_\mu} \left| W_l^{-1/2} L_l \prod_{\mu=1}^{v} (I - \omega_\mu W_l^{-1} L_l)\, [u_l^0 - L_l^{-1} f_l] \right|_0, \end{aligned} \tag{3.3.18 b}$$

where $|u_l|_0 := \langle u_l, u_l\rangle^{1/2}$.

3.3.7 Other Smoothing Iterations

In § 3.3.2 we mentioned Jacobi's iteration applied to the squared equation $(L_l^* L_l) u_l = L_l^* f_l$. Equation $L_l u_l = f_l$ with u_l replaced by $L_l^* v_l$ yields

$$L_l L_l^* v_l = f_l. \tag{3.3.19}$$

The iteration of Kaczmarz [1],

$$\begin{aligned} &\textbf{for } i := 1 \ (1) \ n_l \ \textbf{do } u_l := u_l - a_i (a_i^T u_l - f_{l,i})/(a_i^T a_i), \\ &a_i^T := i^{\text{th}} \text{ row of } L_l, \end{aligned} \tag{3.3.20}$$

converges for any regular matrix L_l. It can be regarded as a Gauß-Seidel iteration applied to (19).

The (damped) Jacobi iteration applied to (19) is known as the method of Cimmino [1].

Another iterative method somewhat related to iteration (20) is the 'distributed relaxation' described in § 11.3. Quite another method (*ILU*, *ILLU*) will be mentioned in § 10.1.3.

3.4 Prolongation

Here we consider only the finite difference case since the finite element discretisation will be discussed in § 3.6. As the grid structures determine the meaning of the corresponding grid functions, we first look at the grids Ω_l and Ω_{l-1}. For the sake of simplicity only the two-dimensional case is treated.

3.4.1 Fine and Coarse Grid

The most frequent choice of Ω_l and Ω_{l-1} is depicted in Fig. 3.4.1. Ω_l and Ω_{l-1} are square grids with mesh sizes h_l and $h_{l-1} = 2 h_l$. We may also assume that the mesh sizes h_l^x and h_l^y in x- and y-direction are different as long as $h_{l-1}^x / h_l^x = h_{l-1}^y / h_l^y = 2$ holds.

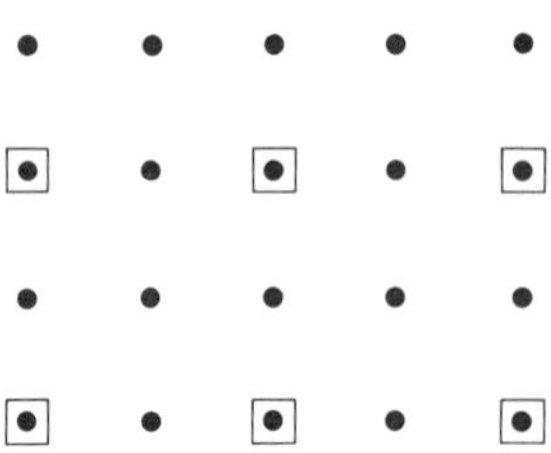

Fig. 3.4.1. Uniform mesh coarsening by factor 2

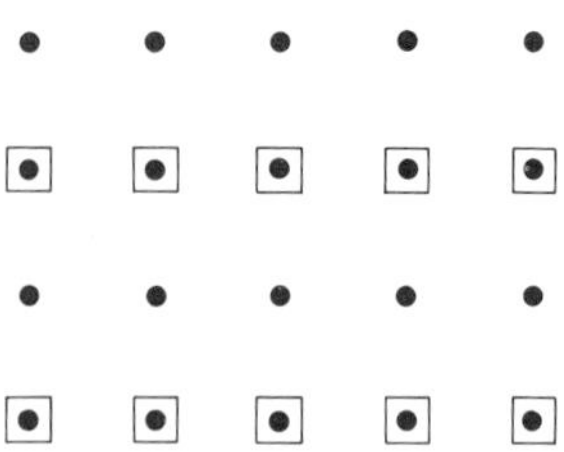

Fig. 3.4.2. Nonuniform mesh coarsening

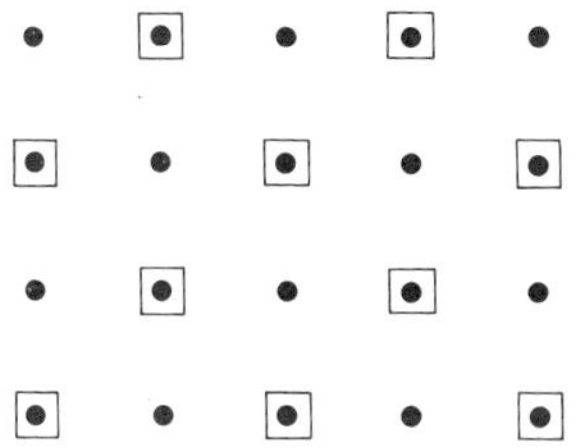

Fig. 3.4.3. Uniform mesh coarsening by factor $\sqrt{2}$

Let Ω_l be a rectangular mesh with step sizes h_l^x, h_l^y. Define the coarse grid by means of $h_{l-1}^x = h_l^x$ and $h_{l-1}^y = 2h_l^y$ (cf. Fig. 3.4.2). Note that a further coarsening with respect to the x-direction ($h_{l-2}^x = 2h_{l-1}^x$, $h_{l-2}^y = h_{l-1}^y$) would yield the same grid at level $l-2$ as obtained in Fig. 3.4.1 at level $l-1$.

Instead of doubling the grid size h_l, it can also be trebled. A larger factor h_{l-1}/h_l would save computational work at the coarse grids, but on the other hand more smoothing is required at the fine grid. It even turns out that a ratio h_{l-1}/h_l less than 2 can be advantageous. Figure 3.4.3 shows how $h_{l-1}/h_l = \sqrt{2}$ can be accomplished. Ω_{l-1} is a rotated square grid (cf. § 4.4.4).

A further situation with $\Omega_l \cap \Omega_{l-1} = \emptyset$ is mentioned in § 3.8.4.

3.4.2 Piecewise Linear Interpolation as Prolongation

It will turn out that the prolongation p can be chosen as piecewise linear interpolation as long as the differential equation is of second order (i.e. $m = 1$ in (1.6)).

Consider first the situation of Fig. 3.4.1 and assume $(0,0)$, $(0,2h)$, $(2h,0)$, $(2h,2h) \in \Omega_l \cap \Omega_{l-1}$. For given v_{l-1} we have to define $v_l = p v_{l-1}$. At the coarse-grid points the values remain unchanged:

$$\begin{aligned} v_l(0,0) &= v_{l-1}(0,0); \qquad & v_l(0,2h) &= v_{l-1}(0,2h); \\ v_l(2h,0) &= v_{l-1}(2h,0); \qquad & v_l(2h,2h) &= v_{l-1}(2h,2h); \end{aligned} \tag{3.4.1 a}$$

where $h = h_l$. For the points $(0,h)$, $(h,0)$, $(2h,h)$, $(h,2h)$ lying directly between coarse-grid points linear interpolation results in

$$\begin{aligned} v_l(0,h) &= \tfrac{1}{2} v_{l-1}(0,0) + \tfrac{1}{2} v_{l-1}(0,2h); \\ v_l(h,0) &= \tfrac{1}{2} v_{l-1}(0,0) + \tfrac{1}{2} v_{l-1}(2h,0); \\ v_l(2h,h) &= \tfrac{1}{2} v_{l-1}(2h,0) + \tfrac{1}{2} v_{l-1}(2h,2h); \\ v_l(h,2h) &= \tfrac{1}{2} v_{l-1}(0,2h) + \tfrac{1}{2} v_{l-1}(2h,2h). \end{aligned} \tag{3.4.1 b}$$

It remains to define v_l at the centre (h,h). One possibility is

$$v_l(h,h) = \tfrac{1}{4} v_{l-1}(0,0) + \tfrac{1}{4} v_{l-1}(0,2h) + \tfrac{1}{4} v_{l-1}(2h,0) + \tfrac{1}{4} v_{l-1}(2h,2h), \tag{3.4.1 c$_1$}$$

another is

$$v_l(h,h) = \tfrac{1}{2} v_{l-1}(0,0) + \tfrac{1}{2} v_{l-1}(2h, 2h). \tag{3.4.1 c$_2$}$$

Analogously v_l is defined in the other cells $[2\nu h, 2\nu h + 2h] \times [2\mu h, 2\mu h + 2h]$.

Nine-point prolongation. The interpolation p by (1 a, b, c$_1$) is called nine-point prolongation (cf. Wesseling [3]) and symbolised by the stencil

$$\begin{bmatrix} \frac{1}{4} & \frac{1}{2} & \frac{1}{4} \\ \frac{1}{2} & 1 & \frac{1}{2} \\ \frac{1}{4} & \frac{1}{2} & \frac{1}{4} \end{bmatrix}. \tag{3.4.2}$$

The reason for the name and the symbol is as follows. Let v_{l-1} be the unit vector being 1 at $(x, y) \in \Omega_{l-1}$ and 0 elsewhere. Its prolongation $v_l = p v_{l-1}$ has exactly nine nonzero values of the pattern (2) at $\{(x', y') \colon x' = x - h_l, x, x + h_l, y' = y - h_l, y, y + h_l\}$.

Note 3.4.1. The nine-point prolongation corresponds to the bilinear finite elements. For example, for $(x, y) \in \Omega_l$, $0 \leqq x$, $y \leqq 2h_l$, the values $v_l(x, y)$ of $v_l = p v_{l-1}$ coincide with the bilinear function $v(x, y) = \alpha_0 + \alpha_1 x + \alpha_2 y + \alpha_3 x y$ interpolating v_{l-1} at $(0,0)$, $(2h, 0)$, $(0, 2h)$, $(2h, 2h)$.

Seven-point prolongation. The interpolation (1 a, b, c$_2$) is called the seven-point prolongation and symbolised by the stencil

$$\begin{bmatrix} 0 & \frac{1}{2} & \frac{1}{2} \\ \frac{1}{2} & 1 & \frac{1}{2} \\ \frac{1}{2} & \frac{1}{2} & 0 \end{bmatrix}. \tag{3.4.3}$$

Again the numbers (3) represent the result of p applied to a unit vector.

Note 3.4.2. The seven-point prolongation corresponds to the linear finite elements over triangles. For $0 \leqq x \leqq y \leqq 2h_l$ the values $v_l(x, y)$ of $v_l = p v_{l-1}$ coincide with the linear function $v(x, y) = \alpha_0 + \alpha_1 x + \alpha_2 y$ interpolating v_{l-1} at $(0,0)$, $(0, 2h_l)$, $(2h_l, 2h_l)$.

Although the star (3) is sparser than (2) and Eq. (1 c$_2$) seems to be simpler than (1 c$_1$), the nine-point prolongation is easier to calculate. For example, consider the grid $\Omega_{l-1} = \{(\nu h_{l-1}, \mu h_{l-1}) \in \bar{\Omega} \colon 0 \leqq \nu, \mu \leqq 1/h_{l-1}\}$ in $\bar{\Omega} = [0,1] \times [0,1]$ and define Ω_l analogously with $h_l = h_{l-1}/2$. The following programmes interpolate the function v which is assumed to be defined on Ω_{l-1}.

nine-point prolongation: (3.4.4)

for $x := 0\ (2h_l)\ 1$ **do for** $y := h_l\ (2h_l)\ 1 - h_l$ **do**
$v(x, y) := [v(x, y - h_l) + v(x, y + h_l)]/2;$
for $y := 0\ (h_l)\ 1$ **do for** $x := h_l\ (2h_l)\ 1 - h_l$ **do**
$v(x, y) := [v(x - h_l, y) + v(x + h_l, y)]/2;$

seven-point prolongation: (3.4.5)

for $x := 0\ (2h_l)\ 1$ **do for** $y := h_l\ (2h_l)\ 1 - h_l$ **do**
$v(x, y) := [v(x, y - h_l) + v(x, y + h_l)]/2;$
for $y := 0\ (2h_l)\ 1$ **do for** $x := h_l\ (2h_l)\ 1 - h_l$ **do**
$v(x, y) := [v(x - h_l, y) + v(x + h_l, y)]/2;$
for $y := h_l\ (2h_l)\ 1 - h_l$ **do for** $x := h_l\ (2h_l)\ 1 - h_l$ **do**
$v(x, y) := [v(x - h_l, y - h_l) + v(x + h_l, y + h_l)]/2;$

The number of operations is the same; however the first prolongation requires only two double-loops. Here we made use of the fact that Eq. ($1c_1$) can be rewritten as

$$v_l(h, h) = \tfrac{1}{2}[v_l(0, h) + v_l(2h, h)] = \tfrac{1}{2}[v_l(h, 2h) + v_l(h, 0)]. \qquad (3.4.1\,c_1')$$

So far we have not taken into account the boundary and its influence upon the interpolation. In § 3.1.2 we proposed to eliminate the variables on the boundary by means of the discretised boundary condition (cf. (1.20)). Therefore, all corrections should satisfy the *homogeneous* boundary condition.

Assume that Dirichlet data are given on Γ and consider the situation of Fig. 3.1.1. Suppose $(x + h_l, y) \in \Omega_{l-1}$, $(x, y) \in \Omega_l \backslash \Omega_{l-1}$, whereas $(x - h_l, y)$ is beyond the boundary. Given the value of v_{l-1} at $(x + h_l, y)$, the value of $v_l = p v_{l-1}$ at (x, y) can be obtained again by piecewise linear interpolation, more precisely by interpolation between $v_{l-1}(x + h_l, y)$ at $(x + h_l, y) \in \Omega_{l-1}$ and the vanishing value at $(x - s h_l, y) \in \Gamma$:

$$v_l(x, y) = \frac{s}{1 + s} v_{l-1}(x + h_l, y).$$

Thus $v_l(x', y')$ can be defined on all lines $y' = \nu h_l$ (ν even). Afterwards the remaining values at $y' = \nu h_l$ (ν odd) are determined by means of linear interpolation with respect to the y-direction. This approach corresponds to (4).

Until now we considered only the uniform doubling of the grid size. In the other situations of § 3.4.1 (cf. Fig. 3.4.2, 3.4.3) the interpolation is still easier since there are less additional fine-grid points $\Omega_l \backslash \Omega_{l-1}$. The extension of the piecewise linear interpolation into three or more dimensions is obvious and need not be defined explicitly.

3.4.3 Interpolation of Higher Order

A boundary value problem of order $2m$ (cf. (1.6)) requires an interpolation of degree m. Since, besides $m = 1$, the only interesting value is $m = 2$, piecewise quadratic prolongation suffices always. Nonetheless, cubic interpolation is often preferred as it is not more involved than quadratic interpolation.

The cubic analogue of the nine-point prolongation can be computed as in (4.4). For $(x, y) = (\nu h_l, \mu h_l)$ (ν odd, μ even) the points $(x \pm h_l, y)$

and $(x \pm 3h_l, y)$ are coarse-grid points. The interpolated value $(p v_{l-1})(x, y)$ is given by

$$v_l(x, y) = -\tfrac{1}{16} v_{l-1}(x - 3h_l, y) + \tfrac{9}{16} v_{l-1}(x - h_l, y) + \tfrac{9}{16} v_{l-1}(x + h_l, y) - \tfrac{1}{16} v_{l-1}(x + 3h_l, y). \tag{3.4.6}$$

Then we can repeat the interpolation process with respect to the y-direction at (x, y) with $y = \nu h_l$, ν odd. Applying this prolongation to a unit vector being 1 at some (interior) point and vanishing elsewhere, we obtain the star

$$\begin{bmatrix} \frac{1}{16^2} & 0 & -\frac{9}{16^2} & -\frac{1}{16} & -\frac{9}{16^2} & 0 & \frac{1}{16^2} \\ 0 & 0 & 0 & 0 & 0 & 0 & 0 \\ -\frac{9}{16^2} & 0 & \frac{81}{16^2} & \frac{9}{16} & \frac{81}{16^2} & 0 & -\frac{9}{16^2} \\ -\frac{1}{16} & 0 & \frac{9}{16} & 1 & \frac{9}{16} & 0 & -\frac{1}{16} \\ -\frac{9}{16^2} & 0 & \frac{81}{16^2} & \frac{9}{16} & \frac{81}{16^2} & 0 & -\frac{9}{16^2} \\ 0 & 0 & 0 & 0 & 0 & 0 & 0 \\ \frac{1}{16^2} & 0 & -\frac{9}{16^2} & -\frac{1}{16} & -\frac{9}{16^2} & 0 & \frac{1}{16^2} \end{bmatrix}. \tag{3.4.7}$$

Note that any bicubic function $\sum_{\nu=0}^{3} \sum_{\mu=0}^{3} \alpha_{\nu\mu} x^\nu y^\mu$ is interpolated exactly by (6).

Formula (6) cannot be applied if one or two neighbours are not inside $\bar{\Omega}$. In the case of Fig. 3.4.4 it is more convenient to use the following quadratic formulae at x and $x + 2h_l$:

$$v_l(x, y) = \frac{s}{1+s}\frac{3}{2} v_{l-1}(x + h_l, y) - \frac{s}{3+s}\frac{1}{2} v_{l-1}(x + 3h_l, y),$$

$$v_l(x + 2h_l, y) = \frac{2+s}{1+s}\frac{1}{2} v_{l-1}(x + h_l, y) + \frac{2+s}{3+s}\frac{1}{2} v_{l-1}(x + 3h_l, y)$$

based on the homogeneous Dirichlet condition $v_{l-1}(x - s h_l, y) = 0$.

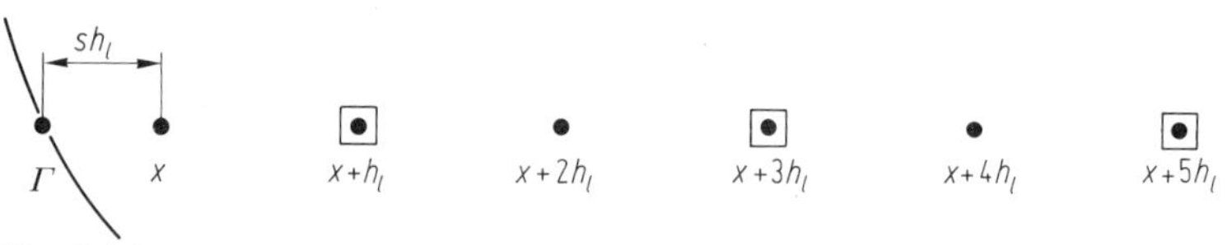

Fig. 3.4.4

The one-sided quadratic interpolation in the interior domain yields

$$v_l(x, y) = -\tfrac{1}{8} v_{l-1}(x - 3h_l, y) + \tfrac{3}{4} v_{l-1}(x - h_l, y) + \tfrac{3}{8} v_{l-1}(x + h_l, y) \tag{3.4.8}$$

leading to the star

$$\begin{bmatrix} 0 & 0 & -\frac{3}{64} & -\frac{1}{8} & -\frac{3}{32} & 0 & \frac{1}{64} \\ 0 & 0 & 0 & 0 & 0 & 0 & 0 \\ 0 & 0 & \frac{9}{32} & \frac{3}{4} & \frac{9}{16} & 0 & -\frac{3}{32} \\ 0 & 0 & \frac{3}{8} & 1 & \frac{3}{4} & 0 & -\frac{1}{8} \\ 0 & 0 & \frac{9}{64} & \frac{3}{8} & \frac{9}{32} & 0 & -\frac{3}{64} \\ 0 & 0 & 0 & 0 & 0 & 0 & 0 \\ 0 & 0 & 0 & 0 & 0 & 0 & 0 \end{bmatrix}. \tag{3.4.9}$$

3.4.4 Modifications

The differential equation can aid the interpolation. We begin with a very simple application. Assume L_l to be a nine-point formula (1.16). The definition (1 c_1 or c_2) of $v_l(h_l, h_l)$ (cf. Fig. 3.4.1) can be changed into

$$(L_l v_l)(h_l, h_l) = 0 \tag{3.4.10}$$

i.e.

$$v_l(h_l, h_l) = \left[\sum_{\substack{i,j=-1 \\ (i,j)\neq(0,0)}}^{+1} - c_{ij} v_l((i+1)h_l, (j+1)h_l) \right] \Big/ c_{00}.$$

Note that for the five-point star (1.15) the Eqs. (1 c_1) and (10) are equivalent. The definition (10) applies to all grid points of the set $\Omega_l^2 = \{(\nu h_l, \mu h_l): \nu, \mu \text{ odd}\}$, which was introduced in § 3.3.1 for the four-colour ordering.

Note 3.4.3. Assume that the pre-smoothing part ends with a Gauß-Seidel step with respect to grid points of Ω_l^2. Apply the coarse-grid correction with p satisfying (10). Then, after the coarse-grid correction the defect of u_l vanishes on Ω_l^2. Hence, a Gauß-Seidel step on Ω_l^2 can be saved in the subsequent smoothing step.

Proof. Since all neighbours of $(x, y) \in \Omega_l^2$ do not belong to Ω_l^2, the Gauß-Seidel step on Ω_l^2 yields $\bar{u}_l$ with vanishing defect on Ω_l^2. The coarse-grid correction changes $\bar{u}_l$ into $\bar{u}_l - p v_{l-1}$. Because of (10), $L_l v_{l-1}$ vanishes on Ω_l^2, too. This proves Note 3.4.3. □

The assertion of Note 3.4.3 can also be stated as follows. Assume that the pre-smoothing part ends with a Gauß-Seidel step on Ω_l^2. Apply the coarse-grid correction with p defined arbitrarily on Ω_l^2. An additional subsequent post-smoothing step on Ω_l^2 re-establishes the result obtained by (10) without post-smoothing.

A special application is given in the case of Fig. 3.4.3 (red-black coarsening), provided that L_l is some five-point scheme. There are two possibilities:

1) After pre-smoothing and coarse-grid correction apply one Gauß-Seidel step on $\Omega_l \backslash \Omega_{l-1}$ as post-smoothing. Then, the definition (and evaluation) of p at $\Omega_l \backslash \Omega_{l-1}$ is not needed at all.

2) Let the pre-smoothing end with a Gauß-Seidel step on $\Omega_l \backslash \Omega_{l-1}$ and apply the coarse-grid correction with p subject to (10) on $\Omega_l \backslash \Omega_{l-1}$. Then post-smoothing on $\Omega_l \backslash \Omega_{l-1}$ can be saved since it does not change the function.

Further modifications will be described in § 10.3.

3.5 Restriction

As long as coarse-grid points are fine-grid points, too, one can define the 'trivial injection' r_{inj}:

$$(r_{\text{inj}}\, d_l)(\mathbf{x}) = d_l(\mathbf{x}) \qquad \text{for all } \mathbf{x} \in \Omega_{l-1} \subset \Omega_l. \tag{3.5.1}$$

With regard to the computational work, the restriction $r = r_{\text{inj}}$ is the optimal one. The evaluation of $r_{\text{inj}}\, d_l$ needs no arithmetical operations; further, the defect $d_l = L_l u_l - f_l$ need not be computed on $\Omega_l \backslash \Omega_{l-1}$. The latter fact reduces the work by 75 pc in the two-dimensional case with $h_{l-1} = 2h_l$.

However, the use of r_{inj} contains some disadvantage and dangers. We give a simple example. Let L_l be a five-point formula and choose Gauß-Seidel's iteration with red-black ordering. There are two possibilities: The smoothing step terminates with a Gauß-Seidel step with respect to the red points (case I) or black points (case II). Since Ω_{l-1} is contained in the set of red points (cf. Fig. 3.3.3), the defect d_l vanishes on Ω_{l-1} in case I. Consequently, $d_{l-1} = r_{\text{inj}}\, d_l$ vanishes and the coarse-grid correction is of no avail. On the other hand, in case II the defect $d_{l-1} = r_{\text{inj}}\, d_l \neq 0$ turns out to be too large. The definition $r = \frac{1}{2} r_{\text{inj}}$ would remedy the situation.

The difficulties can be avoided by using special weighted restrictions as we did in § 2. For this purpose define the scalar product

$$\langle u_l, v_l \rangle = \sum_{\mathbf{x} \in \Omega_l} h_l^d\, u_l(\mathbf{x})\, \overline{v_l(\mathbf{x})} \tag{3.5.2}$$

for all levels (cf. (1.3.10)). d is the dimension: $\Omega \subset \mathbb{R}^d$. For a grid Ω_l with different mesh widths $h_l^{x_1}, \ldots, h_l^{x_d}$ (cf. Fig. 3.4.2) replace the scaling factor h_l^d by $h_l^{x_1} h_l^{x_2} \ldots h_l^{x_d}$. Having defined scalar products, we are able to define adjoint mappings (cf. (1.3.14)).

Let p be one of the prolongations (coarse-to-fine transfers) from the previous section. Its adjoint p^* is a fine-to-coarse transfer and can serve as restriction:

$$r = p^*. \tag{3.5.3}$$

The adjoint of the nine-point prolongation (4.2) is named the *nine-point restriction*. Its symbol is

$$\begin{bmatrix} \frac{1}{16} & \frac{1}{8} & \frac{1}{16} \\ \frac{1}{8} & \frac{1}{4} & \frac{1}{8} \\ \frac{1}{16} & \frac{1}{8} & \frac{1}{16} \end{bmatrix} = \frac{1}{16}\begin{bmatrix} 1 & 2 & 1 \\ 2 & 4 & 2 \\ 1 & 2 & 1 \end{bmatrix}. \tag{3.5.4}$$

By

$$r = \begin{bmatrix} \sigma_{-1,1} & \sigma_{0,1} & \sigma_{1,1} \\ \sigma_{-1,0} & \sigma_{0,0} & \sigma_{1,0} \\ \sigma_{-1,-1} & \sigma_{0,-1} & \sigma_{1,-1} \end{bmatrix} \tag{3.5.5 a}$$

we denote the 'weighted' restriction

$$(r d_l)(x, y) = \sum_{\alpha,\beta=-1}^{+1} \sigma_{\alpha\beta}\, d_l(x + \alpha h_l, y + \beta h_l). \tag{3.5.5 b}$$

Note that the symbols (4.2) and (4) differ by a factor 4, reflecting the fact that the transposed matrix p^T equals $4p^*$. The reason is the different scaling of the scalar products $\langle \cdot, \cdot \rangle_{l-1}$ and $\langle \cdot, \cdot \rangle_l$ by h_{l-1}^d and h_l^d, respectively (cf. Example 1.3.3 and Exercise 2.7.2).

The *seven-point restriction* is the adjoint of the seven-point prolongation (4.3); its symbol is

$$\begin{bmatrix} 0 & \frac{1}{8} & \frac{1}{8} \\ \frac{1}{8} & \frac{1}{4} & \frac{1}{8} \\ \frac{1}{8} & \frac{1}{8} & 0 \end{bmatrix} = \frac{1}{8}\begin{bmatrix} 0 & 1 & 1 \\ 1 & 2 & 1 \\ 1 & 1 & 0 \end{bmatrix}. \tag{3.5.6}$$

In detail we discussed the modifications of prolongations near the boundary. For restrictions such extra considerations are not necessary. Having defined a prolongation p (also near the boundary) we gain a complete description of the restriction r by (3).

For theoretical considerations it is very convenient that $r = p^*$ holds for r and p used in two- (multi-)grid iteration. In practice, however, similar results are observed when $r = \hat{p}^*$ is the adjoint of a prolongation $\hat{p}$ different from p. For example, one may construct a multi-grid iteration with seven-point restriction r and nine-point prolongation p.

In general one should satisfy the condition stated in

Note 3.5.1. Let $2m$ be the order of the differential operator L_Ω (cf. (1.6), (1.9)). Let p be an interpolation of order m_p (this assumption holds if p interpolates polynomials of degree $m_p - 1$ exactly). Further, suppose $r = \hat{p}^*$, where $\hat{p}$ is an interpolation of order m_r. Then, one should satisfy

$$m_p + m_r > 2m. \tag{3.5.7}$$

A justification of this Note can be found in § 6.3.2.2. We give some examples of Inequality (7).

Example 3.5.2. Let $2m = 2$ (second order equation) and let p and $\hat{p} := r^*$ be piecewise linear interpolations (seven or nine-point prolongation). Then $m_p = m_r = 2$ yields $m_p + m_r = 4 > 2 = 2m$.

Example 3.5.3. For $2m = 4$ choose p as piecewise quadratic interpolation and set $r = p^*$. (7) holds because of $m_p + m_r = 3 + 3 = 6 > 4 = 2m$.

A more convenient choice is stated in

Example 3.5.4. For $2m = 4$ choose p as piecewise quadratic (or cubic) interpolation and set $r = \hat{p}^*$, where $\hat{p}$ is the linear interpolation (4.2) or (4.3). As $m_p = 3$ ($m_p = 4$) and $m_r = 2$, inequality (7) is valid.

The trivial injection r_{inj} is the adjoint of $\hat{p} := r_{\text{inj}}^*$ with symbol $\begin{bmatrix} 0 & 0 & 0 \\ 0 & 4 & 0 \\ 0 & 0 & 0 \end{bmatrix}$. Formally, $\hat{p}$ can be regarded as prolongation of order 0 (i.e. $m_r = 0$). According to Note 3.5.1 the order m_p of p should be $m_p = 3$ for second order problems. This is just the choice of Bachvalov [1].

3.6 Canonical Prolongations and Restrictions for Finite Element Equations

Assume that the boundary value problem is described by a bilinear form $a(\cdot,\cdot)$ in $\mathscr{H} \times \mathscr{H}$ with $\mathscr{H} \subset H^m(\Omega)$ (cf. (1.24)). A hierarchy $\mathscr{H}_0 \subset \mathscr{H}_1 \subset \ldots \subset \mathscr{H}_l \subset \ldots \subset \mathscr{H}$ of subspaces is given (cf. (1.26)). The solution of the discrete problem (1.27) is denoted by $u^{\mathscr{H}}$ and belongs to $\mathscr{H}_l$. For the numerical computation one has to fix some basis $\{\varphi_{l,v}: v = 1, \ldots, n_l\}$ of $\mathscr{H}_l$. Every function $u \in \mathscr{H}_l$ is associated with a coefficient vector u_l (cf. (1.29)) belonging to a vector space called $\mathscr{U}_l$. Hereby, a bijection $P_l: \mathscr{U}_l \to \mathscr{H}_l$ is defined (cf. Note 3.1.1). Its adjoint $R_l = P_l^*$ with respect to the scalar products[3] $\langle\cdot,\cdot\rangle$ on $\mathscr{U}_l$ and $(\cdot,\cdot)_{L^2(\Omega)}$ on $\mathscr{H}_l$ is a mapping from $\mathscr{H}'$ (dual of $\mathscr{H}$) onto $\mathscr{U}_l$.

P_l gives rise to a *canonical* choice of p and r. Let $v_{l-1} \in \mathscr{U}_{l-1}$. It corresponds to $v := P_{l-1} v_{l-1} \in \mathscr{H}_{l-1}$. We have to define $v_l = p v_{l-1}$, or equivalently, $v' = P_l v_l \in \mathscr{H}_l$. The canonical choice is $v' := v$ using the inclusion $\mathscr{H}_{l-1} \subset \mathscr{H}_l$. $p v_{l-1}$ is the coefficient vector of $v' = v$ with respect to the basis of $\mathscr{H}_l$. Since P_l^{-1} exists on $\mathscr{H}_l$, we formally have

Definition 3.6.1. The canonical choice of $p: \mathscr{U}_{l-1} \to \mathscr{U}_l$ is $p = P_l^{-1} P_{l-1}$. The canonical restriction is $r = p^*$.

Note 3.6.2. p: $\mathscr{U}_{l-1} \to \mathscr{U}_l$ and r: $\mathscr{U}_l \to \mathscr{U}_{l-1}$ are uniquely defined by

$$P_{l-1} = P_l p, \quad r R_l = R_{l-1}. \tag{3.6.1}$$

3 More generally, the scalar product $(\cdot,\cdot)$ on $\mathscr{H}_l$ must coincide with $(\cdot,\cdot)$ from the right side of (1.24), (1.27)

The first equation expresses that the diagram

$$\begin{array}{ccc} \mathscr{U}_{l-1} & \xrightarrow{p} & \mathscr{U}_l \\ \downarrow P_{l-1} & & \downarrow P_l \\ \mathscr{H}_{l-1} & \xrightarrow{\text{inclusion}} & \mathscr{H}_l \end{array}$$

is commutative.

Usually, the product $R_l P_l$ does not equal the identity I, but it is invertible:

Lemma 3.6.3. $(R_l P_l)^{-1}: \mathscr{U}_l \to \mathscr{U}_l$ exists.

Proof. Since P_l is injective, $u_l \neq 0$ implies $P_l u_l \neq 0$. Hence, $\langle u_l, R_l P_l u_l \rangle = (P_l u_l, P_l u_l) > 0$ shows that $R_l P_l$ is not singular. □

Note that R_l (not P_l) depends on the choice of the discrete scalar product $\langle \cdot, \cdot \rangle$. In Exercise 3.9.7 we choose $\langle \cdot, \cdot \rangle$ such that $R_l P_l = I$.

Thanks to Lemma 3.6.3, the definitions

$$\hat{P}_l := P_l (R_l P_l)^{-1}, \quad \hat{R}_l := (R_l P_l)^{-1} R_l \tag{3.6.2}$$

make sense.

Note 3.6.4. $\hat{P}_l$ and $\hat{R}_l$ satisfy

$$\hat{R}_l = \hat{P}_l^*, \quad R_l \hat{P}_l = \hat{R}_l P_l = I. \tag{3.6.3}$$

The *orthogonal projection* onto $\mathscr{H}_l$ (with respect to the $L^2(\Omega)$ norm) is given by

$$Q_l := \hat{P}_l R_l = P_l \hat{R}_l, \tag{3.6.4}$$

i.e.

$$\| v - Q_l v \|_{L^2(\Omega)} = \inf_{w \in \mathscr{H}_l} \| v - w \|_{L^2(\Omega)}. \tag{3.6.5}$$

Proof. (3) is trivial. Q_l is an orthogonal projection since $Q_l^2 = Q_l = Q_l^*$. □

A further equivalent definition of the canonical prolongations and restrictions is presented in

Note 3.6.5. $p = \hat{R}_l P_{l-1}$, $r = R_{l-1} \hat{P}_l$. (3.6.6)

Proof. p from (6) satisfies condition (1): $P_l p = P_l \hat{R}_l P_{l-1} = P_{l-1}$, since the range of P_{l-1} is $\mathscr{H}_{l-1} \subset \mathscr{H}_l$ which is fixed under $Q_l = P_l \hat{R}_l$. The second part of (6) follows from $r = p^*$. □

With a view to the next section we state the characteristic relation between p, r, and the stiffness matrices L_l, L_{l-1}.

Note 3.6.6. Let L_l and L_{l-1} be the stiffness matrices of § 3.1.3 and let p and r be canonical. Then Eq. (7) holds:

$$L_{l-1} = r L_l p. \tag{3.6.7}$$

Proof. Equation (1.32) yields $L_l = R_l L P_l$ and $L_{l-1} = R_{l-1} L P_{l-1}$. Equation (1) implies

$$L_{l-1} = (r R_l)\, L(P_l p) = r(R_l L P_l)\, p = r L_l p. \qquad \square$$

3.7 Coarse-Grid Matrix

The goal of the two-grid iteration is the solution of $L_l u_l = f_l$ for some fixed l. The equations $L_k u_k = f_k$ $(k \neq l)$ corresponding to other levels need not be defined. Therefore, we are free to decide about a suitable choice of L_{l-1} in step (2.2 c_2) of the two-grid iteration.

There are two strategies for determining L_{l-1}.

First approach. Let the equations $L_l u_l = f_l$ for all levels $l \in \{0, 1, \ldots\}$ be defined by the same discretisation. If, e.g. the Poisson equation $-\Delta u = f$ is discretised by the usual five-point scheme, use the matrix L_{l-1} of the five-point scheme at level $l-1$.

Second approach (*Galerkin approach*). Given L_l, prolongation p, and restriction r, define L_{l-1} by

$$L_{l-1} = r L_l p. \tag{3.7.1}$$

This definition is called the Galerkin approximation since Eq. (1) is true for the stiffness matrices of the Galerkin method (cf. Note 3.6.6). For examples of L_{l-1} compare Exercises 3.9.4–5.

The advantages of the first approach are:

(i) The definition of L_{l-1} is independent of the fact that we want to solve an equation at level l. This will also be important for the multi-grid iteration, where L_{l-1} is needed for the solution of $L_k u_k = f_k$ with $k > l$.

(ii) In the finite difference case this approach needs no additional computations for defining L_{l-1}.

Correspondingly, we have to note the disadvantage of the Galerkin approach:

(iii) The definition of L_{l-1} depends on L_l. If the solution of $L_l u_l = f_l$ is followed by a multi-grid iteration for solving $L_{l+1} u_{l+1} = f_{l+1}$ (cf. § 5.1), we need the matrices $L_l' := r L_{l+1} p$, $L_{l-1}' := r L_l' p$, which possibly differ from L_l and L_{l-1} used before.

(iv) A preprocessing phase is needed to compute L_{l-1} (and $L_{l-2}, \ldots, L_0$ for the multi-grid iteration).

On the other hand there are some advantages:

(v) The possibly irregular difference formulae at points near the boundary are automatically generated.

(vi) If L_l is symmetric (positive definite), and if $r = p^*$, then $L_{l-1} = rL_l p$ is also symmetric (positive definite).

(vii) The usual definition of consistency describes the approximation of the differential operator L by L_l. For the coarse-grid correction we need the *relative consistency* describing the approximation of L_l in terms of L_{l-1} (cf. § 5.2). For the Galerkin approach the relative consistency depends mainly on the orders of p and r (cf. Note 6.3.37).

The sparsity of the matrices L_k ($k = l, l-1, \ldots, 0$) is very important for the efficiency of the two- or multi-grid iteration, since otherwise the smoothing step is too expensive. The following note states that the nine-point pattern is preserved by the Galerkin approach.

Note 3.7.1. Assume $\Omega_{l-1} \subset \Omega_l$, $h_{l-1} = 2h_l$ and let L_l, p and r have a nine-point pattern as, e.g., in (1.16), (4.2), and (5.5a), respectively. Then, L_{l-1} is also a nine-point formula (1.16).

Proof. Let u_{l-1} be a unit vector: $u_{l-1}(\mathbf{x}) = 1$ for some $\mathbf{x} \in \Omega_{l-1}$, $u_{l-1}(\mathbf{x}') = 0$ elsewhere. We have to show $(L_{l-1}u_{l-1})(\mathbf{x}') = 0$ for all $\mathbf{x}' \in \Omega_{l-1}$ with $\|\mathbf{x} - \mathbf{x}'\|_\infty \geqq 2h_{l-1}$, where $\|(x, y)\|_\infty = \max\{|x|, |y|\}$. pu_{l-1} vanishes outside $\{\mathbf{x}' \in \Omega_l\colon \|\mathbf{x}' - \mathbf{x}\|_\infty \leqq h_l\}$, $L_l p u_{l-1}$ equals zero outside $\{\mathbf{x}' \in \Omega_l\colon \|\mathbf{x}' - \mathbf{x}\|_\infty \leqq 2h_l\}$; hence $L_{l-1}u_{l-1} = rL_l p u_{l-1}$ vanishes for all $\mathbf{x}' \in \Omega_{l-1}$ with $\|\mathbf{x}' - \mathbf{x}\|_\infty \geqq 4h_l = 2h_{l-1}$. □

Usually, the five- or seven-point star is not preserved but becomes a nine-point formula (cf. Exercise 3.9.4). A counter-example is given in Exercise 3.9.5. An extension of Note 3.7.1 is mentioned in Exercise 3.9.6.

3.8 Comments

3.8.1 Separated Discrete Boundary Conditions

In the classical formulation of the boundary value problem the differential Eq. (1.9) and the boundary condition (1.3) are well-separated. The weak formulation (1.11) however does not distinguish between both components (cf. Babuška-Aziz [1, pp. 74–76]).

For the discrete problem we proposed a discretisation $L_l u_l = f_l$ with f_l containing the information of f_Ω and f_Γ. The discretised boundary conditions are used to eliminate extra grid points. An alternative approach is the separation of the difference equations and discrete boundary conditions.

Ω_l was defined as the set of grid points serving as midpoints of the difference formulae. Let $\bar{\Omega}_l$ be the set consisting of the points of Ω_l and their neighbours. Often, these extra neighbours are points on the boundary Γ with irregular distance from Ω_l (cf. Fig. 3.1.1), or the extra points are lying outside of $\bar{\Omega}$. Let

$$n_l^\Omega = \text{number of points of } \Omega_l,$$
$$n_l = \text{number of points of } \bar{\Omega}_l.$$

The grid functions u_l are defined on $\bar{\Omega}_l$; thus, they have n_l components. For each of the n_l^Ω (so-called interior) grid points $\mathbf{x} \in \Omega_l$ a difference analogue of $L_\Omega u = f_\Omega$ is given, resulting in the system

$$L_l^\Omega u_l = f_l^\Omega \quad (n_l^\Omega \text{ equations}). \tag{3.8.1 a}$$

For the remaining n_l^Γ points of $\bar{\Omega}_l \backslash \Omega_l$,

$$n_l^\Gamma = n_l - n_l^\Omega,$$

one has to pose n_l^Γ discrete boundary conditions

$$L_l^\Gamma u_l = f_l^\Gamma \quad (n_l^\Gamma \text{ equations}). \tag{3.8.1 b}$$

Setting

$$L_l = \begin{bmatrix} L_l^\Omega \\ L_l^\Gamma \end{bmatrix}, \quad f_l = \begin{bmatrix} f_l^\Omega \\ f_l^\Gamma \end{bmatrix} \tag{3.8.1 c}$$

we regain Eq. (1.19),

$$L_l u_l = f_l \quad (n_l = n_l^\Omega + n_l^\Gamma \text{ equations}). \tag{3.8.1'}$$

Note that the system (1′) is larger than the previous system (1.19), because in § 3.1 the dimension n_l equals n_l^Ω. However, (1′) and (1.19) are algebraically equivalent.

Example 3.8.1. Consider the one-dimensional equation

$$-u''(x) = f_\Omega(x), \quad u(0) = f_\Gamma(0), \quad \frac{du}{dx}(1) = f_\Gamma(1),$$

where $\Omega = (0, 1)$, $\Gamma = \{0, 1\}$. Let $h_l = 2^{-1-l}$. For $l = 1$ we have $n_1 = 5$, $n_1^\Omega = 3$, $n_1^\Gamma = 2$, $\Omega_1 = \{\frac{1}{4}, \frac{1}{2}, \frac{3}{4}\}$, $\bar{\Omega}_1 = \{0, \frac{1}{4}, \frac{1}{2}, \frac{3}{4}, 1\}$,

$$L_1^\Omega = h_1^{-2} \begin{bmatrix} -1 & 2 & -1 & 0 & 0 \\ 0 & -1 & 2 & -1 & 0 \\ 0 & 0 & -1 & 2 & -1 \end{bmatrix}, \quad f_1^\Omega = (f_\Omega(\tfrac{1}{4}), f_\Omega(\tfrac{1}{2}), f_\Omega(\tfrac{3}{4}))^T,$$

$$L_1^\Gamma = \begin{bmatrix} 1 & 0 & 0 & 0 & 0 \\ 0 & 0 & 0 & -1/h_1 & 1/h_1 \end{bmatrix}, \quad f_1^\Gamma = (f_\Gamma(0), f_\Gamma(1))^T.$$

The elimination of $u_1(0)$, $u_1(1)$ be means of $L_1^\Gamma u_1 = f_1^\Gamma$ would yield $L_1 u_1 = f_1$ with

$$L_1 = h_1^{-2}\begin{bmatrix} 2 & -1 & 0 \\ -1 & 2 & -1 \\ 0 & -1 & 1 \end{bmatrix}, \qquad f_1 = f_1^\Omega + (h_1^{-2} f_\Gamma(0),\ 0,\ h_1^{-1} f_\Gamma(1))^T.$$

For separated boundary conditions we can apply formally the same two-grid iteration (2.2), provided that the smoothing procedure, prolongation, and restriction are suitably defined.

Smoothing procedure. Jacobi-like iterations as in § 3.3.2 do not work for Eq. (1′). Blockwise iterations (cf. § 3.3.3–4) with blocks including the corresponding boundary conditions eliminate (at least partially) the discrete boundary conditions. If a pointwise smoothing is used in the interior, it must be completed by a smoothing of the boundary data. It is not necessary to satisfy the boundary conditions exactly, it suffices to produce smooth errors. However, an analysis of this approach does not exist.

Prolongation. In § 3.4 we emphasized that the interpolation at points near the boundary has to take into account homogeneous boundary conditions. For example, the Dirichlet values may not be changed since they are correct. In the present case of separated boundary conditions the interpolation has to change the boundary values at $\mathbf{x} \in \bar{\Omega}_l \backslash \Omega_l$, too.

Restriction. In the present situation the definition $r = p^*$ makes no sense. The restriction has to be defined separately for f_l^Ω and f_l^Γ:

$$f_{l-1}^\Omega = r^\Omega f_l^\Omega, \quad f_{l-1}^\Gamma = r^\Gamma f_l^\Gamma.$$

r^Ω can be chosen as r from § 3.5. r^Γ describes the restriction of the boundary data. If Ω is a d-dimensional cube, Γ consists of $(d-1)$-dimensional cubes, where r^Γ can be defined as a weighted restriction $(p^\Gamma)^*$ with p^Γ being the piecewise linear (quadratic, etc.) interpolation on Γ. In general domains, however, the discrete boundary $\bar{\Omega}_l \backslash \Omega_l$ has a very irregular structure and it is not easy to find a reasonable definition of r^Γ.

3.8.2 Construction of a Finite Element Hierarchy

In § 3.4.1 we described the mesh coarsening for regular grids. An essential advantage of finite element discretisation is the possibility of using non-uniform elements (triangles, quadrangles, etc.). For a multi-grid approach we need a hierarchy of subspaces $\mathscr{H}_l \subset \mathscr{H}$ such that at some level k $\mathscr{H}_k$ has the desired non-uniform structure. The favourable inclusion $\mathscr{H}_{l-1} \subset \mathscr{H}_l$ (cf. (1.26)) is often violated, but it should hold for most of the elements.

The following construction of a hierarchy of triangulations is due to Bank-Sherman [1], [3], Bank [4]. Let $\mathscr{T}_0$ be a given coarsest conforming triangulation of Ω. $\mathscr{T}$ is called 'conforming', if for all triangles $t, t' \in \mathscr{T}$ one of the following conditions hold: (i) $t \cap t' = \emptyset$, (ii) $t = t'$, (iii) $t \cap t'$ is a common vertex or a common side of t and t'. The union $\bigcup_{t \in \mathcal{T}} t$ equals $\bar{\Omega}$. The corresponding subspace $\mathcal{H}_0$ is given by $\{u \in \mathcal{H} : u$ restricted to t is affine for all $t \in \mathcal{T}_0\}$.

For given $\mathcal{T}_0$ the refined triangulation $\mathcal{T}_1$ at level $l = 1$ is obtained by the following steps 0 to 4.

Step 0: $\mathcal{T}_1' := \mathcal{T}_0$

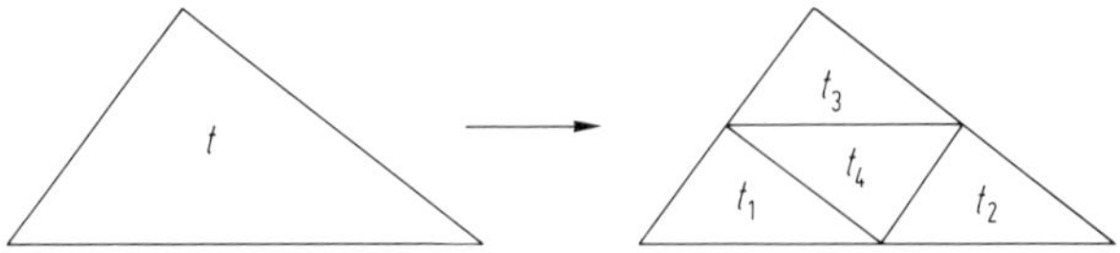

Fig. 3.8.1. Regular subdivision

Step 1: If $t \in \mathcal{T}_1'$ should be refined, replace $\mathcal{T}_1'$ by $\mathcal{T}_1' := (\mathcal{T}_1' \setminus \{t\}) \cup \{t_1, t_2, t_3, t_4\}$ with $t, t_1, \ldots, t_4$ as in Fig. 3.8.1. Repeat Step 1 until $\mathcal{T}_1'$ remains unchanged. Go to Step 2.

Whether t is subdivided or not can be controlled explicitly by the user (e.g. by the condition diameter $(t) \leqq d_l(\mathbf{x})$, $\mathbf{x}$ = midpoint of t) or implicitly by a-posteriori error estimates of a solution over $\mathcal{T}_0$ (cf. Babuška-Rheinboldt [1], Rheinboldt [1], Bank-Weiser [1], Bank-Sherman-Weiser [1]).

Step 1 is repeated as the desired diameter may be smaller than half of diameter (t), $t \in \mathcal{T}_0$. After Step 1 $\mathcal{T}_1'$ may contain triangles with a vertex not being a vertex of the neighbouring triangle (cf. Figs. 3.8.2–3).

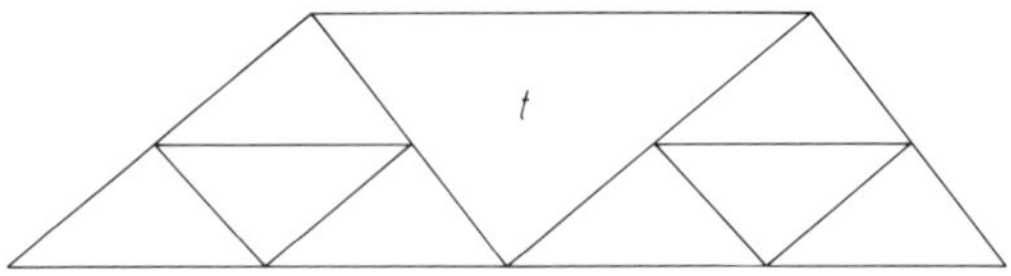

Fig. 3.8.2. Two subdivided neighbours

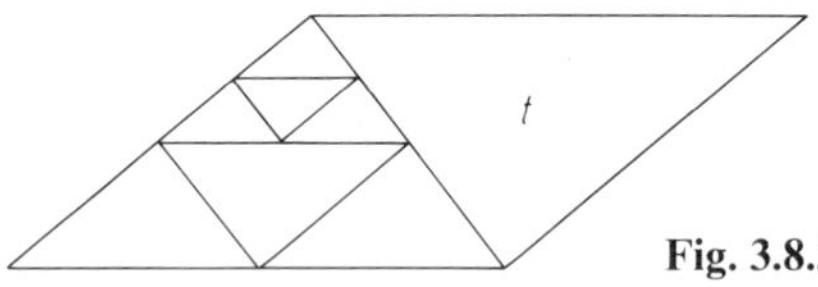

Fig. 3.8.3. One neighbour divided twice

Step 2: Divide any $t \in \mathcal{T}_1'$ as in Step 1 if t has two or three divided neighbours (Fig. 3.8.2) or one neighbour divided twice (Fig. 3.8.3). Repeat Step 2 until $\mathcal{T}_1'$ remains unchanged. Go to Step 3.

After Step 2 there remain triangles with only one neighbour divided once (left part of Fig. 3.8.4).

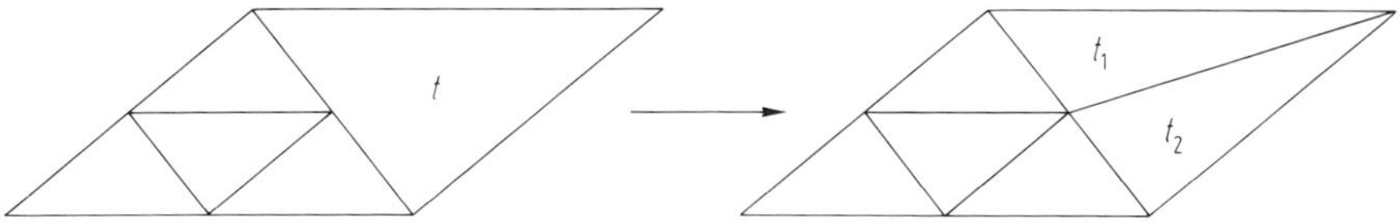

Fig. 3.8.4. Subdivision by halving

Step 3: If $t \in \mathcal{T}_1'$ has a divided neighbour (by virtue of Step 2 there is at most one refined neighbour; it is divided only once), replace $\mathcal{T}_1'$ by $\mathcal{T}_1' := (\mathcal{T}_1' \setminus \{t\}) \cup \{t_1, t_2\}$ with t, t_1, t_2 as in Fig. 3.8.4. Go to Step 4.

Step 4: $\mathcal{T}_1 := \mathcal{T}_1'$.

The further triangulations $\mathcal{T}_2, \mathcal{T}_3, \ldots$ are obtained as follows:

Step 0': $\mathcal{T}_l' := \mathcal{T}_{l-1}$. Cancel the subdivisions made previously by Step 3.

Step 1–4: as above with $\mathcal{T}_l$, $\mathcal{T}_l'$ instead of $\mathcal{T}_1$, $\mathcal{T}_1'$.

Finite element approximations may become poor if there are triangles of $\mathcal{T}_l$ with very small interior angles.

The construction of Bank and Sherman avoids such difficulties because of

Note 3.8.2. Let τ_l be $\tan \alpha_{\min}$, where $\alpha_{\min}$ is the smallest interior angle of any $t \in \mathcal{T}_l$. The numbers τ_l are bounded by $\tau_0/3$ from below.

Proof. Step 1 and Step 2 do not change $\tan \alpha_{\min}$. By Step 3 the value is divided at most by 3. Since Step 3 is removed in the next Step 0', $\tau_l \geqq \tau_0/3$ holds for all levels. □

Note 3.8.3. The construction does not ensure the nestedness $\mathcal{H}_0 \subset \mathcal{H}_1 \subset \ldots \subset \mathcal{H}_l$ (cf. (1.26)).

Proof. Let $t', t'' \in \mathcal{T}_{l-1}$ be obtained by a subdivision according to Fig. 3.8.4 (in Step 3). Step 0' yields $t = t' \cup t'' \in \mathcal{T}_l'$. A regular refinement of t in Step 1 or 2 implies $\mathcal{H}_{l-1} \not\subset \mathcal{H}_l$, since $t' \in \mathcal{T}_{l-1}$ is not a union of triangles of $\mathcal{T}_l$. □

As a consequence the canonical choice of p and r cannot be used. A natural definition of p is

$$(p u_{l-1})(\mathbf{x}) = (P_{l-1} u_{l-1})(\mathbf{x}) \quad \text{at all vertices } \mathbf{x} \text{ of triangles } t \in \mathcal{T}_l \tag{3.8.2}$$

(cf. Exercise 3.9.8). Since $P_{l-1} u_{l-1}$ is a piecewise linear (affine) function, formula (2) describes the piecewise linear interpolation of u_{l-1}.

It is not necessary to violate the nestedness $\mathcal{H}_0 \subset \ldots \subset \mathcal{H}_l$ of the subspaces. The refinement process described by Rivara [1], [2] guarantees (1.26). The refinement strategy reads thus:

Step A: $\mathcal{T}_1' := \mathcal{T}_0$.

As in Step 1 there exists a subset $\mathcal{T}_1^{\text{refine}} \subset \mathcal{T}_1' = \mathcal{T}_0$ of triangles which are to be refined according to a-posteriori estimates etc.

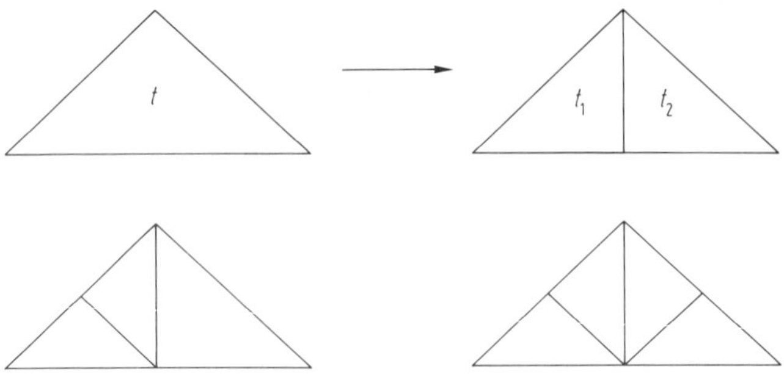

Fig. 3.8.5. Possible refinement of $t \in \mathscr{T}_{l-1}$ in $\mathscr{T}_l$

Step B: Bisect all $t \in \mathscr{T}_1^{\text{refine}}$ by their longest side, i.e. set $\mathscr{T}_1' := (\mathscr{T}_1' \backslash \{t\}) \cup \{t_1, t_2\}$, where t_1, t_2 are generated by bisecting t as depicted in the upper part of Fig. 3.8.5.

A triangle $t \in \mathscr{T}_1'$ contains a 'non-conforming point' $P \in \partial t$, if P is a vertex of a neighbouring triangle but not of t (cf. left part of Fig. 3.8.4). Such a triangle t is called 'non-conforming'.

Step C: Define $\mathscr{T}_1^{\text{non}} := \{t \in \mathscr{T}_1' : t \text{ non-conforming}\}$. If $\mathscr{T}_1^{\text{non}} = \emptyset$ go to Step D. For all $t \in \mathscr{T}_1^{\text{non}}$ do:

(i) Bisect t by its longest side as described in Step B. Let Q be the new bisecting vertex.

(ii) If Q does not coincide with the non-conforming point P of t, join Q and P.

(iii) Repeat Step C with the new triangulation $\mathscr{T}_1'$.

Step D: $\mathscr{T}_1 := \mathscr{T}_1'$.

Analogously, $\mathscr{T}_l$ can be obtained from $\mathscr{T}_{l-1}$.

The example of a sequence of triangulations in Fig. 3.8.6 is cited from Rivara [1]. The singularity at the re-entrant corner requires an increasingly refined grid (cf. Schatz-Wahlbin [1]).

Note 3.8.4 (Rivara [1]). Each $t \in \mathscr{T}_{l-1}$ is refined into at most four triangles of $\mathscr{T}_l$; the possible cases are depicted in Fig. 3.8.5. $\mathscr{H}_0 \subset \mathscr{H}_1 \subset \ldots \subset \mathscr{H}_l$ is guaranteed. The triangulation cannot degenerate, i.e. $\alpha \geqq \alpha_0 > 0$ holds for all interior angles α of $t \in \mathscr{T}_l$, $l \geqq 0$, where α_0 depends only on the initial triangulation $\mathscr{T}_0$.

3.8.3 A Further Variant of the Two-Grid Iteration

The coarse-grid correction (2.2 c_{1-3}) requires the computation of the defect $d_l = L_l u_l - f_l$. The computation of $L_l u_l - f_l$ is also the main part of many smoothing iterations. For instance, the modified Jacobi iteration (3.7) is $u_l^{j+1} = u_l^j - \omega h_l^{2m} d_l$. Thereby, the following two-grid iteration (3) with an

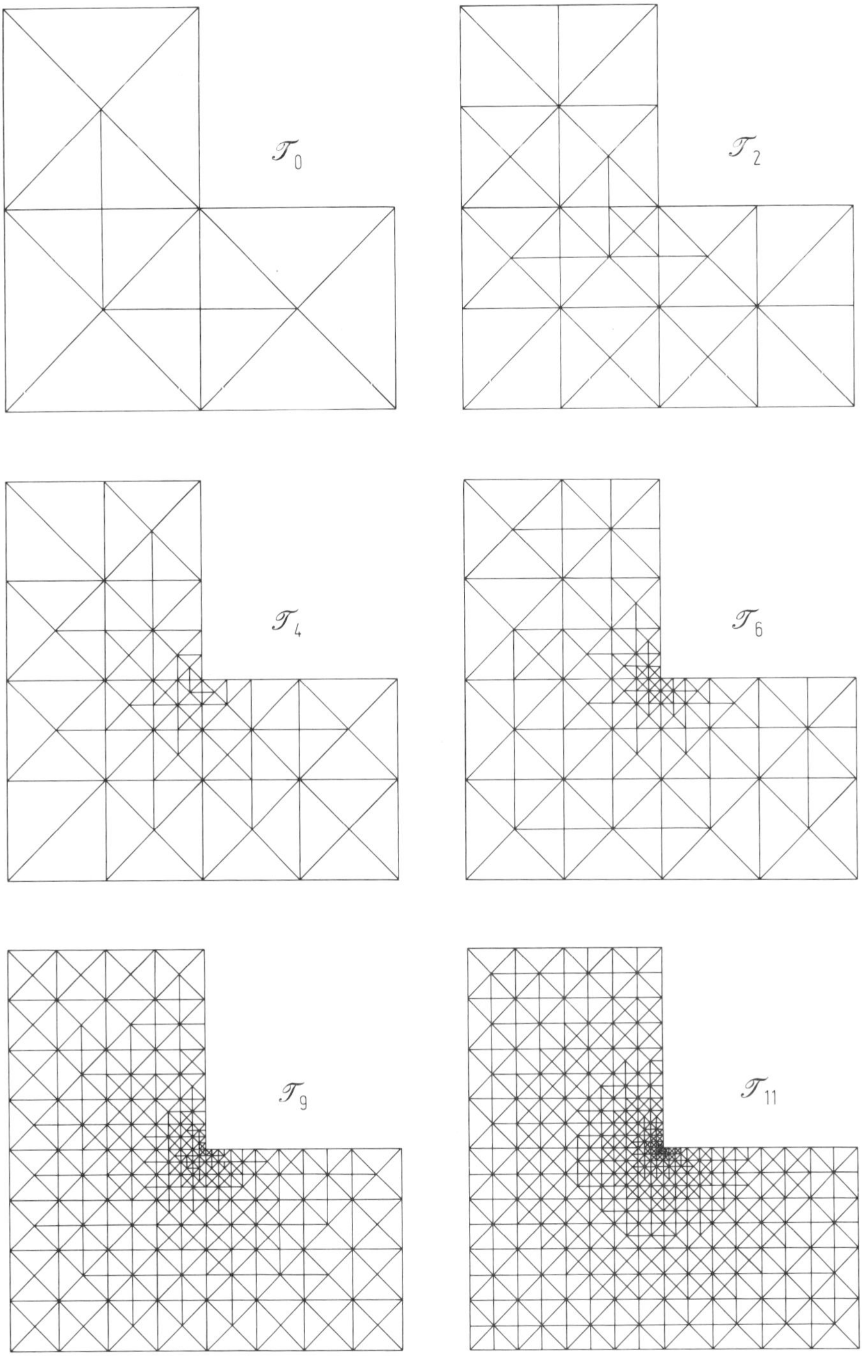

Fig. 3.8.6. Triangulations from the sequence $\{\mathscr{T}_0, \mathscr{T}_1, \ldots\}$ adaptively generated by the refinement strategy of Rivara [1]

extra smoothing iteration takes almost the same computational work as the original iteration (2.2).

$$u_l' := \mathscr{S}_l^{(\nu_1)}(u_l^j, f_l); \qquad \text{(pre-smoothing)} \tag{3.8.3a}$$

$$d_l := L_l u_l' - f_l; \tag{3.8.3b}$$

$$d_{l-1} := r d_l; \tag{3.8.3c}$$

$$u_l'' := u_l' - \omega h_l^{2m} d_l; \qquad \text{(additional smoothing)} \tag{3.8.3d}$$

$$v_{l-1} := L_{l-1}^{-1} d_{l-1}; \qquad \text{(coarse-grid correction)} \tag{3.8.3e}$$

$$u_l''' := u_l'' - p v_{l-1}; \tag{3.8.3f}$$

$$u_l^{j+1} := \mathscr{S}_l^{(\nu_2)}(u_l''', f_l); \qquad \text{(post-smoothing).} \tag{3.8.3g}$$

Seemingly, the added smoothing step disturbs the coarse-grid correction: $p v_{l-1}$ is the correction to u_l', but it is subtracted from u_l''. However, $p v_{l-1}$ has to compensate the low frequency components of the error $u_l' - u_l$. These components are almost equal to those of $u_l'' - u_l$. But numerical experiments with Poisson's equation show that the additional step (d) hardly improves the rate of convergence.

A similar version is mentioned by Brandt [10] for parallel computations. A further ν' smoothing iterations

$$u_l'' = \mathscr{S}_l'^{(\nu')}(u_l', f_l) \tag{3.8.3d'}$$

in place of only one Jacobi iteration (3d) can be performed in parallel to the coarse-grid correction. A good choice of the smoothing procedure $\mathscr{S}_l'$ is the semi-iteration (3.15a) with $\mathscr{S}_l'(u_l, f_l; \omega_\mu) = u_l - \omega_\mu h_l^{2m}(L_l u_l - f_l)$ $(\mu = 1, \ldots, \nu')$ such that the polynomial $\prod_{\mu=1}^{\nu'} (1 - \omega_\mu C_L \xi)$ approximates 1 for $0 \leqq \xi \leqq \frac{1}{2}$, 0 for $\frac{1}{2} \leqq \xi \leqq 1$. C_L is defined by (3.8b).

3.8.4 The Case of $\Omega_{l-1} \not\subset \Omega_l$

In § 3.4.1 we considered various sequences of grids, all satisfying $\Omega_{l-1} \subset \Omega_l$. But there are some situations, where the choice of staggered grids with $\Omega_{l-1} \not\subset \Omega_l$ is very natural.

An example is the discretisation of a Neumann problem in a cube or as in Fig. 3.8.7 of an one-dimensional problem

$$-u_{xx} + u = f_\Omega(x), \quad -u_x(0) = f_\Gamma(0), \quad u_x(1) = f_\Gamma(1). \tag{3.8.4}$$

A simple second order discretisation of Neumann's condition is

$$[u_l(-h_l/2) - u_l(h_l/2)]/h_l = f_\Gamma(0), \quad [u_l(1 + h_l/2) - u_l(1 - h_l/2)]/h_l = f_\Gamma(1) \tag{3.8.5}$$

by central differences. Using (5) in a staggered grid for eliminating $u_l(-h_l/2)$ and $u_l(1 + h_l/2)$, we are led to a symmetric system of equations, whereas in a

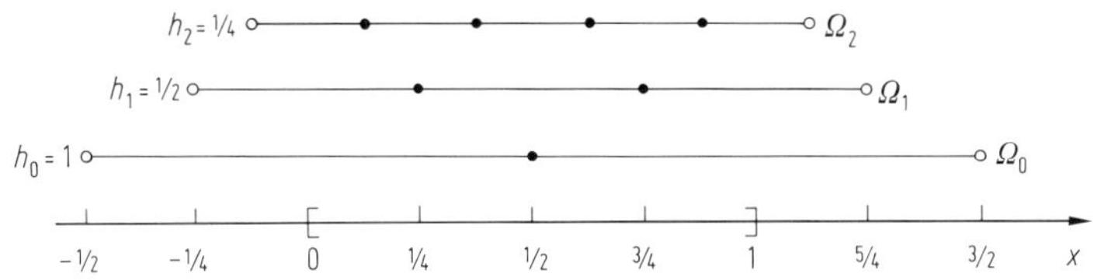

Fig. 3.8.7. Staggered grids

usual grid the central difference $[u_l(-h_l) - u_l(h_l)]/(2h_l) = f_\Gamma(0)$ yields a non-symmetric system.

In the case of Fig. 3.8.7 the *piecewise linear interpolation* is defined by

$$(p v_1)(\tfrac{3}{8}) = \tfrac{3}{4} v_1(\tfrac{1}{4}) + \tfrac{1}{4} v_1(\tfrac{3}{4}), \quad (p v_1)(\tfrac{5}{8}) = \tfrac{1}{4} v_1(\tfrac{1}{4}) + \tfrac{3}{4} v_1(\tfrac{3}{4}). \tag{3.8.6 a}$$

Setting, formally $v_1(\tfrac{5}{4}) = v_1(\tfrac{3}{4})$, because of the (homogeneous) Neumann condition, we obtain

$$(p v_1)(\tfrac{7}{8}) = v_1(\tfrac{3}{4}), \quad (p v_1)(\tfrac{1}{8}) = v_1(\tfrac{1}{4}). \tag{3.8.6 b}$$

It is plain how to extend the prolongation into a two-dimensional piecewise bilinear interpolation.

A very cheap prolongation is the *piecewise constant interpolation.* For Fig. 3.8.7 it reads as

$$(p v_1)(\tfrac{1}{8}) = (p v_1)(\tfrac{3}{8}) = v_1(\tfrac{1}{4}), \quad (p v_1)(\tfrac{5}{8}) = (p v_1)(\tfrac{7}{8}) = v_1(\tfrac{3}{4}). \tag{3.8.7}$$

Its order is 1. By Note 3.5.1 this prolongation should not be used together with $r = p^*$ for second order problems since $2 = m_p + m_r \not> 2m = 2$. A possible choice of r is the adjoint of the piecewise linear interpolation (6).

3.8.5 Comment on Weighted Restrictions

Brandt [3] regards the weighted restriction (5.5 b) as an improvement on trivial injection. The coefficients in (5.5 b) are chosen in order to optimise the efficiency of the coarse-grid correction or of the complete two-grid iteration.

Our choice of the (weighted) restriction $r = p^*$ has another purpose. p acts on the space of grid functions u_l, whereas r is applied to functions f_l of the right-hand side. A natural choice of a norm of f_l is the dual of the norm of u_l. For example, the energy norm $|\cdot|_1$ (cf. (2.6.4)) can be used for u_l, while the dual norm $|\cdot|_{-1}$ (cf. (1.4.5)) is suited for f_l. Estimates of r with respect to the dual norm correspond to estimates of r^* with respect to the original norm. Therefore, r^* should be a sufficiently regular transfer from the coarse grid to the fine grid. The trivial injection is a counter-example since r^*_{inj} is not stable with respect to the energy norm: $|r^*_{\text{inj}}|_{1 \leftarrow 1} = O(h_l^{-1})$.

Accordingly, we do not distinguish between the non-weighted restriction r_{inj} and all other (weighted) restrictions, but between (i) restrictions r such that r^* is a reasonable prolongation and (ii) restrictions with irregular r^*. r_{inj} is an example of the second class of restrictions. A further example is the five-point restriction

$$r = \begin{bmatrix} 0 & \frac{1}{8} & 0 \\ \frac{1}{8} & \frac{1}{2} & \frac{1}{8} \\ 0 & \frac{1}{8} & 0 \end{bmatrix} \tag{3.8.8}$$

due to Trottenberg (adapted from the total reduction method of Schröder-Trottenberg [1]). It is named 'half-weighting' since it is between r_{inj} (no weighting) and the nine-point restriction (5.4) which is also called 'full-weighting'. The reason for the weights $\frac{1}{2}$ and $\frac{1}{8}$ is that 'red' and 'black' points are equally weighted in the case of a red-black ordering:

$$\tfrac{1}{2} = 4 \times \tfrac{1}{8}.$$

In some cases the five-point restriction (8) is as cheap as the trivial injection r_{inj}:

Note 3.8.5. Suppose L_l to be a five-point scheme, $\mathscr{S}_l$ to be the Gauß-Seidel iteration with red-black ordering, and r to be the five-point restriction (8). Then $r d_l$ coincides with $\frac{1}{2} r_{\mathrm{inj}} d_l$, since the defect d_l vanishes at all 'black' points weighted by $\frac{1}{8}$.

3.9 Exercises

Exercise 3.9.1. Suppose L_l to be symmetric. Show that one step of iteration (3.10) is equivalent to two steps of iteration (3.7) with alternating parameters $\omega_1 = +\omega$, $\omega_2 = -\omega$ (ω from (3.10)). This is a special case of (3.15a).

Exercise 3.9.2. Let L_l be a five-point formula (i.e. $c_{\pm 1, \pm 1} = 0$ in (1.16)). Prove that the Gauß-Seidel iterations (3.3) with respect to the chequer-board ordering (Fig. 3.3.3) and with respect to the four-colour ordering (Fig. 3.3.5) are identical.

Exercise 3.9.3. a) Let $\mathscr{H}_l$ be the space of piecewise bilinear finite elements on a square grid and take the nodal values as coefficients. Show that the canonical prolongation and restriction coincide with the nine-point prolongation (4.2) and nine-point restriction (5.4) (cf. Note 3.4.1).

b) Let $\mathscr{H}_l$ be the space of piecewise linear functions over triangles $\{(x, y), (x + h_l, y), (x + h_l, y + h_l)\}$ and $\{(x, y), (x, y + h_l), (x + h_l, y + h_l)\}$ for $(x, y) = (\nu h_l, \mu h_l)$, $\nu, \mu \in \mathbb{Z}$. Prove that the canonical p and r coincide with the seven-point prolongation (4.3) and the seven-point restriction (5.6).

Exercise 3.9.4. Let p be the nine-point prolongation (4.2), r the nine-point restriction (5.4), and L_l the five-point star (1.15) in an infinite grid.

a) Show: $L_{l-1} = r L_l p$ is not a five- but a nine-point formula.

b) Define recursively $L_{k-1} := r L_k p$ for $k = l, l-1, l-2, \dots$. Prove: $h_k^2 L_k$ $(k < l)$ are nine-point schemes tending to the star

$$\frac{1}{3}\begin{bmatrix} -1 & -1 & -1 \\ -1 & 8 & -1 \\ -1 & -1 & -1 \end{bmatrix} \tag{3.9.1}$$

for $k \to -\infty$.

c) Let $L_\Omega = -\Delta$ and discretise by piecewise bilinear finite elements in a square grid of size h_l. Prove that the stiffness matrix is the star (1).

Exercise 3.9.5. Let L_l be the five-point star (1.15) and p and r the seven-point transfers (4.3) and (5.6), respectively. Show that $L_{l-1} = r L_l p$ is again the five-point scheme (1.15) at level $l - 1$.

Exercise 3.9.6. Suppose $\Omega \subset \mathbb{R}^d$ and let p be a $(1 + 2m_p)^d$-point prolongation, L_l a $(1 + 2m_L)^d$-point scheme, and r a $(1 + 2m_r)^d$-point restriction. Prove that $L_{l-1} := r L_l p$ is a $(1 + 2m)^d$-point scheme with $m = [(m_p + m_L + m_r)/2]$, where $[x] = \max\{n \in \mathbb{Z}: n \leqq x\}$.

Exercise 3.9.7. Let P_l be defined as in § 3.6 and set $\langle u_l, v_l \rangle := (P_l u_l, P_l v_l)_{L_2(\Omega)}$. Prove $R_l P_l = I$.

Exercise 3.9.8. Prove: If $\mathscr{H}_{l-1} \subset \mathscr{H}_l$, definition (8.2) yields the canonical prolongation p.

Exercise 3.9.9. Assume (1.26) and let $P_l, R_l, \hat{P}_l, \hat{R}_l, p$, and r as in § 3.6. Prove that the left inverse of p is $\hat{r} := \hat{R}_{l-1} P_l$: $\mathscr{U}_l \to \mathscr{U}_{l-1}$, while the right inverse of r is $\hat{p} := R_l \hat{P}_{l-1} = \hat{r}^*$: $\mathscr{U}_{l-1} \to \mathscr{U}_l$; i.e. $\hat{r} p = r \hat{p} = I$.

4. General Multi-Grid Iteration

The extension of the two-grid iteration to the multi-grid iteration is already described in § 2.5 for the one-dimensional problem. In § 4.1 the algorithm is described in various ways. Typical convergence properties of the multi-grid iteration are discussed in § 4.2. The convergence rates have to be related to the computational work, which is estimated in § 4.3. Numerical results are reported in § 4.4 for the model problem of the Poisson equation in the square and in general domains.

4.1 Multi-Grid Algorithm

We recall that the two-grid iteration is replaced by the multi-grid version not to improve the convergence but to avoid the exact solution of the coarse-grid Eq. (3.2.2 c_2), $L_{l-1}v_{l-1} = d_{l-1}$. Since this equation is of the same form as the original equation $L_l u_l = f_l$, the method can be defined recursively. The $(l+1)$-grid method for solving $L_l u_l = f_l$ is described as follows. The two-grid iteration can be applied to $L_1 u_1 = f_1$. Given the l-grid algorithm for the solution of $L_{l-1}u_{l-1} = f_{l-1}$, the $(l+1)$-grid iteration is the two-grid method (3.2.2) at the levels l, $l-1$ with the exact solution of the coarse-grid equation replaced by γ steps of the l-grid iteration applied to $L_{l-1}v_{l-1} = d_{l-1}$.

The multi-grid version of the two-grid iteration (3.2.2) reads as follows.

multi-grid iteration $MGM^{(\nu_1,\nu_2)}$ for solving $L_l u_l = f_l$ (4.1.1)

procedure $MGM(l,u,f)$; **integer** l; **array** u,f;
if $l=0$ **then** $u := L_0^{-1} * f$ **else** (4.1.1 a)
begin array d,v; **integer** j;
  $u := \mathscr{S}_l^{(\nu_1)}(u,f)$; (4.1.1 b)
  $d := r * (L_l * u - f)$; (4.1.1 c)

$v := 0;$ (4.1.1 d$_1$)
for $j := 1$ **step** 1 **until** γ **do** $MGM(l-1, v, d)$; (4.1.1 d$_2$)
$u := u - p * v;$ (4.1.1 e)
$u := \mathscr{S}_l^{(\nu_2)}(u, f)$ (4.1.1 f)
end;

. . . .

One step $u_l^j \mapsto u_l^{j+1}$ of the multi-grid iteration is performed by

$$u_l := u_l^j; \quad MGM(l, u_l, f_l); \quad u_l^{j+1} := u_l;$$

The only difference from the two-grid iteration (3.2.2) is that the statement (3.2.2 c$_2$), $v := L_{l-1}^{-1} d$, is replaced by (1 d$_{1,2}$). The exact solution is replaced by γ multi-grid iterations applied to the initial guess $v^0 = 0$. The usual value of γ is $\gamma = 1$ (*V*-cycle) or $\gamma = 2$ (*W*-cycle).

Note 4.1.1. When $l = 1$ one may use $\gamma = 1$ in (1 d$_2$) without changing the algorithm.

The multi-grid iteration (2.5.2) is a special case of algorithm (1) with $\nu_1 = \nu$, $\nu_2 = 0$ (no post-smoothing). Figures 2.5.1–2 are still valid for $\nu_2 > 0$ if we add the operation *S* (smoothing) after each correction *P*.

A flow-chart of algorithm (1) is given in Fig. 4.1.1.

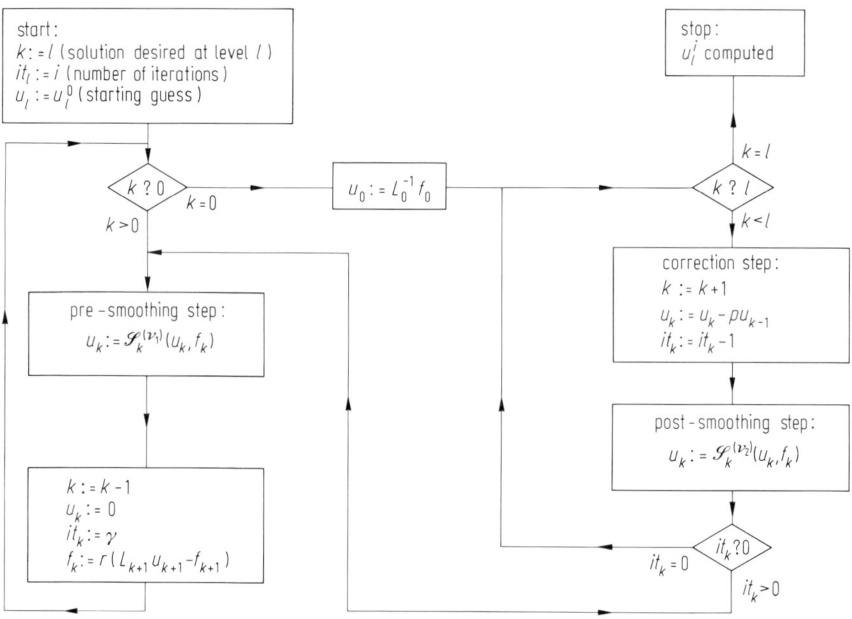

Fig. 4.1.1. Flow-chart of *i* steps of the multi-grid iteration

For readers more familiar with FORTRAN than ALGOL we describe a FORTRAN version of $MGM^{(\nu_1,\nu_2)}$. For simplicity we assume that the functions $\{u_1, u_2, \ldots, u_l\}$ correspond to the array $u(*,*)$. $u(k,j)$ is the jth component of the function u_k at level k. Analogously, $(f_k)_{k=1,\ldots,l}$ corresponds to $f(*,*)$. According to the FORTRAN convention, the lowest level is $l = 1$ (not $l = 0$). We need three subroutines:

heading of subroutine	result of subroutine
subroutine $s(\nu, l, u, f)$ **dimension** $u(*,*),\ f(*,*)$	$u(l,*) := \mathscr{S}_l^{(\nu)}(u(l,*),\ f(l,*))$
subroutine defect (l, u, f) **dimension** $u(*,*),\ f(*,*)$	$f(l-1,*) := r(L_l u(l,*) - f(l,*))$
subroutine $p(l,u)$ **dimension** $u(*,*)$	$u(l,*) := u(l,*) - p u(l-1,*)$

For solving $L_l u_l = f_l$ by i multi-grid iterations one has to call the FORTRAN subroutine *mgm* defined below:

$u(l,*) = u_l^0$ (definition of the starting guess)
$f(l,*) = f_l$ (definition of the right-hand side)
call $mgm(i, l, u, f)$

The ith iterate u_l^i is stored on $u(l,*)$. To denote the componentwise assignment $u(l,j) = u_{l,j}$ $(j = 1, \ldots, n_l)$ we write $u(l,*) = u_l$. The subroutine *mgm* reads as follows:

FORTRAN subroutine for solving $L_l u_l = f_l$ (4.1.2)

```
      subroutine mgm(i,l,u,f)
      dimension u(*,*), f(*,*), it(*)
      k = l
      it(k) = i
C     beginning of the iteration at level k
   10 if (k .eq. 1) goto 30
C     pre-smoothing
   20 call s(ν1,k,u,f)
      call defect (k,u,f)
      k = k − 1
      u(k,*) = 0
```

```
      it(k) = γ
      goto 10
   30 u(1,*) = L_1^{-1} * f(1,*)
   40 if (k .eq. l) return
C     correction step
      k = k + 1
      call p(k,u)
C     post-smoothing
      call s(ν_2,k,u,f)
      it(k) = it(k) - 1
      if (it(k) .eq. 0) goto 40
      goto 20
      end
```

The recursive structure of the ALGOL program (1) can be imitated in FORTRAN, if we write separate subroutines *mgm* 1, *mgm* 2,... for all levels $l = 1, 2, \ldots$. For example, *mgm* 4 would read

```
      subroutine mgm 4(i,u4,f4)
      dimension u3(*), f3(*), u4(*), f4(*)
      do 4 j = 1,i
      call s(ν_1,4,u4,f4)
      call defect (4,u4,f4,f3)
      u3 = 0
      call mgm 3(γ,u3,f3)
      call p(3,u4,u3)
    4 call s(ν_2,4,u4,f4)
      return
      end
```

with subroutines *s*, defect, *p* similar to those in (2).

We recall that in the case $\gamma = 1$ ("*V*-cycle") the algorithm can be written non-recursively:

Routines for the *V*-cycle (4.1.3)

ALGOL	FORTRAN
procedure $MGM(l, u_l, f_l)$;	**subroutine** $mgm(l, u, f)$
begin	**dimension** $u(*,*)$, $f(*,*)$
for $k := l$ **step** -1 **until** 0 **do**	**do** 10 $kk = 2, l$

```
    begin u_k := S_k^(ν1)(u_k, f_k);
          f_k := r(L_k u_k − f_k);
      u_{k−1} := 0
    end;
    u_0 := L_0^{−1} * f_0;
    for k := 1 step 1 until l do
    begin u_k := u_k − p * u_{k−1};
          u_k := S_k^(ν2)(u_k, f_k)
    end
end;
```

```
   k = 2 + l − kk
   call s(ν1, k, u, f)
   call defect (k, u, f)
10 u(k − 1, *) = 0
   u(1, *) = L_1^{−1} * f(1, *)
   do 20 k = 2, l
   call p(k, u)
20 call s(ν2, k, u, f)
   return
   end
```

Note that the routines (3) differ. According to the different index of the coarsest grid, they use $l+1$ (ALGOL) and l (FORTRAN) levels.

There are numerous variants of the multi-grid algorithm (1):

(i) The number of smoothing iterations can be determined a-posteriori during the iteration. E.g., the pre-smoothing statement (1b) may be replaced by

$$\begin{aligned}&\text{norm1} := \| L_l * u - f \|_{\mathscr{F}}; \quad \nu := 0;\\ &\text{smoothing:}\\ &u := \mathscr{S}_l(u, f); \quad \nu := \nu + 1;\\ &\text{norm0} := \text{norm1}; \quad \text{norm1} := \| L_l * u - f \|_{\mathscr{F}};\\ &\textbf{if } \text{norm1} \leqq \kappa * \text{norm0} \textbf{ and } \nu < \nu\text{max} \textbf{ then goto } \text{smoothing};\end{aligned} \tag{4.1.4}$$

with a suitable number $\kappa \in (0,1)$, e.g. $\kappa = \frac{1}{2}$, and some norm $\|\cdot\|_{\mathscr{F}}$. The intention is to smooth as long as it is efficient. One expects a rapid decrease of $\| L_l u_l^i - f_l \|_{\mathscr{F}}$ as long as non-smooth error components are dominating. Asymptotically, the ratio norm1/norm0 approaches the convergence rate of $\mathscr{S}_l$ which should exceed κ. However, a rapid decrease of the error $u_l^i - u_l$ ($u_l = L_l^{-1} f_l$) is not necessarily accompanied by a monotone decrease of the defect.

(ii) The number γ may depend on the level l. For instance, γ from (1 d$_2$) can be replaced by

$$\gamma_l = \begin{cases} 1 & \text{for } l \text{ even,} \\ 2 & \text{for } l \text{ odd.} \end{cases} \tag{4.1.5}$$

Sometimes a very coarse grid at level $l = 0$ causes poor convergence of the two-grid iteration at level 1. In this case we can choose $\gamma_l = \gamma$ ($l \geqq 3$) and $\gamma_2 > \gamma$.

(iii) Pre- and post-smoothing may involve different smoothing iterations $\mathscr{S}_l$ and $\mathscr{S}_l'$.

(iv) The modification mentioned in § 3.8.3 yields a further multi-grid variant.

(v) The exact solution at level $l = 0$ (cf. (1 a)) can be replaced by a suitable approximation. Let $u_0^{j+1} = \mathscr{I}_0(u_0^j, f_0)$ be some iteration and change algorithm (1) as follows:

$$u := \mathscr{I}_0(u, f) \qquad \text{instead of } u = L_0^{-1} * f \text{ in (1 a);}$$
$$\gamma_l = \left.\begin{cases} \gamma & \text{for } l \geqq 2 \\ \gamma_1 & \text{for } l = 1 \end{cases}\right\} \text{instead of } \gamma \text{ in } (1\,\mathrm{d}_2). \tag{4.1.6}$$

Then $u_0 = L_0^{-1} f_0$ is approximated by γ_1 iterations of $\mathscr{I}_0$ applied to the starting guess $u_0^0 = 0$. The simplest choice of $\mathscr{I}_0$ is

$$\mathscr{I}_0 = \mathscr{S}_0 \tag{4.1.7}$$

since no extra programming is needed.

4.2 Convergence of the Multi-Grid Iteration

All smoothing iterations of § 3.3 satisfy

$$u_l = \mathscr{S}_l(u_l, f_l) \qquad (\text{where } u_l = L_l^{-1} f_l) \tag{4.2.1}$$

i.e. the discrete solution is a fixed point of the smoothing iteration. Exercise 1.5.1 proves

Note 4.2.1. Under condition (1) $u_l = L_l^{-1} f_l$ is a fixed point of the multi-grid iteration at level l. Convergence of the multi-grid process implies $u_l^j \to u_l$ as $j \to \infty$.

This result is true even if (1) is violated at other levels.

As stated in (3.2.3) the contraction number and the convergence rate of the two-grid iteration are bounded by $C_{TGM}\,\varrho_L(v)$, $v = v_1 + v_2$, where $\varrho_L(v) \to 0$ as $v \to \infty$. It will be shown in § 7 that a similar result holds for the multi-grid iteration also:

$$\left.\begin{array}{l}\text{multi-grid contraction number} \\ \text{multi-grid convergence rate}\end{array}\right\} \leqq C_{MGM}\,\varrho_L(v), \tag{4.2.2}$$

where $C_{MGM} \geqq C_{TGM}$. Usually, the multi-grid rate is only slightly worse than the two-grid rate. Note that the bound on the right-hand side of (2) is independent not only of the step size h_l but also of the total number of levels involved in the multi-grid process.

The following numbers illustrate the two- and multi-grid convergence. The multi-grid algorithm (4.2) described below solves the discrete Poisson equation in a unit square. Using the parameters $\nu_1 = 2$, $\nu_2 = 0$, $h_0 = \frac{1}{2}$, $h_1 = \frac{1}{4}, \ldots, h_4 = \frac{1}{32}$, we observe the following convergence rates at level $l = 4$:

$$\begin{array}{ll} \text{rate of multi-grid iteration with } \gamma = 1: & 0.1682 \\ \text{rate of multi-grid iteration with } \gamma = 2: & 0.0736 \\ \text{rate of multi-grid iteration with } \gamma = 3: & 0.0735 \\ \text{rate of two-grid iteration:} & 0.0735 \end{array} \tag{4.2.3}$$

Often, the multi-grid convergence rate is better for $\gamma = 2$ (W-cycle) than for $\gamma = 1$ (V-cycle). On the other hand, $\gamma = 2$ requires more computational work as stated in Proposition 4.3.1 below. The choice $\gamma = 3$ is unproductive, since usually the rates corresponding to $\gamma = 3$ are only insignificantly better than those of $\gamma = 2$.

4.3 Computational Work

The convergence rate has to be compared with the computational work per iteration. To estimate the work of the multi-grid iteration (1.1), we assume the following costs of the basic parts of the algorithm:

statement	number of arithmetical operations		
$u_l := \mathscr{S}_l(u_l, f_l)$	$\leqq C_S n_l$	for all $l \geqq 1$	(4.3.1 a)
$d_{l-1} := r(L_l u_l - f_l)$	$\leqq C_D n_l$	for all $l \geqq 1$	(4.3.1 b)
$u_l := u_l - p v_{l-1}$	$\leqq C_C n_l$	for all $l \geqq 1$	(4.3.1 c)
$u_0 := L_0^{-1} f_0$	$\leqq C_0$		(4.3.1 d)

We recall that n_l is the number of unknowns and equations at level l. The bounds are proportional to n_l since the matrices L_l are (uniformly) sparse.

In (1 d) C_0 is assumed to be a constant. This is true as long as the coarsest step size h_0 (hence n_0) is fixed when l increases. Another situation arises if the finest size h is fixed and sequences $h_k = 2^{l-k} h$ $(k \leqq l)$ with changing $h_0 = 2^l h$ are considered.

The constant

$$C_H := \sup_{l \geqq 1} n_{l-1}/n_l \tag{4.3.2}$$

takes the value

$$C_H = 2^{-d} \tag{4.3.3 a}$$

for d-dimensional grids $\Omega_l \subset \mathbb{R}^d$ with mesh size $h_{l-1} = 2h_l$. An exception is the case of Fig. 3.4.3 ($h_{l-1} = \sqrt{2}\,h_l$, $d = 2$), where

$$C_H = \tfrac{1}{2}. \tag{4.3.3 b}$$

The following proposition states that the work of one multi-grid iteration is also proportional to n_l.

Proposition 4.3.1. Suppose

$$\vartheta := \gamma C_H < 1 \qquad (\gamma \text{ from } (1.1\,\mathrm{d}_2)). \tag{4.3.4}$$

Then one step of the multi-grid iteration (1.1) requires $C_l n_l$ operations, where

$$\begin{aligned} C_l &\leqq \frac{\nu C_S + C_D + C_C}{1-\vartheta} + \vartheta^l \left[\frac{C_0}{\gamma C_H^l n_l} - \frac{\nu C_S + C_D + C_C}{1-\vartheta}\right] \\ &< \frac{\nu C_S + C_D + C_C}{1-\vartheta} + \vartheta^{l-1} C_0' \quad \text{with} \quad C_0' = C_0/n_1. \end{aligned} \tag{4.3.5}$$

Proof. One iteration at level l involves γ^{l-k} iterations at level k ($1 \leqq k \leqq l$). According to Note 4.1.1 γ^{l-1} systems are to be solved at level 0. Hence the number of operations is bounded by

$$\begin{aligned} C_l n_l &\leqq (\nu C_S + C_D + C_C)(n_l + \gamma n_{l-1} + \ldots + \gamma^{l-1} n_1) + \gamma^{l-1} C_0 \\ &\leqq (\nu C_S + C_D + C_C)\, n_l (1 + \vartheta + \ldots + \vartheta^{l-1}) + \vartheta^l C_0/(\gamma C_H^l) \end{aligned}$$

proving the first part of (5). The second part follows from $C_H^{l-1} n_l \geqq n_l$. □

In standard cases the condition $\vartheta < 1$ is satisfied:

Corollary 4.3.2. If $h_{l-1} = 2h_l$ the condition (4) holds for all

$$\gamma \leqq 2^d - 1.$$

In the two-dimensional case ($d = 2$) inequality (5) becomes

$$C_l < \begin{cases} \frac{4}{3}[\nu C_S + C_D + C_C] + (\frac{1}{4})^{l-1} C_0' & \text{for } \gamma = 1, \\ 2[\nu C_S + C_D + C_C] + (\frac{1}{2})^{l-1} C_0' & \text{for } \gamma = 2. \end{cases} \tag{4.3.6}$$

However, the *W*-cycle (i.e. $\gamma = 2$) in the case $d = 1$ or the *V*-cycle ($\gamma = 1$) in the case of (3 b) implies $\vartheta = 1$. Then the work is no longer proportional to n_l.

Proposition 4.3.3. If $\vartheta = 1$, the number of operations is bounded by

$$[(\nu C_S + C_D + C_C)\, l + C_0']\, n_l, \tag{4.3.7}$$

where $C_0' = C_0/n_1$. Bound (7) is of order $O(n_l \log_2 n_l)$ if $h_{l-1} = 2h_l$. For $\vartheta > 1$ the bound increases exponentially as $O(\vartheta^l n_l)$. In the case $h_{l-1} = 2h_l$ the latter expression equals $O(n_l^{1+(\log_2 \vartheta)/d})$.

The estimates (5) and (6) show that the *W*-cycle ($\gamma = 2$) takes nearly 50% more work than the *V*-cycle ($\gamma = 1$). Note that the number $\nu C_S + C_D + C_C$ corresponds to the work of the two-grid iteration when the solution of the coarse-grid equation is not counted.

The proof of Proposition 4.3.1 exhibits the fact that the work spent at the coarser grids $l-1, l-2, \ldots$ is fast decreasing, though an increasing number of γ^{l-k} auxiliary equations at level k has to be solved.

Note 4.3.4. There are still some possibilities of saving computational work:

(i) For smoothing by Gauß-Seidel (with red-black, zebra, or four-colour ordering of the grid points) the defect $L_l u_l - f_l$ vanishes at some grid points. Hence the computation of $r(L_l u_l - f_l)$ requires less work than in the general case.

(ii) The correction $u_l - p v_{l-1}$ need not be computed at points that are immediately redefined by the subsequent Gauß-Seidel step (either by $\mathscr{S}_l^{\nu_2}$ during the same iteration, or by $\mathscr{S}_l^{\nu_1}$ in a next iteration).

(iii) The iterations at the auxiliary levels $k < l$ are started with initial guess $u_k = 0$. Hence, the first smoothing step $\mathscr{S}_k(u_k, f_k) = \mathscr{S}_k(0, f_k)$ can be performed with a reduced number of operations.

For case of the alternating choice (1.5), $\gamma_l = 1, 2$, the leading terms of formulae (5) and (6) become $[\nu C_S + C_D + C_C]\ (1 + C_H)/(1 - 2C_H^2)$ and $\frac{10}{7}\,[\nu C_S + C_D + C_C]$, if l is an even integer.

Formula (5) indicates that the work mainly depends on the number of smoothing steps. Often, the computation of the defect is comparable with the performance of a smoothing iteration, whereas the term C_C can be neglected:

$$\nu C_S + C_D + C_C \approx (\nu + 1)\, C_S. \tag{4.3.8}$$

What is the optimal ν? To compare iterations of different rates and operation counts, one has to introduce the 'efficiency'

$$E = -\frac{\text{operations per iteration}}{\log(\text{contraction number})} \tag{4.3.9}$$

which is the work required for an error reduction by a factor $1/e$. In § 2.4 the contraction number was found to be approximately

$$\zeta_\nu \approx C/(\nu + 1). \tag{4.3.10}$$

(10) is not restricted to the model problem of § 2 as will be shown in § 6.2. By (8) and (10) the number $E = E(\nu)$ equals $C_S(\nu + 1)/\log((\nu + 1)/C)$. The (real) value ν minimising $E(\nu)$ depends on the constant C only:

$$\nu_{\text{opt}} = Ce - 1 \quad \text{yielding} \quad \zeta_{\nu_{\text{opt}}} = 1/e \approx 0.368.$$

But note that this result is only valid if the contraction number really behaves as $C/(\nu + 1)$ at $\nu = \nu_{\text{opt}}$. If ζ_ν decreases more rapidly, the optimal ν will be larger. An example is given in Table 4.3.1, where $E(\nu)/C_S = -(\nu + 1)/\log(\zeta_\nu)$ and $E_S(\nu)/C_S := -(\nu + 1)/\log(\varrho_\nu)$ are listed for the values ζ_ν and ϱ_ν from Table 2.4.1. Here, $\nu = 3$ is optimal.

Table 4.3.1. Efficiency for different values of ν

ν	1	2	3	4	5
$-(\nu+1)/\log(\zeta_\nu)$	2.89	2.16	2.11	2.31	2.55
$-(\nu+1)/\log(\varrho_\nu)$	2.89	2.16	1.92	2.01	2.22

Finally, we mention the storage requirements of the multi-grid iteration (1.1). Assume first that L_l has constant coefficients which therefore need not be stored. Algorithm (1.1) requires storage for u_k, f_k, $k = 0, \ldots, l$; thus

$$S_l^{MGM} := \text{total storage} = 2 \sum_{k=0}^{l} n_k,$$

whereas an iteration using only the first grid requires the storage

$$S_l = 2 n_l.$$

Note 4.3.5. Let $C_H < 1$ be defined by Eq. (2). Then

$$S_l^{MGM} \leqq S_l/(1 - C_H). \tag{4.3.11}$$

In the standard case of $C_H = 2^{-2}$ (cf. (3a)), the estimate (11) becomes $S_l^{MGM} = \frac{4}{3} S_l$. Only 33% additional storage is needed. In the three-dimensional case the respective percentage is less than 14.3%.

If the matrices L_l have to be stored, the total storage S_l^{MGM} increases, but as S_l increases by the same factor, inequality (11) remains valid.

Special techniques for saving storage are described by Brandt [3, p. 362], [5, p. 81] and Hackbusch [14].

4.4 Examples of Multi-Grid Iterations and Numerical Results

4.4.1 A Multi-Grid Program for Solving Poisson's Equation

The program (2) completely documented below follows the structure of algorithm (1.1). The underlying problem is the five-point discretisation (3.1.15) of Poisson's equation in the square $\Omega = (0,1) \times (0,1)$. A square is chosen since for a general domain an easily readable program cannot be expected because of the irregular structure at points near the boundary. All components of program (2) are chosen as simply as possible:

(i) *sequence of grids*. The grid sizes are $h_l = 2^{-l-1}$ for $l \geqq 0$. The array declarations of program (2) have to be changed if $l > 5$.

(ii) *pre-smoothing*. We apply ν steps of the Gauß-Seidel iteration with red-black ordering. One call of s corresponds to a quarter Gauß-Seidel step. Each step involves one of the four subgrids $\Omega_l^1, \ldots, \Omega_l^4$ described in Fig. 3.3.5. $s(\ldots, i, j)$, $1 \leqq i, j \leqq 2$, refers to the subgrid with mesh points $(\mu_1 h_l, \mu_2 h_l)$, where $\mu_1 \equiv i(\mathrm{mod}\ 2)$ and $\mu_2 \equiv j(\mathrm{mod}\ 2)$. The combination of $s(\ldots, 1, 1)$ and $s(\ldots, 2, 2)$ yields a half Gauß-Seidel step for the red points of $\Omega_l^r = \Omega_l^1 \cup \Omega_l^2$, while $s(\ldots, 1, 2)$ together with $s(\ldots, 2, 1)$ performs the half step on Ω_l^b.

(iii) *restriction*. r is chosen as nine-point restriction (3.5.4).

(iv) *prolongation*. p is the nine-point prolongation (3.4.2).

(v) *post-smoothing*: none, i.e. $\nu_2 = 0$.

(vi) *solution at level* $l = 0$. The coarsest grid Ω_0 contains only one point $(\frac{1}{2}, \frac{1}{2})$. Therefore, the call of $s(\ldots, 1, 1)$ produces the exact solution of $L_0 u_0 = f_0$.

We apply the program to

$$\Delta u(x, y) = 4 \quad \text{in} \quad \Omega = (0,1) \times (0,1), \quad u(x, y) = x^2 + y^2 \quad \text{on} \quad \Gamma. \tag{4.4.1}$$

One may easily verify that the solution $u(x, y) = x^2 + y^2$ is also the solution of the five-point scheme. We recall that the choice of the right-hand side in (1) has no influence on the convergence rate. A non-smooth or random right side $f(x, y)$ in $\Delta u = f$ would lead to the same multi-grid convergence.

. . . .

Multi-grid program for Poisson's equation in a square (4.4.2)

. . . .

```
begin real h2; integer l, i, j, n, n1, nn, ny, gamma, m, j1, k;
array w, u, f [0:64, 0:64], h2l[0:5]; integer array nl[0:5];

real procedure rhs(x, y); real x, y; rhs := - 4;

real procedure dir(x, y); real x, y; dir := x * x + y * y;

procedure start;
begin for i := 0 step 1 until l do
   begin nl[i] := 2 ** (i + 1); h2l[i] := 1/(nl[i] * nl[i]) end;
   n := nl[l]; n1 := n - 1;
   for i := 1 step 1 until n1 do for j := 1 step 1 until n1 do u[i, j] := 0;
   for i := 1 step 1 until n1 do
   begin u[0, i] := dir(0.0, i/n); u[i, 0] := dir(i/n, 0.0);
         u[n, i] := dir(1.0, i/n); u[i, n] := dir(i/n, 1.0)
   end Dirichlet data defined;
    for i := 1 step 1 until n1 do
    for j := 1 step 1 until n1 do f[i, j] := rhs(i/n, j/n)
end start;
```

```
procedure comp;
begin real s; n:= nl[l]; n1 := n - 1; s:= 0.0;
   for i:= 1 step 1 until n1 do for j:= 1 step 1 until n1 do
   s:= s + (u[i,j] - dir(i/n, j/n)) ** 2;
   print('absolute error: ', sqrt(h2l[l] * s))
end comparison with exact solution;

procedure s(l, u, f, ip, jp); integer l, ip, jp; array u, f;
begin n:= nl[l]; n1 := n - 1; h2:= h2l[l];
   for i:= ip step 2 until n1 do for j:= jp step 2 until n1 do
  u[i,j]:= (h2 * f[i,j] + u[i + 1,j] + u[i - 1,j] + u[i,j - 1] + u[i,j + 1])/4
end a quarter Gauß-Seidel step;

procedure defect(l, d, u, f); integer l; array u, f, d;
begin n:= nl[l]; n1 := n - 1; h2:= h2l[l];
   for i:= 1 step 1 until n1 do for j:= 1 step 1 until n1 do w[i,j]:=
      (4 * u[i,j] - u[i,j - 1] - u[i - 1,j] - u[i + 1,j] - u[i,j + 1])/
      h2 - f[i,j];
   for i:= 2 step 2 until n - 2 do for j:= 2 step 2 until n - 2 do
      d[i'/'2, j'/'2]:= (((w[i - 1, j - 1] + w[i + 1, j - 1] + w[i - 1, j + 1]
                        + w[i + 1, j + 1])/2 + w[i,j - 1] + w[i + 1, j]
                        + w[i - 1,j] + w[i,j + 1])/2 + w[i,j])/4
end computation of r(Lu - f);

procedure p(l, u, v); integer l; array u, v;
begin nn:= nl[l - 1]; n:= nl[l]; n1 := n - 1;
   for i:= 1 step 2 until n1 do
      w[i,0]:= w[0,i]:= w[n,i]:= w[i,n]:= 0.0;
   for i:= 0 step 1 until nn do for j:= 0 step 1 until nn do
      w[2 * i, 2 * j]:= v[i,j];
   for i:= 1 step 2 until n1 do for j:= 2 step 2 until n - 2 do
      w[i,j]:= (w[i - 1,j] + w[i + 1,j])/2;
   for i:= 1 step 1 until n1 do for j:= 1 step 2 until n1 do
      w[i,j]:= (w[i,j - 1] + w[i,j + 1])/2;
   for i:= 1 step 1 until n1 do for j:= 1 step 1 until n1 do
      u[i,j]:= u[i,j] - w[i,j]
end computation of u - pv;

procedure mgm(l, u, f); integer l; array u, f;
if l = 0 then s(l, u, f, 1, 1) else
begin integer i, j; array v, d[0:nl[l - 1], 0:nl[l - 1]];
   comment pre-smoothing;
   for i:= 1 step 1 until ny do
   begin s(l, u, f, 2, 2); s(l, u, f, 1, 1); s(l, u, f, 1, 2); s(l, u, f, 2, 1) end;
   comment coarse-grid correction;
```

```
    defect (l, d, u, f);
    for i := 0 step 1 until nl[l − 1] do for j := 0 step 1 until nl[l − 1] do
        v[i, j] := 0;
    for i := 1 step 1 until gamma do mgm(l − 1, v, d);
    p(l, u, v)
end multi-grid iteration;

l := 4; gamma := 2; ny := 2; start; comp;
for k := 1 step 1 until 10 do begin mgm(l, u, f); comp end
end
```

. . . .

Table 4.4.1 shows the development of the errors $\| u_l^j - u_l \|_{\mathscr{U}}$. $u_l(x, y) = x^2 + y^2$ is the discrete solution, u_l^j is the result of program (2) with $v = \gamma = 2$ applied to the starting value $u_l^0(x, y) = 0$. The level number is chosen as $l = 4$ corresponding to $h_l = \frac{1}{32}$. The Euclidean norm $\| u_l \|_{\mathscr{U}} = \langle u_l, u_l \rangle^{1/2}$ is defined by (3.5.2). The ratios $\| u_l^{j+1} - u_l \|_{\mathscr{U}} / \| u_l^j - u_l \|_{\mathscr{U}}$ listed in the last row converge to the convergence rate $\varrho_l = \varrho(M_l'(v, 0))$ (M_l' from (7.1.4)). Further computations show $\varrho_l = 0.0736$.

Computing the convergence rates ϱ_l for all levels $l = 1, \ldots, 5$, we obtain the numbers of Table 4.4.2. Obviously, the rates do not tend to 1 as classical iterations do. For the two-grid iteration we shall prove $\varrho_l < 0.0741$ for all l (cf. §8.1.1). Note that for increasing γ the multi-grid rates converge to the two-grid rates (cf. (2.3)).

Table 4.4.1. The iteration errors for $h_l = \frac{1}{32}$, $v = \gamma = 2$

j	0		1		2		3		4		5		6
$\| u_l^j - u_l \|_{\mathscr{U}}$	7.48 − 1[a]		4.20 − 2		1.93 − 3		1.03 − 4		5.76 − 6		3.36 − 7		2.03 − 8
ratio		0.0561		0.0460		0.0533		0.0560		0.0583		0.0603	

[a] Read $7.48_{10} - 1$ and other values analogously

Table 4.4.2. The convergence rates at different levels for $v = 2$

l	1	2	3	4	5
h_l	$\frac{1}{4}$	$\frac{1}{8}$	$\frac{1}{16}$	$\frac{1}{32}$	$\frac{1}{64}$
ϱ_l for $\gamma = 1$	0.0625	0.1263	0.1587	0.1682	0.174
ϱ_l for $\gamma = 2$	0.0625	0.0642	0.0738	0.0736	0.073

Table 4.4.3. The convergence rate depending on v for $h_l = \frac{1}{32}$

v	1	2	3	4	5
ϱ_l for $\gamma = 1$	0.310	0.168	0.115	0.087	0.0706
ϱ_l for $\gamma = 2$	0.244	0.074	0.052	0.040	0.033

It remains to investigate the dependence of ϱ_l on the number v of smoothing iterations. $\varrho_l(v)$ is listed in Table 4.4.3 for $\gamma = 1$ and $\gamma = 2$. As for the very simple example of § 2 the rates behave like C_γ/v.

To judge the efficiency of the multi-grid iteration we must also take account of the computational work. Counting the operations $+, -, *, /$ (except those involved in indices) in program (2), we obtain the values

$$C_S = 6, \quad C_D = 9.75, \quad C_C = 2.5, \quad C_0 = 6 \tag{4.4.3 a}$$

for the constants of (3.1). Following the advice of Note 4.3.4(i), we can reduce the work:

$$C_S = 6, \quad C_D = 4.5, \quad C_C = 2.5, \quad C_0 = 6. \tag{4.4.3 b}$$

If we replace f_l by $h_l^2 f_l$ one multiplication can be saved in *s* and one division in *defect*:

$$C_S = 5, \quad C_D = 4, \quad C_C = 2.5, \quad C_0 = 5. \tag{4.4.3 c}$$

For simplicity we neglect the further reduction due to Note 4.3.4(ii), (iii). Thus an optimised version of program (2) would require C_l operations per grid point, where C_l can approximately be estimated by

$$\bar{C} = \bar{C}(v, \gamma) := (v\, C_S + C_D + C_C)/(1 - \gamma/4).$$

The values of $\bar{C}$ are listed in Table 4.4.4. Replacing the efficiency E from (3.9) by

$$\bar{E} = \bar{E}(v, \gamma) = -\bar{C}(v, \gamma)/\log(\varrho_l(v, \gamma)),$$

we obtain the values in Table 4.4.5 from the data of Tables 4.4.3–4.

For a comparison with the SOR method we mention that the efficiency of SOR with optimal overrelaxation parameter equals $7/(2\pi h_l) \approx 1.11/h_l$.

Table 4.4.4. The number $\bar{C}$ of operations per grid point in the case of (3c)

v	1	2	3	4	5
$\bar{C}$ for $\gamma = 1$	$15\frac{1}{3}$	22	$28\frac{2}{3}$	$35\frac{1}{3}$	42
$\bar{C}$ for $\gamma = 2$	23	33	43	53	63

Table 4.4.5. Efficiency $\bar{E}$

γ \ v	1	2	3	4	5
1	13.1	12.3	13.3	14.5	15.8
2	16.3	12.7	14.5	16.5	18.5

Finally we mention an improvement of the multi-grid program (2) (cf. Hackbusch [2]). Exploiting Note 4.3.4 we define the pre-smoothing by

$$\mathscr{S}_l^{\text{pre}} = \mathscr{S}_{l1} \tag{4.4.4a}$$

and the post-smoothing on the finest grid by

$$\mathscr{S}_l^{\text{post}} = \mathscr{S}_{l1}\mathscr{S}_{l4}\mathscr{S}_{l3}\mathscr{S}_{l1}\mathscr{S}_{l2}\mathscr{S}_{l4}\mathscr{S}_{l3} \qquad (l = l_{\max}), \tag{4.4.4b}$$

where $\mathscr{S}_{lj}$ denotes the quarter Gauß-Seidel step on Ω_l^j (cf. Fig. 3.3.5). Thus, one iteration requires $\frac{8}{4}$ Gauß-Seidel steps at the maximum level $l_{\max}$. At lower levels $l < l_{\max}$ the post-smoothing reads as

$$\mathscr{S}_l^{\text{post}} = \mathscr{S}_{l2}\mathscr{S}_{l1}\mathscr{S}_{l4}\mathscr{S}_{l3}\mathscr{S}_{l1}\mathscr{S}_{l2}\mathscr{S}_{l4}\mathscr{S}_{l3} \qquad (l < l_{\max}). \tag{4.4.4c}$$

The step sizes, the restrictions and prolongations are chosen as before, but the auxiliary matrices L_l $(l < l_{\max})$ are defined by the Galerkin approach (3.7.1). Further, the V-cycle $\gamma = 1$ is used.

The computational work per iteration is comparable with program (2) for $\gamma = 1$, $\nu = 2$, since $\bar{C} = 2.75$, however the convergence rate is improved; e.g. for $h_4 = \frac{1}{32}$ we obtain $\varrho_4 = 0.061$ instead of $\varrho_4 = 0.168$ (cf. Table 4.4.3). The efficiency becomes $\bar{E} = 9.8$.

4.4.2 Poisson's Equation in a General Domain

In a general domain $\Omega \subset \mathbb{R}^2$ the Shortley-Weller scheme (cf. (3.1.22)) replaces the five-point formula (3.1.15). The nine-point prolongation has to be modified near the boundary as mentioned at the end of § 3.4.2. Having defined p (irregularly near the boundary), we still define the restriction by $r = p^*$.

The multi-grid iteration is modified as described previously. Testing the algorithm for the domains Ω depicted in Fig. 4.4.1 we observe the rates listed in the left part of Table 4.4.6.

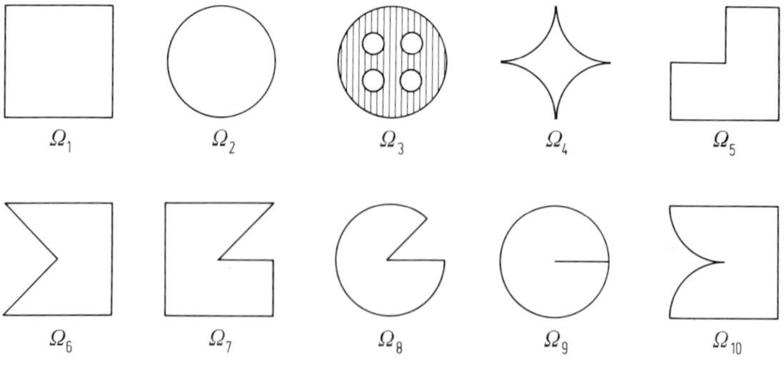

Fig. 4.4.1. Regions Ω_i (redrawn from Computing 20 (1978) p. 298)

Table 4.4.6. Averaged reduction factors [1] of the multi-grid iteration applied to the five-point Poisson discretisation

Ω	pointwise Gauß-Seidel	x-line Gauß-Seidel	Angle of the re-entrant corner
Ω_1	0.044–0.052	0.040–0.041	π
Ω_2	0.030–0.035	0.030–0.034	π
Ω_3	0.042–0.056	0.028–0.032	π
Ω_4	0.025–0.031	0.027–0.028	π
Ω_5	0.073–0.085	0.043–0.044	$3\pi/2$
Ω_6	0.066–0.078	0.041–0.042	$3\pi/2$
Ω_7	0.096–0.116	0.045–0.054	$7\pi/4$
Ω_8	0.093–0.116	0.044–0.054	$7\pi/4$
Ω_9	0.113–0.147	0.053–0.079	2π
Ω_{10}	0.108–0.139	0.053–0.077	2π

Obviously, the rates depend on the angle of the re-entrant corner(s). This angle determines the regularity of the boundary value problem (cf. Definition 6.3.5). This result is in accordance with the theoretical results of § 6.

The pointwise Gauß-Seidel iteration can be replaced by the linewise version (zebra-line Gauß-Seidel). We choose

$$\mathscr{S}_l^{\text{pre}} = \mathscr{S}_l^b \qquad (l \leqq l_{\max}), \tag{4.4.5a}$$

$$\mathscr{S}_l^{\text{post}} = \mathscr{S}_l^r \mathscr{S}_l^b \mathscr{S}_l^r \qquad (l = l_{\max}), \tag{4.4.5b}$$

$$\mathscr{S}_l^{\text{post}} = \mathscr{S}_l^r \mathscr{S}_l^b \mathscr{S}_l^r \mathscr{S}_l^b \qquad (l < l_{\max}), \tag{4.4.5c}$$

where $\mathscr{S}_l^r$ and $\mathscr{S}_l^b$ are the half Gauß-Seidel steps on the respective subsets Ω_l^r and Ω_l^b (cf. § 3.3.1 and Fig. 3.3.4). The observed rates are listed on the right side of Table 4.4.6 (cf. Hackbusch [3], [8]). The results still depend on the angle of the re-entrant corner (i.e. on the regularity of the problem) but the dependence is very slight.

4.4.3 Other Boundary Conditions

A multi-grid program for a general linear second order differential equation

$$a(x,y)\,u_{xx} + b(x,y)\,u_{xy} + c(x,y)\,u_{yy} + d(x,y)\,u_x + e(x,y)\,u_y + f(x,y)\,u = g(x,y) \tag{4.4.6}$$

on a rectangle $\Omega = (x_1, x_2) \times (y_1, y_2)$ subject to Dirichlet, Neumann, mixed, or periodic boundary conditions is described by Hackbusch [6]. It uses the

1 The averaged reduction factor is defined by

$$\tilde{\varrho} = [\| u_l^{i+j} - u_l \|_{\mathscr{U}} / \| u_l^i - u_l \|_{\mathscr{U}}]^{1/j}, \quad j \text{ sufficiently large.}$$

For $j \to \infty$, $\tilde{\varrho}$ approaches the convergence rate ϱ

Table 4.4.7. Errors $\| u_l^j - u_l \|$ with $h_l = \frac{1}{64}$ for Poisson's equation

j	1	2	3	4	5	6	7	8	ARF[a]
Dirichlet	$7.1_{10}-1$	$1.2_{10}-2$	$4.5_{10}-4$	$1.4_{10}-5$	$6.6_{10}-7$	$2.7_{10}-8$	$1.5_{10}-9$	$7.8_{10}-11$	0.038
Neumann	$7.9_{10}-1$	$7.1_{10}-3$	$8.2_{10}-5$	$3.4_{10}-6$	$4.7_{10}-8$	$3.0_{10}-9$	$5.0_{10}-10$	$2.8_{10}-12$	0.023
mixed	$7.1_{10}-1$	$1.0_{10}-2$	$3.5_{10}-4$	$1.1_{10}-5$	$5.1_{10}-7$	$2.4_{10}-8$	$1.4_{10}-9$	$7.5_{10}-11$	0.038
periodic	$4.8_{10}-1$	$1.5_{10}-2$	$4.9_{10}-4$	$1.6_{10}-5$	$5.0_{10}-7$	$1.6_{10}-8$	$7.5_{10}-10$	$1.6_{10}-11$	0.032

[a] ARF, Averaged reduction factor

Galerkin approach (3.7.1). The pre- and post-smoothing are specified by (5a–c). Table 4.4.7 shows the errors $\| u_5^j - u_5 \|_{\mathcal{U}}$ obtained for the Poisson equation $\Delta u = f$ in $\Omega = (0,1) \times (0,1)$ with $h_5 = \frac{1}{64}$. The boundary conditions, the right side f, and the discrete solution u_5 are

Dirichlet: $u = x^2 + y^2$ on Γ, $f = 4$, $u_5 = x^2 + y^2$;

Neumann: $u_x = 2x$ for $x = 0, 1$, $u_y = 2y$ for $y = 0, 1$, $f = 4$, $u_5 = x^2 + y^2$;

mixed: $u = x^2 + y^2$ on Γ if $x \neq 0$, $-u_x + u = y^2$ if $x = 0$, $f = 4$, $u_5 = x^2 + y^2$;

periodic: $u(x,0) = u(x,1)$, $u_y(x,0) = u_y(x,1)$, $u(0,y) = u(1,y)$, $u_x(0,y) = u_x(1,y)$, $f = -8 \cdot 64^2 \cdot \sin^2(\pi/64) \cdot u_5$, $u_5 = \sin(2\pi x)\sin(2\pi y)$.

For the Neumann and periodic conditions the solution is determined up to a constant. Concerning this problem we refer to Remark 12.1.5. The numerical results show that the multi-grid convergence speed is rather improved by the choice of other boundary conditions than made worse.

For numerical results of Eq. (6) with variable coefficients we refer to Hackbusch [6], [8] and § 10.2.2.

4.4.4 Further Poisson Solvers

The program *MG00* (cf. Foerster-Witsch [1], [2]; also mentioned by Stüben-Trottenberg [2]) is very close to program (2). The only difference is that post-smoothing is also performed (e.g. $v_1 = 2$, $v_2 = 1$). Depending on the version the nine-point restriction is replaced by the 'half weighting' (3.8.8).

Ries-Trottenberg-Winter [1] and Winter [1] proposed a multi-grid program ("*MGR*") which is especially adapted to Poisson's equation. Instead of the standard coarsening (cf. Fig. 3.4.1) it uses the rotated grid of size $h_{l-1} = \sqrt{2}\, h_l$ depicted in Fig. 3.4.3. One can exploit the following facts:

- the Poisson equation can be discretised by the rotated five-point scheme;
- the restriction r becomes $(r\,u)\,(\mathbf{x}) = \frac{1}{2}u(\mathbf{x})$, $\mathbf{x} \in \Omega_{l-1}$, after a half Gauß-Seidel step on $\Omega_l \backslash \Omega_{l-1}$ (cf. Note 4.3.4(i));
- the prolongation p need not be performed if a half Gauß-Seidel step on $\Omega_l \backslash \Omega_{l-1}$ follows (cf. Note 4.3.4(ii) and § 3.4.4).

One V-cycle ($v = 1$, $\gamma = 1$) requires $12.75 \times \frac{4}{3} = 17$ operations (cf. Ries-Trottenberg-Winter [1, Table 2]). The three-grid convergence rate (involving h_l, $h_{l-1} = \sqrt{2}\,h_l, h_{l-2}$) equals 0.074; the five-grid rate is 0.093. Since $\bar{E} \approx 7.2$ this MGR-version is more efficient than the previous programs.

Another multi-grid program of Wesseling will be described in § 10.4.3.2. A multi-grid Poisson solver with optimal performance on a special vector computer is developed by Barkai-Brandt [1].

5. Nested Iteration Technique

5.1 Algorithm

Given some iterative process, the natural approach is to start with a more or less accurate initial value u_l^0 and to perform several steps of the iteration. For the multi-grid iteration the program would be

$$\begin{aligned}&\tilde{u}_l := u_l^0;\\ &\textbf{for } j := 1 \textbf{ step } 1 \textbf{ until } i \textbf{ do } MGM(l, \tilde{u}_l, f_l);\end{aligned} \tag{5.1.1}$$

The error of $\tilde{u}_l$ satisfies

$$\| \tilde{u}_l - u_l \| \mathscr{U} \leqq \zeta^i \| u_l^0 - u_l \|_{\mathscr{U}} \tag{5.1.2a}$$

where ζ is the contraction number of the iteration with respect to $\| \cdot \|_{\mathscr{U}}$ (cf. (1.3.16b)). In particular, the simplest choice $u_l^0 = 0$ yields an estimate of the relative error:

$$\| \tilde{u}_l - u_l \|_{\mathscr{U}} / \| u_l \|_{\mathscr{U}} \leqq \zeta^i \qquad (\text{if } u_l^0 = 0). \tag{5.1.2b}$$

In order to obtain a fixed (relative) error ε, one needs $i \geqq \log(\varepsilon)/\log(\zeta) = O(|\log(\varepsilon)|)$ iterations. Often, ε is not given in advance and one has to choose a suitable value. Usually, it is useless to take ε smaller than the discretisation error (i.e. the difference between u_l and the continuous solution u). Unfortunately, the quantitative size of the discretisation error is not known a priori, but only its asymptotic behaviour $O(h_l^\kappa)$ (κ: consistency order). From $\varepsilon = O(h_l^\kappa)$ one concludes that $i = O(|\log(h_l)|)$ iterations are required to obtain an iterate $\tilde{u}_l$ with error of the size of the discretisation error. The corresponding number of operations is $O(n_l |\log(h_l)|) = O(h_l^{-d} |\log(h_l)|)$.

The nested iteration described below (also called 'full multi-grid method', cf. Brandt [12]) has several advantages:

(i) Although no a priori knowledge of the discretisation error $\varepsilon_{\text{disc}}$ is needed, the nested iteration produces approximations $\tilde{u}_l$ with error $O(\varepsilon_{\text{disc}})$.

(ii) The nested iteration is cheaper than the simple approach (1). An approximation with error $O(\varepsilon_{\text{disc}})$ is calculated by $O(n_l)$ operations.

(iii) Besides $\tilde{u}_l$, the coarser-grid solutions $\tilde{u}_{l-1}, \tilde{u}_{l-2}, \ldots$ are also approximated and are at one's disposal.

The nested iteration can be combined with any iterative process and has often been used together with successive overrelaxation iterations. The idea is to provide a good starting guess u_l^0 by means of iterating on the coarser grid. The quasi-ALGOL formulation reads as follows:

nested iteration (5.1.3)

$\tilde{u}_0 := L_0^{-1} f_0$; (5.1.3a)
for $k := 1$ **step** 1 **until** l **do**
begin $\tilde{u}_k := \tilde{p}\,\tilde{u}_{k-1}$; (5.1.3b)
for $j := 1$ **step** 1 **until** i **do** $MGM(k, \tilde{u}_k, f_k)$ (5.1.3c)
end;

At each level $k = 1, 2, \ldots, l$ i iterations are performed. The starting value is obtained from the coarse-grid approximation $\tilde{u}_{l-1}$. $\tilde{p}$ denotes some interpolation. From the programming point of view the optimal choice is $\tilde{p} = p$ (p from multi-grid algorithm), since no additional subroutine is needed. However, interpolations $\tilde{p}$ of higher order than p should be taken into considerations, too, as pointed out in § 5.2.

An appropriate choice of i is discussed in the next section. The same value i can be chosen for all levels, since the contraction numbers of the multi-grid iteration are bounded independently of the level $0 \leqq k \leqq l$. The analogous nested iteration, e.g., applied to the successive overrelaxation would require increasing numbers $i = i_k$.

Finally, we mention three modifications of algorithm (3). The exact solution of $L_0 u_0 = f_0$ in (3a) is not necessary. One may replace (3a) by

$$\tilde{u}_0 \approx L_0^{-1} f_0 \tag{5.1.4a}$$

provided $\| \tilde{u}_0 - u_0 \|_{\mathscr{U}}$ is small enough.

The second generalisation concerns the interpolation $\tilde{p}$ in (3b). Sometimes, simple high order interpolations can be found that employ the difference equation. We give an example for the discrete Poisson equation. From $\tilde{u}_{l-1}$ at the coarse-grid points (x, y), $(x + 2h_l, y)$, $(x, y + 2h_l)$, $(x + 2h_l, y + 2h_l)$ one can compute the value at $(x + h_l, y + h_l)$ by

$$\begin{aligned} \tilde{u}_l(x + h_l, y + h_l) := {} & \tfrac{1}{4}[\tilde{u}_{l-1}(x, y) + \tilde{u}_{l-1}(x + 2h_l, y) + \tilde{u}_{l-1}(x, y + 2h_l) \\ & + \tilde{u}_{l-1}(x + 2h_l, y + 2h_l)] + \tfrac{1}{2} h_l^2 f_l(x + h_l, y + h_l) \end{aligned}$$

according to the diagonal five-point formula $(2h_l^2)^{-1}\begin{bmatrix} -1 & 0 & -1 \\ 0 & 4 & 0 \\ -1 & 0 & -1 \end{bmatrix} \tilde{u}_l = f_l$.

The values of $\tilde{u}_l$ at the remaining grid points can be computed by the usual five-point formula. This interpolation, which may be written

$$\tilde{u}_k := \tilde{p}(\tilde{u}_{k-1}, f_k) \tag{5.1.4b}$$

with $\tilde{p}$ linear in $(\tilde{u}_{k-1}, f_k)$, is of order four. Cf. Kronjö-Dahlquist [1], Hyman [1].

Thirdly, we mention that the number i in (3c) may be different for different levels: $i = i_k$. Since most of the computational work is spent at level l, it might be advantageous to choose $i_l < i_{l-1} = i_{l-2} = \ldots = i_1$ (cf. § 5.3).

5.2 Analysis of the Nested Iteration

The nested iteration (1.3) requires the specification of the iteration number i. The following analysis suggests how to choose i. It will often turn out that $i = 1$ is satisfactory.

Let ζ_k be the contraction number of the multi-grid iteration employed at level k:

$$\| u_k^{j+1} - u_k \|_{\mathscr{U}} \leqq \zeta_k \, \| u_k^j - u_k \|_{\mathscr{U}} \tag{5.2.1}$$

(cf. (1.3.16)). As pointed out before, the numbers ζ_k are uniformly bounded by some $\zeta < 1$. Set

$$\zeta := \max_{1 \leqq k \leqq l} \zeta_k \tag{5.2.2}$$

where l is the maximum level from (1.3).

The discretisation error is the (suitably defined) difference between the discrete solution $u_k = L_k^{-1} f_k$ and the continuous solution u. The difference between u_k and $u_{k-1} = L_{k-1}^{-1} f_{k-1}$ is often called the *relative discretisation error* (error of u_{k-1} relative to u_k). We may estimate it by

$$\| \tilde{p} u_{k-1} - u_k \|_{\mathscr{U}} \leqq C_1 h_k^{\kappa} \quad (1 \leqq k \leqq l). \tag{5.2.3}$$

For a proof of (3) see Exercise 5.7.1. Note that the exponent κ in (3) depends on the (relative) consistency order *and* on the interpolation order of $\tilde{p}$. Therefore, we are led to

Note 5.2.1. The interpolation order of $\tilde{p}$ should at least equal the (relative) consistency order.

Consider the standard case of a second order discretisation ($\kappa = 2$) of a second order differential equation ($2m = 2$). By Note 5.2.1, $\tilde{p}$ should be at least piecewise linear. Hence, $\tilde{p}$ may coincide with the standard prolongation of § 3.4.2.

To indicate the levels involved we write $\tilde{p} = \tilde{p}_{k \leftarrow k-1}$. Let $\|\cdot\|_{\mathscr{U} \leftarrow \mathscr{U}}$ be defined as in § 1.3.2 and set

$$C_{20} := \max_{1 \leq k \leq l} \|\tilde{p}_{k \leftarrow k-1}\|_{\mathscr{U} \leftarrow \mathscr{U}}. \tag{5.2.4a}$$

Together with

$$C_{21} := \max_{1 \leq k \leq l} (h_{k-1}/h_k)^\kappa \tag{5.2.4b}$$

we define the constant

$$C_2 := C_{20} C_{21}. \tag{5.2.4c}$$

Note 5.2.2. In Exercise 5.7.2 it will be shown that $C_{20} = 1$ for the most frequent choices of $\tilde{p}$ and $\|\cdot\|_{\mathscr{U}}$. Moreover, $C_{21} = 2^\kappa$ holds for the usual sequence $h_k = h_0/2^k$. Hence, the value of C_2 is well-known: $C_2 = 2^\kappa$.

These values are sufficient to describe the errors of the results of the nested iteration (1.3).

Theorem 5.2.3. Assume (3) and

$$C_2 \zeta^i < 1 \tag{5.2.5}$$

with C_2 from (4a–c), ζ from (2), and i from (1.3c). Then the nested iteration (1.3) with i multi-grid iterations per level results in $\tilde{u}_k (0 \leqq k \leqq l)$ satisfying the error estimate

$$\|\tilde{u}_k - u_k\|_{\mathscr{U}} \leqq C_3(\zeta, i)\, C_1\, h_k^\kappa \qquad (0 \leqq k \leqq l) \tag{5.2.6}$$

where $u_k = L_k^{-1} f_k$ and

$$C_3(\zeta, i) := \zeta^i/(1 - C_2 \zeta^i). \tag{5.2.7}$$

Proof. (6) holds for $k = 0$ because $\tilde{u}_0 = u_0$. Let (6) be valid for $k-1$. Denote the iterates of (1.3c) by u_k^j, $j = 0, \ldots, i$. The definition of ζ implies

$$\|\tilde{u}_k - u_k\|_{\mathscr{U}} = \|u_k^i - u_k\|_{\mathscr{U}} \leqq \zeta^i \|u_k^0 - u_k\|_{\mathscr{U}}.$$

The starting guess is $u_k^0 = \tilde{p}\tilde{u}_{k-1}$. Its error can be estimated by

$$\begin{aligned}\|u_k^0 - u_k\|_{\mathscr{U}} &= \|\tilde{p}\tilde{u}_{k-1} - u_k\|_{\mathscr{U}} = \|[\tilde{p}u_{k-1} - u_k] + \tilde{p}[\tilde{u}_{k-1} - u_{k-1}]\|_{\mathscr{U}} \\ &\leqq \|\tilde{p}u_{k-1} - u_k\|_{\mathscr{U}} + \|\tilde{p}\|_{\mathscr{U} \leftarrow \mathscr{U}} \|\tilde{u}_{k-1} - u_{k-1}\|_{\mathscr{U}}.\end{aligned}$$

From (6) (for $k-1$ instead of k) and (4b) we conclude that

$$\|\tilde{u}_{k-1} - u_{k-1}\|_{\mathscr{U}} \leqq C_3(\zeta, i)\, C_1\, h_{k-1}^\kappa \leqq C_3(\zeta, i)\, C_{21}\, C_1\, h_k^\kappa.$$

The last three inequalities together with (3) and (4a) yield

$$\begin{aligned}\|\tilde{u}_k - u_k\|_{\mathscr{U}} &\leqq \zeta^i [C_1 h_k^\kappa + C_{20} C_{21} C_3(\zeta, i)\, C_1\, h_k^\kappa] \\ &= [1 + C_2 C_3(\zeta, i)]\, \zeta^i C_1 h_k^\kappa = C_3(\zeta, i)\, C_1\, h_k^\kappa\end{aligned}$$

proving that if (6) holds for $k-1$, then it also holds for k. □

Similar results for the modifications (1.4a) and (1.4b) are discussed in Exercises 5.7.3–4.

Theorem 5.2.3 ensures that the errors at *all* levels $k = 0, 1, \ldots, l$ differ from the (bound of the) relative discretisation error $C_1 h_k^\kappa$ only by a factor $C_3(\zeta, i)$ which is well-known because of Note 5.2.2. The only condition on i is (5), $C_2 \zeta^i < 1$. The nested iteration is as cheap as possible if $i = 1$ satisfies (5). With $C_2 = 2^\kappa$ from Note 5.2.2 this condition becomes $2^\kappa \zeta < 1$. Assuming, in addition, the standard case of $\kappa = 2$, we obtain

Note 5.2.4. Assume $C_2 = 4$ and $\zeta < \frac{1}{4}$. Then the estimates (6) of Theorem 5.2.3 hold for $i = 1$ (only one multi-grid iteration per level).

Even if the condition $C_2 \zeta^i < 1$ is violated, the nested iteration produces reasonable results $\tilde{u}_k$ as studied in Exercise 5.7.5, but the error is no longer proportional to the relative discretisation error $C_1 h_k^\kappa$.

Hitherto, the errors $\| \tilde{u}_k - u_k \|_{\mathscr{U}}$ were compared only with the relative discretisation error. To get a realistic comparison with the (usual) discretisation error, assume an error expansion

$$u_k = \tilde{R}_k u + h_k^\kappa \tilde{R}_k e + E_k, \quad \| E_k \|_{\mathscr{U}} = o(h_k^\kappa), \tag{5.2.8}$$

(cf. Pereyra [2]) where $(\tilde{R}_k u)(\mathbf{x}) = u(\mathbf{x})\ (\mathbf{x} \in \Omega_k)$ is the trivial restriction of the continuous solution u to the grid Ω_k. The function e is assumed to be independent of k. The interpolation of u_{k-1} by $\tilde{p}$ yields

$$\tilde{p} u_{k-1} = \tilde{p} \tilde{R}_{k-1} (u + h_{k-1}^\kappa e) + o(h_{k-1}^\kappa) = \tilde{R}_k (u + h_{k-1}^\kappa e) + o(h_k^\kappa),$$

if the order of interpolation exceeds κ. Hence, the relative discretisation error is

$$\| \tilde{p} u_{k-1} - u_k \|_{\mathscr{U}} = (h_{k-1}^\kappa - h_k^\kappa) \| \tilde{R}_k e \|_{\mathscr{U}} + o(h_k^\kappa).$$

For $h_{k-1} = 2 h_k$ the constant C_1 of inequality (3) equals almost $\max \{(2^\kappa - 1) \| \tilde{R}_k e \|_{\mathscr{U}} : 1 \leqq k \leqq l\}$. One can prove

Theorem 5.2.5. Assume the validity of expansion (8), let the interpolation order of $\tilde{p}$ exceed κ, and suppose $h_{k-1} = 2 h_k$. Under condition (5), $C_2 \zeta^i < 1$, the results $\tilde{u}_k$ of the nested iteration (1.3) satisfy

$$\| \tilde{u}_k - u_k \|_{\mathscr{U}} \leqq C_4(\zeta, i)\, h_k^\kappa \max_{1 \leqq k \leqq l} \| \tilde{R}_k e \|_{\mathscr{U}} + o(h_k^\kappa) \tag{5.2.9}$$

with $C_4(\zeta, i) = (2^\kappa - 1)\, C_3(\zeta, i) = (2^\kappa - 1)\, \zeta^i / (1 - C_2 \zeta^i)$.

Since $h_k^\kappa \| \tilde{R}_k e \|_{\mathscr{U}} + o(h_k^\kappa)$ is the discretisation error, estimate (9) may be interpreted to mean

iteration error $\leqq C_4(\zeta, i) *$ discretisation error.

For $\kappa = 2$ and $C_2 = 2^\kappa = 4$ (cf. Note 5.2.2) the factor $C_4(\zeta, i) = 3 \zeta^i / (1 - 4 \zeta^i)$ is listed in Table 5.2.1. $i = 1$ may be chosen if $\zeta < \frac{1}{4}$ (cf. Note 5.2.4).

Table 5.2.1. Factors $C_4(\zeta, i)$ for $i = 1, 2, 3$

ζ	$C_4(\zeta,1)$	ζ	$C_4(\zeta,2)$	ζ	$C_4(\zeta,3)$
0.200	3.00	0.44	2.57	0.60	4.76
0.182	2.00	0.40	1.33	0.58	3.00
0.175	1.75	0.38	1.00	0.57	2.00
0.143	1.00	0.33	0.60	0.52	1.00
0.100	0.50	0.30	0.42	0.50	0.75
0.071	0.30	0.25	0.25	0.46	0.50
0.053	0.20	0.20	0.143	0.40	0.25

Typical values of multi-grid contraction numbers are $\zeta \approx \frac{1}{7} \approx 0.143$ yielding $C_4(\zeta, 1) \approx 1.0$. Hence the iteration error $\tilde{u}_k - u_k$ is of nearly the same size as the discretisation error. If $\zeta > 0.18$ one should choose $i = 2$.

Theorem 5.2.5 entails a higher interpolation order of $\tilde{p}$. Otherwise, the interpolation error, say $w_k h_k^\kappa = O(h_k^\kappa)$, is superposed on $h_k^\kappa \tilde{R}_k e$ and the maximum in (9) becomes $\max\{\| \tilde{R}_k e \|_{\mathscr{U}} + \| w_k \|_{\mathscr{U}}/(2^\kappa - 1) \colon 1 \leqq k \leqq l\}$. But often $\| w_k \|_{\mathscr{U}} \ll (2^\kappa - 1) \| \tilde{R}_k e \|_{\mathscr{U}}$ will be valid so that inequality (9) still holds approximately.

Although the estimate (9) seems to be very favourable, it is too pessimistic for some cases. Assume the expansion (8) with $E_k = O(h_k^\tau)$, $\tau > \kappa$, and let the interpolation order of $\tilde{p}$ be τ. Then

$$u_k = \tilde{p}\, u_{k-1} + h_k^\kappa \tilde{e}_k + h_k^\tau \tilde{E}_k \tag{5.2.10}$$

holds with $\tilde{e}_k = (h_{k-1}^\kappa / h_k^\kappa - 1) \tilde{R}_k e$. Denote the multi-grid iteration matrix by M_k (cf. (1.3.2)). By definition of ζ, $\| M_k \tilde{e}_k \|_{\mathscr{U}} \leqq \zeta \| \tilde{e}_k \|_{\mathscr{U}}$ is valid. However, in regular cases the function $\tilde{e}_k$ is smooth and the coarse-grid correction might yield $\| M_k \tilde{e}_k \|_{\mathscr{U}} = O(h_k^{\tau-\kappa})$. Then the condition $\zeta^i C_2' < 1$ $(C_2' = C_{20} C_{21}', C_{21}' = \max(h_{k-1}/h_k)^\tau)$ implies

$$\| \tilde{u}_k - u_k \| \mathscr{U} = O(h_k^\tau), \tag{5.2.11}$$

i.e. the iteration error is of even better order than the discretisation error.

5.3 Computational Work and Efficiency

The computational work of one multi-grid iteration at level k is approximately

$$W_k = (v\, C_s + C_D + C_C)\, n_k / (1 - \vartheta) \qquad (\vartheta = \gamma\, C_H < 1)$$

as can be seen from (4.3.5), when the term $O(\vartheta^l)$ is neglected. The inequality $n_{k-1} \leqq C_H n_k$ yields $W_k \leqq C_H^{l-k} W_l$ $(1 \leqq k \leqq l)$. Hence, the computational work of the nested iteration (1.3) equals

$$i\,W_1 + i\,W_2 + \ldots + i\,W_l \leqq i \sum_{k=1}^{l} C_H^{l-k}\, W_l < \frac{i}{1 - C_H}\, W_l .$$

Here we do not take into account the solution of $L_0 u_0 = f_0$ and the work required by $\tilde{p}$, since both are much less than the work taken by the iterations.

Note 5.3.1. The nested iteration (1.3) with i iterations per level is only more expensive than the simple iteration (1.1) by a factor $1/(1 - C_H)$. In the two-dimensional case ($d = 2$) with $h_{k-1} = 2h_k$, this factor equals $\frac{4}{3}$; hence the total work of the nested iteration is less than

$$\frac{4}{3}\, i\, W_l = \frac{16}{3(4-\gamma)}\, i(v\,C_S + C_D + C_C)\, n_l . \tag{5.3.1}$$

The factor $16/[3(4-\gamma)]$ equals 1.78 for $\gamma = 1$ (V-cycle) and 2.67 for $\gamma = 2$ (W-cycle).

One concludes that an approximation $\tilde{u}_l$ with an error $O(h_l^\kappa)$ can be obtained by only $O(n_l)$ operations (instead of $O(n_l\,|\log h_l|)$ for iteration (1.1)).

What choice of i is optimal? The answer depends on the desired accuracy and on ζ. Assume for instance that one aims to obtain $C_4(\zeta, i) = 1$, i.e. $\|\tilde{u}_l - u_l\|_{\mathscr{U}} \approx$ discretisation error. This may be difficult for certain values of ζ. Consider e.g. $\zeta = 0.2$. From Table 5.2.1 we see that $i = 1$ yields $C_4(0.2, 1) = 3 \gg 1$, whereas $i = 2$ gives $C_4(0.2, 2) = 0.143$ indicating that $\tilde{u}_l$ is more accurate than necessary. A compromise is the choice of different i's at different levels: $i_l = 1, i_{l-1} = i_{l-2} = \ldots = i_1 = 2$. As shown in Exercise 5.7.6 the corresponding factor C_4 at level l equals 0.714 which is fairly close to 1. The computational work of the nested iteration is $\frac{4}{3} W_l$ for $i = 1$, $\frac{8}{3} W_l$ for $i = 2$, and only $\frac{5}{3} W_l$ for $i_l = 1$, $i_{l-1} = \ldots = i_1 = 2$.

5.4 Nested Iteration with Extrapolation

In § 5.2 we used an asymptotic expansion

$$u_k = \tilde{R}_k u + h_k^\kappa \tilde{R}_k e + O(h_k^\tau), \qquad \tau > \kappa, \tag{5.4.1}$$

to obtain quantitative estimates of the discretisation error. On the other hand an asymptotic expansion (1) with e independent of h_k enables us to apply Richardson's extrapolation (cf. § 14.1). An obvious strategy would be as follows: Compute $\tilde{u}_k$, $0 \leqq k \leqq l$, by the nested iteration (1.3) and extrapolate to $h = 0$. However, Richardson's extrapolation needs $\tilde{u}_k$ with errors $\tilde{u}_k - u_k$ not exceeding the remainder $O(h_k^\tau)$ of (1), whereas $\tilde{u}_k - u_k = O(h_k^\kappa)$ (cf. Theorem 5.2.5). A remedy would be to perform $i' - i$ further iterations with i' such that $\zeta^{i'-i} \leqq \text{const} * h_l^{\tau-\kappa}$ (l: fixed highest level). Then the iteration errors are of the same order $O(h_k^\tau)$ but may still be of a different size to the remainder in (1). The same situation arises in the case (2.11).

In the following we describe a modified nested iteration that yields iteration errors of the size of the remainder in (1). The algorithm uses an extrapolation to the finer grid size.

Assume $\Omega_{k-2} \subset \Omega_{k-1} \subset \Omega_k$ and $h_{k-2} = 2h_{k-1} = 4h_k$. For every $\mathbf{x} \in \Omega_{k-2}$ the expansion (1) implies

$$u_k(\mathbf{x}) = u_{k-1}(\mathbf{x}) + \frac{2^\kappa - 1}{4^\kappa - 2^\kappa}(u_{k-1}(\mathbf{x}) - u_{k-2}(\mathbf{x})) + O(h_k^\tau). \tag{5.4.2}$$

Kronsjö and Dahlquist [1] proposed to use the successive overrelaxation within a nested iteration based on (2). The algorithm is similar to (1.3),

nested iteration with extrapolation (5.4.3)

$\tilde{u}_0 := L_0^{-1} f_0;\quad \tilde{u}_1 := L_1^{-1} f_1;$ (5.4.3 a)
for $k := 2$ **step** 1 **until** l **do**
begin $\tilde{u}_k := \pi(\tilde{u}_{k-1}, \tilde{u}_{k-2});$ (5.4.3 b)
  for $j := 1$ **step** 1 **until** i **do** $MGM(k, \tilde{u}_k, f_k)$ (5.4.3 c)
end;

where the extrapolation is hidden in π. Several suggestions for the choice of π will be given below. For simplicity only the exact solutions are required in (3 a).

The difficulty of describing the extrapolation π is caused by the fact that formula (2) applies to $\mathbf{x} \in \Omega_{k-2}$ only, whereas $\tilde{u}_k$ has to be defined for all $\mathbf{x} \in \Omega_k$. The first version of π is based on an extrapolation on Ω_{k-2} followed by an interpolation $\tilde{p}_{k \leftarrow k-2}$ from level $k-2$ to k:

$$\pi(\tilde{u}_{k-1}, \tilde{u}_{k-2}) := \tilde{p}_{k \leftarrow k-2}\left[\hat{r}\tilde{u}_{k-1} + \frac{2^\kappa - 1}{4^\kappa - 2^\kappa}(\hat{r}\tilde{u}_{k-1} - \tilde{u}_{k-2})\right]. \tag{5.4.4}$$

$\hat{r}$ is the trivial injection onto $\Omega_{k-2} \subset \Omega_{k-1}$. The values of $\tilde{u}_{k-1}$ at $\mathbf{x} \in \Omega_{k-1} \backslash \Omega_{k-2}$ are not involved in (4) but are used in

$$\pi(\tilde{u}_{k-1}, \tilde{u}_{k-2}) := \tilde{p}_{k \leftarrow k-1}\left[\tilde{u}_{k-1} + \frac{2^\kappa - 1}{4^\kappa - 2^\kappa}(\tilde{u}_{k-1} - \tilde{p}_{k-1 \leftarrow k-2}\tilde{u}_{k-2})\right]. \tag{5.4.5}$$

An intermediate formula is

$$\pi(\tilde{u}_{k-1}, \tilde{u}_{k-2}) := \tilde{p}_{k \leftarrow k-1}\left[\tilde{u}_{k-1} + \frac{2^\kappa - 1}{4^\kappa - 2^\kappa}\tilde{\tilde{p}}_{k-1 \leftarrow k-2}(\hat{r}\tilde{u}_{k-1} - \tilde{u}_{k-2})\right], \tag{5.4.6}$$

where $\tilde{p}$ and $\tilde{\tilde{p}}$ are distinguished because of the following note:

Note 5.4.1. In order to guarantee $\pi(u_{k-1}, u_{k-2}) = u_k + O(h_k^\tau)$, the interpolation order of $\tilde{p}$ in (4–6) must be at least τ, whereas the order of $\tilde{\tilde{p}}$ has to be $\geqq \tau - \kappa$.

For instance, when of $\kappa = 2$, $\tau = 4$ one should use linear interpolation $\tilde{\tilde{p}}$ and cubic interpolation $\tilde{p}$. Again, the interpolation $\tilde{p}$ can be replaced by the more general procedure (1.4 b) (cf. "Experiment B" of Kronsjö-Dahlquist [1]).

The counterparts of the assumptions (2.3), (2.4 a–c) are

$$\| u_k - \pi(u_{k-1}, u_{k-2}) \|_{\mathscr{U}} \leqq C_1 h_k^\tau \quad (2 \leqq k \leqq l), \tag{5.4.7}$$

$$C_{20} := \max_{2 \leqq k \leqq l} (h_{k-1}/h_k)^\tau, \tag{5.4.8 a}$$

$$C_{21} := \max_{2 \leqq k \leqq l} \| \pi_1 \|_{\mathscr{U} \leftarrow \mathscr{U}}, \quad C_{22} := \max_{2 \leqq k \leqq l} \| \pi_2 \|_{\mathscr{U} \leftarrow \mathscr{U}}, \tag{5.4.8 b}$$

$$C_2 := C_{20} C_{21} + C_{20}^2 C_{22}, \tag{5.4.8 c}$$

where π_1 and π_2 are defined by $\pi(v_{k-1}, v_{k-2}) = \pi_1 v_{k-1} + \pi_2 v_{k-2}$.

As in Note 5.2.2 we obtain

$$C_2 = \frac{4^\kappa - 1}{4^\kappa - 2^\kappa} 2^\tau + \frac{2^\kappa - 1}{4^\kappa - 2^\kappa} 4^\tau \tag{5.4.9}$$

in the standard case. For the most frequent pairs (κ, τ) the value of C_2 is

$$C_2 = 14 (\kappa = 1, \tau = 2), \quad C_2 = 26 (\kappa = 2, \tau = 3), \quad C_2 = 84 (\kappa = 2, \tau = 4). \tag{5.4.10}$$

The first and third case correspond to discretisations by unsymmetric and symmetric difference formulae, respectively. An expansion with $\kappa = 2$, $\tau = 3$ appears if unsymmetric formulae are used only at points near the boundary (cf. Meis-Marcowitz [1]).

The following theorem guarantees $\tilde{u}_k = u_k + O(h_k^\tau)$ if $i > \log(C_2)/|\log(\zeta)|$.

Theorem 5.4.2. Let C_1 and C_2 be defined by (7) and (8). Choose the number i in (3 c) such that (2.5), $C_2 \zeta^i < 1$, holds. Then the nested iteration (3) results in approximations $\tilde{u}_k$ satisfying

$$\| \tilde{u}_k - u_k \|_{\mathscr{U}} \leqq C_3(\zeta, i)\, C_1 h_k^\tau \quad (0 \leqq k \leqq l) \tag{5.4.11}$$

with $C_3(\zeta, i) = \zeta^i/(1 - C_2 \zeta^i)$ as in (2.7).

The proof is similar to that of Theorem 5.2.3. The values (10) show that C_2 is much larger than in the case of the nested iteration (1.3) without extrapolation. Accordingly, the iteration number i must be larger. $i = 1$ works only for very fast multi-grid iterations (e.g. $\zeta < \frac{1}{14}$ in the case of $2\kappa = \tau = 2$).

Using $\tilde{u}_k$ for extrapolation to $h \to 0$ one obtains approximations of accuracy $O(h_k^\tau)$. The computational work is still $O(n_k)$ since i is independent of k.

5.5 Numerical Example

We apply the nested iteration (1.3) to the Dirichlet problem

$$\begin{aligned} -\Delta u &= 3\,y/(x+1) - y^3/(x+1)^3 \quad \text{in } \Omega = (0,1)\times(0,1), \\ u &= 0.5\,y^3/(x+1) \quad \text{on } \Gamma. \end{aligned} \tag{5.5.1}$$

The multi-grid iteration MGM in (1.3c) is chosen as program (4.4.2) with $\nu = \gamma = 2$. In Table 5.5.1 we present the results of the nested iteration, where $\tilde{p}$ is the cubic interpolation (3.4.6/7) in the interior and the quadratic interpolation (3.4.8) near the boundary. To illustrate the development of the iteration errors $\|\tilde{u}_k^j - u_k\|_{\mathscr{U}}$, the total errors $\|\tilde{u}_k^j - u\|_{\mathscr{U}}$ and the discretisation error $\|u_k - u\|_{\mathscr{U}}$ are displayed (u_k: exact discrete solution; $\|\cdot\|_{\mathscr{U}}$: Euclidean norm; $u = 0.5\,y^3/(x+1)$: exact continuous solution). $\tilde{u}_k^j$ are the intermediate results from (1.3c): $\tilde{u}_k^0$ equals $\tilde{u}_k$ from (1.3b), while $\tilde{u}_k^i$ is the final result $\tilde{u}_k$ of (1.3c).

The rate of the multi-grid iteration is $\varrho \approx 0.07$ (cf. Table 4.4.2) but the iteration errors indicate a contraction number $\zeta \approx 0.1$. According to Table 5.2.1 the nested iteration with $i = 1$ should yield $\tilde{u}_l$ with an iteration error being half of the discretisation error. The asymptotic behaviour is confirmed by the results for $h = \frac{1}{64}$, $i = 1$ (iteration error: $2.59_{10}-7$, discretisation

Table 5.5.1. Nested iteration with cubic interpolation $\tilde{p}$ for Eq. (1)

	nested iteration with $i=1$				nested iteration with $i=2$		
h_k		total error $\|\tilde{u}_k^j - u\|_{\mathscr{U}}$	iteration error $\|\tilde{u}_k^j - u_k\|_{\mathscr{U}}$	discr. error $\|u_k - u\|_{\mathscr{U}}$		total error $\|\tilde{u}_k^j - u\|_{\mathscr{U}}$	iteration error $\|u_k^j - u_k\|_{\mathscr{U}}$
$\frac{1}{2}$	$\tilde{u}_0$	$1.45_{10}-4$	0	$1.45_{10}-4$	$\tilde{u}_0$	$1.45_{10}-4$	0
$\frac{1}{4}$	$\tilde{u}_1^0$	$9.85_{10}-3$	$9.81_{10}-3$	$8.81_{10}-5$	$\tilde{u}_1^0$	$9.85_{10}-3$	$9.81_{10}-3$
	$\tilde{u}_1^1$	$3.06_{10}-4$	$2.59_{10}-4$		$\tilde{u}_1^1$	$3.06_{10}-4$	$2.59_{10}-4$
					$\tilde{u}_1^2$	$9.04_{10}-5$	$4.14_{10}-6$
$\frac{1}{8}$	$\tilde{u}_2^0$	$1.15_{10}-3$	$1.14_{10}-3$	$2.64_{10}-5$	$\tilde{u}_2^0$	$9.98_{10}-4$	$9.89_{10}-4$
	$\tilde{u}_2^1$	$1.21_{10}-4$	$1.08_{10}-4$		$\tilde{u}_2^1$	$9.34_{10}-5$	$8.00_{10}-5$
					$\tilde{u}_2^2$	$2.86_{10}-5$	$4.62_{10}-6$
$\frac{1}{16}$	$\tilde{u}_3^0$	$1.75_{10}-4$	$1.72_{10}-4$	$6.89_{10}-6$	$\tilde{u}_3^0$	$1.00_{10}-4$	$9.77_{10}-5$
	$\tilde{u}_3^1$	$2.00_{10}-5$	$1.70_{10}-5$		$\tilde{u}_3^1$	$1.22_{10}-5$	$8.71_{10}-6$
					$\tilde{u}_3^2$	$6.98_{10}-6$	$4.65_{10}-7$
$\frac{1}{32}$	$\tilde{u}_4^0$	$2.37_{10}-5$	$2.29_{10}-5$	$1.74_{10}-6$	$\tilde{u}_4^0$	$1.13_{10}-5$	$1.02_{10}-5$
	$\tilde{u}_4^1$	$2.93_{10}-6$	$2.19_{10}-6$		$\tilde{u}_4^1$	$2.01_{10}-6$	$8.34_{10}-7$
					$\tilde{u}_4^2$	$1.74_{10}-6$	$4.14_{10}-8$
$\frac{1}{64}$	$\tilde{u}_5^0$	$3.16_{10}-6$	$2.91_{10}-6$	$4.36_{10}-7$	$\tilde{u}_5^0$	$1.91_{10}-6$	$1.52_{10}-6$
	$\tilde{u}_5^1$	$5.14_{10}-7$	$2.59_{10}-7$		$\tilde{u}_5^1$	$4.47_{10}-7$	$7.62_{10}-8$
					$\tilde{u}_5^2$	$4.36_{10}-7$	$4.03_{10}-9$

error: $4.36_{10} - 7$). The right part of Table 5.5.1 shows that the choice $i = 2$ yields hardly better results with respect to the total error.

Table 5.5.2 contains the analogous results for the linear interpolation $\tilde{p}$ in (1.3b). With $i = 1$ one obtains a very accurate result, however the iteration error is still larger than the discretisation error by a factor of 10 at least. To make both errors nearly equal, one has to choose $i = 2$. That the linear interpolation $\tilde{p}$ requires $i = 2$ is due to the fact that the interpolation error $\| \tilde{p} u_{k-1} - u_k \|_{\mathscr{U}}$ is much larger than the discretisation error $\| u_{k-1} - u \|_{\mathscr{U}}$. The choice $i = 1$ suffices to ensure $\| \tilde{u}_k - u \|_{\mathscr{U}} = O(h_k^2)$, but two iterations are needed to reach the size of the discretisation error. The ratio of the interpolation and discretisation errors depends on the problem. A slight change of Eq. (1) yields

$$\begin{aligned} -\Delta u &= 3y/(x+0.1) - y^3/(x+0.1)^3 \quad \text{in } \Omega = (0,1) \times (0,1), \\ u &= 0.5\, y^3/(x+0.1) \quad \text{on } \Gamma. \end{aligned} \tag{5.5.2}$$

Then the interpolation error $\| \tilde{p} u_{k-1} - u_k \|_{\mathscr{U}}$ is comparable with the discretisation error $\| u_{k-1} - u \|_{\mathscr{U}}$ and the choice $i = 1$ suffices for the nested iteration as can be seen from Table 5.5.3.

In (3.1) we described the number of operations neglecting the interpolation $\tilde{p}$. Because of the following smoothing step the additional work of $\tilde{p}$ can be

Table 5.5.2. Nested iteration with linear interpolation $\tilde{p}$ for Eq. (1)

	nested iteration with $i = 1$				nested iteration with $i = 2$		
h_k		total error $\| \tilde{u}_k^j - u \|_{\mathscr{U}}$	iteration error $\| \tilde{u}_k^j - u_k \|_{\mathscr{U}}$	discr. error $\| u_k - u \|_{\mathscr{U}}$		total error $\| \tilde{u}_k^j - u \|_{\mathscr{U}}$	iteration error $\| \tilde{u}_k^j - u_k \|_{\mathscr{U}}$
$\frac{1}{2}$	$\tilde{u}_0$	$1.45_{10} - 4$	0	$1.45_{10} - 4$	$\tilde{u}_0$	$1.45_{10} - 4$	0
$\frac{1}{4}$	$\tilde{u}_1^0$	$2.23_{10} - 2$	$2.23_{10} - 2$	$8.81_{10} - 5$	$\tilde{u}_1^0$	$2.23_{10} - 2$	$2.23_{10} - 2$
	$\tilde{u}_1^1$	$1.57_{10} - 3$	$1.49_{10} - 3$		$\tilde{u}_1^1$	$1.57_{10} - 3$	$1.49_{10} - 3$
					$\tilde{u}_1^2$	$1.72_{10} - 4$	$9.16_{10} - 5$
$\frac{1}{8}$	$\tilde{u}_2^0$	$7.26_{10} - 3$	$7.24_{10} - 3$	$2.64_{10} - 5$	$\tilde{u}_2^0$	$6.49_{10} - 3$	$6.47_{10} - 3$
	$\tilde{u}_2^1$	$6.19_{10} - 4$	$5.98_{10} - 4$		$\tilde{u}_2^1$	$5.13_{10} - 4$	$4.92_{10} - 4$
					$\tilde{u}_2^2$	$5.23_{10} - 5$	$2.86_{10} - 5$
$\frac{1}{16}$	$\tilde{u}_3^0$	$2.10_{10} - 3$	$2.09_{10} - 3$	$6.89_{10} - 6$	$\tilde{u}_3^0$	$1.73_{10} - 3$	$1.73_{10} - 3$
	$\tilde{u}_3^1$	$1.34_{10} - 4$	$1.30_{10} - 4$		$\tilde{u}_3^1$	$1.03_{10} - 4$	$9.91_{10} - 5$
					$\tilde{u}_3^2$	$9.89_{10} - 6$	$4.91_{10} - 6$
$\frac{1}{32}$	$\tilde{u}_4^0$	$5.18_{10} - 4$	$5.17_{10} - 4$	$1.74_{10} - 6$	$\tilde{u}_4^0$	$4.45_{10} - 4$	$4.43_{10} - 4$
	$\tilde{u}_4^1$	$2.59_{10} - 5$	$2.54_{10} - 5$		$\tilde{u}_4^1$	$1.87_{10} - 5$	$1.82_{10} - 5$
					$\tilde{u}_4^2$	$2.02_{10} - 6$	$8.52_{10} - 7$
$\frac{1}{64}$	$\tilde{u}_5^0$	$1.24_{10} - 4$	$1.23_{10} - 4$	$4.36_{10} - 7$	$\tilde{u}_5^0$	$1.12_{10} - 4$	$1.12_{10} - 4$
	$\tilde{u}_5^1$	$4.81_{10} - 6$	$4.76_{10} - 6$		$\tilde{u}_5^1$	$3.32_{10} - 6$	$3.25_{10} - 6$
					$\tilde{u}_5^2$	$4.63_{10} - 7$	$1.47_{10} - 7$

Table 5.5.3. Nested iteration ($i = 1$) with linear interpolation for Eq. (2)

h_k		total error $\|\tilde{u}_k^j - u\|_{\mathscr{U}}$	iteration error $\|\tilde{u}_k^j - u_k\|_{\mathscr{U}}$	discr. error $\|u_k - u\|_{\mathscr{U}}$
$\frac{1}{2}$	$\tilde{u}_0$	$4.11_{10}-2$	0	$4.1_{10}-2$
$\frac{1}{4}$	$\tilde{u}_1^0$ $\tilde{u}_1^1$	$1.22_{10}-1$ $7.53_{10}-2$	$5.71_{10}-2$ $5.29_{10}-3$	$7.1_{10}-2$
$\frac{1}{8}$	$\tilde{u}_2^0$ $\tilde{u}_2^1$	$1.26_{10}-1$ $5.40_{10}-2$	$8.10_{10}-2$ $8.33_{10}-3$	$4.7_{10}-2$
$\frac{1}{16}$	$\tilde{u}_3^0$ $\tilde{u}_3^1$	$7.70_{10}-2$ $2.27_{10}-2$	$5.92_{10}-2$ $5.37_{10}-3$	$1.8_{10}-2$
$\frac{1}{32}$	$\tilde{u}_4^0$ $\tilde{u}_4^1$	$3.11_{10}-2$ $6.70_{10}-3$	$2.60_{10}-2$ $1.99_{10}-3$	$5.5_{10}-3$
$\frac{1}{64}$	$\tilde{u}_5^0$ $\tilde{u}_5^1$	$9.35_{10}-3$ $1.65_{10}-3$	$8.11_{10}-3$ $5.18_{10}-4$	$1.4_{10}-3$

i	1	2
$\tilde{p}$: linear	$45\frac{1}{3}$	$89\frac{1}{3}$
$\tilde{p}$: cubic	$47\frac{1}{3}$	$91\frac{1}{3}$

Table 5.5.4. Operation count for nested iteration (1.3) with program (4.4.2) ($v = \gamma = 2$)

reduced to 1 operation ($\tilde{p}$: linear), or 2.5 operations ($\tilde{p}$: cubic) per grid point. The resulting operation numbers (per grid point of the finest grid Ω_l) of the nested iteration (1.3) using program (4.4.2) with $v = \gamma = 2$ are listed in Table 5.5.4. We recall that program (4.4.2) is not optimal. With other multi-grid iterations (cf. § 4.4) smaller operation counts can be achieved.

5.6 Comments

In § 5.1 we clearly distinguished the simple iteration (1.1) from the nested iteration (1.3) and proved different error estimates. However, a paradox arises in Exercise 5.7.7 which shows that one multi-grid iteration can be regarded as nested iteration and vice versa. The solution of this problem is hidden in the following considerations.

The iteration (1.1) requires the right-hand side f_l at level l only and the matrices L_k, $0 \leqq k \leqq l$, whereas the nested iteration needs in addition the right-hand sides f_k for $k < l$. Therefore, a problem arises if the system $L_l u_l = f_l$ is only given at level l. In § 3.7 the construction of coarse-grid matrices L_k ($k < l$) is studied. What f_k should be used for $k < l$? The obvious definition $f_{l-1} = r f_l$,

$f_{l-2} = r f_{l-1}, \ldots$ is possible for homogeneous boundary conditions. Otherwise, the nested iteration does not yield the desired accuracy. This is due to the violation of condition (2.3). The right-hand side f_l can be written as $f_l = R_l^\Omega f_\Omega + R_l^\Gamma f_\Gamma$ (f_Ω, f_Γ from (3.1.1/2)), where R_l^Ω and R_l^Γ are implicitly given by the discretisation. A possible choice of f_k $(k < l)$ would be $f_k = R_k^\Omega f_\Omega + R_k^\Gamma f_\Gamma$, where f_Ω and f_Γ are condensed separately. $r R_l^\Gamma f_\Gamma$ is not a correct approximation to $R_{l-1}^\Gamma f_\Gamma$. The reader may construct simple examples to verify this fact. Hence $r f_l$ cannot serve as f_{l-1} in the nested iteration.

5.7 Exercises

Exercise 5.7.1. Define the discretisation error by $\tilde{R}_k u - u_k$, where $\tilde{R}_k$ is some (possibly weighted) restriction onto the grid Ω_k. Assume some $\tilde{r}$ such that $\tilde{R}_{k-1} = \tilde{r} \tilde{R}_k$, and define the interpolation error of $\tilde{p}$ by $\tilde{p} \tilde{r} u_k - u_k$. Show that

$$\| \tilde{R}_k u - u_k \|_{\mathscr{U}} = O(h_k^\kappa), \quad \| \tilde{p} \|_{\mathscr{U} \leftarrow \mathscr{U}} = O(1), \quad \| \tilde{p} \tilde{r} u_k - u_k \|_{\mathscr{U}} = O(h_k^\tau)$$

(κ: consistency order; τ: interpolation order) imply

$$\| \tilde{p} u_{k-1} - u_k \|_{\mathscr{U}} \leqq C_1 h_k^{\min(\kappa, \tau)}$$

for the sufficiently smooth solutions u and u_k of the continuous and discrete problems.

Exercise 5.7.2. Let Ω_l be the infinite grid $\{(\nu h_l, \mu h_l) \colon \nu, \mu \in \mathbb{Z}\}$, $h_{l-1} = 2 h_l$, and define the Euclidean norm by $\| u_l \|_{\mathscr{U}} = [h_l^2 \Sigma_{\Omega_l} | u_l(x, y) |^2]^{1/2}$. The piecewise (i) bilinear, (ii) biquadratic, and (iii) bicubic prolongations $\tilde{p}$ are defined as follows. For all $(x, y) \in \Omega_{l-1}$ the grid function $v_l = \tilde{p} v_{l-1}$ coincides at $(x + \nu h_l, y + \mu h_l)$ $(\nu, \mu = 0, 1, 2)$ with the polynomial $v(\xi, \eta) = \sum_{\alpha, \beta \leqq \sigma} a_{\alpha\beta} \xi^\alpha \eta^\beta$ interpolating v_{l-1} at $(\xi, \eta) \in \Omega_{l-1}$ subject to

(i) $x \leqq \xi \leqq x + h_{l-1}$, $y \leqq \eta \leqq y + h_{l-1}$ with $\sigma = 1$,
(ii) $x \leqq \xi \leqq x + 2h_{l-1}$, $y \leqq \eta \leqq y + 2h_{l-1}$ with $\sigma = 2$,
(iii) $x - h_{l-1} \leqq \xi \leqq x + 2h_{l-1}$, $y - h_{l-1} \leqq \eta \leqq y + 2h_{l-1}$ with $\sigma = 3$,
respectively. Prove $\| \tilde{p} \|_{\mathscr{U} \leftarrow \mathscr{U}} = 1$ for all the cases (i), (ii), (iii).

Hint. The assertion can be reduced to the one-dimensional case.

Exercise 5.7.3. Consider the nested iteration with (1.3a) replaced by (1.4a). Show:

(i) If $\| \tilde{u}_0 - u_0 \|_{\mathscr{U}} \leqq C_3(\zeta, i)\, C_1 h_0^\kappa$, the estimate (2.6) is still valid.
(ii) If $\| \tilde{u}_0 - u_0 \|_{\mathscr{U}} = C_3(\zeta, i)\, C_1 h_0^\kappa + \delta$, $\delta > 0$, inequality (2.6) has to be replaced by

$$\| \tilde{u}_k - u_k \|_{\mathscr{U}} \leqq C_3(\zeta, i)\, C_1 h_k^\kappa + (C_{20} \zeta^i)^k \delta \qquad (0 \leqq k \leqq l)$$

with C_{20} from (2.4a).

(iii) (2.5) implies $(C_{20}\zeta^i)^k = o(h_k^\kappa)$. In the standard case of $C_{20} = 1$, $C_{21} = 2^\kappa$, this term equals $O(h_k^\sigma)$ with $\sigma = i\,|\log_2(\zeta)| > \kappa$.

Exercise 5.7.4. Consider the nested iteration with (1.3 b) replaced by (1.4 b). Show that Theorem 5.2.3 remains valid if C_{20} is defined by

$$C_{20} := \max_{1 \leq k \leq l} \sup_{v_{k-1} \neq 0} \|\tilde{p}(v_{k-1}, 0)\|_{\mathcal{U}} / \|v_{k-1}\|_{\mathcal{U}}$$

and if inequality (2.3) is replaced by

$$\|\tilde{p}(u_{k-1}, f_k) - u_k\|_{\mathcal{U}} \leqq C_1 h_k^\kappa \quad (1 \leqq k \leqq l).$$

The following exercise studies the case where the condition $C_2\zeta^i < 1$ of Theorem 5.2.3 is violated.

Exercise 5.7.5. Assume (2.3), $h_{k-1} = 2h_k$, and $1 \leqq C_2\zeta^i < 2^\kappa = C_{21}$. Prove the following estimates of the results of the nested iteration (1.3):

$$\|\tilde{u}_k - u_k\|_{\mathcal{U}} \leqq k\zeta^i C_1 h_k^\kappa \quad \text{if } C_2\zeta^i = 1 \quad (0 \leqq k \leqq l),$$
$$\|\tilde{u}_k - u_k\|_{\mathcal{U}} \leqq C_k\zeta^i C_1 h_k^\alpha \quad \text{if } 1 < C_2\zeta^i < 2^\kappa,$$

where $\alpha = \kappa - \log_2(C_2\zeta^i) > 0$, $C_k = h_0^{\kappa-\alpha} \min\{1/(C_2\zeta^i - 1),\, k/(C_2\zeta^i)\}$.

Exercise 5.7.6. Replace the number i in (1.3 c) by i_k. Prove: If $i_{l-1} = i_{l-2} = \ldots = i_1$, the Theorems 5.2.3 and 5.2.5 remain valid for $k < l$ with $i := i_{l-1}$. For $k = l$ the factors C_3 and C_4 must be replaced by $C_*(\zeta, i_l, i_{l-1}) = C_*(\zeta, i_{l-1})\,\zeta^{i_l - i_{l-1}}$ $(* = 3, 4)$.

Exercise 5.7.7. (i) Consider the following iteration: Given the problem $L_l w_l = g_l$ and an approximation w_l^j, apply the nested iteration onto $L_l u_l = f_l := L_l w_l^j - g_l$ with suitably defined $f_k := \tilde{r}_{k\leftarrow l} f_l$ and set $w_l^{j+1} := w_l^j - u_l$. Show that the nested iteration can be regarded as one step of this iteration applied to $w_l^0 = 0$.

(ii) Modify the nested iteration (1.3) as follows: Let MGM be the multi-grid iteration (4.1.1) with post-smoothing only (i.e. $v_1 = 0$, $v_2 > 0$) and with $\gamma \geqq 1$. Take $\tilde{p} = p$ (p from (4.1.1 e)), $i = \gamma - 1$ ($i = 0$ permitted), and precede (1.3 c) by $\tilde{u}_k := \mathcal{S}_k^{v_2}(\tilde{u}_k, f_k)$. The mapping $\tilde{r}_{k\leftarrow l}$ mentioned in (i) is defined by $\tilde{r}_{l-1\leftarrow l} = r$, $\tilde{r}_{k\leftarrow l} = r\,\tilde{r}_{k+1\leftarrow l}$ (r from (4.1.1 c)). According to (i) the modified nested iteration gives rise to a new iteration. Prove: The generated iteration is the same multi-grid iteration (4.1.1).

6. Convergence of the Two-Grid Iteration

6.1 Sufficient Conditions for Convergence

6.1.1 The Two-Grid Iteration Matrix

The two-grid iteration $TGM^{(\nu_1,\nu_2)}$ from (3.2.2) is a linear iteration having a representation (1.3.2),

$$u_l^{j+1} = M_l u_l^j + N_l f_l,$$

where the iteration matrix M_l depends on the numbers ν_1 and ν_2: $M_l = M_l(\nu_1, \nu_2)$.

We recall that $S_l^{(\nu)}$ are the iteration matrices of the smoothing procedures $\mathscr{S}_l^{(\nu)}$. $S_l^{(\nu)} = S_l^{\nu}$ holds if $\mathscr{S}_l^{(\nu)}$ is the ν-fold application of $\mathscr{S}_l$. Using Exercise 1.5.1 one proves

Lemma 6.1.1. a) The two-grid iteration matrix equals

$$M_l(\nu_1, \nu_2) = S_l^{(\nu_2)}(I - p\, L_{l-1}^{-1}\, r\, L_l)\, S_l^{(\nu_1)}. \tag{6.1.1}$$

b) If the solution of $L_l u_l = f_l$ is a fixed point of $\mathscr{S}_l^{(\nu)}$, i.e.

$$\mathscr{S}_l^{(\nu)}(u_l, f_l) = u_l \quad \text{for } L_l u_l = f_l, \quad \nu \geqq 0, \tag{6.1.2}$$

then it is also a fixed point of the two-grid iteration.

The two-grid iteration converges if and only if $\varrho(M_l(\nu_1, \nu_2)) < 1$ (cf. Lemma 1.3.1). In Exercise 2.7.7 we have already proved the following note.

Note 6.1.2. If $S_l^{(\nu)} = S_l^{\nu}$, the spectral radius $\varrho(M_l(\nu_1, \nu_2))$ depends on $\nu := \nu_1 + \nu_2$, only. In particular, $\varrho(M_l(\nu_1, \nu_2)) = \varrho(M_l(\nu))$ holds, where

$$M_l(\nu) := M_l(\nu, 0). \tag{6.1.3}$$

Unfortunately, knowledge of the two-grid spectral radius does not help to determine the multi-grid spectral radius. For this reason we need norm estimates.

Let $\mathscr{U}_l$ be the vector space of the grid functions u_l endowed with the norm $\|\cdot\|_{\mathscr{U}}$. For instance, $\|\cdot\|_{\mathscr{U}}$ may be the Euclidean norm generated by the scalar product (3.5.2). In addition, we introduce the vector space $\mathscr{F}_l$ consisting of the grid functions f_l equipped with the norm $\|\cdot\|_{\mathscr{F}}$. The standard case will be $\mathscr{U}_l = \mathscr{F}_l$, $\|\cdot\|_{\mathscr{U}} = \|\cdot\|_{\mathscr{F}} = |\cdot|_0 :=$ Euclidean norm. The associated matrix norms $\|\cdot\|_{\mathscr{U}\leftarrow\mathscr{U}}$, $\|\cdot\|_{\mathscr{F}\leftarrow\mathscr{U}}$, etc are defined in § 1.3.

6.1.2 A Factorisation of M_l

The calculation of $\varrho(M_l)$ or $\|M_l\|_{\mathscr{U}\leftarrow\mathscr{U}}$ is rather involved even in simple model cases (cf. § 8). Nevertheless, a fruitful analysis can be based on a suitable factorisation of M_l.

First we concentrate on the case of $\nu_2 = 0$ (i.e. only pre-smoothing; cf. (3)). We split M_l into the factors

$$M_l(\nu) = [L_l^{-1} - p\,L_{l-1}^{-1}\,r][L_l\,S_l^{(\nu)}] \tag{6.1.4}$$

and conclude that

$$\|M_l(\nu)\|_{\mathscr{U}\leftarrow\mathscr{U}} \leqq \|L_l^{-1} - p\,L_{l-1}^{-1}\,r\|_{\mathscr{U}\leftarrow\mathscr{F}}\,\|L_l\,S_l^{(\nu)}\|_{\mathscr{F}\leftarrow\mathscr{U}} \tag{6.1.5}$$

(cf. (1.3.12a)). The factorisation $M_l = [I - p\,L_{l-1}^{-1}\,r\,L_l]\cdot S_l^{(\nu)}$ might seem more natural. It leads to the estimate

$$\|M_l\|_{\mathscr{U}\leftarrow\mathscr{U}} \leqq \|I - p\,L_{l-1}^{-1}\,r\,L_l\|_{\mathscr{U}\leftarrow\mathscr{V}}\,\|S_l^{(\nu)}\|_{\mathscr{V}\leftarrow\mathscr{U}} \tag{6.1.6}$$

with a further norm $\|\cdot\|_{\mathscr{V}}$ on $\mathscr{U}_l$. However, the approaches (5) and (6) are equivalent when $\|\cdot\|_{\mathscr{F}}$ and $\|\cdot\|_{\mathscr{V}}$ are connected by

$$\|f_l\|_{\mathscr{F}} = \|L_l^{-1} f_l\|_{\mathscr{V}} \quad \text{or} \quad \|u_l\|_{\mathscr{V}} = \|L_l\,u_l\|_{\mathscr{F}}.$$

We prefer the factorisation (5) since for standard problems it allows one to choose $\|\cdot\|_{\mathscr{U}}$ and $\|\cdot\|_{\mathscr{F}}$ as the simple Euclidean norm.

Usually, the right-hand side of inequality (5) is larger than the left-hand side, so that we obtain somewhat pessimistic estimates. However, for an appropriate choice of $\|\cdot\|_{\mathscr{F}}$ the estimate becomes arbitrarily sharp, since

$$\|M_l\|_{\mathscr{U}\leftarrow\mathscr{U}} = \inf_{\|\cdot\|_{\mathscr{F}}} \|L_l^{-1} - p\,L_{l-1}^{-1}\,r\|_{\mathscr{U}\leftarrow\mathscr{F}}\,\|L_l\,S_l^{(\nu)}\|_{\mathscr{F}\leftarrow\mathscr{U}}$$

holds for the infimum taken over all norms (cf. Exercise 6.6.2).

6.1.3 Smoothing and Approximation Properties

The second factor $\|L_l\,S_l^{(\nu)}\|_{\mathscr{F}\leftarrow\mathscr{U}}$ in (5) has already been discussed in § 2.6.3. It describes how effectively $\mathscr{S}_l^{(\nu)}$ is smoothing. The smaller the other factor $\|L_l^{-1} - p\,L_{l-1}^{-1}\,r\|_{\mathscr{U}\leftarrow\mathscr{F}}$, the better the coarse-grid solutions approximate u_l. Hence, the two essential parts of the two-grid iteration, namely the smoothing step and the coarse-grid correction can be analysed separately.

Extending the definition (2.6.6) we expect the estimate $\|L_l\,S_l^{(\nu)}\|_{\mathscr{F}\leftarrow\mathscr{U}} \leqq \varrho_L(\nu)\,\|L_l\|_{\mathscr{F}\leftarrow\mathscr{U}}$ with $\varrho_L(\nu)\to 0$ as $\nu\to\infty$. If $\|\cdot\|_{\mathscr{U}} = \|\cdot\|_{\mathscr{F}}$ and if the elliptic equation is of order $2m$ (cf. Eq. (3.1.6)), the factor $\|L_l\|_{\mathscr{F}\leftarrow\mathscr{U}}$ is at least of order $O(h_l^{-2m})$. But even a simple scaling $\|\cdot\|_{\mathscr{F}} = h_l^{\beta}\,\|\cdot\|_{\mathscr{U}}$ can change the exponent. In the sequel we shall replace $\|L_l\|_{\mathscr{F}\leftarrow\mathscr{U}}$ by $O(h_l^{-\alpha})$.

Definition 6.1.3. Let $\|\cdot\|_{\mathscr{U}}$ and $\|\cdot\|_{\mathscr{F}}$ be given. $S_l^{(\nu)}$ is said to possess the 'smoothing property' if there exist functions $\eta(\nu)$ and $\bar{\nu}(h)$ and a number α such that

$$\| L_l S_l^{(\nu)} \|_{\mathscr{F} \leftarrow \mathscr{U}} \leqq \eta(\nu) h_l^{-\alpha} \quad \text{for all } 1 \leqq \nu < \bar{\nu}(h_l), \quad l \geqq 1, \tag{6.1.7 a}$$

$$\eta(\nu) \to 0 \quad \text{as} \quad \nu \to \infty, \tag{6.1.7 b}$$

$$\bar{\nu}(h) = \infty \quad \text{or} \quad \bar{\nu}(h) \to \infty \quad \text{as} \quad h \to 0. \tag{6.1.7 c}$$

The functions η and $\bar{\nu}$ are independent of l or h_l. The condition

$$\| L_l S_l^{(\nu)} \|_{\mathscr{F} \leftarrow \mathscr{U}} \leqq \eta(\nu) h_l^{-\alpha} \quad \text{for all } \nu \geqq 1, \ l \geqq 1, \tag{6.1.7 a'}$$

is described formally by $\bar{\nu}(h) = \infty$. Replacement of $\|\cdot\|_{\mathscr{F}}$ by $h_l^{\alpha} \|\cdot\|_{\mathscr{F}}$ proves

Note 6.1.4. By a suitable scaling of $\|\cdot\|_{\mathscr{F}}$ one can achieve $\alpha = 0$.

Lemma 6.1.5. Suppose L_l to be non-singular and $S_l^{(\nu)} = S_l^{\nu}$. Then (7 a') and (7 b) imply that $\mathscr{S}_l$ is a convergent iteration for all $l \geqq 1$.

Proof. $\| S_l^{(\nu)} \|_{\mathscr{U} \leftarrow \mathscr{U}} \leqq \| L_l^{-1} \|_{\mathscr{U} \leftarrow \mathscr{F}} \| L_l S_l^{(\nu)} \|_{\mathscr{F} \leftarrow \mathscr{U}} \leqq C_1 \eta(\nu) \to 0 \quad (\nu \to \infty)$ and $\varrho(S_l) = \inf_{\nu} \| S_l^{(\nu)} \|_{\mathscr{U} \leftarrow \mathscr{U}}^{1/\nu}$ prove $\varrho(S_l) < 1$ for all l. □

By Lemma 6.1.5 the estimate (7 a') cannot hold for divergent iterations $\mathscr{S}_l$; nonetheless, there are divergent iterations which are applicable as a smoothing iteration. Therefore, the more general condition (7 a) is needed.

Note that in general the estimate of $L_l S_l^{\nu}$ by $\| L_l S_l^{\nu} \|_{\mathscr{F} \leftarrow \mathscr{U}} \leqq \| L_l \|_{\mathscr{F} \leftarrow \mathscr{U}} \| S_l^{\nu} \|_{\mathscr{U} \leftarrow \mathscr{U}}$ does not imply (7 a, b). For all S_l from § 3.3 there is no better bound on $\| S_l \|_{\mathscr{U} \leftarrow \mathscr{U}}$ than $1 - c h_l^{\beta}$ $(\beta > 0)$ yielding $\eta(\nu) = C(1 - c h_l^{\beta})^{\nu}$. But η has to be independent of l; hence, $\eta(\nu) := C \sup_l (1 - c h_l^{\beta})^{\nu} = C$ is the lowest uniform bound, but violates (7 b).

The other factor in (5) is considered in

Definition 6.1.6. The 'approximation property' holds if there is some constant C_A such that

$$\| L_l^{-1} - p L_{l-1}^{-1} r \|_{\mathscr{U} \leftarrow \mathscr{F}} \leqq C_A h_l^{\alpha} \quad \text{for all } l \geqq 1 \tag{6.1.8}$$

with α from (7 a).

Implicitly, the estimate (8) contains the condition that L_l is non-singular for all $l \geqq 0$. Again, the scaling mentioned in Note 6.1.4 yields $\alpha = 0$ in (8).

The smoothing and approximation properties are discussed in detail in § 6.2 and § 6.3. In the next subsection we shall see that these conditions are sufficient for two-grid convergence, but it is even more important that they are also almost sufficient for multi-grid convergence (cf. § 7).

6.1.4 A General Convergence Theorem

Theorem 6.1.7 (two-grid convergence of $TGM^{(\nu, 0)}$). Suppose the smoothing property (7 a–c) and the approximation property (8). Let $\varrho > 0$ be a fixed number.

a) In the case of $\bar{\nu}(h) = \infty$ (cf. (7 c)) there is a number $\underline{\nu}$ such that the two-grid contraction number satisfies

$$\| M_l(\nu) \|_{\mathscr{U} \leftarrow \mathscr{U}} \leqq C_A \eta(\nu) \leqq \varrho \tag{6.1.9}$$

whenever $\nu \geqq \underline{\nu}$ and $l \geqq 1$.

b) In the remaining case of $\bar{\nu}(h) \to \infty$ there are $\bar{h} > 0$ and $\underline{\nu}$ such that inequality (9) holds for all $\underline{\nu} \leqq \nu < \bar{\nu}(h_l)$ and all l with $h_l \leqq \bar{h}$. For such l the interval $[\underline{\nu}, \bar{\nu}(h_l))$ is not empty.

c) Under condition (2) and for $\varrho < 1$ one obtains convergence:

$$u_l^j \to u_l = L_l^{-1} f_l.$$

Proof. a) Let $\bar{\nu} = \infty$. (7 a) and (8) yield $\| M_l \|_{\mathscr{U} \leftarrow \mathscr{U}} \leqq C_A \eta(\nu)$ (cf. (5)). By (7 b) there is $\underline{\nu}$ with $C_A \eta(\nu) \leqq \varrho$ for all $\nu \geqq \underline{\nu}$.

b) Determine $\underline{\nu}$ as above. By $\bar{\nu}(h) \to \infty$ there is $\bar{h}$ such that $\bar{\nu}(h) > \underline{\nu}$ for all $h \leqq \bar{h}$. Hence, $[\underline{\nu}, \bar{\nu}(h))$ is not empty. For $\nu \in [\underline{\nu}, \bar{\nu}(h))$ inequality (7 a) is valid and yields (9).

c) Use Lemma 1.3.1 and Lemma 6.1.1. □

Corollary 6.1.8. The errors of the iterates u_l^j of $TGM^{(\nu,0)}$ satisfy

$$\| u_l^j - u_l \|_{\mathscr{U}} \leqq \| M_l(\nu) \|_{\mathscr{U} \leftarrow \mathscr{U}}^j \| u_l^0 - u_l \|_{\mathscr{U}} \tag{6.1.10}$$

(cf. (1.3.16)).

Note that the right-hand side $C_A \eta(\nu)$ of (9) is independent of l. This result coincides with the estimate (2.4.11) for the one-dimensional example of § 2.

The smoothing number $\varrho_L(\nu)$ from (2.6.6) is

$$\varrho_L(\nu) := \sup \{ \| L_l S_l^{(\nu)} \|_{\mathscr{F} \leftarrow \mathscr{U}} / \| L_l \|_{\mathscr{F} \leftarrow \mathscr{U}} : l \geqq 1 \}. \tag{6.1.11}$$

The estimate

$$h_l^\alpha \| L_l \|_{\mathscr{F} \leftarrow \mathscr{U}} \leqq C_L \tag{6.1.12}$$

holds at least with $C_L = \eta(0)$, if (7a) is extended to $\nu = 0$ (cf. Note 6.2.4 and Proposition 6.3.28). The convergence result (9) can be expressed by means of $\varrho_L(\nu)$:

Corollary 6.1.9. Assume (12). In Theorem 6.1.7 the estimate (9) may be replaced by (13) with $C_{TGM} := C_A C_L$:

$$\| M_l(\nu) \|_{\mathscr{U} \leftarrow \mathscr{U}} \leqq C_{TGM} \, \varrho_L(\nu). \tag{6.1.13}$$

6.1.5 The Case of $\nu_2 \neq 0$

Hitherto, we have considered only the case of no post-smoothing ($\nu_2 = 0$). In the reverse case of no pre-smoothing ($\nu_1 = 0$) define the matrix

$$\hat{M}_l(\nu) = L_l M_l(0, \nu) L_l^{-1}, \tag{6.1.14}$$

which is similar to $M_l(0,\nu)$; hence $\varrho(\hat{M}_l(\nu)) = \varrho(M_l(0,\nu))$. $\|\hat{M}_l(\nu)\|_{\mathscr{F}\leftarrow\mathscr{F}} < 1$ would be sufficient for convergence. Since

$$\hat{M}_l(\nu) = [L_l S_l^{(\nu)}][L_l^{-1} - p L_{l-1}^{-1} r], \tag{6.1.14'}$$

one concludes

$$\|\hat{M}_l(\nu)\|_{\mathscr{F}\leftarrow\mathscr{F}} \leqq \|L_l S_l^{(\nu)}\|_{\mathscr{F}\leftarrow\mathscr{U}} \|L_l^{-1} - p L_{l-1}^{-1} r\|_{\mathscr{U}\leftarrow\mathscr{F}}$$

with the same right-hand side as in (5). This proves

Theorem 6.1.10 (two-grid convergence of $TGM^{(0,\nu)}$). If $\nu_1 = 0$ and $\nu_2 = \nu$, Theorem 6.1.7 with (9) replaced by

$$\|\hat{M}_l(\nu)\|_{\mathscr{F}\leftarrow\mathscr{F}} \leqq C_A \eta(\nu) \leqq \varrho \tag{6.1.15}$$

remains valid.

Corollary 6.1.8 becomes

Corollary 6.1.11. The defects of the iterates u^l of $TGM^{(0,\nu)}$ satisfy

$$\|L_l u_l^j - f_l\|_{\mathscr{F}} \leqq \|\hat{M}_l(\nu)\|^j_{\mathscr{F}\leftarrow\mathscr{F}} \|L_l u_l^0 - f_l\|_{\mathscr{F}}. \tag{6.1.16}$$

The general case of $\nu_1 \neq 0$, $\nu_2 \neq 0$ will be treated in § 6.4. A simple but non-optimal approach is the estimation of $M_l(\nu_1, \nu_2)$ by

$$\|M_l(\nu_1,\nu_2)\|_{\mathscr{U}\leftarrow\mathscr{U}} \leqq \|S_l^{(\nu_2)}\|_{\mathscr{U}\leftarrow\mathscr{U}} \|M_l(\nu_1,0)\|_{\mathscr{U}\leftarrow\mathscr{U}}. \tag{6.1.17}$$

A better approach is based on the splitting

$$\|M_l(\nu_1,\nu_2)\|_{\mathscr{U}\leftarrow\mathscr{U}} \leqq \|S_l^{(\nu_2)}\|_{\mathscr{U}\leftarrow\mathscr{V}} \|L_l^{-1} - p L_{l-1}^{-1} r\|_{\mathscr{V}\leftarrow\mathscr{F}} \|L_l S_l^{(\nu_1)}\|_{\mathscr{F}\leftarrow\mathscr{U}} \tag{6.1.18}$$

with a suitable normed space $\mathscr{V}_l$ or based on

Lemma 6.1.12. If $L_{l-1} = p L_l r$, then

$$M_l(\nu_1,\nu_2) = M_l(0,\nu_2) M_l(\nu_1,0). \tag{6.1.19}$$

Proof. Set $C_l := L_l^{-1} - p L_{l-1}^{-1} r$ and note that $C_l L_l C_l = C_l$. □

6.2 The Smoothing Property

It is the goal of this section to demonstrate that the smoothing property holds for a large class of iterative methods. In § 6.2.1 we formulate some lemmata and introduce special norms. § 6.2.2 contains a criterion showing that it suffices to check the smoothing property of the principal part of the discrete operator. Special smoothing iterations are studied in the subsequent subsections.

6.2.1 Preparatory Lemmata

Let $\mathscr{U}_l$ be a Hilbert space with scalar product $\langle\cdot,\cdot\rangle_{\mathscr{U}}$ and norm (1.3.9). Often the estimation of $L_l S_l^\nu$ reduces to an estimation of an expression of the form $A(I-A)^\nu$ for a positive definite matrix $O \leqq A \leqq I$. In such cases the following lemma applies.

Lemma 6.2.1. Let $O \leqq A = A^* \leqq I$. Then

$$\|A(I-A)^\nu\|_{\mathscr{U}\leftarrow\mathscr{U}} \leqq \eta_0(\nu), \tag{6.2.1 a}$$

where

$$\eta_0(\nu) := \nu^\nu/(\nu+1)^{\nu+1}. \tag{6.2.1 b}$$

Proof. Lemma 1.3.5 (iii) implies $\|A(I-A)^\nu\|_{\mathscr{U}\leftarrow\mathscr{U}} = \max\{|f(\lambda)| : \lambda \in \sigma(A)\}$ with $f(x) = x(1-x)^\nu$. $O \leqq A \leqq I$ implies $0 \leqq \lambda \leqq 1$; hence $\|A(I-A)^\nu\|_{\mathscr{U}\leftarrow\mathscr{U}} \leqq \max\{f(x)\colon 0 \leqq x \leqq 1\}$. f attains its maximum at $x_0 := 1/(\nu+1)$ yielding $\eta_0(\nu) = f(x_0)$. □

Corollary 6.2.2. The asymptotic behaviour of $\eta_0(\nu)$ is

$$\eta_0(\nu) = \frac{1}{e\,\nu} + O(\nu^{-2}) \quad \text{as} \quad \nu \to \infty. \tag{6.2.2 a}$$

Particular estimates are

$$\eta_0(\nu) \leqq \eta_0(\mu)(\mu+c)/(\nu+c) \quad \text{for } \nu \geqq \mu \geqq 0, \quad c \geqq \tfrac{1}{2}, \tag{6.2.2 b}$$

$$\eta_0(\nu) \leqq \frac{1}{\nu+c}\max\left\{\frac{1}{e}, \eta_0(\mu)(\mu+c)\right\} \quad \text{for } \nu \geqq \mu \geqq 0, \quad 0 \leqq c < \tfrac{1}{2}, \tag{6.2.2 c}$$

$$\eta_0(\nu) \leqq \frac{1}{e(\nu+c(\mu))} \quad \text{with} \quad c(\mu) := \frac{1}{e\,\eta_0(\mu)} - \mu \quad \text{for } \nu \geqq \mu, \tag{6.2.2 d}$$

$$\eta_0(\nu) \leqq \frac{\frac{3}{8}}{\nu+\frac{1}{2}} \qquad (\nu \geqq 1), \tag{6.2.2 e}$$

$$\eta_0(\nu) \leqq \frac{2/(3\sqrt{3})}{\nu+\frac{1}{2}} \qquad (\nu \geqq \tfrac{1}{2}), \tag{6.2.2 f}$$

$$\frac{1}{e(\nu+\frac{1}{2})} \leqq \eta_0(\nu) \leqq \frac{1}{e\,\nu+1} \quad (\nu \geqq 0). \tag{6.2.2 g}$$

Proof. Consider $\varphi(\nu) := (\nu+c)\,\eta_0(\nu)$ with $\varphi'(\nu) = \eta_0\psi$, where $\psi(\nu) = 1 - (\nu+c)\log(1+1/\nu)$. Let $c \geqq \frac{1}{2}$. A simple analysis yields $\alpha(\zeta) := \exp(\zeta/(1+c\zeta)) \leqq 1+\zeta$ $(\zeta > 0)$, which is equivalent to $\psi(\nu) = \psi(1/\zeta) < 0$ for $\zeta = 1/\nu$. Hence, $\varphi(\nu)$ is decreasing and (2 b) follows. The analysis of the case $c \in [0, \frac{1}{2}]$ shows that there is a unique positive root ζ_0 of $\alpha(\zeta) = 1+\zeta$. Thus, $\varphi(\nu)$ has only one minimum at $\nu_0 = 1/\zeta_0$ and is bounded in $[\mu, \infty)$ by $\max\{\varphi(\mu), \varphi(\infty)\} = \max\{(\mu+c)\,\eta_0(\mu), 1/e\}$ proving (2 c). (2 d) is a special case of (2 c) since $c(\mu) \leqq \frac{1}{2}$ (cf (2 g)). (2 e), (2 f), and the second part of (2 g) are particular examples of (2 c, d). □

Corollary 6.2.3. Under the conditions of Lemma 6.2.1 the estimate

$$\| A^\alpha (I - A)^\beta \|_{\mathscr{U} \leftarrow \mathscr{U}} \leqq \left(\eta_0 \left(\frac{\beta}{\alpha}\right)\right)^\alpha \quad (\alpha > 0, \beta \geqq 0) \tag{6.2.3}$$

holds for all real $\alpha > 0$, $\beta \geqq 0$.

In the following analysis the basic scalar product of $\mathscr{U}_l$ is denoted by $\langle \cdot, \cdot \rangle$ (e.g. from (3.5.2)) with associated norm

$$|u_l|_0 = \langle u_l, u_l \rangle^{1/2}. \tag{6.2.4 a}$$

Given a positive definite matrix $\Lambda_l = \Lambda_l^*$, one may define new norms

$$|u_l|_s := |\Lambda_l^s u_l|_0 \tag{6.2.4 b}$$

for all real s. Thus, the spaces $(\mathscr{U}_l, |\cdot|_s)$ form a Hilbert scale (cf. § 1.4.4). If $L_l = L_l^* > 0$, the natural choice of Λ_l will be

$$\Lambda_l := L_l^{1/2}, \tag{6.2.5}$$

According to (1.3.11) the associated matrix norms are defined by

$$|A_l|_{s \leftarrow t} := \sup \{ |A_l u_l|_s / |u_l|_t : 0 \neq u_l \in \mathscr{U}_l \}$$

for any $s, t \in \mathbb{R}$. We introduce the abbreviation $\| \cdot \|$ for the spectral norm:

$$\| \cdot \| := | \cdot |_{0 \leftarrow 0}. \tag{6.2.6 a}$$

An equivalent description of $| \cdot |_{s \leftarrow t}$ is

$$|A_l|_{s \leftarrow t} = \| \Lambda_l^s A_l \Lambda_l^{-t} \|. \tag{6.2.6 b}$$

The norms $\| \cdot \|_{\mathscr{U}}$ and $\| \cdot \|_{\mathscr{F}}$ which appear in § 6.1 will be chosen as

$$\| \cdot \|_{\mathscr{U}} := | \cdot |_{s_{\mathscr{U}}}, \quad \| \cdot \|_{\mathscr{F}} := | \cdot |_{s_{\mathscr{F}}} \tag{6.2.7}$$

with appropriate numbers $s_{\mathscr{U}}$ and $s_{\mathscr{F}}$.

Note 6.2.4. Let $\| \cdot \|_{\mathscr{F}}$ and $\| \cdot \|_{\mathscr{U}}$ be given by (7). Then the smoothing property (1.7 a) becomes

$$\| \Lambda_l^{s_{\mathscr{F}}} L_l S_l^{(\nu)} \Lambda_l^{-s_{\mathscr{U}}} \| \leqq \eta(\nu) h_l^{-\alpha}. \tag{6.2.8 a}$$

In the case of (5) with $L_l = L_l^* > 0$, Inequality (8 a) reduces to

$$\| L_l^{1 + s_{\mathscr{F}}/2} S_l^{(\nu)} L_l^{-s_{\mathscr{U}}/2} \| \leqq \eta(\nu) h_l^{-\alpha}. \tag{6.2.8 b}$$

If in addition $S_l^{(\nu)}$ commutes with L_l, one obtains

$$\| L_l^{1 + (s_{\mathscr{F}} - s_{\mathscr{U}})/2} S_l^{(\nu)} \| \leqq \eta(\nu) h_l^{-\alpha}. \tag{6.2.8 c}$$

The natural value of α is

$$\alpha = (2 + s_{\mathscr{F}} - s_{\mathscr{U}}) m, \tag{6.2.9}$$

since $\| L_l \|_{\mathscr{F} \leftarrow \mathscr{U}} = \| L_l^{1 + (s_{\mathscr{F}} - s_{\mathscr{U}})/2} \| = \| L_l \|^{1 + (s_{\mathscr{F}} - s_{\mathscr{U}})/2} = O(h_l^{-2m - (s_{\mathscr{F}} - s_{\mathscr{U}})m})$ provided that

$$\| L_l \| \leqq C_L h_l^{-2m}, \tag{6.2.10}$$

where $2m$ is the order of the differential operator, and

$$s_{\mathscr{U}} \leqq s_{\mathscr{F}} + 2 \tag{6.2.11}$$

(cf. Lemma 1.3.5 (ii)). Inequality (10) will be proved in § 6.3.1.5 for finite element discretisations.

Sometimes, it is possible to estimate $L_l S_l^{(v)}$ and $S_l^{(v)}$, but an elementary estimate of $L_l^s S_l^{(v)}$ is not available. Then the following interpolation theorem can be applied.

Lemma 6.2.5. Suppose $\Lambda_1 = \Lambda_1^* > 0$, $\Lambda_2 = \Lambda_2^* > 0$. Then for any matrix A and any scalars $\alpha \leqq \beta \leqq \gamma$ the following inequality holds:

$$\| \Lambda_2^\beta A \Lambda_1^{-\beta} \| \leqq \| \Lambda_2^\alpha A \Lambda_1^{-\alpha} \|^{(\gamma - \beta)/(\gamma - \alpha)} \| \Lambda_2^\gamma A \Lambda_1^{-\gamma} \|^{(\beta - \alpha)/(\gamma - \alpha)}.$$

Proof. For completeness we give the proof, though this lemma is a special case of Lemma 1.4.3. $g(s) := \| \Lambda_2^s A \Lambda_1^{-s} \|$ is a continuous function of s and satisfies

$$g(s)^2 = \| \Lambda_2^s A \Lambda_1^{-2s} A^* \Lambda_2^s \| = \varrho(\Lambda_2^r A \Lambda_1^{-r} \Lambda_1^{-t} A^* \Lambda_2^t) \leqq g(r) g(t)$$

for all r and t with $r + t = 2s$. Hence $G(s) := \log g(s)$ is convex. The inequality

$$G(\beta) \leqq \frac{\gamma - \beta}{\gamma - \alpha} G(\alpha) + \frac{\beta - \alpha}{\gamma - \alpha} G(\gamma)$$

proves the lemma. □

Lemma 6.2.6. Let $| \cdot |_s (s \in \mathbb{R})$ be defined by (4). Then the following estimates are valid for all matrices A:

$$|A|_{r \leftarrow r} \leqq |A|_{s \leftarrow s}^{(t - r)/(t - s)} |A|_{t \leftarrow t}^{(r - s)/(t - s)} \qquad (s \leqq r \leqq t), \tag{6.2.12 a}$$

$$|A|_{-r \leftarrow r} \leqq |A|_{0 \leftarrow 0}^{1 - r/t} |A|_{-t \leftarrow t}^{r/t} \qquad (0 \leqq r \leqq t \text{ or } 0 \geqq r \geqq t), \tag{6.2.12 b}$$

$$|A|_{-s \leftarrow r} \leqq |A|_{0 \leftarrow 0}^{1 - (r+s)/t} |A|_{0 \leftarrow t}^{r/t} |A|_{-t \leftarrow 0}^{s/t} \qquad (r \geqq 0, s \geqq 0, r + s \leqq t). \tag{6.2.12 c}$$

Proof. a) The quantity $a(p, q) := |A|_{q \leftarrow p}$ equals $|\Lambda_l^q A \Lambda_l^{-p}|_{0 \leftarrow 0}$. Lemma 6.2.5 with $\Lambda_1 = \Lambda_2 = \Lambda_l$, $\alpha = s$, $\beta = r$, $\gamma = t$ yields (12 a).

b) (12 b) follows from Lemma 6.2.5 with $\Lambda_1 = \Lambda_l$, $\Lambda_2 = \Lambda_l^{-1}$.

c) Lemma 6.2.5 with $\Lambda_1 = I$, $\Lambda_2 = \Lambda_l$ proves $a(r + s, 0) \leqq a(0, 0)^{1 - \sigma} a(t, 0)^\sigma$ where $\sigma = (r + s)/t$. Similary, $a(0, -r - s) \leqq a(0, 0)^{1 - \sigma} a(0, -t)^\sigma$ follows. Applying Lemma 6.2.5 to $\Lambda_l^{-r-s} A$ instead of A, one obtains $a(r, -s) = a(r + s, 0)^{r/(r+s)} a(0, -r - s)^{s/(r+s)}$. Inserting the estimates of $a(r + s, 0)$ and $a(0, -r - s)$ we are led to (12 c). □

6.2.2 Criteria

Often a discrete operator L_l can be split into $L_l' + L_l''$, where L_l' is the principal part of L_l satisfying the smoothing property, whilst L_l'' is a term of lower order. Let $S_l'^{(\nu)}$ be the smoothing iteration matrix associated with L_l' and define $S_l''^{(\nu)} := S_l^{(\nu)} - S_l'^{(\nu)}$. Then the assumptions (15), (16) of the next criterion are very natural.

Criterion 6.2.7. Let $L_l = L_l' + L_l''$ and $S_l^{(\nu)} = S_l'^{(\nu)} + S_l''^{(\nu)}$ and assume that L_l' and $S_l'^{(\nu)}$ satisfy the smoothing property

$$\| L_l' S_l'^{(\nu)} \|_{\mathscr{F} \leftarrow \mathscr{U}} \leqq \eta'(\nu) h_l^{-\alpha} \quad \text{for } 0 \leqq \nu < \bar{\nu}'(h_l), \quad l \geqq 1, \tag{6.2.13 a}$$

$$0 < \eta'(\nu) \to 0 \quad \text{as } \nu \to \infty, \tag{6.2.13 b}$$

$$\bar{\nu}'(h) = \infty \quad \text{or} \quad \bar{\nu}'(h) \to \infty \quad \text{as } h \to 0. \tag{6.2.13 c}$$

Further, suppose $S_l'^{(0)} = I$ and

$$\| S_l'^{(\nu)} \|_{\mathscr{U} \leftarrow \mathscr{U}} \leqq C_S'(\nu) \quad \text{for all } l \geqq 1, \quad \nu < \bar{\nu}'(h_l) \tag{6.2.14}$$

with some $C_S'(\nu)$ and

$$\lim_{l \to \infty} h_l^{\alpha} \| L_l'' \|_{\mathscr{F} \leftarrow \mathscr{U}} = 0, \tag{6.2.15}$$

$$\lim_{l \to \infty} \| S_l''^{(\nu)} \|_{\mathscr{U} \leftarrow \mathscr{U}} = 0 \quad \text{for all } \nu. \tag{6.2.16}$$

Then the smoothing property (1.7 a–c) holds for $L_l S_l^{(\nu)}$ with the same α. $\eta(\nu)$ can be chosen as $\eta(\nu) = (1 + \varepsilon)\eta'(\nu)$ for any $\varepsilon > 0$. $\bar{\nu}(h) \leqq \bar{\nu}'(h)$ satisfies (1.7 c). Furthermore, there is a function $C_S(\nu) := C_S'(\nu) + \text{constant}$ such that

$$\| S_l^{(\nu)} \|_{\mathscr{U} \leftarrow \mathscr{U}} \leqq C_S(\nu) \quad \text{for } l \geqq 1, \quad \nu \leqq \bar{\nu}(h_l). \tag{6.2.17}$$

Proof. Let $\varepsilon > 0$ be given and set

$$\Theta(\nu) = \varepsilon \eta'(\nu)/(2\eta'(0)).$$

By (16) there is a decreasing bound $h_\Theta(\nu) > 0$ with

$$\| S_l''^{(\nu)} \|_{\mathscr{U} \leftarrow \mathscr{U}} \leqq \Theta(\nu) \quad \text{for } h_l \leqq h_\Theta(\nu).$$

Let $\nu_1(h) := \sup\{\nu \in \mathbb{N} \cup \{\infty\} : \nu \leqq \bar{\nu}'(h), h \leqq h_\Theta(\nu)\}$. As $0 > h_\Theta(\nu) \to 0$, either $\nu_1(h) = \infty$ or $\nu_1(h) \to \infty$ $(h \to 0)$ holds, and $\nu \leqq \nu_1(h)$ implies $h \leqq h_\Theta(\nu_1(h)) \leqq h_\Theta(\nu)$. Let $\Theta_0 := \max\{\Theta(\nu) : \nu \in \mathbb{N}\} < \infty$. The inequalities (14) and $\| S_l''^{(\nu)} \|_{\mathscr{U} \leftarrow \mathscr{U}} \leqq \Theta_0$ for $\nu \leqq \nu_1(h_l)$ prove (17) with $C_S(\nu) := C_S'(\nu) + \Theta_0$.

The splitting $L_l S_l^{(\nu)} = L_l' S_l'^{(\nu)} + L_l'' S_l^{(\nu)} + L_l' S_l''^{(\nu)}$ implies

$$\begin{aligned} & h_l^{\alpha} \| L_l S_l^{(\nu)} \|_{\mathscr{F} \leftarrow \mathscr{U}} \\ & \quad \leqq h_l^{\alpha} [\| L_l' S_l'^{(\nu)} \|_{\mathscr{F} \leftarrow \mathscr{U}} + \| L_l'' \|_{\mathscr{F} \leftarrow \mathscr{U}} \| S_l^{(\nu)} \|_{\mathscr{U} \leftarrow \mathscr{U}} + \| L_l' \|_{\mathscr{F} \leftarrow \mathscr{U}} \| S_l''^{(\nu)} \|_{\mathscr{U} \leftarrow \mathscr{U}}] \\ & \quad \leqq \eta'(\nu) + C_S(\nu) h_l^{\alpha} \| L_l'' \|_{\mathscr{F} \leftarrow \mathscr{U}} + \eta'(0)\Theta(\nu) \end{aligned}$$

$(\nu \leqq \nu_1(h_l) \leqq \bar{\nu}'(h_l))$ by means of (13 a) and (17). The last term $\eta'(0)\Theta(\nu)$ equals

$(\varepsilon/2)\,\eta'(v)$ by the definition of $\Theta(v)$. Using (15) there is a bound $v_2(h) \leqq \bar{v}'(h)$ with $v_2(h) \to \infty$ such that

$$C_S(v)\,h_l^\alpha \,\| L_l'' \|_{\mathscr{F} \leftarrow \mathscr{U}} \leqq \frac{\varepsilon}{2}\,\eta'(v) \quad \text{for } v \leqq v_2(h_l), \qquad l \geqq 1 .$$

Set $\bar{v}(h) := \min\{v_1(h),\ v_2(h)\} \leqq \bar{v}'(h)$. (1.7 a) holds for $v \leqq \bar{v}(h)$ with $\eta(v) = (1+\varepsilon)\,\eta'(v)$. □

Usually, $S_l^{(v)}$ is the v-fold application of $\mathscr{S}_l$, and $S_l^{(v)}$ equals S_l^v. Then the conditions (14) and (16) can be modified:

Criterion 6.2.8. Assume $S_l^{(v)} = S_l^v$, $S_l'^{(v)} = S_l'^{\,v}$ and $L_l = L_l' + L_l''$, $S_l = S_l' + S_l''$. Under the conditions (13), (15), and

$$\| S_l' \|_{\mathscr{U} \leftarrow \mathscr{U}} \leqq C_S' \quad \text{for all } l \geqq 1, \tag{6.2.18 a}$$

$$\lim_{l \to \infty} \| S_l'' \|_{\mathscr{U} \leftarrow \mathscr{U}} = 0, \tag{6.2.18 b}$$

the smoothing property (1.7 a–c) holds for $L_l S_l^v$.

Proof. (14) is valid with $C_S'(v) := C_S'^{\,v}$. $C_S := C_S' + \max\{\| S_l'' \|_{\mathscr{U} \leftarrow \mathscr{U}} : l \geqq 1\}$ is a bound of $\| S_l \|_{\mathscr{U} \leftarrow \mathscr{U}}$. Hence,

$$\begin{aligned} \| S_l''^{(v)} \|_{\mathscr{U} \leftarrow \mathscr{U}} &= \| S_l^v - S_l'^{\,v} \|_{\mathscr{U} \leftarrow \mathscr{U}} = \left\| \sum_{\mu=0}^{v-1} S_l^\mu (S_l - S_l') S_l'^{\,v-1-\mu} \right\|_{\mathscr{U} \leftarrow \mathscr{U}} \\ &= \left\| \sum_{\mu=0}^{v-1} S_l^\mu S_l'' S_l'^{\,v-1-\mu} \right\|_{\mathscr{U} \leftarrow \mathscr{U}} \leqq \left(\sum_{\mu=0}^{v-1} C_S^\mu C_S'^{\,v-1-\mu} \right) \| S_l'' \|_{\mathscr{U} \leftarrow \mathscr{U}} \\ &\leqq v\, C_S^{v-1} \, \| S_l'' \|_{\mathscr{U} \leftarrow \mathscr{U}} \end{aligned}$$

implies (16). □

In some cases $S_l^{(v)} \neq S_l^v$. For instance, when $S_l^{(v)}$ is a semi-iterative process, $S_l^{(v)}$ may be some polynomial $\pi_v(S_l)$.

Corollary 6.2.9. Assume $S_l^{(v)} = \pi_v(S_l)$, $S_l'^{(v)} = \pi_v(S_l')$, where $\pi_v(z) = \sum_{\mu=0}^{v} \alpha_{\mu,v} z^\mu$. Then Criterion 6.2.8 is also valid.

Proof. $S_l''^{(v)}$ equals $\sum_\mu \alpha_{\mu,v}(S_l^\mu - S_l'^{\,\mu})$. The proof of Criterion 6.2.8 shows $\| S_l''^{(v)} \|_{\mathscr{U} \leftarrow \mathscr{U}} \leqq (\sum_\mu \mu\, |\alpha_{\mu,v}|\, C_S^{\mu-1})\, \| S_l'' \|_{\mathscr{U} \leftarrow \mathscr{U}}$. Hence, (16) holds again. □

The condition (14) is weaker than the requirement

$$\| S_l'^{(v)} \|_{\mathscr{U} \leftarrow \mathscr{U}} \leqq C_S' \quad \text{for all } v < \bar{v}'(h_l), \qquad l \geqq 1, \tag{6.2.19}$$

which expresses the 'stability' of $\{S_l'^{(v)} : v < \bar{v}'(h_l), l \geq 1\}$ (cf. Richtmyer-Morton [1]). Note that (19) can be satisfied by a divergent iteration. The proof of Criterion 6.2.7 showed $C_S(v) = C_S'(v) + \Theta_0 = C_S' + \Theta_0$ with Θ_0 proportional to ε. Therefore, the bound C_S on $\| S_l^{(v)} \|_{\mathscr{U} \leftarrow \mathscr{U}}$ is independent of v, too, and we obtain

Corollary 6.2.10. Inequality (19) and Eq. (16) imply

$$\| S_l^{(v)} \|_{\mathscr{U} \leftarrow \mathscr{U}} \leqq C_S \quad \text{for } v < \bar{v}(h_l), \qquad l \geqq 1 . \tag{6.2.20}$$

If $S_l^{(\nu)} = S_l^\nu$ or $S_l^{(\nu)} = \pi_\nu(S_l)$ (π_ν: polynomial), condition (16) may be replaced by (18 b).

Even if L_l' satisfies the smoothing property with $\bar{\nu}'(h) = \infty$, the new bound $\bar{\nu}(h)$ may be finite. We shall see that the damped Jacobi iteration (3.3.7′) allows $\bar{\nu}'(h) = \infty$, whenever L_l' is positive definite. Let λ be larger than the first eigenvalue of L_l'. $L_l := L_l' - \lambda I$ is a perturbation of L_l' by a low-order term. But since L_l is indefinite, the damped Jacobi iteration diverges and $\bar{\nu}(h) < \infty$ follows from Lemma 6.1.5.

6.2.3 The Smoothing Property of the Jacobi Iterations

6.2.3.1 The Simple Damped Jacobi Iteration

First we study the damped Jacobi iteration (3.3.7′), $S_l = I - \omega_l h_l^{2m} L_l$, with $0 < \omega_l \leqq \| h_l^{2m} L_l \|^{-1}$. For a positive definite $L_l = L_l^*$ the norms $\|\cdot\|$, $|\cdot|_{t \leftarrow s}$ etc. are defined in (4), (5), (6).

Proposition 6.2.11. Suppose $L_l = L_l^* \geqq O$ and $0 < \omega_l \leqq \| h_l^{2m} L_l \|^{-1}$. Then the damped Jacobi iteration (3.3.7′) satisfies

$$\| L_l S_l^\nu \| \leqq \frac{1}{\omega_l} h_l^{-2m} \eta_0(\nu) \quad \text{for all } \nu \tag{6.2.21}$$

with $\eta_0(\nu)$ from (1 b). If in addition $1/C_0 \leqq \omega_l$, the smoothing property (1.7) holds with $\|\cdot\|_{\mathscr{U}} = \|\cdot\|_{\mathscr{F}} = |\cdot|_0$ and with

$$\eta(\nu) := C_0 \eta_0(\nu) \leqq \frac{3C_0}{8} \Big/ (\nu + \tfrac{1}{2}), \qquad \alpha = 2m, \qquad \bar{\nu}(h) = \infty .$$

Proposition 6.2.11 is a particular case of

Proposition 6.2.12. Let L_l and ω_l be as in Proposition 6.2.11 and define $|\cdot|_s$ and $|\cdot|_t$ by (4), (5) with $s - t < 2$. Then the damped Jacobi iteration satisfies

$$|L_l S_l^\nu|_{t \leftarrow s} \leqq \left[\frac{1}{\omega_l} \eta_0\left(\frac{2\nu}{2 + t - s} \right) h_l^{-2m} \right]^{1 + (t-s)/2} . \tag{6.2.22 a}$$

Hence, if $1/C_0 \leqq \omega_l$, the smoothing property (1.7) holds for $\|\cdot\|_{\mathscr{U}} = |\cdot|_s$ and $\|\cdot\|_{\mathscr{F}} = |\cdot|_t$ with

$$\eta(\nu) := \left[C_0 \eta_0 \left(\frac{2m\nu}{\alpha} \right) \right]^{\alpha/2m}, \qquad \alpha = (2 - s + t)m, \qquad \bar{\nu}(h) = \infty . \tag{6.2.22 b}$$

Proof. $|L_l S_l^\nu|_{t \leftarrow s}$ equals $\| L_l^{\alpha/2m} S_l^\nu \|$ as stated in (8 c). Define $A_l := \omega_l h_l^{2m} L_l$. The choice of ω_l ensures $0 \leqq A_l \leqq I$. Hence, Corollary 6.2.3 yields

$$\| L_l^{\alpha/2m} S_l^\nu \| = (\omega_l h_l^{2m})^{-\alpha/2m} \| A_l^{\alpha/2m} (I - A_l)^\nu \| \leqq \omega_l^{-\alpha/2m} h_l^{-\alpha} \eta_0 \left(\frac{2m\nu}{\alpha} \right)^{\alpha/2m}$$

for all ν (i.e. for $\bar{\nu}(h) = \infty$). □

The condition $s - t < 2$ (hence $\alpha > 0$) is needed to ensure $\eta(v) \to 0$. Values of ω_l exceeding $\| h_l^{2m} L_l \|^{-1}$ are considered in Exercise 6.6.3.

Note 6.2.13. The function η from (22 b) behaves like $\eta(v) = O(v^{-\alpha/2m})$. If $0 < \alpha \leqq 2m$, an upper bound is

$$\eta(v) \leqq \frac{m}{2\alpha}\left[\frac{3}{2} C_0 \Big/ \left(v + \frac{1}{2}\right)\right]^{\alpha/2m}, \qquad \alpha = 2m - s + t.$$

For $s = t = 0$ we regain $\eta(v) = O(1/v)$, which corresponds to the analogous behaviour (2.4.12) of the contraction numbers for the model problem of § 2.

The smoothing property can also be established for matrices with complex eigenvalues. However, the bounds are more complicated. Examples are studied in Exercise 6.6.4.

6.2.3.2 The Damped Jacobi Iteration (3.3.6)

The damped form of the original Jacobi iteration is (3.3.6′) with $S_l = I - \Theta D_l^{-1} L_l$, where D_l is the diagonal (or block-diagonal) of L_l.

Proposition 6.2.14. Assume $L_l = L_l^*$, $D_l = D_l^*$, and

$$O \leqq \Theta L_l \leqq D_l \leqq C_D h_l^{-2m} I \tag{6.2.23 a}$$

for some constant C_D. Then, (3.3.6′) satisfies

$$\| L_l S_l^v \| \leqq C_D \Theta^{-1} \eta_0(v) h_l^{-2m}. \tag{6.2.24 b}$$

Hence, the smoothing property (1.7) holds for $\|\cdot\|_{\mathcal{U}} = \|\cdot\|_{\mathcal{F}} = |\cdot|_0$ with $\eta(v) := C_D \eta_0(v)/\Theta$, $\alpha = 2m$, $\bar{v}(h) = \infty$.

Proof. (23 a) implies $O \leqq A_l \leqq I$ for $A_l := \Theta D_l^{-1/2} L_l D_l^{-1/2}$. Set $X_l := \Theta D_l^{-1/2} L_l S_l^v D_l^{-1/2}$. From Lemma 6.2.1 one concludes $\| X_l \| = \| A_l (I - A_l)^v \| \leqq \eta_0(v)$. Equation (1.3.19 a) yields

$$\begin{aligned} \| L_l S_l^v \| &= \Theta^{-1} \| D_l^{1/2} X_l D_l^{1/2} \| \leqq \Theta^{-1} \| D_l^{1/2} \| \, \| X_l \| \, \| D_l^{1/2} \| \\ &\leqq \Theta^{-1} \eta_0(v) \| D_l \| . \end{aligned}$$

□

Note 6.2.15. If D_l is the diagonal of L_l and if L_l is weakly diagonally dominant, then inequality (23 a) holds for $0 < \Theta \leqq \frac{1}{2}$.

6.2.3.3 Jacobi's Iteration with Squared Matrix

In (3.3.10) we described a modification of Jacobi's iteration with $S_l = I - \omega_l^2 h_l^{4m} L_l^* L_l$. The following propositions holds for a general L_l.

Proposition 6.2.16. If $1/C_0 \leqq \omega_l \leqq 1/\| h_l^{2m} L_l \|$, the iteration (3.3.10) satisfies

$$\| L_l S_l^v \| \leqq C_0 h_l^{-2m} \sqrt{\eta_0(2v)}. \tag{6.2.24 a}$$

Hence, the smoothing property (1.7) holds for $\|\cdot\|_{\mathcal{U}} = \|\cdot\|_{\mathcal{F}} = |\cdot|_0$ with

$$\eta(v) := C_0 \sqrt{\eta_0(2v)} \leqq C_0 \frac{\sqrt{2}}{3} \Big/ \sqrt{v + \tfrac{1}{2}}, \qquad \alpha = 2m, \qquad \bar{v}(h) = \infty. \tag{6.2.24 b}$$

Table 6.2.1. Smoothing numbers of (3.3.7′) and (3.3.10)

v	1	2	3	4
$\eta_0(v)$	0.250	0.148	0.105	0.082
$\sqrt{\eta_0(2v)}$	0.385	0.286	0.238	0.208

Proof. Set $A_l := \omega_l^2 h_l^{4m} L_l^* L_l$. Since $S_l = S_l^* = I - A_l$, one obtains $\| L_l S_l^v \| = \| (L_l S_l^v)^* (L_l S_l^v) \|^{1/2} = \| S_l^v L_l^* L_l S_l^v \|^{1/2} = (\omega_l h_l^{2m})^{-1} \| A_l (I - A_l)^{2v} \|^{1/2}$. Since $\varrho(L_l^* L_l) = \| L_l \|^2$ (cf. (1.3.15 a)), the choice of ω_l yields $O \leqq A_l \leqq I$ and Lemma 6.2.1 proves (24 a). □

Set $\omega_l = 1/\| h_l^{2m} L_l \|$. The comparison of (21) and (24 a) shows that the smoothing effect is weaker in the latter case. Table 6.2.1 contains the smoothing numbers

$$\varrho_L(v) := \sup_{l \geq 0} \| L_l S_l^v \|_{\mathscr{F} \leftarrow \mathscr{U}} / \| L_l \|_{\mathscr{F} \leftarrow \mathscr{U}} \tag{6.2.25}$$

(cf. (1.11)), which are $\eta_0(v)$ for (21) and $\sqrt{\eta_0(2v)}$ for (24 a), provided that $L_l = L_l^* \geqq 0$.

6.2.3.4 The Case of Non-Hermitian L_l

Propositions 6.2.11, 6.2.12, and 6.2.14 require a positive definite L_l. Otherwise one has to apply Criterion 6.2.8. The symmetric part of L_l is $(L_l + L_l^*)/2$. For suitable λ the matrix

$$L_l' := \tfrac{1}{2}(L_l + L_l^*) + \lambda I \tag{6.2.26}$$

should be positive definite. Usually, the remainder

$$L_l'' := \tfrac{1}{2}(L_l - L_l^*) - \lambda I \tag{6.2.27}$$

consists of terms of lower order; i.e. $\| L_l'' \| = O(h_l^{-\beta})$, $\beta < 2m$. In particular, $\lim_{l \to \infty} \| h_l^{2m} L_l'' \| = 0$ can be expected.

Proposition 6.2.17. Assume that L_l' from (26) satisfies $O \leqq L_l' \leqq h_l^{-2m} C_L' I$, C_L' constant, while L_l'' from (27) satisfies $h_l^{2m} \| L_l'' \| \to 0\,(l \to \infty)$. Choose $\omega_l \in [1/C_0, 1/\| h_l^{2m} L_l' \|]$. The damped Jacobi iteration (3.3.7′) satisfies the smoothing property (1.7 a–c) with $\alpha = 2m$, $\| \cdot \|_{\mathscr{U}} = \| \cdot \|_{\mathscr{F}} = |\cdot|_0$. Further, inequality (20) is valid: $\| S_l^v \| \leqq C_S$ $(0 < v < \bar{v}(h_l))$.

Proof. Proposition 6.2.11 proves the smoothing property (13 a–c) of $L_l' S_l'^{(v)}$, where $S_l' := I - \omega_l h_l^{2m} L_l'$. As $S_l'' = -\omega_l h_l^{2m} L_l''$, both (15) and (18 b) are satisfied. (18 a) holds with $C_S' = 1$. Hence Criterion 6.2.8 yields the smoothing property of $L_l S_l^v$, while Corollary 6.2.10 proves the estimate $\| S_l^v \| \leqq C_S$. □

Since L_l' is assumed to be positive definite, the norms (4), $|\cdot|_s$, can be defined by

$$|u_l|_s := |\Lambda_l^s u_l|_0, \qquad \Lambda_l := L_l'^{1/2}. \tag{6.2.28}$$

If $\|\cdot\|_{\mathscr{U}} = |\cdot|_{s_{\mathscr{U}}}$ and $\|\cdot\|_{\mathscr{F}} = |\cdot|_{s_{\mathscr{F}}}$ are given with $s_{\mathscr{U}} \neq 0$ or $s_{\mathscr{F}} \neq 0$, Proposition 6.2.17 can be extended.

Corollary 6.2.18. Let L_l' and ω_l be as in Proposition 6.2.17, assume $s_{\mathscr{U}} - s_{\mathscr{F}} < 2$ and

$$\lim_{l\to\infty} h_l^\alpha \|L_l''\|_{\mathscr{F}\leftarrow\mathscr{U}} = \lim_{l\to\infty} h_l^{2m} \|L_l''\|_{\mathscr{U}\leftarrow\mathscr{U}} = 0. \tag{6.2.29}$$

Then the damped Jacobi iteration (3.3.7′) satisfies the smoothing property (1.7a–c). Further, inequality (20), $\|S_l^\nu\|_{\mathscr{U}\leftarrow\mathscr{U}} \leqq C_S$, is valid for $0 < \nu < \bar{\nu}(h_l)$.

Proof. (13a–c) follows from Proposition 6.2.12 and (18a) holds with $C_S' = 1$. Hence, Criterion 6.2.8 proves the corollary. □

If $h_l^\alpha \|L_l''\|_{\mathscr{F}\leftarrow\mathscr{U}} \to 0$ (first part of (29)) and $0 \leqq s_{\mathscr{U}} - s_{\mathscr{F}}$, the second part of Eq. (29) holds, too (cf. Exercise 6.6.5). For the finite element case, Eq. (29) is verified in § 6.3.1.4.

6.2.4 The Smoothing Property of the Gauß-Seidel Iteration

6.2.4.1 The 2-Cyclic Gauß-Seidel Iteration

Let L_l be decomposed into $L_l = D_l - A_l - B_l$ (cf. (3.3.2) or (3.3.11)), where A_l and B_l are strictly lower and upper triangular matrices, respectively. D_l is diagonal in the case of (3.3.1) and block-diagonal for the block-Gauß-Seidel iteration (3.3.12). The point- or blockwise Gauß-Seidel iterations are characterised by

$$S_l = (D_l - A_l)^{-1} B_l = I - (D_l - A_l)^{-1} L_l. \tag{6.2.30}$$

We recall some properties of the Gauß-Seidel-iteration matrix in

Note 6.2.19. (i) The identity (31) holds:

$$L_l S_l = \hat{S}_l L_l \quad \text{with} \quad \hat{S}_l := B_l (D_l - A_l)^{-1}. \tag{6.2.31}$$

(ii) In the symmetric case of $B_l = A_l^*$, $D_l = D_l^*$, the matrix $\hat{S}_l^*$ is the iteration matrix of the Gauß-Seidel iteration with reversed ordering.

(iii) Let $L_l = L_l^*$ be positive definite and either D_l diagonal or $D_l = D_l^* > 0$. Then the Gauß-Seidel iteration converges and satisfies

$$\varrho(S_l) \leqq \|L_l^{1/2} S_l L_l^{-1/2}\| < 1 \tag{6.2.32 a}$$

as known from Ostrowski's theorem (cf. Varga [1, p. 77], Meis-Marcowitz [1, p. 297]). (32 a) follows from

$$L_l^{-1/2} S_l^* L_l S_l L_l^{-1/2} = I - L_l^{1/2} (D_l - B_l)^{-1} D_l (D_l - A_l)^{-1} L_l^{1/2}. \tag{6.2.32 b}$$

6.2.4.2 The Lexicographical Gauß-Seidel Iteration

The lexicographical Gauß-Seidel iteration cannot be treated as the 2-cyclic iteration. However, a particular case is easy to analyse. Consider, e.g., the Helmholtz equation $-\Delta u + cu = f$ in the infinite plane $\Omega = \mathbb{R}^2$. The lexicographical ordering leads to the stencils

$$L_l = h_l^{-2}\begin{bmatrix} & -1 & \\ -1 & 4+ch_l^2 & -1 \\ & -1 & \end{bmatrix}, \quad D_l = h_l^{-2}\begin{bmatrix} & 0 & \\ 0 & 4+ch_l^2 & 0 \\ & 0 & \end{bmatrix},$$
$$A_l = h_l^{-2}\begin{bmatrix} & 0 & \\ -1 & 0 & 0 \\ & -1 & \end{bmatrix}, \quad B_l = A_l^* = h_l^{-2}\begin{bmatrix} & -1 & \\ 0 & 0 & -1 \\ & 0 & \end{bmatrix}. \tag{6.2.38}$$

The matrices L_l, A_l, B_l, D_l (and also S_l) commute. In particular, the assumptions of the following proposition are met.

Proposition 6.2.25. Let $L_l = D_l - A_l - A_l^*$ be symmetric. Assume that A_l is normal and that

$$D_l = d_l I, \qquad d_l > 0, \qquad \|A_l\| \leqq d_l/2.$$

Then

$$\|L_l S_l^\nu\| \leqq d_l \cdot \begin{cases} 2/3^\nu & \text{if } \nu = 1,2; \\ \frac{1}{2}\sqrt{\eta_0(\nu-2)\,\eta_0(\nu-1)} & \text{if } \nu \geqq 3. \end{cases} \tag{6.2.39}$$

Hence, if $\|\cdot\|_{\mathscr{U}} = \|\cdot\|_{\mathscr{F}} = |\cdot|_0$ and $d_l \leqq C_D h_l^{-2m}$, the smoothing property is satisfied. Further, S_l^ν is bounded by

$$\|S_l^\nu\| = \|\hat{S}_l^\nu\| \leqq 1 \quad \text{for all } \nu. \tag{6.2.40}$$

Proof. Since $L_l S_l^\nu = (d_l I - A_l - A_l^*)(d_l I - A_l)^{-\nu} A_l^{*\nu}$ is normal, its norm equals $\varrho(L_l S_l^\nu)$. Hence,

$$\|L_l S_l^\nu\| = \max_{\lambda\in\sigma(A_l)} (d_l - 2\operatorname{Re}\lambda)\,|\bar{\lambda}/(d_l-\lambda)|^\nu \leqq \max_{|\lambda|\leqq d_l/2} \Phi(\lambda),$$

where $\sigma(A_l) = \{\lambda\colon \text{eigenvalues of } A_l\}$ and $\Phi(\lambda) := (d_l - 2\operatorname{Re}\lambda)\,|\bar{\lambda}/(d_l-\lambda)|^\nu$. Introducing $x = \operatorname{Re}\lambda$ and $y = |\lambda|$, we can rewrite Φ as

$$\Phi(\lambda) = \varphi(x,y) := (d_l - 2x)\, y^\nu (d_l^2 - 2xd_l + y^2)^{-\nu/2}$$
$$= (d_l - 2x)\Big/\left(1 + \frac{d_l^2 - 2xd_l}{y^2}\right)^{\nu/2}.$$

φ has to be maximised over $x \in [-y, y]$, $y \in [0, d_l/2]$. The estimate $\varphi(x,y) \leqq \varphi(x, d_l/2) = d_l\left(1 - \frac{2x}{d_l}\right)\Big/\left(5 - \frac{8x}{d_l}\right)^{\nu/2}$ is immediate. The maximum is attained at $x = -d_l/2$ if $\nu \leqq 2$, and at $x = d_l\left(\frac{1}{2} - \frac{1}{4(\nu-2)}\right)$ if $\nu \geqq 3$, resulting in the numbers of (39). The estimate (40) is a consequence of $|\bar{\lambda}/(d_l - \lambda)| \leqq 1$. □

For bounded domains Ω the matrices A_l are no longer normal. That case is considered in § 6.4.2, § 8.2.2, and Hackbusch [34].

6.2.4.3 The Symmetric Gauß-Seidel Iteration

Some symmetry properties of L_l carry over to S_l, when the symmetric Gauß-Seidel process (3.3.5) is applied. One step of the iteration consists of one usual iteration $u_l \mapsto (D_l - A_l)^{-1}(B_l u_l + f_l) =: \hat{u}_l$ and one iteration with reversed ordering: $\hat{u}_l \mapsto (D_l - B_l)^{-1}(A_l \hat{u}_l + f_l)$. The resulting iteration matrix is

$$S_l = (D_l - B_l)^{-1} A_l (D_l - A_l)^{-1} B_l = I - [(D_l - A_l) D_l^{-1} (D_l - B_l)]^{-1} L_l. \tag{6.2.41}$$

Proposition 6.2.25 extends to this iteration with ν on the right side of (39) replaced by 2ν. However, a more general result can be obtained. The symmetric Gauß-Seidel iteration is a special case of an iteration of the type

$$W_l(u_l^j - u_l^{j+1}) = L_l u_l^j - f_l. \tag{6.2.42}$$

Note 6.2.26. (i) The symmetric Gauß-Seidel iteration (41) is equivalent to (42) with

$$W_l = (D_l - A_l) D_l^{-1} (D_l - B_l). \tag{6.2.43}$$

(ii) If $L_l = L_l^* > 0$ and $D_l = D_l^* > 0$, the matrix W_l from (43) is symmetric and positive definite. Furthermore, it satisfies

$$O < L_l \leqq W_l. \tag{6.2.44}$$

The following two propositions apply to the iteration (42). Note that the norms involved are different from (7).

Proposition 6.2.27. Let the iteration (42) satisfy $L_l = L_l^*$, $W_l = W_l^*$, and (44); define the norms by $\|u_l\|_{\mathscr{U}} = |W_l^{1/2} u_l|_0$, $\|f_l\|_{\mathscr{F}} = |W_l^{-1/2} f_l|_0$. Then the smoothing property (1.7) holds with $\eta(\nu) = \eta_0(\nu)$, $\alpha = 0$, $\bar{\nu}(h) = \infty$.

Proof. (44) implies $O \leqq X \leqq I$ for $X := W_l^{-1/2} L_l W_l^{-1/2}$; hence, Lemma 6.2.1 yields

$$\begin{aligned} \|L_l S_l^\nu\|_{\mathscr{F} \leftarrow \mathscr{U}} &= \|W_l^{-1/2} L_l (I - W_l^{-1} L_l)^\nu W_l^{-1/2}\| \\ &= \|X(I - X)^\nu\| \leqq \eta_0(\nu). \end{aligned}$$ □

Proposition 6.2.28. Under the conditions of Proposition 6.2.27, but with $\|u_l\|_{\mathscr{U}} = |L_l^{1/2} u_l|_0$ and $\|f_l\|_{\mathscr{F}} = |W_l^{-1/2} f_l|_0$, the smoothing property (1.7) holds with $\eta(\nu) = \sqrt{\eta_0(2\nu)}$, $\alpha = 0$, $\bar{\nu}(h) = \infty$.

Proof. (44) implies $W_l^{-1} \leqq L_l^{-1}$ and $O < Y := L_l^{1/2} W_l^{-1} L_l^{1/2} \leqq I$. The assertion follows from

$$\begin{aligned} \|L_l S_l^\nu\|_{\mathscr{F} \leftarrow \mathscr{U}} &= \|W_l^{-1/2} L_l S_l^\nu L_l^{-1/2}\| = \|W_l^{-1/2} L_l^{1/2} (I - Y)^\nu\| \\ &= \|Y^{1/2}(I - Y)^\nu\| \leqq \sqrt{\eta_0(2\nu)}. \end{aligned}$$ □

6.2.4.4 Estimates with Respect to Other Norms

In § 6.2.4.1 we proved the smoothing property with respect to $\|\cdot\|_{\mathcal{U}} = \|\cdot\|_{\mathcal{F}} = |\cdot|_0$. In this chapter we consider the more general norms (4 b). The following criteria apply to arbitrary smoothing iterations. The smoothing property in a general norm is derived from the one with respect to $|\cdot|_0$.

Criterion 6.2.29. Let $|\cdot|_s$, $s \in \mathbb{R}$, be defined by (4) with some $\Lambda_l = \Lambda_l^* > 0$, and define $\|\cdot\|_{\mathcal{U}}$ and $\|\cdot\|_{\mathcal{F}}$ by (7) with $0 \leqq s_{\mathcal{U}} < 1$, $s_{\mathcal{F}} = -s_{\mathcal{U}}$. Assume

$$|L_l|_{-1\leftarrow 1} \leqq C_L, \quad |S_l^{(v)}|_{1\leftarrow 1} \leqq C_S, \quad |L_l S_l^{(v)}|_{0\leftarrow 0} \leqq \eta^*(v)\, h_l^{-2m}$$

for $1 \leqq v < \bar{v}(h)$, $l \geqq 1$, where $\eta^*(v)$ and $\bar{v}$ satisfy (1.7 b, c). Then the smoothing property $\|L_l S_l^{(v)}\|_{\mathcal{F}\leftarrow\mathcal{U}} = |L_l S_l^{(v)}|_{-s_{\mathcal{U}}\leftarrow s_{\mathcal{U}}} \leqq \eta(v)\, h_l^{-\alpha}$ holds with $\alpha = 2m(1-s_{\mathcal{U}})$, $\eta(v) = (C_L C_S)^{s_{\mathcal{U}}}\, \eta^*(v)^{\alpha/2m}$.

Proof. Set $X := L_l S_l^{(v)}$. (12 b) implies $\|X\|_{\mathcal{F}\leftarrow\mathcal{U}} \leqq |X|_{0\leftarrow 0}^{\alpha/2m}\, |X|_{-1\leftarrow 1}^{s_{\mathcal{U}}}$. Insert $|X|_{0\leftarrow 0} \leqq \eta^*(v)\, h_l^{-2m}$ and $|X|_{-1\leftarrow 1} \leqq |L_l|_{-1\leftarrow 1}\, |S_l^{(v)}|_{1\leftarrow 1} \leqq C_L C_S$. □

Criterion 6.2.30. Let $\|\cdot\|_{\mathcal{U}}$ and $\|\cdot\|_{\mathcal{F}}$ be defined by (7) with $\alpha = (2 + s_{\mathcal{F}} - s_{\mathcal{U}})m > 0$, $-2 \leqq s_{\mathcal{F}} \leqq 0 \leqq s_{\mathcal{U}} \leqq 2$. Assume

$$|L_l|_{0\leftarrow 2} \leqq C_L, \quad |L_l^*|_{0\leftarrow 2} \leqq C_L^*,$$

$$|S_l^{(v)}|_{0\leftarrow 0} \leqq C_S, \quad |\hat{S}_l^{(v)}|_{0\leftarrow 0} \leqq C_S^*, \quad |L_l S_l^{(v)}|_{0\leftarrow 0} \leqq \eta^*(v)\, h_l^{-2m}$$

for $1 \leqq v < \bar{v}(h)$, $l \geqq 1$ (η^*, $\bar{v}$ satisfying (1.7 b, c)), where $\hat{S}_l^{(v)} := L_l S_l^{(v)} L_l^{-1}$ (cf. (31)). Then the smoothing property (1.7), $\|L_l S_l^{(v)}\|_{\mathcal{F}\leftarrow\mathcal{U}} \leqq \eta(v)\, h_l^{-\alpha}$, holds with $\eta(v) = (C_S^* C_L)^{s_{\mathcal{U}}/2}\, (C_L^* C_S)^{-s_{\mathcal{F}}/2}\, \eta^*(v)^{\alpha/2m}$.

Proof. $X := L_l S_l^{(v)}$ satisfies $|X|_{0\leftarrow 2} = |\hat{S}_l^{(v)} L_l|_{0\leftarrow 2} \leqq C_S^* C_L$, $|X|_{0\leftarrow 0} \leqq \eta^*(v))\, h_l^{-2m}$ and

$$|X|_{-2\leftarrow 0} \leqq |L_l|_{-2\leftarrow 0}\, |S_l^{(v)}|_{0\leftarrow 0} = |L_l^*|_{0\leftarrow 2}\, |S_l^{(v)}|_{0\leftarrow 0} \leqq C_L^* C_S\,.$$

The assertion follows from (12 c). □

All assumptions of these criteria are satisfied for the 2-cyclic Gauß-Seidel iteration with Hermitian, positive definite L_l.

Proposition 6.2.31. Assume $L_l = L_l^* > 0$ and define $|\cdot|_s$ by (4 a, b), (5). Then the 2-cyclic Gauß-Seidel iteration (30) satisfies

$$|L_l S_l^v|_{-r\leftarrow r} \leqq [\|D_l\|\, \eta_0(v - \tfrac{1}{2})]^{\alpha/2m}, \quad \alpha = 2m(1-r), \quad \text{for } 0 \leqq r < 1\,,$$

and

$$|L_l S_l^v|_{-s\leftarrow r} \leqq [2 \operatorname{cond}(D_l)]^{\frac{r+s}{4}}\, [\|D_l\|\, \eta_0(v - \tfrac{1}{2})]^{\alpha/2m}$$

for $\alpha = (2 - r - s)m > 0$, $s \geqq 0$, $r \geqq 0$. Hence, under assumption (36) the smoothing property (1.7) holds for $\|\cdot\|_{\mathcal{U}}$, $\|\cdot\|_{\mathcal{F}}$ from (7) with $s_{\mathcal{F}} \leqq 0 \leqq s_{\mathcal{U}}$, $\alpha := (2 - s_{\mathcal{U}} + s_{\mathcal{F}})m > 0$.

Proof. Criterion 6.2.29 applies with $C_L = 1$, $C_S = 1$ (cf. (32)), $\eta^*(v) = \|D_l\| \eta_0(v - \frac{1}{2})$, while the bounds in Criterion 6.2.30 are $C_L = C_L^* = 1$, $C_S = C_S^* = [2\,\mathrm{cond}\,(d_l)]^{1/2}$ (cf. (35)), and η^* as above. □

6.2.4.5 The Case of Non-Hermitian L_l

The previous proofs are restricted to $L_l = L_l^* > 0$. In the general case of non-Hermitian L_l split L_l into $L_l' + L_l''$ according to (26), (27).

Proposition 6.2.32. Let $L_l = D_l - A_l - B_l$ be a 2-cyclic decomposition (33). Assume $L_l = L_l' + L_l''$ with a corresponding decomposition $L_l'' = D_l'' - A_l'' - B_l''$, while $L_l' = L_l'^* > O$ and $D_l' = D_l'^*$. If

$$\lim_{l \to \infty} h_l^{2m} \|L_l''\| = 0, \quad \|L_l'\| \leqq C_L h_l^{-2m}, \quad \|D_l^{-1}\| \leqq C_D h_l^{2m}, \quad \|D_l'^{-1}\| \leqq C_D' h_l^{2m},$$

the smoothing property holds for $L_l S_l^v$ with respect to $\|\cdot\|_{\mathscr{U}} = \|\cdot\|_{\mathscr{F}} = |\cdot|_0$.

Proof. By assumption the smoothing property holds for $L_l' S_l'^{(v)}$ (cf. Proposition 6.2.22). All assumptions of Criterion 6.2.8 besides (18 b) are trivial. $S_l'' := S_l - S_l'$ equals $(D_l - A_l)^{-1} [(D_l'' - A_l'') S_l' - L_l'']$. Since the decomposition is 2-cyclic, the assumptions imply $\|(D_l - A_l)^{-1}\| = O(h_l^{2m})$, whereas $\|D_l''\| \leqq \|L_l''\|$ and $\|A_l''\| \leqq \|L_l''\|$. Since $\|S_l'\| \leqq$ const by (35), inequality (18 b) is proved and Criterion 6.2.8 implies the assertion. □

If the smoothing property has to be proved for other norms, there are two approaches. One can apply Criterion 6.2.8 directly with respect to these norms, as is done in Corollary 6.2.18 for Jacobi's iteration. Alternatively, one may use Proposition 6.2.32 to prove the smoothing property with respect to $|\cdot|_0$. Then the criteria 6.2.29 or 6.2.30 yield the desired result.

6.2.5 The Smoothing Property of Semi-Iterative Methods

6.2.5.1 Richardson's Iteration

The choice $\omega_l \leqq 1/\|h_l^{2m} L_l\|$ for the Jacobi iteration (3.3.7) is not optimal as already mentioned in § 2.6.3. More generally, one can use different factors ω_μ ($\mu = 1, \ldots, v$) for each of the v smoothing iterations. The iteration matrix of the resulting Richardson iteration is

$$S_l^{(v)} = \pi_v(h_l^{2m} L_l) \quad \text{with} \quad \pi_v(\zeta) = \prod_{\mu=1}^{v} (1 - \omega_\mu \zeta) \tag{6.2.45 a}$$

(cf. (3.3.15 b)). The same smoothing procedure is generated by

$$S_l^{(v)}(u_l, f_l) = u_l - \tilde{\pi}_{v-1}(h_l^{2m} L_l) h_l^{2m} (L_l u_l - f_l), \tag{6.2.45 b}$$

(cf. § 3.3.5), where $\tilde{\pi}_{v-1}$ is the polynomial

$$\tilde{\pi}_{v-1}(\zeta) = (1 - \pi_v(\zeta))/\zeta. \tag{6.2.45 c}$$

By Lemma 1.3.5 the norm of $L_l S_l^{(\nu)}$ equals

$$\| L_l S_l^{(\nu)} \| = h_l^{-2m} \max_{\lambda \in \sigma(h_l^{2m} L_l)} |\lambda \pi_\nu(\lambda)| \leqq h_l^{-2m} \max_{0 \leqq \lambda \leqq C_L} |\lambda \pi_\nu(\lambda)|,$$

where $L_l = L_l^* > O$ and $C_L = \sup \| h_l^{2m} L_l \|$. The usual choice of the Čebyšev polynomial π_ν minimises $|\pi_\nu(\lambda)|$ in $[0, C_L]$. This corresponds to a minimum error. When we minimise $|\lambda \pi_\nu(\lambda)|$, the defect becomes a minimum.

Proposition 6.2.33. The optimisation of $\lambda \pi_\nu$ yields the estimate

$$\| L_l S_l^{(\nu)} \| \leqq \frac{\| L_l \|}{\nu + 1} \frac{\sin(\pi/(2\nu + 2))}{1 + \cos(\pi/(2\nu + 2))} \leqq \| L_l \| \frac{2}{1 + \sqrt{2}} \frac{1}{(\nu + 1)^2}, \tag{6.2.46 a}$$

which is attained for the transformed Čebyšev polynomial

$$\zeta \pi_\nu(\zeta) = \frac{\| h_l^{2m} L_l \|}{\nu + 1} \frac{\sin(\pi/(2\nu + 2))}{1 + \cos(\pi/(2\nu + 2))} \times \cos\left((\nu + 1) \arccos\left(\frac{\zeta}{\| h_l^{2m} L_l \|} - \left(1 - \frac{\zeta}{\| h_l^{2m} L_l \|}\right) \cos \frac{\pi}{2\nu + 2}\right)\right) \tag{6.2.46 b}$$

corresponding to the following $\omega_1, \ldots, \omega_\nu$:

$$\omega_\mu = \frac{1}{\| h_l^{2m} L_l \|} \left(1 + \cos \frac{\pi}{2\nu + 2}\right) \Big/ \left(\cos \frac{\pi}{2\nu + 2} - \cos \frac{(2\mu + 1)\pi}{2\nu + 2}\right). \tag{6.2.46 c}$$

The norm $\| L_l \|$ in (46 a, c) may be replaced by $h_l^{-2m} C_L$, where $C_L \geqq \| h_l^{2m} L_l \|$. Hence, the smoothing property is satisfied for $\| \cdot \|_{\mathscr{U}} = \| \cdot \|_{\mathscr{F}} = | \cdot |_0$ with $\alpha = 2m$, $\bar{\nu}(h) = \infty$, and

$$\eta(\nu) = \frac{C_L}{\nu + 1} \frac{\sin(\pi/(2\nu + 2))}{1 + \cos(\pi/(2\nu + 2))} \leqq \frac{2 C_L}{1 + \sqrt{2}} \frac{1}{(\nu + 1)^2}. \tag{6.2.46 d}$$

Note that the asymptotic behaviour of $\eta(\nu)$ is $O(\nu^{-2})$ instead of $O(\nu^{-1})$ as for the previous iterations. The smoothing numbers (25) and the factors (46 c) are listed below.

Table 6.2.3. Optimal smoothing numbers for Richardson's iteration

ν	1	2	3	4
$\varrho_L(\nu)$	0.207	0.0893	0.0497	0.0317
$\omega_\mu C_L$	1.207	2.1547	3.5549	5.3708
		1.0774	1.4725	2.0515
			1.0412	1.2679
				1.0257

6.2.5.2 The Accelerated Symmetric Gauß-Seidel Iteration

Semi-iterative methods can also be based on the symmetric Gauß-Seidel iteration (cf. Young [3]). Let $\mathscr{S}_l$ denote the smoothing procedure by the symmetric Gauß-Seidel iteration (3.3.5). The semi-iterative process $\mathscr{S}_l^{(\nu)}(u_l, f_l; \boldsymbol{\omega})$, $\boldsymbol{\omega} = (\omega_1, \ldots, \omega_\nu)$, is defined by

$$u_l^0 := u_l;$$

$$\textbf{for} \quad \mu := 1(1)\nu \quad \textbf{do} \quad u_l^\mu := u_l^{\mu-1} + \omega_\mu(\mathscr{S}_l(u_l^{\mu-1}, f_l) - u_l^{\mu-1}); \tag{6.2.47 a}$$

$$\mathscr{S}_l^{(\nu)}(u_l, f_l; \boldsymbol{\omega}) := u_l^\nu;$$

Its iteration matrix is

$$S_l^{(\nu)}(\boldsymbol{\omega}) = \prod_{\mu=1}^{\nu} (I + \omega_\mu(S_l - I)), \quad S_l \quad \text{from (41).} \tag{6.2.47 b}$$

Iteration (47) is a special case of the semi-iterative variant of (42):

$$u_l^0 := u_l;$$

$$\textbf{for} \quad \mu := 1(1)\nu \quad \textbf{do} \quad u_l^\mu := u_l^{\mu-1} - \omega_\mu W_l^{-1}(L_l u_l^{\mu-1} - f_l); \tag{6.2.48 a}$$

$$\mathscr{S}_l^{(\nu)}(u_l, f_l; \boldsymbol{\omega}) := u_l^\nu;$$

The corresponding iteration matrix is

$$S_l^{(\nu)}(\boldsymbol{\omega}) = \prod_{\mu=1}^{\nu} (I - \omega_\mu W_l^{-1} L_l). \tag{6.2.48 b}$$

Proposition 6.2.34. Let $S_l^{(\nu)}(\omega)$ be given by (48 b). Under the assumptions of Proposition 6.2.27 the optimal estimate

$$\| L_l S_l^{(\nu)}(\boldsymbol{\omega}) \|_{\mathscr{F} \leftarrow \mathscr{U}} \leqq \frac{\sin(\pi/(2\nu+2))}{1 + \cos(\pi/(2\nu+2))} \bigg/ (\nu + 1) < 0.83/(\nu+1)^2 \tag{6.2.49 a}$$

is attained for

$$\omega_\mu = \left(1 + \cos\frac{\pi}{2\nu+2}\right) \bigg/ \left(\cos\frac{\pi}{2\nu+2} - \cos\frac{(2\mu+1)\pi}{2\nu+2}\right), \quad 1 \leqq \mu \leqq \nu. \tag{6.2.49 b}$$

(49 a, b) and (46 a, c) differ only by the scaling factor. The left-hand side of (49 a) equals $X\pi_\nu(X)$ with $X = W_l^{-1/2} L_l W_l^{-1/2}$. The optimal π_ν is (46 b) with $\| h_l^{2m} L_l \|$ replaced by 1. Another polynomial arises in the next proposition, where a different $|\cdot|_{\mathscr{U}}$ norm is used.

Proposition 6.2.35. Under the assumptions of Proposition 6.2.28 the optimal estimate

$$\| L_l S_l^{(\nu)}(\boldsymbol{\omega}) \|_{\mathscr{F} \leftarrow \mathscr{U}} \leqq \frac{1}{2\nu + 1} \tag{6.2.50 a}$$

is attained for

$$\omega_\mu = 2\Big/\left[1 + \cos\left(\frac{2\mu - 1}{2\nu + 1}\pi\right)\right], \quad \mu = 1, \ldots, \nu. \tag{6.2.50 b}$$

Proof. $\| L_l S_l^{(\nu)}(\omega) \|_{\mathcal{F} \leftarrow \mathcal{U}}$ equals $\| Y^{1/2} \pi_\nu(Y) \|$ with $O \leqq Y := L_l^{1/2} W_l^{-1} L_l^{1/2} \leqq I$ and $\pi_\nu(x) = \prod_{\mu=1}^{\nu} (1 - \omega_\mu x)$. The optimal polynomial is

$$\pi_\mu(x) = [(1 + T_{2\nu+1}(2x - 1))/(2x)]^{1/2}/(2\nu + 1),$$

$$T_k(x) = \cos(k \cdot \arccos x). \qquad \square$$

Conjugate gradient methods are closely related to § 6.2.3.1–2. They will be discussed in Exercise 6.6.8.

6.2.5.3 The Alternating Direction Method

The iteration matrix of the alternating direction implicit iteration (ADI) is

$$S_l(\omega) = (\omega h_l^{-2m} I + B_l)^{-1} (\omega h_l^{-2m} I - A_l)(\omega h_l^{-2m} I + A_l)^{-1} (\omega h_l^{-2m} I - B_l),$$

where $L_l = A_l + B_l$ (cf. (3.3.13)). The convergence of ADI can only easily be analysed if A_l and B_l commute (cf. Varga [1, p. 219]). Under similar assumptions the smoothing property can be discussed.

Lemma 6.2.36. Let $L_l = A_l + B_l, O \leqq A_l = A_l^*, O \leqq B_l = B_l^*, \| A_l \| \leqq C h_l^{-2m}$, $\| B_l \| \leqq C h_l^{-2m}$, and suppose that A_l and B_l commute. The parameters $\varrho_\mu := \omega_\mu / C$ have to be chosen so that the right-hand side of (51) becomes a minimum:

$$\| L_l S_l^{(\nu)}(\omega) \| \leqq C h_l^{-2m} \max_{0 \leqq \alpha, \beta \leqq 1} (\alpha + \beta) \prod_{\mu=1}^{\nu} \left| \frac{\varrho_\mu - \alpha}{\varrho_\mu + \alpha} \frac{\varrho_\mu - \beta}{\varrho_\mu + \beta} \right|. \tag{6.2.51}$$

The ω_μ's minimising (51) are different from the usual optimal ADI parameters, as can be seen from the result for $\nu = 1$ mentioned in § 3.3.4.

6.3 The Approximation Property

The technique for proving the approximation property (1.8) is different for finite element discretisations and finite difference schemes. The former case is studied in § 6.3.1, while the latter will be found in § 6.3.2. We recall that the approximation property is closely related to the smoothing property. Together they imply convergence, provided they refer to the same norms $\| \cdot \|_{\mathcal{U}}$, $\| \cdot \|_{\mathcal{F}}$, and to the same α (cf. Theorem 6.1.7). Therefore, the approximation property has to be discussed for the norms that have been used in § 6.2.

6.3.1 The Approximation Property of Finite Element Equations

6.3.1.1 The Definition of the Problem

Equation (3.1.11) is a particular example of a variational formulation of a boundary value problem. In general, it can be described by a sesqui-linear form $a\colon \mathscr{H}^1 \times \mathscr{H}^1 \to \mathbb{C}$ and a functional $f \in (\mathscr{H}^1)' =: \mathscr{H}^{-1}$ (cf. § 1.4.5; in particular, Example 1.4.6)[1]. The Hilbert spaces $\mathscr{V}$ and $\mathscr{U}$ of § 1.4.5 will be denoted by $\mathscr{H}^1$ and $\mathscr{H}^0$, respectively. The common choice is $\mathscr{H}^0 = L^2(\Omega)$, $H_0^m(\Omega) \subset \mathscr{H}^1 \subset H^m(\Omega)$. We define $\mathscr{H}^{-1}$ as the dual space of $\mathscr{H}^1$ and assume

$$\mathscr{H}^{-1} := (\mathscr{H}^1)' \supset \mathscr{H}^0 \supset \mathscr{H}^1 \quad \text{with continuous and dense embedding.} \tag{6.3.1 a}$$

As shown in § 1.4 there is a uniquely defined Hilbert scale $\{\mathscr{H}^s\colon -1 \leqq s \leqq 1\}$ connecting $\mathscr{H}^{-1}$, $\mathscr{H}^0$, and $\mathscr{H}^1$. Whenever $\mathscr{H}^s$, $|s| \leqq 1$, appears, it is defined by (1.4.8 b, c). If for instance $\mathscr{H}^1 = H_0^m(\Omega)$, the intermediate spaces are $\mathscr{H}^s = H_0^{sm}(\Omega)$, $|s| \leqq 1$, with exceptions explained in (1.4.11 b).

Moreover, we shall need spaces $\mathscr{H}^s$ for certain $s > 1$; e.g. $\mathscr{H}^2 = H^{2m}(\Omega) \cap \mathscr{H}^1$ might appear in (5 a). The norm of $\mathscr{H}^s$ is denoted by $|\cdot|_s$. Generally, we assume

$$\mathscr{H}^s \supset \mathscr{H}^t \quad \text{for } s \leqq t, \quad \mathscr{H}^{-s} := (\mathscr{H}^s)', \tag{6.3.1 b}$$

and define $|\cdot|_{-s}$, $s > 1$, as dual norm of $|\cdot|_s$ since it is already true for $|s| < 1$. The operator norms $\|\cdot\|_{\mathscr{H}_s \leftarrow \mathscr{H}_t}$ are abbreviated by $|\cdot|_{s \leftarrow t}$. (1 b) implies

Note 6.3.1. $|A|_{s \leftarrow t} = |A^*|_{-t \leftarrow -s}$ holds for any $A \in L(\mathscr{H}^t, \mathscr{H}^s)$.

The variational formulation of a boundary value problem reads as

$$a(u, v) = f(v) \quad \text{for all } v \in \mathscr{H}^1 \tag{6.3.2 a}$$

with $f \in \mathscr{H}^{-1}$, while the form a is bounded and possibly $\mathscr{H}^1$-coercive relative to $\mathscr{H}^0$:

$$|a(u, v)| \leqq C_L^* |u|_1 |v|_1 \quad \text{for all } u, v \in \mathscr{H}^1, \tag{6.3.3 a}$$

$$\operatorname{Re} a(v, v) + \lambda_0 |v|_0^2 \geqq \varepsilon |v|_1^2, \quad \varepsilon > 0, \quad \lambda_0 \in \mathbb{R}, \quad \text{for all } v \in \mathscr{H}^1. \tag{6.3.3 b}$$

Lemma 6.3.2. Let $\mathscr{H}^1$ be compactly embedded in $\mathscr{H}^0$ or assume $\sup\{|a(v, u)|\colon v \in \mathscr{H}^1, |v|_1 = 1\} > 0$ for all $0 \neq u \in \mathscr{H}^1$. Then problem (2a) has a unique solution for arbitrary $f \in \mathscr{H}^{-1}$ if and only if

$$\sup\{|a(u, v)|\colon v \in \mathscr{H}^1, |v|_1 = 1\} \geqq |u|_1 / C_S \quad \text{for all } u \in \mathscr{H}^1 \tag{6.3.3 c}$$

is valid for some $C_S \in R$. Trivially, (3 c) holds for $\mathscr{H}^1$-elliptic forms (cf. (1.4.17)).

The proof of Lemma 6.3.2 is evident by the following rewriting of (2 a), (3 a–c). According to Proposition 1.4.7 the form a corresponds to an associated

---

1 The space $\mathscr{H}^1$ was called $\mathscr{H}$ in § 3

operator $L: \mathscr{H}^1 \to \mathscr{H}^{-1}$. An equivalent formulation of (2a), (3 a–c) is

$$Lu = f, \tag{6.3.2a'}$$

$$|L|_{-1\leftarrow 1} \leqq C_L^*, \tag{6.3.3 a'}$$

$$|[\tfrac{1}{2}(L+L^*) + \lambda_0 I]^{-1}|_{1\leftarrow -1} \leqq 1/\varepsilon, \tag{6.3.3 b'}$$

$$|L^{-1}|_{1\leftarrow -1} \leqq C_S. \tag{6.3.3 c'}$$

For the conforming finite element method one has to replace $\mathscr{H}^1$ in Eq. (2 a) by a finite dimensional subspace:

$$a(u, v) = f(v) \equiv (f, v) \quad \text{for all } v \in \mathscr{H}_l. \tag{6.3.2 b}$$

As in § 3.6 we assume that there exists a hierarchy

$$\mathscr{H}_0 \subset \mathscr{H}_1 \subset \ldots \subset \mathscr{H}_{l-1} \subset \mathscr{H}_l \subset \ldots \subset \mathscr{H}^1 \tag{6.3.4}$$

of subspaces. Each $\mathscr{H}_l$ is connected with a 'step size' h_l by

$$\inf\{|u-v|_1 : v \in \mathscr{H}_l\} \leqq C_a h_l^{(t-1)m} |u|_t \quad \text{for all } u \in \mathscr{H}^t \subset \mathscr{H}^1 \tag{6.3.5 a}$$

with fixed $t > 1$. $\mathscr{H}^t$ denotes a Hilbert space of functions of higher order of differentiability. Usually, $\mathscr{H}^t$ equals $H^{tm}(\Omega) \cap \mathscr{H}^1$. The parameters h_l have to satisfy

$$h_0 > h_1 > \ldots > h_{l-1} > h_l > \ldots,$$
$$\lim_{l\to\infty} h_l = 0, \quad h_{l-1}/h_l \leqq C_h \quad \text{for all } l \geqq 1. \tag{6.3.5 b}$$

Lemma 6.3.3. If $\{\mathscr{H}^s : 1 \leqq s \leqq t\}$ is a Hilbert scale, Inequality (5 a) is equivalent to

$$\inf\{|u-v|_1 : v \in \mathscr{H}_l\} \leqq C_a^{(s-1)/(t-1)} h_l^{(s-1)m} |u|_s \quad \text{for all } u \in \mathscr{H}^s, \quad 1 \leqq s \leqq t. \tag{6.3.5 a'}$$

Proof. The left side of (5 a) equals $|u - Q_l u|_1$, where $Q_l: \mathscr{H}^1 \to \mathscr{H}_l$ is the $\mathscr{H}^1$-orthogonal projection onto $\mathscr{H}_l$; hence $|I - Q_l|_{1\leftarrow 1} = 1$. (5 a) can be written as $|I - Q_l|_{1\leftarrow t} \leqq C_a h_l^{(t-1)m}$. (1.4.12) and (1.4.10 b) prove (5 a'). □

Besides (5 a), another common condition is the 'inverse assumption'

$$|u|_1 \leqq C_i h_l^{-m} |u|_0 \quad \text{for all } u \in \mathscr{H}_l. \tag{6.3.5 c}$$

The prolongations p and the restrictions r are the canonical ones defined in § 3.6 by means of

$$p = P_l^{-1} P_{l-1} : \mathscr{U}_{l-1} \to \mathscr{U}_l, \quad r = p^* = R_{l-1} R_l^{-1} : \mathscr{U}_l \to \mathscr{U}_{l-1},$$

where $P_l: \mathscr{U}_l \to \mathscr{H}_l \subset \mathscr{H}^1$ and $R_l = P_l^*: \mathscr{H}^{-1} \to \mathscr{U}_l$ depend on the choice of the basis and on the scalar product $\langle \cdot, \cdot \rangle$ of $\mathscr{U}_l$. The norm $|u_l|_0 = \langle u_l, u_l \rangle^{1/2}$ of $\mathscr{U}_l$ has to correspond to that of $\mathscr{H}^0$. Therefore,

$$\underline{C}_P^{-1} |u_l|_0 \leqq |P_l u_l|_0 \leqq \bar{C}_P |u_l|_0 \quad \text{for all } u_l \in \mathscr{U}_l, \quad l \geqq 0 \tag{6.3.6}$$

is a natural assumption.

The stiffness matrices of the discrete equations $L_l u_l = f_l$ satisfy

$$L_l = R_l L P_l, \quad L_{l-1} = r L_l p$$

(cf. (3.1.32), (3.6.7)). To ensure the unique solvability of $L_l u_l = f_l$ we need the discrete analogue of condition (3 c):

$$\sup\{|a(u,v)|: v \in \mathscr{H}_l, |v|_1 = 1\} \geqq |u|_1 / C_S^* \quad \text{for all } u \in \mathscr{H}_l. \tag{6.3.7}$$

Remark 6.3.4. If $a(\cdot,\cdot)$ is $\mathscr{H}^1$-elliptic, i.e. $a(u,u) \geqq \varepsilon |u|_1^2$, Inequality (7) holds with $C_S^* = 1/\varepsilon$. (7) implies that L_l is a non-singular matrix.

6.3.1.2 The Standard Case

First we shall prove the approximation property (1.8) for the norms $\|\cdot\|_{\mathscr{U}} = \|\cdot\|_{\mathscr{F}} = |\cdot|_0$ with $\alpha = 2m$:

$$\|L_l^{-1} - p L_{l-1}^{-1} r\| \leqq C_A h_l^{2m}.$$

These norms and the same α have also been used in most of the previous proofs of the smoothing property. The proof of the approximation property requires a certain regularity of the boundary value problem.

Definition 6.3.5. A differential operator L, e.g. the operator associated with the sesqui-linear form a, is called $\mathscr{H}^t$-regular, if

$$|L^{-1}|_{t \leftarrow t-2} \leqq C_R \quad \text{for some } C_R \in \mathbb{R}. \tag{6.3.8}$$

For instance, the associated operator of $a(\cdot,\cdot)$ is $\mathscr{H}^2$-regular, if the solution of Eq. (2a) satisfies $|u|_2 \leqq C_R |f|_0$ for any $f \in \mathscr{H}^0$. Furthermore, the adjoint operator L^* is $\mathscr{H}^2$-regular, if the solution u^* of $a(v,u^*) = (v,f)$ for all $v \in \mathscr{H}^1$ satisfies $|u^*|_2 \leqq C_R |f|_0$ (cf. (1.4.16)).

Remark 6.3.6. The $\mathscr{H}^1$-regularity of L is already stated in (3 c). By (1 b) and Note 6.3.1, the $\mathscr{H}^1$-regularity of L and L^* are equivalent. If the coefficients of L as well as the boundary Γ of Ω are sufficiently smooth, $\mathscr{H}^1$-regularity implies $\mathscr{H}^t$-regularity with $\mathscr{H}^t = H^{tm}(\Omega) \cap \mathscr{H}^1$ for all t (cf. Lions-Magenes [1], Nečas [2]).

In the case of Dirichlet boundary conditions (i.e. $\mathscr{H}^1 = H_0^m(\Omega)$) the following result is known (cf. also Nečas [2, p. 201], Ladyženskaja-Ural'ceva [1, p. 188]).

Theorem 6.3.7 (Kadlec [1]). Suppose that $\Omega \subset \mathbb{R}^n$ is convex. Let $a: H_0^1(\Omega) \times H_0^1(\Omega) \to \mathbb{R}$ be defined by

$$a(u,v) = \int_\Omega \sum_{i,j=1}^n a_{ij}(\mathbf{x}) \frac{\partial u(\mathbf{x})}{\partial x_i} \frac{\partial v(\mathbf{x})}{\partial x_j} d\mathbf{x} \tag{6.3.9 a}$$

with uniformly Lipschitz continuous coefficients a_{ij} satisfying

$$\sum_{i,j} a_{ij}(\mathbf{x}) \xi_i \xi_j \geqq \varepsilon |\xi|^2, \varepsilon > 0 \quad \text{for all } \mathbf{x} \in \Omega, \ \xi \in \mathbb{R}^n$$

Proof. Let $E_l := L^{-1} - P_l L_l^{-1} R_l$ and note that

$$P_l(L_l^{-1} - p\,L_{l-1}^{-1}\,r)\,R_l = P_l L_l^{-1} R_l - P_{l-1} L_{l-1}^{-1} R_{l-1} = E_{l-1} - E_l$$

(cf. (3.6.1)). Since $\hat{R}_l P_l = R_l \hat{P}_l = I$, one concludes from (12 a) and (5 b) that

$$\| L_l^{-1} - p\,L_{l-1}^{-1}\,r \| = \| \hat{R}_l (E_{l-1} - E_l)\hat{P}_l \| \leqq \underline{C}_P^2 (\| E_{l-1} \| + \| E_l \|).$$

(11 c) proves (13) with $C_A = C_A^* \underline{C}_P^2 (C_h^{2m} + 1)$. □

A better bound $C_A = C_A^* \underline{C}_P^2 C_h^{2m}$ can be proved if $L_l = L_l^* > 0$ (cf. Exercise 6.6.9). The previous results are collected in

Theorem 6.3.15. Assume: (1 a, b), (3 a, c′), (4), (5 a) for $t = 2$, (5 b), (6), (7), L and L^* $\mathscr{H}^2$-regular. Then the approximation property (11 c) holds.

Besides for $\|\cdot\|_{\mathscr{U}} = \|\cdot\|_{\mathscr{F}} = |\cdot|_0$, the smoothing property has been proved for $\|u_l\|_{\mathscr{U}} = |W_l^{1/2} u_l|_0$, $\|f_l\|_{\mathscr{F}} = |W_l^{-1/2} f_l|_0$ and $\alpha = 0$ (cf. Proposition 6.2.27). Thus, the corresponding approximation property becomes

$$\| L_l^{-1} - p\,L_{l-1}^{-1}\,r \|_{\mathscr{U} \leftarrow \mathscr{F}} = \| W_l^{1/2} (L_l^{-1} - p\,L_{l-1}^{-1}\,r)\,W_l^{1/2} \| \leqq C_A. \tag{6.3.14}$$

An obvious sufficient condition is stated in

Proposition 6.3.16. Assume (13) and $\| W_l \| \leqq C_W h_l^{-2m}$ for all $l \geqq 1$. Then the approximation property (14) holds.

6.3.1.3 The Less Regular Case

Unfortunately, the conditions of Theorem 6.3.15 are not always satisfied. For instance, the $\mathscr{H}^2$-regularity is lacking for boundary value problems with domain Ω having a re-entrant corner. If the boundary is still Lipschitz continuous, one can guarantee a weaker $\mathscr{H}^t$-regularity for $t\,m \in (m - \frac{1}{2}, m + \frac{1}{2})$, where $\mathscr{H}^t = H_0^{tm}(\Omega)$. The original formulation of the following theorem applies to problems of general order $2m$.

Theorem 6.3.17 (Nečas [1]). Let t and κ be numbers with $1 < t < \frac{3}{2}$. $\kappa > t - 1$. Assume Ω to be a Lipschitz domain and suppose that the coefficients of L from (9 b) satisfy $b_i, d \in L^\infty(\Omega)$, $a_{ij}, c_i \in \bar{C}^\kappa(\Omega)$, a_{ij} real, $\sum_{i,j} a_{ij}(\mathbf{x})\,\xi_i \xi_j \geqq \varepsilon |\xi|^2$ for all $\mathbf{x} \in \Omega$, $\xi \in \mathbb{R}^n$. If (3 c′) holds (i.e. if the inverse exists), L is $\mathscr{H}^t$-regular. Under the same assumption, but with b_i and c_i interchanged, L^* is $\mathscr{H}^t$-regular.

Another difficulty arises, e.g., for fourth order problems (i.e. $2m = 4$, $\mathscr{H}^1 = H_0^2(\Omega)$, $\mathscr{H}^2 = H_0^2(\Omega) \cap H^4(\Omega)$), which are discretised by quadratic finite elements. The estimate (5 a) does not hold for $t = 2$ but for $t = \frac{3}{2}$, where $\mathscr{H}^{3/2} = H_0^2(\Omega) \cap H^3(\Omega)$.

In both cases it is not possible to establish inequality (11 c), whereas (11 b) is still available for certain $s, s^* \in [-1, 0)$ with $s + s^* > -2$. Hence, one has to prove the approximation property with respect to norms $\|\cdot\|_{\mathscr{U}}$, $\|\cdot\|_{\mathscr{F}}$, which are discrete counterparts of those of $\mathscr{H}^{-s^*}$ and $\mathscr{H}^s$, respectively. If $L_l = L_l^* > 0$, one can choose $\|\cdot\|_{\mathscr{U}} = |\cdot|_{-s^*}$ and $\|\cdot\|_{\mathscr{F}} = |\cdot|_s$ with

$$|u_l|_s := |\Lambda_l^s u_l|_0, \qquad s \in \mathbb{R}, \tag{6.3.15 a}$$

where $\Lambda_l = L_l^{1/2}$ (cf. (2.4), (2.5)). In the general case, Λ_l can be constructed according to

Definition 6.3.18. Suppose that the sesqui-linear form a is $\mathscr{H}^1$-coercive relative to $\mathscr{H}^0$; take λ_0 from (3 b) and set

$$L' := \tfrac{1}{2}(L + L^*) + \lambda_0 I, \quad \Lambda := L'^{1/2}, \quad L_l' = R_l L' P_l, \quad \Lambda_l = L_l'^{1/2}. \tag{6.3.16}$$

Define the norm $|\cdot|_s$ on $\mathscr{U}_l$ and $\mathscr{F}_l$ by (15 a) with Λ_l from (16).

L_l' is not exactly of the form (2.26) as discussed below in § 6.3.1.5. If $L = L^* > O$, (3 b) holds with $\lambda_0 = 0$ and we regain $L_l' = L_l$.

Lemma 6.3.19. Suppose (1 a, b), (3 a, b), and define Λ and Λ_l by (16).

i) Λ^2 is a homomorphism of $\mathscr{H}^{-1}$ onto $\mathscr{H}^1$, i.e. $|\Lambda^2|_{-1\leftarrow 1} \leqq C$, $|\Lambda^{-2}|_{1\leftarrow -1} \leqq C$.

ii) The spaces $\mathscr{H}^s$ of the Hilbert scale $\{\mathscr{H}^s\colon -1 \leqq s \leqq 1\}$ can be endowed with the equivalent norms

$$|u|_s = |\Lambda^s u|_0 \quad \text{for } u \in \mathscr{H}^s, \quad s \in [-1, 1]. \tag{6.3.15 b}$$

Proof. (i) (1 a) implies $|I|_{-1\leftarrow 1} \leqq C_I$ for some C_I. Thus Note 6.3.1 and (3 a′) show $|L^*|_{-1\leftarrow 1} = |L|_{-1\leftarrow 1} \leqq C_L^*$ and $|\Lambda^2| \leqq C_L^* + |\lambda_0| C_I$. The estimate of Λ^{-2} is already stated in (3 b′).

(ii) Using part i), $|\Lambda u|_0$ and $|u|_1$ are equivalent norms. The same result holds for the dual norms $|\Lambda^{-1} u|_0$ and $|u|_{-1}$, while $|\Lambda^0 u|_0 = |u|_0$. By interpolation, $\mathscr{H}^s$ must have a norm equivalent to $|\Lambda^s u|_0$ for $|s| \leqq 1$ (cf. (1.4.7 b)). □

By virtue of the preceding lemma we may assume that the norm of $\mathscr{H}^s$, $-1 \leqq s \leqq 1$, is defined by (15 b); in particular, $|u|_{\pm 1} = |\Lambda^{\pm 1} u|_0$. In the sequel, this will be the general assumption.

The proof of Proposition 6.3.14 can be repeated for the actual norms if the estimates of (12 a) can be generalised to

$$|\hat{P}_l|_{s\leftarrow s} \leqq \hat{C}_p, \quad |\hat{P}_l|_{s^*\leftarrow s^*} \leqq \hat{C}_p^*. \tag{6.3.17}$$

From the proof mentioned above we have

$$|L_l^{-1} - p L_{l-1}^{-1} r|_{-s^*\leftarrow s} \leqq |\hat{R}_l|_{-s^*\leftarrow -s^*} |E_{l-1} - E_l|_{-s^*\leftarrow s} |\hat{P}_l|_{s\leftarrow s}.$$

Since $\hat{R}_l^* = \hat{P}_l$, Note 3.6.1 shows $|\hat{R}_l|_{-s^*\leftarrow -s^*} = |\hat{P}_l|_{s^*\leftarrow s^*} \leqq \hat{C}_p^*$. This proves

Lemma 6.3.20. (5 b), (11 b) and (17) imply the approximation property (1.8) with respect to $\|\cdot\|_{\mathscr{U}} = |\cdot|_{-s^*}$, $\|\cdot\|_{\mathscr{F}} = |\cdot|_s$ and with $\alpha = (2 + s + s^*)m$:

$$|L_l^{-1} - p L_{l-1}^{-1} r|_{-s^*\leftarrow s} \leqq C_A h_l^{(2+s+s^*)m}. \tag{6.3.18}$$

In the remaining part of this chapter we show that the desired estimates (17) with $s, s^* \in [-1, 0]$ are a consequence of the assumptions of Lemma 6.3.11 and the inverse assumption (5 c), which has not been used before. The result will be

Theorem 6.3.21. Let the norms be defined by (2.7) with $-1 \leqq s_{\mathscr{U}} \leqq 0 \leqq s_{\mathscr{F}} \leqq 1$. The approximation property (1.8) holds under following conditions: (1 a, b), (3 a, c′), (4), (5 a) for $t := 2 + s_{\mathscr{F}}$ and $t := 2 - s_{\mathscr{U}}$, (5 b, c), (6), (7), L $\mathscr{H}^{2+s_{\mathscr{F}}}$-regular, L^* $\mathscr{H}^{2-s_{\mathscr{U}}}$-regular.

The proof is prepared by a sequence of lemmata.

Lemma 6.3.22. Let the assumptions of Theorem 6.3.21 be valid. Then for $s = s_{\mathscr{F}}$ and $s = -s_{\mathscr{U}}$ there are projections Q_l onto $\mathscr{H}_l$ such that

$$|Q_l|_{1 \leftarrow 1} \leqq C, \quad |I - Q_l|_{-s \leftarrow 1} \leqq C h_l^{(s+1)m}. \tag{6.3.19}$$

Proof. a) Set $X_l := L(L^{-1} - P_l L_l^{-1} R_l)$. (3 a′) and Lemma 6.3.10 yield $|X_l|_{-1 \leftarrow s} \leqq C h_l^{(s+1)m}$ for $s = s_{\mathscr{F}}$; hence $|X_l^*|_{-s \leftarrow 1} \leqq C h_l^{(s+1)m}$. X_l^* equals $I - Q_l$, where $Q_l := P_l L_l^{*\,-1} R_l L^*$ is the required projection. For $s = -s_{\mathscr{U}}$ use $X_l := L^*(L^{*\,-1} - P_l L_l^{*\,-1} R_l)$.

b) An upper bound of $|Q_l|_{1 \leftarrow 1}$ is $|P_l|_{1 \leftarrow 1} |L_l^{*\,-1}|_{1 \leftarrow -1} |R_l|_{-1 \leftarrow -1} |L^*|_{-1 \leftarrow 1}$. (3 a′) implies $|L^*|_{-1 \leftarrow 1} \leqq C_l^*$, while (7) is equivalent to

$$|L_l^{-1}|_{1 \leftarrow -1} \leqq C_S^* \quad (C_S^* \text{ from (7)}). \tag{6.3.20 a}$$

It remains to show

$$|P_l|_{1 \leftarrow 1} = |R_l|_{-1 \leftarrow -1} = 1, \tag{6.3.20 b}$$

which follows from

$$|P_l u_l|_1^2 = |\Lambda P_l u_l|_0^2 = \langle u_l, R_l \Lambda^2 P_l u_l \rangle = \langle u_l, \Lambda_l^2 u_l \rangle = |\Lambda_l u_l|_0^2 = |u_l|_1^2. \quad \square$$

Lemma 6.3.23. Let a projection Q_l onto $\mathscr{H}_l$ satisfy (19) for some $s \in (-1, 1]$ and assume (5 c). Then the $\mathscr{H}^0$-orthogonal projection Q_l^0 onto $\mathscr{H}_l$ satisfies

$$|I - Q_l^0|_{t \leftarrow 1} \leqq C h_l^{(1-t)m} \quad \text{for } |t| \leqq 1, \tag{6.3.21 a}$$

$$|Q_l^0|_{1 \leftarrow 1} \leqq C. \tag{6.3.21 b}$$

Proof. a) Assume first $0 \leqq s \leqq 1$. (19) implies $|I - Q_l|_{t \leftarrow 1} \leqq C' h_l^{(1-t)m}$ for $t = 1$ and $t = -s \leqq 0$. By interpolation (cf. (1.4.10 b)) we obtain $|I - Q_l|_{0 \leftarrow 1} \leqq C' h_l^m$; hence, (19) holds also for $s = 0$.

b) By virtue of a), we may assume that (19) holds for some $s \in (-1, 0]$. Let Q_l^t be the $\mathscr{H}^t$-orthogonal projection onto $\mathscr{H}_l$. Since $|I - Q_l^{-s}|_{-s \leftarrow 1} \leqq |I - Q_l|_{-s \leftarrow 1}$, Proposition 1.4.5 can be applied with $\eta := C h_l^{(s+1)m}$ and yields (21 a) for $-1 \leqq t \leqq -s$.

c) Let $u \in \mathscr{H}^1$ and set $v := (Q_l^0 - Q_l) u \in \mathscr{H}_l$. $|v|_{-s} \leqq |(I - Q_l) u|_{-s} + |(I - Q_l^0) u|_{-s} \leqq C' h_l^{(s+1)m} |u|_1$ and (5 c) prove $|v|_1 \leqq C'' |u|_1$, since (5 c) implies $|u|_1 \leqq (C_i h_l^{-m})^{1+s} |u|_{-s}$, $u \in \mathscr{H}_l$ (cf. (1.4.9 d)). (21 b) follows from $|Q_l^0 u|_1 \leqq |Q_l u|_1 + |v|_1 \leqq (C + C'') |u|_1$. It is equivalent to (21 a) with $t = 1$. Interpolation between $t = 1$ and $t = -s$ completes the proof. $\square$

The projection Q_l^0 can be represented by $Q_l^0 = P_l \hat{R}_l$. From

$$|Q_l^0|_{1\leftarrow 1}^2 = \|\Lambda P_l \hat{R}_l \Lambda^{-1}\|^2 = \|\Lambda^{-1} \hat{P}_l R_l \Lambda^2 P_l \hat{R}_l \Lambda^{-1}\|$$
$$= \|\Lambda^{-1} \hat{P}_l \Lambda_l^2 \hat{R}_l \Lambda^{-1}\| = \|\Lambda_l \hat{R}_l \Lambda^{-1}\|^2 = |\hat{R}_l|_{1\leftarrow 1}^2$$

one concludes the first part of

Lemma 6.3.24. (i) $|Q_l^0|_{1\leftarrow 1} = |\hat{R}_l|_{1\leftarrow 1}$.
(ii) (21 b) and (6) imply (17) for $-1 \leqq s, s^* \leqq 0$ and

$$\underline{C}_p^{-1} |u_{l-1}|_s \leqq |p\, u_{l-1}|_s \leqq \bar{C}_p |u_{l-1}|_s \quad \text{for all } 0 \leqq s \leqq 1, \quad u_{l-1} \in \mathscr{U}_{l-1}, \tag{6.3.20 c}$$

$$|r|_{s\leftarrow s} \leqq \bar{C}_r, \quad \inf\{|u_l|_s \colon u_l \in \mathscr{U}_l, r\, u_l = u_{l-1}\} \leqq \underline{C}_r |u_{l-1}|_s \quad \text{for all } u_{l-1} \in \mathscr{U}_{l-1}, \quad -1 \leqq s \leqq 0. \tag{6.3.20 d}$$

Proof of (ii). Interpolation of (12 a), $|\hat{R}_l|_{0\leftarrow 0} \leqq \underline{C}_p$ and $|\hat{R}_l|_{1\leftarrow 1} \leqq C$ yields $|\hat{R}_l|_{t\leftarrow t} \leqq C^t \underline{C}_p^{1-t}$ for all $0 \leqq t \leqq 1$ (cf. (1.4.10 b)). Since $|\hat{P}_l|_{-t\leftarrow -t} = |\hat{R}_l|_{t\leftarrow t}$, (17) is valid. For p and r compare Exercise 3.9.9. □

Proof of Theorem 6.3.21. a) In the less interesting case of $s_{\mathscr{F}} = -1$, $s_{\mathscr{U}} = 1$ the result follows from $|r|_{-1\leftarrow -1} = |p|_{1\leftarrow 1} = 1$ and $|L_l^{-1}|_{1\leftarrow -1} = |L_{l-1}^{-1}|_{1\leftarrow -1} \leqq C_S^*$.

b) Otherwise, either $s_{\mathscr{F}} > -1$ or $s_{\mathscr{U}} < 1$ must hold and Lemmata 6.3.22–23 –24–11 prove (18) with $s = s_{\mathscr{F}}$, $s^* = -s_{\mathscr{U}}$. □

6.3.1.4 The Very Regular Case

In the previous subsection L and L^* were assumed to be t-regular, where $1 \leqq t < 2$. The standard case $t = 2$ has been considered in § 6.3.1.2. It remains to discuss the case of $t > 2$. This higher regularity must be accompanied with higher order finite elements.

To prove (17) for $s, s^* \in (0, 1]$, it is sufficient to have

$$|\hat{P}_l|_{1\leftarrow 1} \leqq C. \tag{6.3.22 a}$$

Lemma 6.3.25 (i) $|\hat{P}_l|_{1\leftarrow 1} = |(R_l P_l)^{-1}|_{1\leftarrow 1}$.
(ii) (22 a) and (12 a) imply (17) for all $0 \leqq s, s^* \leqq 1$.
(iii) (22 a) and (21 b) yield (17) for all $-1 \leqq s, s^* \leqq 1$.

Proof. (i) $|\hat{P}_l|_{1\leftarrow 1} = \|\Lambda P_l (R_l P_l)^{-1} \Lambda_l^{-1}\| = \|\Lambda_l (R_l P_l)^{-1} \Lambda_l^{-1}\| = |(R_l P_l)^{-1}|_{1\leftarrow 1}$.
(ii) Apply (1.4.10 b) to $|\hat{P}_l|_{1\leftarrow 1} \leqq C$, $|\hat{P}_l|_{0\leftarrow 0} \leqq \underline{C}_P$.
(iii) (21 b) implies $|\hat{P}_l|_{-1\leftarrow -1} \leqq C$ (cf. Lemma 6.3.24). Continue as in (ii). □

Using part (ii), condition (17) of Lemma 6.3.20 is satisfied and one obtains

Theorem 6.3.26. Let the norms be given by (2.7) with $s_{\mathscr{U}}, s_{\mathscr{F}} \in [-1, +1]$. Under the conditions of Theorem 6.3.21 and the additional assumption (22 a), the approximation property (1.8) holds.

Inequality (22 a) is equivalent to $|u_l|_{-1} \leqq C\,|P_l u_l|_{-1}$ for all $u_l \in \mathscr{U}_l$. To obtain full equivalence of the norms we need in addition

$$|P_l|_{-1\leftarrow -1} \leqq C. \tag{6.3.22 b}$$

The proof of the following lemma is left to the reader.

Lemma 6.3.27. (i) (22 b) is equivalent to $|P_l R_l|_{1\leftarrow 1} \leqq C$.
(ii) (22 a, b) implies (21 b), i.e. $|Q_l^0|_{1\leftarrow 1} \leqq C$.
(iii) (22 a, b) holds if and only if

$$C^{-1}\,|u_l|_s \leqq |P_l u_l|_s \leqq C\,|u_l|_s \quad \text{for all } u_l \in \mathscr{U}_l, \quad |s| \leqq 1, \quad l \geqq 0. \tag{6.3.22 c}$$

(iv) (22 c) implies (20 c, d) even for $s \in [-1, 1]$.

6.3.1.5 The Smoothing Property Revisited

The results of this chapter are also useful for the analysis of the smoothing property. For instance, we often required $\|h_l^{2m} L_l\|^{-1} \geqq \omega_l \geqq 1/\text{const}$ (cf. Proposition 6.2.11), which is equivalent to

$$\|L_l\| \leqq C_L h_l^{-2m}. \tag{6.3.23}$$

Proposition 6.3.28. (3 a), (5 c) and (12 a) imply (23) with $C_L := C_L^* C_i^2 \bar{C}_P^2$. More generally, $|L_l|_{s\leftarrow t} \leqq C_L^* (C_i \bar{C}_P h_l^{-m})^{2+s-t}$ holds for any $s, t \in [-1, 1]$.

Proof. a) $|L_l|_{-1\leftarrow 1}$ equals $\sup\{|\langle u_l, L_l v_l\rangle| : |u_l|_1 = |v_l|_1 = 1\}$. (3 a) proves that $|\langle u_l, L_l v_l\rangle| = |a(P_l u_l, P_l v_l)| \leqq C_L^* |P_l u_l|_1 |P_l v_l|_1 = C_L^* |u_l|_1 |v_l|_1 = C_L^*$.

b) Let I be the identity on $\mathscr{U}_l$. The desired estimate follows from $|L_l|_{s\leftarrow t} = |I|_{s\leftarrow -1} |L_l|_{-1\leftarrow 1} |I|_{1\leftarrow t}$ by means of the next lemma. □

Lemma 6.3.29. $|I|_{s\leftarrow t} \leqq (C_i \bar{C}_P h_l^{-m})^{s-t}$, $-1 \leqq t \leqq s \leqq 1$, holds for the identity $I: \mathscr{U}_l \to \mathscr{U}_l$ with C_i from (5 c), $\bar{C}_P$ from (12 a).

Proof. a) (5 c) shows $|u_l|_1 = |P_l u_l|_1 \leqq C_i h_l^{-m} |P_l u_l|_0 \leqq C_i \bar{C}_P h_l^{-m} |u_l|_0$; hence the estimate holds for $s = 1$, $t = 0$.

a) The case $s = 1, t = -1$ follows from $|I|_{1\leftarrow -1} \leqq |I|_{1\leftarrow 0} |I|_{0\leftarrow -1} = |I|_{1\leftarrow 0}^2$.

c) Interpolation between $|I|_{1\leftarrow -1} \leqq \dots$ and $|I|_{r\leftarrow r} = 1$ for $r = (t+s)/(2+t-s)$ yields the estimate of $|I|_{s\leftarrow t}$. □

For a general L we constructed Λ and Λ_l to define the basic norms. In § 6.2.3.4 we used $\Lambda_l = L_l'^{1/2}$ with $L_l' = (L_l + L_l^*)/2 + \lambda I$ (cf. (2.26), whereas in § 6.3.1.3 the choice $L_l' = R_l L' P_l = (L_l + L_l^*)/2 + \lambda_0 R_l P_l$ (cf. (16)) was more convenient. Although the norms involved in the smoothing and approximation property must coincide, this difference does not matter because of

Lemma 6.3.30. Assume inequality (6). The norms $|u_l|_s^I = |\Lambda_{l,I}^s u_l|_0$ and $|u_l|_s^{II} = |\Lambda_{l,II}^s u_l|_0$ with $\Lambda_{l,I}^2 = \frac{1}{2}(L_l + L_l^*) + \lambda I$ and $\Lambda_{l,II}^2 = \frac{1}{2}(L_l + L_l^*) + \lambda_0 R_l P_l$ are uniformly equivalent:

$$C^{-1}\,|u_l|_s^I \leqq |u_l|_s^{II} \leqq C\,|u_l|_s^I \quad \text{for all } -1 \leqq s \leqq 1, \quad u_l \in \mathscr{U}_l, \quad l \geqq 0.$$

Proof. a) $|u_l|_1^I \leqq C_1 |u_l|_1^{II}$ is equivalent to $\Lambda_{l,I}^2 \leqq C_1 \Lambda_{l,II}^2$ or $\Lambda_{l,II}^{-1} \Lambda_{l,I}^2 \Lambda_{l,II}^{-1} \leqq C_1$. Since $\Lambda_{l,II}^{-1} \Lambda_{l,I}^2 \Lambda_{l,II}^{-1} = I - \Lambda_{l,II}^{-1}(\lambda_0 R_l P_l - \lambda I) \Lambda_{l,II}^{-1}$ and $\|\Lambda_{l,II}^{-2}\| \leqq \underline{C}_P^2/\lambda_0$, the desired inequality holds with $C_1 = 1 + (\lambda/\lambda_0 + \bar{C}_P^2)\,\underline{C}_P^2$.

b) Analogously, one proves $|u_l|_1^{II} \leqq C_2 |u_l|_1^I$ with $C_2 = 2 + \lambda_0 \bar{C}_P^2/\lambda$.

c) From a), b) one deduces $C_1^{-1}|u_l|_{-1}^{II} \leqq |u_l|_{-1}^{I} \leqq C_2 |u_l|_{-1}^{II}$ for the dual norms.

d) Interpolation yields the assertion of the lemma with $C = \max\{C_1, C_2\}$. □

Finally we discuss the estimate (2.29) of $L_l'' = \frac{1}{2}(L_l - L_l^*) - \lambda_0 R_l P_l$. This inequality guarantees the smoothing property of the damped Jacobi iteration (cf. Corollary 6.2.18) and it is the main condition for the smoothing property of the Gauß-Seidel iteration (cf. Proposition 6.2.32). If the coefficients of the sesqui-linear form a are smooth enough, the anti-symmetric part $b := \frac{1}{2}(a - a^*) - \lambda_0$ corresponds to an operator L'' containing only derivatives of order $\leqq 2m - 1$. Therefore, $|L''|_{-1\leftarrow 1-1/m}$ is bounded and implies $|L''|_{-1\leftarrow 1-1/m} \leqq C$. Lemma 6.3.29 shows that $\|L_l''\| \leqq C h_l^{1-2m}$ holds for $L_l'' = \frac{1}{2}(L_l - L_l^*) - \lambda_0 R_l P_l$ as well as for $L_l'' = \frac{1}{2}(L_l - L_l^*) - \lambda I$. Hence condition (2.29) is satisfied.

6.3.2 The Approximation Property of Finite Difference Schemes

The previous technique of proof used the error estimate $|u - P_l u_l|_{s_u} \leqq C h_l^{(2+s_f-s_u)m} |f|_{s_f}$, e.g. $|u - P_l u_l|_0 \leqq C h_l^{2m} |f|_0$, which is equivalent to (11 b). This is not the common form of error estimates of finite difference solutions u_l, although as a result of the following theory such estimates can be proved.

The key to the proof is the discrete regularity of L_l, the discrete counterpart of the regularity from Definition 6.3.5. In the foregoing section the difference $L_l^{-1} - p L_{l-1}^{-1} r$ was estimated by means of $L^{-1} - P_l L_l^{-1} R_l$ and $L^{-1} - P_{l-1} L_{l-1}^{-1} R_{l-1}$. In the following we shall not use the triangle inequality, but compare L_l^{-1} and $p L_{l-1}^{-1} r$ directly.

6.3.2.1 The Discrete Regularity

The uniform boundedness of $\|L_l^{-1}\|$ is well-known as 'stability'. However, too little attention is paid to the fact that even $|L_l^{-1}|_{s+2\leftarrow s}$ may be uniformly bounded, where $|\cdot|_s$ is a norm involving differences of order $s \cdot m$. It is not necessary to define a scale of norms $\{|\cdot|_s : s \in [-t, t]\}$. Only three or four different values of s will appear.

For $s = 0$ the norm $|\cdot|_0$ can be chosen to be the Euclidean norm associated with the scalar product (3.5.2):

$$|u_l|_0 := [h_l^d \sum_{\mathbf{x} \in \Omega_l} |u_l(\mathbf{x})|^2]^{1/2}, \tag{6.3.24 a}$$

where Ω_l is the set of grid points.

$\mathscr{H}^1$ equals $H_0^1(\Omega)$ if we consider Dirichlet boundary conditions and if $m=1$. Then $|u_l|_1$ can easily be defined by

$$|u_l|_1 := [\sum_{|\alpha|\leqq 1} |\partial^\alpha \tilde{u}_l|_0^2]^{1/2}; \quad \tilde{u}_l := \begin{cases} u_l & \text{in} \quad \Omega_l, \\ 0 & \text{in} \quad Q_l \backslash \Omega_l, \end{cases} \tag{6.3.24 b}$$

where $|\cdot|_0$ is the norm in the infinite grid Q_l from (3.1.13). The sum is taken over multi-indices $\alpha = (\alpha_1, \ldots, \alpha_d)$ with $\alpha_j \geqq 0$ and $|\alpha| = \alpha_1 + \ldots + \alpha_d$. ∂^α and T^α are difference and transition operators:

$$T^\alpha = T_1^{\alpha_1} \ldots T_d^{\alpha_d}, \quad T_j^k u(x_1, \ldots, x_j, \ldots, x_d) = u(x_1, \ldots, x_j + k h_l, \ldots, x_d), \tag{6.3.25 a}$$

$$\partial^\alpha = \partial_1^{\alpha_1} \ldots \partial_d^{\alpha_d}, \quad \partial_j := h^{-1}(I - T_j^{-1}). \tag{6.3.25 b}$$

Replacing $|\alpha| \leqq 1$ in (24 b) by $|\alpha| \leqq k$, one obtains the norm $|\cdot|_k$ corresponding to the norm of $H_0^k(\Omega)$. The norm $|\cdot|_s$ corresponding to $H_0^{sm}(\Omega)$, $s \neq \frac{1}{2}$ + integer, can be defined even for all real $s \geqq 0$ by means of a Fourier transform:

$$|u_l|_s := |v|_{L^2([-\pi,\pi]^d)}, \quad \text{with } s \geqq 0, \tag{6.3.24 c}$$

$$v(\xi) := [1 + h_l^{-2} |\xi|^2]^{s/2} \hat{u}(\xi), \quad \hat{u}(\xi) = (h_l/2\pi)^{d/2} \sum_{\mathbf{x}\in\Omega_l} u_l(\mathbf{x}) e^{i\mathbf{x}\xi/h_l},$$

where $\mathbf{x}\xi = x_1\xi_1 + \ldots + x_d\xi_d$.

It is more difficult, e.g., to define a discrete norm $|\cdot|_2$ corresponding to $H^2(\Omega) \cap H_0^1(\Omega)$. The divided differences ∂^α, $|\alpha| = 2$, are to be applied only at grid points $\mathbf{x}$ such that the neighbouring grid points involved in $(\partial^\alpha u_l)(\mathbf{x})$ belong to Ω_l, whereas $\partial^\alpha \tilde{u}_l$, $|\alpha| \leqq 1$, has to be summed over all Ω_l. If irregular boundary points occur (cf. Fig. 3.1.1), special second order differences are needed to define $|\cdot|_2$ (cf. Hackbusch [32]).

For negative index the norm $|\cdot|_{-s}$ is defined as the dual norm of $|\cdot|_s$:

$$|u_l|_{-s} := \sup\{|\langle u_l, v_l\rangle| : v_l \in \mathscr{U}_l, |v_l|_s = 1\}. \tag{6.3.24 d}$$

Definition 6.3.31. A family $\{L_l : l = 0, 1, \ldots\}$ is called t-regular, if there exists some C_R such that

$$|L_l^{-1}|_{t\leftarrow t-2} \leqq C_R \quad \text{for all } l \geqq 0. \tag{6.3.26}$$

In the sequel we shall say that 'L_l is t-regular' for brevity.

The 2-regularity of the five-point scheme (3.1.15) can easily be shown if Ω is a rectangle. Severe problems arise for curvi-linear boundaries. As a typical difference scheme with irregular discretisations near the boundary, we consider the Shortley-Weller formula (cf. (3.1.22)).

Theorem 6.3.32 (Hackbusch [32]). For sufficiently smooth boundary and suitably defined $|\cdot|_2$, the Shortley-Weller discretisation of the Poisson equation yields 2-regular operators L_l.

A similar statement for a special difference scheme in a non-uniform convex grid Ω_l is proved by Dryja [1].

If Γ is not smooth enough, the five-point scheme cannot be 2-regular, since the Laplace operator is not $\mathscr{H}^2$-regular. However, the analogue of Theorem 6.3.17 can still be proved:

Theorem 6.3.33 (Hackbusch [23]). Let Ω be a Lipschitz domain and L_l be the difference scheme

$$L_l = \sum_{|\alpha|,|\beta| \leqq m} \sum_{\gamma,\delta \in \mathbb{Z}^d} \partial^\alpha T^\gamma c_{\alpha\beta\gamma\delta} T^\delta \partial^\beta$$

with T and ∂ from (25). $c_{\alpha\beta\gamma\delta} = c_{\alpha\beta\gamma\delta}(\mathbf{x}, h_l)$ are real coefficients vanishing for almost all indices. Let $1 \leqq t < \frac{3}{2}$. Assume

$$|c_{\alpha\beta\gamma\delta}(\mathbf{x}, h)| \leqq C \quad \text{for all } \mathbf{x} \in \mathbb{R}^d, \quad 0 < h \leqq h_0,$$

$$c_{\alpha\beta\gamma\delta} \in \bar{C}^\kappa(\mathbb{R}^d \times [0, h_0]) \quad \text{if} \begin{Bmatrix} |\alpha| = m \text{ and } t > 1 \\ |\alpha| = |\beta| = m, \ t = 1 \end{Bmatrix},$$

where either $\kappa > t - 1 > 0$ or $\kappa \geqq t - 1 = 0$. Suppose that L_l is elliptic, i.e. $\lambda(\mathbf{x}, \xi) \geqq \varepsilon |\xi|^{2m}$ for all $\mathbf{x} \in \mathbb{R}^d$, $|\xi_j| \leqq \pi$, where $\varepsilon > 0$ and

$$\lambda(\mathbf{x}, \xi) = \sum_{|\alpha| = |\beta| = m} \sum_{\gamma,\delta \in \mathbb{Z}^d} c_{\alpha\beta\gamma\delta}(\mathbf{x}, 0)\, e^{-i\xi(\gamma+\delta)} \prod_{j=1}^{d} (1 - e^{i\xi_j})^{\alpha_j + \beta_j}.$$

For all grid points $\mathbf{x} \in \Omega_l$, $(L_l u_l)(\mathbf{x})$ is defined by $(L_l \tilde{u}_l)(\mathbf{x})$, where $\tilde{u}_l = u_l$ on Ω_l, $\tilde{u}_l = 0$ outside. Then, the stability $\|L_l^{-1}\| \leqq C$ implies t-regularity of L_l.

6.3.2.2 Criteria

If L_{l-1} is defined according to (3.7.1), the quantity

$$\delta_{l-1} := r L_l p - L_{l-1} \tag{6.3.27}$$

vanishes. Otherwise, $\delta_{l-1} u_{l-1}$ should equal $O(h_{l-1}^\kappa)$ (κ: consistency order) for smooth u_{l-1}, since L_l and L_{l-1} are consistent discretisations of the same differential operator. E.g., a reasonable assumption is

$$|\delta_{l-1}|_{s_{\mathscr{U}}-2 \leftarrow s_{\mathscr{F}}+2} \leqq C_\delta h_l^{(s_{\mathscr{F}} - s_{\mathscr{U}} + 2)m} \tag{6.3.28 a}$$

for some $s_{\mathscr{F}}$, $s_{\mathscr{U}} < s_{\mathscr{F}} + 2$, where of course $(s_{\mathscr{F}} - s_{\mathscr{U}} + 2)m$ cannot exceed the consistency order κ. If $\kappa = 2m$, we may, in particular, assume

$$|\delta_{l-1}|_{-2 \leftarrow 2} \leqq C_\delta h_l^{2m}. \tag{6.3.28 a*}$$

Here $|\cdot|_t$ denotes the norms discussed above in § 6.3.2.1. If L_l is $s_{\mathscr{U}}$-regular and $(2 + s_{\mathscr{F}})$-regular, one has

$$|L_l^{-1}|_{s_{\mathscr{U}} \leftarrow s_{\mathscr{U}}-2} \leqq C_R, \quad |L_l^{-1}|_{s_{\mathscr{F}}+2 \leftarrow s_{\mathscr{F}}} \leqq C_R. \tag{6.3.28 b}$$

$s_{\mathscr{U}}$-regularity of L_l is equivalent to $(2 - s_{\mathscr{U}})$-regularity of L_l^* (cf. (1.5.2)). Therefore, 2-regularity of L_l and L_l^* implies

$$|L_l^{-1}|_{0 \leftarrow -2} \leqq C_R, \quad |L_l^{-1}|_{2 \leftarrow 0} \leqq C_R. \tag{6.3.28 b*}$$

Further natural assumptions are (28 c) or (28 c*), respectively:

$$|L_l|_{s_{\mathscr{U}}-2 \leftarrow s_{\mathscr{U}}} \leqq C_L, \tag{6.3.28 c}$$

$$|L_l|_{-2 \leftarrow 0} = |L_l^*|_{0 \leftarrow 2} \leqq C_L. \tag{6.3.28 c*}$$

The interpolation error of p can be described by the difference $I - p r'$ with a suitable restriction $r' : \mathscr{U}_l \to \mathscr{U}_{l-1}$. Assume

$$|I - p r'|_{s_{\mathscr{U}} \leftarrow s_{\mathscr{F}}+2} \leqq C_I h_l^{(s_{\mathscr{F}} - s_{\mathscr{U}} + 2)m}. \tag{6.3.28 d}$$

If, e.g., p is the piecewise linear interpolation from § 3.4.2, if $m = 1$ and $r' = r_{inj}$ from (3.5.1), one has

$$|I - p r'|_{0 \leftarrow 2} \leqq C_I h_l^{2m}. \tag{6.3.28 d*}$$

r' as well as r and p have to satisfy the stability conditions

$$|r|_{s_{\mathscr{U}}-2 \leftarrow s_{\mathscr{U}}-2} \leqq C_r, \quad |r'|_{s_{\mathscr{F}}+2 \leftarrow s_{\mathscr{F}}+2} \leqq C_r', \quad |p|_{s_{\mathscr{U}} \leftarrow s_{\mathscr{U}}} \leqq C_p. \tag{6.3.28 e}$$

For $s_{\mathscr{U}} = s_{\mathscr{F}} = 0$ these inequalities become

$$|r|_{-2 \leftarrow -2} \leqq C_r, \quad |r'|_{2 \leftarrow 2} \leqq C_r', \quad |p|_{0 \leftarrow 0} \leqq C_p \tag{6.3.28 e*}$$

which hold, e.g., for r from (3.5.4) or (3.5.6) and for $r' = r_{inj}$, if $m = 1$.

Remark 6.3.34. Define $\|\cdot\|_{\mathscr{U}} := |\cdot|_{s_{\mathscr{U}}}$, $\|\cdot\|_{\mathscr{F}} := |\cdot|_{s_{\mathscr{F}}}$. The assumptions (28 a–e) imply

$$\|L_{l-1}^{-1} \delta_{l-1} r' L_l^{-1}\|_{\mathscr{U} \leftarrow \mathscr{F}} \leqq C_1 h_l^{\alpha}, \tag{6.3.29 a}$$

$$\|L_{l-1}^{-1} r L_l\|_{\mathscr{U} \leftarrow \mathscr{U}} \leqq C_2, \quad \|p\|_{\mathscr{U} \leftarrow \mathscr{U}} \leqq C_p, \tag{6.3.29 b}$$

$$\|[I - p r'] L_l^{-1}\|_{\mathscr{U} \leftarrow \mathscr{F}} \leqq C_3 h_l^{\alpha} \tag{6.3.29 c}$$

with $C_1 = C_R^2 C_\delta C_r'$, $C_2 = C_R C_r C_L$, $C_3 = C_I C_R$, and

$$\alpha = (s_{\mathscr{F}} - s_{\mathscr{U}} + 2)m. \tag{6.3.29 d}$$

Criterion 6.3.35. Suppose that there exists some $r' : \mathscr{U}_l \to \mathscr{U}_{l-1}$ such that (27), (29 a, b, c) are valid with an $\alpha \geqq 0$ for all l. Then the approximation property (1.8) holds with $C_A = C_h^{\alpha}[(1 + C_p C_2) C_3 + C_p C_1]$.

Proof. From $(I - p L_{l-1}^{-1} r L_l) p = - p L_{l-1}^{-1} \delta_{l-1}$ one concludes that

$$\begin{aligned}
&\|L_l^{-1} - p L_{l-1}^{-1} r\|_{\mathscr{U} \leftarrow \mathscr{F}} = \|(I - p L_{l-1}^{-1} r L_l) L_l^{-1}\|_{\mathscr{U} \leftarrow \mathscr{F}} \\
&= \|(I - p L_{l-1}^{-1} r L_l)(I - p r') L_l^{-1} - p L_{l-1}^{-1} \delta_{l-1} r' L_l^{-1}\|_{\mathscr{U} \leftarrow \mathscr{F}} \\
&\leqq \|I - p L_{l-1}^{-1} r L_l\|_{\mathscr{U} \leftarrow \mathscr{U}} \|(I - p r') L_l^{-1}\|_{\mathscr{U} \leftarrow \mathscr{F}} \\
&\quad + \|p\|_{\mathscr{U} \leftarrow \mathscr{U}} \|L_{l-1}^{-1} \delta_{l-1} r' L_l^{-1}\|_{\mathscr{U} \leftarrow \mathscr{F}} \\
&\leqq [(1 + C_p C_2) C_3 + C_p C_1] h_l^{\alpha}.
\end{aligned}$$

□

Note that α from (29 d) is the same value as required in (6.2.9) for the smoothing property. Remark 6.3.34 implies

Corollary 6.3.36. Sufficient conditions of the approximation property are (27), (28 a–e), $s_{\mathscr{U}} \leqq s_{\mathscr{F}} + 2$. If $L_{l-1} = r L_l p$, the estimate (28 e) of r' is not needed.

The value of α is bounded by the consistency order of the discretisation (cf. (29 a)) and by the order of the interpolation p (cf. (29 c)). However, if $\delta_{l-1} = 0$ the consistency order does not matter:

Note 6.3.37. By the choice $L_{l-1} = r L_l p$ one can achieve the approximation property with α larger than the consistency order of the discretisation.

The smoothing property requires $\alpha = (s_{\mathscr{F}} - s_{\mathscr{U}} + 2)m > 0$ (cf. Proposition 6.2.12, Criterion 6.2.30, or (2.3)). The choice $s_{\mathscr{F}} = -1$, $s_{\mathscr{U}} = 1$ is excluded. Hence, the simple 1-regularity of L_l is not sufficient. Another factor determines the choice of p and r. Assume that p and r^* are piecewise interpolations by polynomials of the respective degrees $m_p - 1$ and $m_r - 1$. Condition (28 e) requires $(2 - s_{\mathscr{U}})m \leqq m_r$ and $s_{\mathscr{U}} m \leqq m_p$. Hence,

$$m_r + m_p \geqq 2m$$

follows. The assumption $m_r + m_p = 2m$ would imply $s_{\mathscr{U}} m = m_p$. The estimation of $\delta_{l-1} = r L_l p - L_{l-1}$ by (28 a) can hardly be proved if $|p|_{s_{\mathscr{F}}+2 \leftarrow s_{\mathscr{F}}+2} \leqq C$ does not hold. The latter inequality requires $m_p \geqq (s_{\mathscr{F}} + 2)m$ in contradiction to $m_p = s_{\mathscr{U}} m = (s_{\mathscr{F}} + 2)m - \alpha < (s_{\mathscr{F}} + 2)m$ as $\alpha > 0$. Another argument [the $(s_{\mathscr{F}} + 2)$-regularity of L_l] can be applied if $\delta_{l-1} = 0$. Thus, the inequality

$$m_r + m_p > 2m$$

should be satisfied as stated in Note 3.5.1.

A dual version of Criterion 6.3.35 can be proved by means of the identity $r[I - L_l p L_{l-1}^{-1} r] = -\delta_{l-1} L_{l-1}^{-1} r$.

Criterion 6.3.38. Suppose that there exists some prolongation $p': \mathscr{F}_{l-1} \to \mathscr{F}_l$ with

$$\| L_l^{-1} p' \delta_{l-1} L_{l-1}^{-1} \|_{\mathscr{U} \leftarrow \mathscr{F}} \leqq C_1^* h_l^{\alpha}, \tag{6.3.30 a}$$

$$\| L_l p L_{l-1}^{-1} \|_{\mathscr{F} \leftarrow \mathscr{F}} \leqq C_2^*, \qquad \| r \|_{\mathscr{F} \leftarrow \mathscr{F}} \leqq C_r, \tag{6.3.30 b}$$

$$\| L_l^{-1} (I - p' r) \|_{\mathscr{U} \leftarrow \mathscr{F}} \leqq C_3^* h_l^{\alpha}, \tag{6.3.30 c}$$

and assume (27), and $\alpha \geqq 0$. Then the approximation property (1.8) holds.

Remark 6.3.39. Let $\| \cdot \|_{\mathscr{U}} = | \cdot |_{s_{\mathscr{U}}}$, $\| \cdot \|_{\mathscr{F}} = | \cdot |_{s_{\mathscr{F}}}$. The conditions (30 a–c) with $\alpha = (s_{\mathscr{F}} - s_{\mathscr{U}} + 2)m$ follow from (28 a, b) and

$$\begin{aligned} &|L_l|_{s_{\mathscr{F}} \leftarrow s_{\mathscr{F}}+2} \leqq C_L, \quad |I - p' r|_{s_{\mathscr{U}}-2 \leftarrow s_{\mathscr{F}}} \leqq C_I h_l^{(s_{\mathscr{F}} - s_{\mathscr{U}} + 2)m}, \\ &|p'|_{s_{\mathscr{U}}-2 \leftarrow s_{\mathscr{U}}-2} \leqq C_p', \quad |p|_{s_{\mathscr{F}}+2 \leftarrow s_{\mathscr{F}}+2} \leqq C_p, \quad |r|_{s_{\mathscr{F}} \leftarrow s_{\mathscr{F}}} \leqq C_r. \end{aligned} \tag{6.3.31}$$

A combination of both criteria is given for the special case of $s_{\mathscr{U}} = 1 - \vartheta$, $s_{\mathscr{F}} = -1 + \vartheta$. Note that Theorem 6.3.33 yields no better regularity than $(1 + \vartheta)$-regularity with $|\vartheta| < 1/(2m)$.

Criterion 6.3.40. Let $|u_l|_s = |\Lambda_l^s u_l|_0$ for $|s| \leqq 1 + \vartheta$ with $\|\Lambda_l\| \leqq C h_l^{-m}$. The approximation property (1.8) holds with $\alpha = 2\vartheta m$ and $\|\cdot\|_{\mathscr{F}} = |\cdot|_{-1+\vartheta}$, $\|\cdot\|_{\mathscr{U}} = |\cdot|_{1-\vartheta}$, if the following conditions are satisfied: $\vartheta \geqq 0$, (27), there are p' and r' such that the quantities $|L_l^{-1}|_{1\pm\vartheta\leftarrow-1\pm\vartheta}$, $|L_l|_{-1\pm\vartheta\leftarrow 1\pm\vartheta}$, $|p'|_{-1\leftarrow-1}$, $|r'|_{1\leftarrow 1}$, $|p|_{1\pm\vartheta\leftarrow 1\pm\vartheta}$, $|r|_{-1\pm\vartheta\leftarrow-1\pm\vartheta}$ are uniformly bounded, while

$$|\delta_{l-1}|_{-1-\vartheta\leftarrow 1+\vartheta} \leqq C_\delta h_l^{2\vartheta m}, \qquad |I - p' r|_{-1\leftarrow\vartheta-1} \leqq C_I h_l^{\vartheta m},$$
$$|I - p r'|_{1-\vartheta\leftarrow 1} \leqq C_I h_l^{\vartheta m}.$$

Proof. Set $C_l := L_l^{-1} - p L_{l-1}^{-1} r$. The assumptions imply $|C_l|_{1\leftarrow-1+\vartheta} \leqq C h_l^{\vartheta m}$ and $|C_l|_{1-\vartheta\leftarrow-1} \leqq C h_l^{\vartheta m}$ by means of Criteria 6.3.35, 6.3.38. The estimate of $|C_l|_{1-\vartheta\leftarrow-1+\vartheta}$ can be obtained from the identity

$$C_l = C_l L_l C_l + p L_{l-1}^{-1} \delta_{l-1} L_{l-1}^{-1} r. \qquad \square$$

For a curvi-linear boundary one needs extra definitions of p and r near the boundary. It is not necessary to require the same order of interpolation. In the case of a Dirichlet problem, even very simple definitions (e.g., $p = 0$ near the boundary) are sufficient. The weak conditions of Criterion 6.3.40 are satisfied if $0 < \vartheta < 1/(2m)$ (cf. Hackbusch [13]).

6.4 The Symmetric Case

6.4.1 Quantitative Analysis

The classical iterations admit a simple convergence proof if L_l is positive definite:

$$L_l = L_l^* > 0 \tag{6.4.1 a}$$

In this case, also the two-grid analysis can be improved, provided that r and p satisfy

$$r = p^* \tag{6.4.1 b}$$

(cf. (3.5.3)). For ease of exposition we assume in addition (3.7.1):

$$L_{l-1} = r L_l p. \tag{6.4.1 c}$$

Remark 6.4.1. If $a(.,.)$ is an $\mathscr{H}^1$-elliptic form, the finite element approach leads to L_l, r, p satisfying (1 a–c).

Also the smoothing iteration must satisfy a symmetry condition. $S_l = S_l^*$ is not required, but instead

$$S_l = I - W_l^{-1} L_l, \qquad W_l = W_l^* \geqq L_l. \tag{6.4.1 d}$$

Remark 6.4.2. The damped Jacobi iterations (3.3.6), (3.3.7), (3.3.10), as well as the symmetric Gauß-Seidel iteration (3.3.5) satisfy (1 d) (cf. § 6.2.4.3).

First we consider the two-grid iteration $TGM^{(\nu_1,\nu_2)}$ from (3.2.2) with $\nu_1 = \nu_2 = \nu/2$; i.e., the number of pre- and post-smoothing steps is equal. The approximation property is required with respect to the norms $\|u_l\|_{\mathscr{W}} := |W_l^{1/2} u_l|_0$, $\|f_l\|_{\mathscr{W}^*} = |W_l^{-1/2} f_l|_0$ instead of $\|\cdot\|_{\mathscr{U}}$ and $\|\cdot\|_{\mathscr{F}}$:

$$\|L_l^{-1} - p L_{l-1}^{-1} r\|_{\mathscr{W}\leftarrow\mathscr{W}^*} := \|W_l^{1/2}(L_l^{-1} - p L_{l-1}^{-1} r) W_l^{1/2}\| \leqq C_A. \tag{6.4.2}$$

Note that the factor h_l^α can be skipped because of $\alpha = 0$ (cf. Proposition 6.2.27). We recall Proposition 6.3.16: If $\|W_l\| \leqq C_W h_l^{-2m}$ (as for the examples mentioned in Remark 6.4.2), Inequality (2) follows from the approximation property $\|L_l^{-1} - p L_{l-1}^{-1} r\| \leqq h_l^{2m} C_A/C_W$, which has been proved in § 6.3.

Remark 6.4.3. Suppose (1 a–c). Inequality (2) is equivalent to

$$O \leqq L_l^{-1} - p L_{l-1}^{-1} r \leqq C_A W_l^{-1}. \tag{6.4.2'}$$

Proof. The right-hand inequality is obvious from (1.5.3 a, c). (1 c) yields $L_{l-1}^{-1/2} r L_l p L_{l-1}^{-1/2} = I$, (1.5.3 d) proves $L_l^{1/2} p L_{l-1}^{-1} r L_l^{1/2} \leqq I$; hence, $p L_{l-1}^{-1} r \leqq L_l^{-1}$ proves the left-hand inequality of (2′). □

In the general case, Theorem 6.1.7 could guarantee convergence only for $\nu \geqq \underline{\nu}$. Now we can prove convergence for any $\nu > 0$.

Theorem 6.4.4. Suppose (1 a–d), (2), $S_l^{(\nu)} = S_l^\nu$, and let $\nu > 0$ be even. Then the two-grid iteration $TGM^{(\nu/2,\nu/2)}$ from (3.2.2) converges. Its spectral radius and contraction number with respect to the energy norm $\|u_l\|_{\mathscr{U}} := |L_l^{1/2} u_l|_0$ satisfy (3) for all $\nu \geqq 0$:

$$\varrho\left(M_l\left(\frac{\nu}{2},\frac{\nu}{2}\right)\right) = \left\|M_l\left(\frac{\nu}{2},\frac{\nu}{2}\right)\right\|_{\mathscr{U}\leftarrow\mathscr{U}} \leqq \begin{cases} 1 & \text{if} \quad \nu = 0, \\ C_A \eta_0(\nu) & \text{if } 0 \leqq C_A \leqq 1+\nu, \\ (1 - 1/C_A)^\nu & \text{if} \quad C_A > 1+\nu. \end{cases} \tag{6.4.3}$$

$\eta_0(\nu)$ is defined in (2.1 b).

Proof. Introduce the transformed matrices

$$\check{M}_l := L_l^{1/2} M_l\left(\frac{\nu}{2},\frac{\nu}{2}\right) L_l^{-1/2}, \qquad \check{S}_l = L_l^{1/2} S_l L_l^{-1/2} = I - X_l,$$

$$X_l := L_l^{1/2} W_l^{-1} L_l^{1/2}, \tag{6.4.4 a}$$

$$Q_l := L_l^{1/2}(L_l^{-1} - p L_{l-1}^{-1} r) L_l^{1/2}. \tag{6.4.4 b}$$

Since $\check{M}_l = \check{S}_l^{\nu/2} Q_l \check{S}_l^{\nu/2}$ (cf. (1.1)) is symmetric, $\varrho\left(M_l\left(\frac{\nu}{2},\frac{\nu}{2}\right)\right) = \varrho(\check{M}_l)$ equals $\|\check{M}_l\| = \left\|M_l\left(\frac{\nu}{2},\frac{\nu}{2}\right)\right\|_{\mathscr{U}\leftarrow\mathscr{U}}$. The first inequality of

$$O \leqq Q_l \leqq C_A X_l, \quad X_l := L_l^{1/2} W_l^{-1} L_l^{1/2}, \tag{6.4.5 a}$$

$$O \leqq Q_l \leqq I, \tag{6.4.5 b}$$

is equivalent to (2′), while (5 b) holds since Q_l is an orthogonal projection. The linear combination of (5 a, b) yields

$$O \leqq Q_l \leqq \alpha C_A X_l + (1-\alpha) I \qquad \text{for all } 0 \leqq \alpha \leqq 1. \tag{6.4.5 c}$$

Multiplication by $\check{S}_l^{\nu/2}$ from the left and the right proves

$$O \leqq \check{M}_l = \check{S}_l^{\nu/2} Q_l \check{S}_l^{\nu/2} \leqq \check{S}_l^{\nu/2} [\alpha C_A X_l + (1-\alpha) I] \check{S}_l^{\nu/2} = f(X_l; \alpha) \tag{6.4.5 d}$$

with $f(\xi;\alpha) := (1-\xi)^\nu (1-\alpha+\alpha C_A \xi)$ for all $0 \leqq \alpha \leqq 1$. Since $O \leqq X_l \leqq I$, Lemma 1.3.5 yields

$$\|\check{M}_l\| \leqq \|f(X_l;\alpha)\| \leqq m(\alpha) := \max\{f(\xi;\alpha) : 0 \leqq \xi \leqq 1\}$$

for all $0 \leqq \alpha \leqq 1$. $m(1)$ equals $C_A \eta_0(\nu)$. If $C_A \geqq 1+\nu$, the value $\alpha^* = \nu/(C_A - 1)$ lies in $[0,1]$ and one calculates that $m(\alpha^*) = f(1/C_A; \alpha^*) = (1 - 1/C_A)^\nu$. Hence (3) is proved. □

We remark that $M_l\left(\frac{\nu}{2}, \frac{\nu}{2}\right)$ can be defined also for odd ν and that such an $M_l\left(\frac{\nu}{2}, \frac{\nu}{2}\right)$ satisfies (3).

If the numbers of pre- and post-smoothing iterations are different, one can apply

Proposition 6.4.5. Suppose (1 a–d), $S_l^{(\nu)}$ = polynomial in S_l, and define $\|\cdot\|_{\mathcal{U}}$ as in Theorem 6.4.4. The contraction number of the two-grid iteration $TGM^{(\nu_1,\nu_2)}$ from (3.2.2) is bounded by

$$\|M_l(\nu_1,\nu_2)\|_{\mathcal{U}\leftarrow\mathcal{U}} \leqq \|M_l(\nu_1,\nu_1)\|_{\mathcal{U}\leftarrow\mathcal{U}}^{1/2} \|M_l(\nu_2,\nu_2)\|_{\mathcal{U}\leftarrow\mathcal{U}}^{1/2}$$
$$\text{for all } \nu_1, \nu_2 \geqq 0. \tag{6.4.6}$$

If $S_l^{(\nu)} = S_l^\nu$, $\varrho(M_l(\nu_1,\nu_2))$ equals $\|M_l(\nu/2,\nu/2)\|_{\mathcal{U}\leftarrow\mathcal{U}}$, where $\nu = \nu_1 + \nu_2$. $\nu_1 + \nu_2 > 0$ implies convergence.

Proof. Define $\check{M}_l(\nu_1,\nu_2) := L_l^{1/2} M_l(\nu_1,\nu_2) L_l^{-1/2} = \check{S}_l^{(\nu_1)} Q_l \check{S}_l^{(\nu_2)}$. Since $Q_l^2 = Q_l$, the equation $\check{M}_l(\nu_1,\nu_2) = \check{M}_l(0,\nu_2)\check{M}_l(\nu_1,0)$ (cf. Lemma 6.1.12) proves that $\|\check{M}_l(\nu_1,\nu_2)\| \leqq \|\check{M}_l(0,\nu_2)\| \, \|\check{M}_l(\nu_1,0)\|$; hence,

$$\|M_l(\nu_1,\nu_2)\|_{\mathcal{U}\leftarrow\mathcal{U}} \leqq \|M_l(0,\nu_2)\|_{\mathcal{U}\leftarrow\mathcal{U}} \|M_l(\nu_1,0)\|_{\mathcal{U}\leftarrow\mathcal{U}}. \tag{6.4.7 a}$$

By $\check{M}_l(\nu,0)^* = \check{M}_l(0,\nu)$ and $\check{M}_l(\nu,\nu) = \check{M}_l(\nu,0)^* \check{M}_l(\nu,0)$,

$$\|M_l(0,\nu)\|_{\mathcal{U}\leftarrow\mathcal{U}} = \|M_l(\nu,0)\|_{\mathcal{U}\leftarrow\mathcal{U}} = \|M_l(\nu,\nu)\|_{\mathcal{U}\leftarrow\mathcal{U}}^{1/2} \tag{6.4.7 b}$$

follows. (7 a) and (7 b) imply (6). Concerning $\varrho(M_l(\nu_1,\nu_2))$ compare Note 6.1.2. □

If we choose semi-iterative smoothers (2.48 b), we can prove similar results. Proposition 6.4.5 applies already to this case. The right-hand side of (3) becomes $\min_{0\leqq s\leqq 1} \max_{0\leqq\xi\leqq 1} (C_A\xi)^s \prod_{\mu=1}^{\nu/2} (1-\omega_\mu\xi)^2 < 1$. For sufficiently large ν the minimum is attained at $s=1$ and the bound equals $C_A(\nu+1)^{-2}$, if ω_μ, $\mu=1,\ldots,\nu/2$, are given by (2.50 b).

Finally, we consider the case of less regular problems. The norm is given by $|u_l|_s = |\Lambda_l^s u_l|_0$, $-1 \leqq s \leqq 1$, where $\Lambda_l = L_l^{1/2}$.

Lemma 6.4.6. Assume (1 a–d), $\|W_l\| \leqq C_W h_l^{-2m}$ and $s \in (-1,0]$. The approximation property

$$|L_l^{-1} - p L_{l-1}^{-1} r|_{-s\leftarrow s} \leqq (C_A' h_l^{2m})^{1+s} \tag{6.4.8}$$

implies (9) with $C_A := C_A' C_W$:

$$O \leqq Q_l \leqq C_A^\alpha X_l^{\alpha} \quad \text{for all } 0 \leqq \alpha \leqq 1+s \quad \text{with} \quad X_l \text{ from (4a)}. \tag{6.4.9}$$

Proof. (9) holds for $\alpha = 0$ because of (5 b). By Exercise 6.6.10 it remains to prove (9) for $\alpha = 1+s$. The inequality

$$|L_l^{-1} - p L_{l-1}^{-1} r|_{-s\leftarrow s} = \|L_l^{-s/2}(L_l^{-1} - p L_{l-1}^{-1} r) L_l^{-s/2}\| \leqq (C_A' h_l^{2m})^{1+s}$$

can be rewritten as

$$L_l^{-s/2}(L_l^{-1} - p L_{l-1}^{-1} r) L_l^{-s/2} = L_l^{-(1+s)/2} Q_l L_l^{-(1+s)/2} \leqq (C_A' h_l^{2m})^{1+s} I$$

and gives

$$O \leqq Q_l \leqq (C_A' h_l^{2m})^{1+s} L_l^{1+s}. \tag{6.4.10 a}$$

$\|W_l\| \leqq C_W h_l^{-2m}$ implies $W_l \leqq C_W h_l^{-2m} I$ and $I \leqq C_W h_l^{-2m} W_l^{-1}$; hence

$$L_l = L_l^{1/2} I L_l^{1/2} \leqq L_l^{1/2}(C_W h_l^{-2m} W_l^{-1}) L_l^{1/2} = C_W h_l^{-2m} X_l. \tag{6.4.10 b}$$

Applying Exercise 6.6.10 to (10 b), one obtains in particular

$$L_l^{1+s} \leqq (C_W h_l^{-2m})^{1+s} X_l^{1+s} \tag{6.4.10 c}$$

since $0 < 1+s \leqq 1$. (10a) and (10 c) imply Inequality (9) for $\alpha = 1+s$. □

The proof of the following result is similar to the proof of Theorem 6.4.4.

Proposition 6.4.7. Assume (1 a–d), $\|W_l\| \leqq C_W h_l^{-2m}$, $S_l^{(\nu)} = S_l^\nu$, and (8) for some $s \in (-1,0]$. Then the two-grid iteration $TGM^{(\nu/2,\nu/2)}$ converges whenever $\nu > 0$. The contraction number with respect to the energy norm $\|u_l\|_{\mathscr{U}} = |L_l^{1/2} u_l|_0$ is given by (11) with $C_A := C_A' C_W$:

$$\varrho\left(M_l\left(\frac{\nu}{2},\frac{\nu}{2}\right)\right) = \left\|M_l\left(\frac{\nu}{2},\frac{\nu}{2}\right)\right\|_{\mathscr{U}\leftarrow\mathscr{U}} \leqq \begin{cases} \left[C_A \eta_0\left(\dfrac{\nu}{1+s}\right)\right]^{1+s} < 1 & \text{if } 0 \leqq C_A \leqq \dfrac{\nu+1+s}{1+s} \\[2ex] (1-1/C_A)^\nu \quad < 1 & \text{if } C_A \geqq \dfrac{\nu+1+s}{1+s}. \end{cases} \tag{6.4.11}$$

Proposition 6.4.5 is still applicable, since it does not require condition (2).

6.4.2 Estimates Without Regularity Assumptions

The approximation property (2′) can be rewritten as $Q_l \leqq C_A X_l$ (cf. (5 a)). Multiplying by the orthogonal projection Q_l from both sides, one obtains the weaker condition

$$Q_l \leqq C_A Q_l X_l Q_l. \tag{6.4.12}$$

Inequality (12) can be proved much more easily than (2′), since it does not require 2-regularity of L_l.

Lemma 6.4.8. The weak approximation property (12) follows from (1 a–d) and

$$\inf\{|W_l^{1/2}(u_l - p u_{l-1})|_0 : u_{l-1} \in \mathscr{U}_{l-1}\} \leqq \sqrt{C_A}\,|u_l|_1 \quad \text{for all } u_l \in \mathscr{U}_l. \tag{6.4.13}$$

Proof. Let $u_l \in \mathscr{U}_l$ be arbitrary and define $w_l := L_l^{-1/2} Q_l u_l$. Choosing the optimal $w_{l-1} \in \mathscr{U}_{l-1}$ according to (13) and noting that $r L_l w_l = 0$, we prove

$$|w_l|_1^2 = \langle L_l w_l, w_l\rangle = \langle L_l w_l, w_l - p w_{l-1}\rangle \leqq |W_l^{-1/2} L_l w_l|_0 \sqrt{C_A}\,|w_l|_1;$$

hence, $|w_l|_1 \leqq \sqrt{C_A}\,|W_l^{-1/2} L_l w_l|_0$, which is equivalent to

$$\langle u_l, Q_l u_l\rangle \leqq C_A \langle u_l, Q_l L_l^{1/2} W_l^{-1} L_l^{1/2} Q_l u_l\rangle, \text{ i.e., to (12).} \qquad \square$$

Note 6.4.9. Let $Q_l^0 = p(rp)^{-1} r$ denote the orthogonal projection onto $p\,\mathscr{U}_{l-1}$. (13) holds with $C_A = C_1 C_2$ if

$$|I - Q_l^0|_{0 \leftarrow 1} \leqq C_1 h_l^m, \qquad \|W_l\| \leqq C_2^2 h_l^{-2m}. \tag{6.4.14}$$

The matrix L_l is involved only indirectly in $|u_l|_1 := |L_l^{1/2} u_l|_0$, which implies the 1-regularity of L_l. Though further conditions are not imposed on L_l, the two-grid convergence is immediate.

Proposition 6.4.10. Define $\|\cdot\|_{\mathscr{U}}$ as the energy norm $|\cdot|_1$. The assumptions (1 a–d), $S_l^{(\nu)} = S_l^\nu$, and (12) imply the estimates

$$\|M_l(\nu_1,\nu_2)\|_{\mathscr{U} \leftarrow \mathscr{U}} \leqq \left(1 - \frac{1}{C_A}\right)^{\kappa} \quad \text{with} \quad \kappa = \begin{cases} \frac{1}{2} & \text{if } \nu_1 \geqq 1 \text{ or } \nu_2 \geqq 1 \\ 1 & \text{if } \nu_1, \nu_2 \geqq 1 \end{cases}, \tag{6.4.15a}$$

$$\varrho(M_l(\nu_1,\nu_2)) \leqq 1 - \frac{1}{C_A} \quad \text{for all } \nu_1 + \nu_2 \geqq 1. \tag{6.4.15b}$$

Proof. By Proposition 6.4.5, the spectral radius equals $\|M_l(\nu,\nu)\|_{\mathscr{U} \leftarrow \mathscr{U}}$ with $\nu = (\nu_1 + \nu_2)/2 \geqq \frac{1}{2}$. Hence (15 b) follows from (15 a), if (15 a) holds for multiples ν_1, ν_2 of $\frac{1}{2}$.

Firstly let $\nu_1 = 0$, $\nu := \nu_2 \geqq \frac{1}{2}$ and set $\check{M}_l = L_l^{1/2} M_l(0,\nu) L_l^{-1/2} = \check{S}_l^\nu Q_l$. (12) and $\check{S}_l \leqq I$ imply

$$\check{M}_l^* \check{M}_l = Q_l \check{S}_l^{2\nu} Q_l \leqq Q_l \check{S}_l Q_l = Q_l (I - X_l) Q_l \leqq \left(1 - \frac{1}{C_A}\right) Q_l \leqq \left(1 - \frac{1}{C_A}\right) I,$$

which gives $\|\check{M}_l\| = \|M_l(0,\nu)\|_{\mathscr{U} \leftarrow \mathscr{U}} \leqq (1 - C_A^{-1})^{1/2}$. The other cases $\nu_1 \geqq \frac{1}{2}$, $\nu_2 = 0$ and $\nu_1, \nu_2 \geqq \frac{1}{2}$ can be treated by means of (7 a, b). $\square$

The symmetry of $\check{S}_l$ is not needed as stated in

Corollary 6.4.11. Assume (1 a–c), $S_l^{(\nu)} = S_l^\nu$, and

$$\check{S}_l^* \check{S}_l \leqq I - X_l, \quad \text{i.e.} \quad S_l^* L_l S_l \leqq L_l - L_l W_l^{-1} L_l \tag{6.4.16a}$$

for some $X_l = L_l^{1/2} W_l^{-1} L_l^{1/2} \leqq I$ satisfying (12). Then (15 a) holds for $M_l(0, \nu)$. The estimation of $M_l(\nu, 0)$ by (15 a) requires

$$\check{S}_l \check{S}_l^* \leqq I - X_l \tag{6.4.16b}$$

instead of (16 a). With (16 a) and (16 b) one proves the estimate (15 a) for $M_l(\nu_1, \nu_2)$, $\nu_1 + \nu_2 \geqq 1$.

Proof. Use $\check{M}_l^* \check{M}_l = Q_l \check{S}_l^{*\nu} \check{S}_l^\nu Q_l \leqq Q_l \check{S}_l^* \check{S}_l Q_l \leqq Q_l(I - X_l) Q_l \leqq \left(1 - \frac{1}{C_A}\right) I$. □

The latter result can be applied to the Gauß-Seidel iteration. Let $L_l = D_l - A_l - B_l$ be the splitting defining $S_l := (D_l - A_l)^{-1} B_l$.

Proposition 6.4.12. Assume (1 a–c), $D_l = D_l^* > 0$, $A_l = B_l^*$, $S_l = (D_l - A_l)^{-1} B_l$,

$$\begin{aligned} &\| I - D_l^{-1/2} A_l D_l^{-1/2} \| \leqq C_1, \\ &\inf \{ |D_l^{1/2} (u_l - p u_{l-1})|_0 : u_{l-1} \in \mathscr{U}_{l-1} \} \leqq C_2 |u_l|_1 \end{aligned} \tag{6.4.17a}$$

or

$$\begin{aligned} &\| (D_l - A_l) D_l^{-1} (D_l - B_l) \| \leqq C_3 h_l^{-2m}, \\ &\| (D_l - B_l) D_l^{-1} (D_l - A_l) \| \leqq C_3 h_l^{-2m}, \\ &\inf \{ |u_l - p u_{l-1}|_0 : u_{l-1} \in \mathscr{U}_{l-1} \} \leqq C_4 h_l^m |u_l|_1 . \end{aligned} \tag{6.4.17b}$$

Then the two-grid iteration with Gauß-Seidel smoothing satisfies (15 a) with $C_A = C_1^2 C_2^2$ or $C_A = C_3 C_4^2$, respectively.

Proof. (2.32 b) shows that $\check{S}_l^* \check{S}_l = I - L_l^{1/2} (D_l - B_l)^{-1} D_l (D_l - A_l)^{-1} L_l^{1/2} \leqq I - C_1^{-2} L_l^{1/2} D_l^{-1} L_l^{1/2}$. Thus, $W_l := C_1^2 D_l$ satisfies (13) with $\sqrt{C_A} = C_1 C_2$ and the results follow from Corollary 6.4.11. The case of (17 b) is analogous. □

6.5 Comments

The smoothing and approximation properties (introduced by Hackbusch [10], [21]) are hidden in all previously published convergence proofs. These proofs are often very involved since the proofs of the smoothing and approximation properties and of Theorem 6.1.7 are intertwined.

A variant of our approach is the following. Let Q_l be a projection, set $Q_l^\perp := I - Q_l$, and abbreviate $I - p L_{l-1}^{-1} r L_l$ by C_l. In

$$\begin{aligned} \| M_l(\nu, 0) \| &= \| C_l S_l^{(\nu)} \| = \| C_l Q_l S_l^{(\nu)} + C_l Q_l^\perp S_l^{(\nu)} \| \\ &= \| C_l Q_l \| \, \| S_l^{(\nu)} \| + \| C_l \| \, \| Q_l^\perp S_l^{(\nu)} \| \end{aligned} \tag{6.5.1}$$

we may assume $\| S_l^{(\nu)} \| \leqq C$, $\| C_l \| \leqq C$. If the range of Q_l consists of sufficiently smooth functions, the inequality $\| C_l Q_l \| \leqq \varepsilon$ (ε small) can be regarded as a kind of approximation property, while $\| Q_l^\perp S_l^{(\nu)} \| \leqq \eta(\nu)$ serves as smoothing property. Inequality (1) implies $\| M_l(\nu) \| \leqq C(\varepsilon + \eta(\nu))$.

A similar approach has been proposed by Wesseling [2]. He uses our approximation property and describes the smoothing property by

$$\| Q_l S_l^{(\nu)} Q_l \| \leqq 1, \quad \| Q_l^\perp S_l^{(\nu)} Q_l^\perp \| \leqq \eta(\nu), \quad \| Q_l S_l^{(\nu)} Q_l^\perp + Q_l^\perp S_l^{(\nu)} Q_l \| \leqq C(\nu)\, h_l^m.$$

The definition of Q_l by Wesseling [2] originates from Bachvalov [1]. Q_l is the orthogonal projection onto the span of all eigenvectors corresponding to eigenvalues $\lambda \leqq \sigma \| L_l \|$, where $\sigma \in (0,1)$ has to be chosen suitably. The optimal choice of σ depends on ν. If we take instead the span of all 'low frequencies', we are led to the local mode analysis described in § 8.2.

The approximation property for finite element equations. We recall that there is a canonical definition of the Hilbert scale $\{\mathscr{H}^s : -1 \leqq s \leqq 1\}$. In addition we need $\mathscr{H}^t$, $t > 1$, (e.g. $\mathscr{H}^2$) to define the t-regularity and to estimate the approximation error by (3.5 a). We never required a scale connecting $\mathscr{H}^0$, $\mathscr{H}^1$, and $\mathscr{H}^t$.

For ease of presentation we defined the t-regularity of L and L^* by means of the same Hilbert space $\mathscr{H}^t$. In general it can happen that the ranges of L^{-1} and L^{*-1} are different. Then one should introduce two Hilbert spaces $\mathscr{H}^t$ and $\mathscr{H}_*^t$, $t > 1$, but still $\mathscr{H}^s = \mathscr{H}_*^s$ coincide for $|s| \leqq 1$. Inequality (3.5 a) has to hold for $|\cdot|_t = \|\cdot\|_{\mathscr{H}^t}$ and $|\cdot|_t = \|\cdot\|_{\mathscr{H}_*^t}$, while the t-regularity of $L(L^*)$ uses $\mathscr{H}^t$ ($\mathscr{H}_*^t$). Then, all conclusions of § 6.3 remain valid. However under the conditions as given in Corollary 6.3.9 one can always choose $\mathscr{H}^t = \mathscr{H}_*^t$.

A finite element hierarchy violating $\mathscr{H}_{l-1} \subset \mathscr{H}_l$. The assumption (3.4), $\mathscr{H}_{l-1} \subset \mathscr{H}_l$, is needed for the canonical definition of p and r (cf. § 3.6) and simplifies the proofs. However, a finite element hierarchy with $\mathscr{H}_{l-1} \not\subset \mathscr{H}_l$ does not destroy the convergence. E.g., $\mathscr{H}_{l-1} \not\subset \mathscr{H}_l$ may arise during the construction of finite element spaces in § 3.8.3 (cf. Note 3.8.2). The prolongation p can still be defined by the finite element interpolation (3.8.2). Thus we are in the same situation as in the finite difference case, where p is chosen in a non-canonical way.

Even if the triangulation involved in $\mathscr{H}_{l-1}$ is divided regularly in the interior, the refinement near the boundary must lead to $\mathscr{H}_{l-1} \not\subset \mathscr{H}_l$ if Γ is curvilinear. Mansfield [1] studies such a case and proves multi-grid convergence for a finite element discretisation with isoparametric elements.

Discrete regularity. A typical error estimate for a difference scheme is $\max\{|u_l(\mathbf{x}) - u(\mathbf{x})| : \mathbf{x} \in \Omega_l\} \leqq C h_l^2 \|u\|_{C^4(\Omega)}$ which is far from being optimal and does not meet our needs. An optimal error estimate $|u_l - R_l u|_0 \leqq C h_l^2 \|u\|_{H^2(\Omega)}$ as known for finite element methods has been proved by Nitsche-Nitsche [1] for a constant difference operator in a square, where Fourier analysis can be applied. In such a case also the multi-grid convergence can be proved by means of the Fourier analysis (cf. § 8.1). Bachvalov [1] already considered finite difference schemes with varying coefficients in a rectangle. As long as Ω is a rectangle (or a cube, etc.) one can apply partial summation to prove regularity results. Regularity and optimal error estimates are closely related. Also the result of Dryja [1] using a convex grid Ω_l belongs to this class of problems: Although Ω is a general convex domain, the grid is chosen non-uniformly so that partial summation still applies. For further results of this kind we refer to D'jakonov [1] and Lapin [1].

The Fourier analysis can also be applied to the interior of a general region Ω. Interior regularity results are proved e.g. by Thomée-Westergren [1]. However, for a proof of the approximation property we need the regularity 'up to the boundary'. For Dirichlet boundary value problems in a general domain Ω some results have been proved by Hackbusch [23], [32] (cf. Theorems 6.3.32, 6.3.33). The treatment of other boundary conditions is still an open problem.

Convergence without regularity assumptions. Papers concerning the symmetric case of § 6.4.1 will be mentioned in § 7.3. The results of § 6.4.2 differ from the other convergence results since we do not require the usual approximation property. The substitute (4.2) is provable without any t-regularity ($t > 1$) of L. It seems that in return one obtains only weaker convergence results. They are still uniform with respect to l, but the estimates do not improve with increasing $\nu = \nu_1 + \nu_2$. Furthermore, a generalisation of this proof to the symmetric multi-grid case has not yet been developed.

The pioneering paper in this respect is a report of Bank-Dupont [1]. A quantitative result has been obtained by Braess [1]. He proved a contraction number $\frac{1}{2}$ for a regular finite element discretisation of Poisson's equation with $h_{l-1} = \sqrt{2}\, h_l$. The main tool is a strengthened Cauchy-Schwarz inequality $|\langle u_l, v_l \rangle| \leqq \gamma\, |u_l|\, |v_l|, \gamma < 1$, for u_l and v_l from different subspaces spanning $\mathscr{U}_l$. The constant γ (as well as our constant C_A from (4.12)) can be calculated by studying the single triangles of the finite element triangulation. Values of γ for further finite element methods are given by Maitre-Musy [1]. Mandel [3] proves the same convergence results for the Gauß-Seidel iteration as smoother and extends the theorem to variational inequalities.

6.6 Exercises

Exercise 6.6.1. Prove that the matrix $N_l(\nu_1, \nu_2)$ (cf. (1.3.2)) of the two-grid iteration (3.2.2) is given by

$$N_l(\nu_1, \nu_2) = T_l^{(\nu_2)} + S_l^{(\nu_2)} [T_l^{(\nu_1)} - p\, L_{l-1}^{-1}\, r (L_l\, T_l^{(\nu_1)} - I)],$$

where $\mathscr{S}_l^{(\nu)}(u_l, f_l) = S_l^{(\nu)} u_l + T_l^{(\nu)} f_l$.

Exercise 6.6.2. Let A and B be two $n \times n$-matrices and suppose that the norm is connected with a scalar product by $\|\cdot\|_{\mathscr{U}} = \langle \cdot, \cdot \rangle^{1/2}$. Prove: For every $\varepsilon > 0$ there is a norm $\|\cdot\|_{\mathscr{F}}$ with $\|AB\|_{\mathscr{U} \leftarrow \mathscr{U}} \leqq \|A\|_{\mathscr{U} \leftarrow \mathscr{F}}\, \|B\|_{\mathscr{F} \leftarrow \mathscr{U}} + \varepsilon$.

Exercise 6.6.3. Consider the Jacobi iteration (3.3.7′) with $L_l = L_l^* \geqq 0$. Prove that $\| L_l S_l^\nu \| \leqq C_L h_l^{-2m} \eta_0(\nu)$ for $\omega_l \in \left[1/C_L, \left(2 - \frac{\log 2\nu}{\nu + 1}\right) \Big/ C_L \right]$ and C_L from (2.10).

Exercise 6.6.4. Consider the iteration (3.3.7′) with some ω_l. Suppose that L_l is normal (i.e. $L_l^* L_l = L_l L_l^*$) and that $\omega_l h_l^{2m} L_l$ has eigenvalues only in a certain domain $\Omega \subset \mathbb{C}$.

(i) Show: $\| L_l S_l^\nu \| \leqq (\omega_l h_l^{2m})^{-1} \max\limits_{\zeta \in \Omega} |\zeta| |1 - \zeta|^\nu$.

(ii) Remark: For any function $\eta^*(\nu) \geqq \eta_0(\nu)$ one can define

$$\Omega_\nu = \{\zeta \in \mathbb{C}: |\zeta| |1 - \zeta|^\nu \leqq \eta^*(\nu)\}.$$

Then $\| L_l S_l^\nu \| \leqq \eta^*(\nu)/(\omega_l h_l^{2m})$ holds if the spectrum of $\omega_l h_l^{2m} L_l$ lies in Ω_ν. Show that $\bigcap_{\nu=1}^{\infty} \Omega_\nu$ is not empty and that the smoothing property holds, if the spectrum is contained in $\bigcap_\nu \Omega_\nu$.

(iii) Set $\eta^*(\nu) := [\nu^2/((1+\nu)(2+\nu))]^{\nu/2}/\sqrt{1+\nu}$. Let $R_\nu(\varphi)$ be the largest root of $r[(1-r)^2 + 4r\sin^2(\varphi/2)]^{\nu/2} = \eta^*(\nu)$. Prove: $\| L_l S_l^\nu \| \leqq \eta^*(\nu)/(\omega_l h_l^{2m})$, if the spectrum of $\omega_l h_l^{2m} L_l$ is contained in $\{z = r e^{i\varphi}: 0 \leqq r \leqq R_\nu(\varphi), |\varphi| \leqq \pi\}$.

Exercise 6.6.5. Let $\| \cdot \|_{\mathscr{U}}$ and $\| \cdot \|_{\mathscr{F}}$ be defined by (2.4), (2.6), (2.7), and (2.28). Assume $\alpha := (2 + s_{\mathscr{F}} - s_{\mathscr{U}})m \leqq 2m$, $\| L_l' \| \leqq C_L' h_l^{-2m}$, $h_l^\alpha \| L_l'' \|_{\mathscr{F} \leftarrow \mathscr{U}} \to 0$. Prove that $h_l^{2m} \| L_l'' \|_{\mathscr{U} \leftarrow \mathscr{U}} \to 0$ as $l \to \infty$ (cf. Corollary 6.2.18).

Exercise 6.6.6. Discretise the equation $-u''(x) = f(x)$ in $\mathbb{R}$ (without boundary conditions) by (2.1.5′) and define $\mathscr{S}_l$ by (2.3.1). Prove the following smoothing property with respect to the supremum norm:

$$\| L_l S_l^\nu \|_\infty \leqq h_l^{-2} 4^{1-\nu} \binom{2\nu + 1}{\nu + k + 1} \frac{2k+1}{2\nu+1}$$

$$\leqq \left[h_l^{-2} \sqrt{\frac{32}{\pi e}} \Big/ (\nu + \tfrac{1}{2}) \right] (1 + O(\nu^{-1/2})),$$

where $k = (\nu+1)/2$ (ν odd), $k = \nu/2$ (ν even), $\| \cdot \|_\infty$ = row-sum matrix norm.

Exercise 6.6.7. The elliptic equation $Lu = L(x)u(x) = f(x)$ in Ω has variable coefficients. Let L_l be a difference operator and S_l the smoothing iteration matrix. For simplicity assume that S_l is sparse (i.e. exclude blockwise iterations). Let the coefficients of the stencils L_l and S_l be bounded and uniformly Hölder continuous. Denote the operators with constant coefficients evaluated at some fixed $\xi \in \Omega$ by $L_l(\xi)$ and $S_l(\xi)$. Define $\| \cdot \|_{\mathscr{U}} = \| \cdot \|_{\mathscr{F}} = | \cdot |_0$ by means of (3.5.2). Prove: The smoothing property (1.7) with $\alpha = 2m$ holds for $L_l S_l^\nu$ if and only if

$$\| L_l(\xi) S_l^\nu(\xi) \|_{\mathscr{F} \leftarrow \mathscr{U}} \leqq \eta^*(\nu) h_l^{-2m} \quad \text{for all } \nu < \bar{\nu}(h_l), \quad l \geqq 1, \quad \xi \in \Omega,$$

with $\bar{\nu}$ independent of ξ and some $\eta^*(\nu) \to 0$ $(\nu \to \infty)$.

Remark. The $c\,g$ methods (3.3.16) and (3.3.17) yield a nonlinear $\mathscr{S}_l^{(\nu)}$. However, the considerations of this chapter remain valid if the smoothing property (1.7 a) is replaced by

$$\|L_l[\mathscr{S}_l^{(\nu)}(u_l,f_l) - L_l^{-1}f_l]\|_{\mathscr{F}} \leqq \eta(\nu)\,h_l^{-\alpha}\,\|u_l - L_l^{-1}f_l\|_{\mathscr{U}}$$
$$\text{for all } u_l \in \mathscr{U}_l, \quad f_l \in \mathscr{F}_l. \tag{6.6.1}$$

Exercise 6.6.8. Suppose $O < L_l = L_l^* \leqq W_l = W_l^*$. Show:

(i) Let $\mathscr{S}_l^{(\nu)}$ be defined by (3.3.16). Inequality (1) holds with $\alpha = 0$, $\eta(\nu) = 1/(2\nu + 1)$ if $\|f_l\|_{\mathscr{F}} = |L_l^{-1/2} f_l|_0$ and $\|u_l\|_{\mathscr{U}} = |W_l^{1/2}u_l|_0$.
(ii) Let $\mathscr{S}_l^{(\nu)}$ be defined by (3.3.17). If $\|f_l\|_{\mathscr{F}} = |W_l^{-1/2} f_l|_0$ and $\|u_l\|_{\mathscr{U}} = |L_l^{1/2}u_l|_0$, α and η are as in (i). If instead $\|u_l\|_{\mathscr{U}} = |W_l^{1/2}u_l|_0$, Inequality (1) holds with $\alpha = 0$ and η equal to the right side of (2.49 a). Hint: Use Note 3.3.5.

Exercise 6.6.9. Let $L_l = L_l^* > O$, P_l, R_l, $\hat{P}_l$, $\hat{R}_l$ be as in § 6.3.1.2. Prove that

$$O \leqq L_l^{-1} - p\,L_{l-1}^{-1}\,r \leqq \hat{R}_l(L^{-1} - P_{l-1}\,L_{l-1}^{-1}\,R_{l-1})\,\hat{P}_l.$$

Exercise 6.6.10. Let A, B, C be matrices with $A = A^* \geqq O$, $B = B^* \geqq O$, $C = C^*$. Prove:

(i) If $C \leqq A^p$, $C \leqq A^q$ for some $p \leqq q$, then $C \leqq A^s$ for all $p \leqq s \leqq q$.
(ii) If $A^p \leqq B^p$ for some $p \geqq 0$, then $A^s \leqq B^s$ for all $s \in [0,p]$.

The stiffness matrix L_l may be perturbed because of integration errors. This situation is considered in

Exercise 6.6.11. Assume that $\tilde{L}_l := L_l + \delta L_l$, where

$$\|\delta L_l\|_{\mathscr{F} \leftarrow \mathscr{U}} \leqq \varepsilon\,h_l^{\alpha}.$$

Let ε be sufficiently small so that $\tilde{L}_l$ is non-singular. Suppose that the norms $\|\cdot\|_{\mathscr{U}}$, $\|\cdot\|_{\mathscr{F}}$ are defined as in § 6.3 with $\alpha = (2 - s_{\mathscr{U}} + s_{\mathscr{F}})m$. Prove that

(i) under the conditions of the respective Theorems 6.3.15, 6.3.21, 6.3.26 the approximation property is still valid for $\tilde{L}_l$;
(ii) the damped Jacobi iteration applied to $\tilde{L}_l$ satisfies the smoothing property.

7. Convergence of the Multi-Grid Iteration

In § 7.1 we shall prove that two-grid convergence, more precisely, the sufficient conditions from Theorem 6.1.7, almost imply the multi-grid convergence. The multi-grid contraction number is bounded by some $\zeta(\nu) < 1$ with $\zeta(\nu) \to 0$ $(\nu \to \infty)$, which is independent of l, i.e. independent of the finest step size h_l and independent of the number of involved levels. However, the estimates are somewhat pessimistic. In particular, the V-cycle (cf. § 2.5) cannot be treated by this approach. § 7.2 provides sharper estimates, which apply to the V-cycle also.

7.1 The General Convergence Theorem

In addition to the suppositions of Theorem 6.1.7 we need

$$\| S_l^{(\nu)} \|_{\mathscr{U} \leftarrow \mathscr{U}} \leqq C_S \qquad \text{for all } l \geqq 1,\ 0 < \nu < \bar{\nu}(h_1) \tag{7.1.1}$$

with $\bar{\nu}(h)$ from (6.1.7 a, c). This inequality has already been considered. Thanks to Corollary 6.2.10, it suffices to prove (1) for the positive definite case $L_l = L_l^* > 0$. The modifications of Jacobi's iteration satisfy (1) with $C_S = 1$. For the Gauß-Seidel iteration compare Proposition 6.2.23.

Further, we require the inequalities

$$\underline{C}_p^{-1} \| u_{l-1} \|_{\mathscr{U}} \leqq \| p u_{l-1} \|_{\mathscr{U}} \leqq \bar{C}_p \| u_{l-1} \|_{\mathscr{U}} \qquad \text{for all } u_{l-1} \in \mathscr{U}_{l-1},\ l \geqq 1 \tag{7.1.2}$$

which have also been stated above for different choices of $\| \cdot \|_{\mathscr{U}}$ (cf. (6.3.12 b), (6.3.20 c, d), Lemma 6.3.27 (iv)).

Remark 7.1.1. Let $\mathscr{U}_l$ be a Hilbert space. An equivalent formulation of (2) reads as follows: There exists some $\hat{r}\colon \mathscr{U}_l \to \mathscr{U}_{l-1}$ with

$$\| \hat{r} \|_{\mathscr{U} \leftarrow \mathscr{U}} \leqq \underline{C}_p, \quad \hat{r} p = I, \quad \| p \|_{\mathscr{U} \leftarrow \mathscr{U}} \leqq \bar{C}_p. \tag{7.1.2'}$$

Condition (2) ensures that the norms $\| \cdot \|_{\mathscr{U}}$ at the different levels $l = 0, 1, \ldots$ are comparable.

With these assumptions the assertions about the two-grid convergence (cf. Theorem 6.1.7) can be established for the multi-grid case also. The symbol $M_l'(\nu_1, \nu_2)$ denotes the iteration matrix of the multi-grid algorithm $MGM^{(\nu_1, \nu_2)}$ described in (4.1.1).

Theorem 7.1.2 (multi-grid convergence of $MGM^{(\nu,0)}$). Suppose $\gamma \geqq 2$, (1), (2), the smoothing property (6.1.7a–c), and the approximation property (6.1.8). Let $\zeta' \in (0,1)$ be a fixed number.

a) In the case of $\bar{\nu}(h) = \infty$ (cf. (6.1.7c)) there is a number $\underline{\nu}$ such that the multi-grid contraction number satisfies

$$\| M_l'(\nu, 0) \|_{\mathscr{U} \leftarrow \mathscr{U}} \leqq \zeta' < 1, \quad \| M_l'(\nu, 0) \|_{\mathscr{U} \leftarrow \mathscr{U}} \leqq \frac{\gamma}{\gamma - 1} C_A \eta(\nu), \tag{7.1.3}$$

whenever $\nu \geqq \underline{\nu}$, independently of $l \geqq 1$.

b) In the remaining case of $\bar{\nu}(h) \to \infty$ there exist $\bar{h} > 0$ and $\underline{\nu}$ such that Inequality (3) holds for all $\nu \in [\underline{\nu}, \bar{\nu}(h_1))$ and all $l \geqq 1$, provided that $h_1 \leqq \bar{h}$. For such h_1 the interval $[\underline{\nu}, \bar{\nu}(h_1))$ is not empty.

c) Under condition (6.1.2) one obtains convergence $u_l^j \to L_l^{-1} f_l$ as $j \to \infty$.

The *proof* of part c) is immediate: $\zeta' < 1$ ensures convergence $\lim_{j \to \infty} u_l^j =: u_l^*$ and $u_l^* = L_l^{-1} f_l$ follows from

Remark 7.1.3. The discrete solution $L_l^{-1} f_l$ is a fixed point of the multi-grid iteration (4.1.1), provided that (6.1.2) holds.

To prove the parts a), b) we have to characterise the multi-grid iteration matrices $M_l'(\nu_1, \nu_2)$. The matrix is defined recursively like the iteration.

Lemma 7.1.4. The iteration matrix $M_l'(\nu_1, \nu_2)$ of the multi-grid algorithm (4.1.1) is defined by

$$M_1'(\nu_1, \nu_2) = M_1(\nu_1, \nu_2) \quad \text{for} \quad l = 1, \tag{7.1.4a}$$

$$M_l'(\nu_1, \nu_2) = M_l(\nu_1, \nu_2) + S_l^{(\nu_2)} p \,[M_{l-1}'(\nu_1, \nu_2)]^\gamma L_{l-1}^{-1} r L_l S_l^{(\nu_1)} \quad \text{for} \quad l \geqq 2, \tag{7.1.4b}$$

where M_l is the two-grid iteration matrix from (6.1.1). One can also start with $M_0 = 0$ and use (4b) for $l \geqq 1$.

Proof. For $l = 1$ the two- and multi-grid algorithms are identical and (4a) follows. Let (4b) hold for $l - 1$ (instead of l). By Exercise 1.5.1, $M_l' = M_l'(\nu_1, \nu_2)$ equals $S_l^{(\nu_2)} C_l' S_l^{(\nu_1)}$, where C_l' represents the coarse-grid correction (4.1.1c–e). To determine C_l' set $f_l' = 0$ and let $u_l \in \mathscr{U}_l$ be arbitrary. Step (4.1.1c) results in $d_{l-1} := r L_l u_l$. In (4.1.1d) γ iterates $v_{l-1}^1, \ldots, v_{l-1}^\gamma$ are computed from $v_{l-1}^0 = 0$. Since the multi-grid iteration at level $l - 1$ can be described by

$$v_{l-1}^{j+1} = M_{l-1}' v_{l-1}^j + N_{l-1}' d_{l-1}, \quad M_{l-1}' := M_{l-1}'(\nu_1, \nu_2),$$

(cf. (1.3.2)), Eq. (1.3.3) shows

$$v_{l-1}^\gamma = \sum_{k=0}^{\gamma-1} (M_{l-1}')^k N_{l-1}' d_{l-1} = \sum_{k=0}^{\gamma-1} (M_{l-1}')^k N_{l-1}' r L_l u_l.$$

By virtue of Remark 7.1.3, Eq. (1.3.6) applies:

$$N'_{l-1} = (I - M'_{l-1})\, L_{l-1}^{-1}.$$

Using $\sum_{k=0}^{\gamma-1} \xi^k (1-\xi) = 1 - \xi^\gamma$, one obtains

$$v_{l-1}^{\gamma} = [I - (M'_{l-1})^{\gamma}]\, L_{l-1}^{-1} r L_l u_l.$$

Hence, the coarse-grid correction (4.1.1e) results in $C'_l u_l = u_l - p v_{l-1}^{\gamma}$ proving

$$C'_l = I - p[I - (M'_{l-1})^{\gamma}]\, L_{l-1}^{-1} r L_l = I - p L_{l-1}^{-1} r L_l + p (M'_{l-1})^{\gamma}\, L_{l-1}^{-1} r L_l$$

and

$$M'_l = S_l^{(\nu_2)} C'_l S_l^{(\nu_1)} = S_l^{(\nu_2)} [I - p L_{l-1}^{-1} r L_l]\, S_l^{(\nu_1)} + S_l^{(\nu_2)} p (M'_{l-1})^{\gamma}\, L_{l-1}^{-1} r L_l S_l^{(\nu_1)}.$$

Since the first term equals $M_l(\nu_1, \nu_2)$ (cf. (6.1.1)), Eq. (4b) is proved by induction. □

By Eq. (4b), the multi-grid iteration matrix can be regarded as the two-grid matrix M_l plus a perturbation, which is hopefully small. This leads to the estimate

$$\begin{aligned} \| M'_l(\nu_1, \nu_2) \|_{\mathscr{U} \leftarrow \mathscr{U}} &\leqq \| M_l(\nu_1, \nu_2) \|_{\mathscr{U} \leftarrow \mathscr{U}} \\ &+ \| S_l^{(\nu_2)} \|_{\mathscr{U} \leftarrow \mathscr{U}} \| p \|_{\mathscr{U} \leftarrow \mathscr{U}} \| M'_{l-1}(\nu_1, \nu_2) \|_{\mathscr{U} \leftarrow \mathscr{U}}^{\gamma} \| L_{l-1}^{-1} r L_l S_l^{(\nu_1)} \|_{\mathscr{U} \leftarrow \mathscr{U}}. \end{aligned} \tag{7.1.5a}$$

Under the conditions of Theorem 7.1.2 we have $\| p \|_{\mathscr{U} \leftarrow \mathscr{U}} \leqq \bar{C}_p$ and $\| S_l^{(\nu_2)} \|_{\mathscr{U} \leftarrow \mathscr{U}} = 1$ because of $\nu_2 = 0$. Hence, (5a) becomes

$$\begin{aligned} \| M'_l(\nu, 0) \|_{\mathscr{U} \leftarrow \mathscr{U}} &\leqq \| M_l(\nu, 0) \|_{\mathscr{U} \leftarrow \mathscr{U}} \\ &+ \bar{C}_p \| M'_{l-1}(\nu, 0) \|_{\mathscr{U} \leftarrow \mathscr{U}}^{\gamma} \| L_{l-1}^{-1} r L_l S_l^{(\nu)} \|_{\mathscr{U} \leftarrow \mathscr{U}}. \end{aligned} \tag{7.1.5b}$$

To estimate the last factor we may apply

Lemma 7.1.5. Suppose (1) and the first part of (2). Then

$$\| L_{l-1}^{-1} r L_l S_l^{(\nu)} \|_{\mathscr{U} \leftarrow \mathscr{U}} \leqq \underline{C}_p (C_S + \| M_l(\nu, 0) \|_{\mathscr{U} \leftarrow \mathscr{U}}).$$

Proof. Use $\| L_{l-1}^{-1} r L_l S_l^{(\nu)} \|_{\mathscr{U} \leftarrow \mathscr{U}} \leqq \underline{C}_p \| p L_{l-1}^{-1} r L_l S_l^{(\nu)} \|_{\mathscr{U} \leftarrow \mathscr{U}}$ and

$$p L_{l-1}^{-1} r L_l S_l^{(\nu)} = S_l^{(\nu)} - [L_l^{-1} - p L_{l-1}^{-1} r]\, L_l S_l^{(\nu)} = S_l^{(\nu)} - M_l(\nu, 0).$$ □

We shall consider only ν with $\| M_l(\nu, 0) \|_{\mathscr{U} \leftarrow \mathscr{U}} < 1$. Hence, Inequality (5b) becomes

$$\| M_l(\nu, 0) \|_{\mathscr{U} \leftarrow \mathscr{U}} \leqq \| M_l(\nu, 0) \|_{\mathscr{U} \leftarrow \mathscr{U}} + C^* \| M'_{l-1}(\nu, 0) \|_{\mathscr{U} \leftarrow \mathscr{U}}^{\gamma} \tag{7.1.5c}$$

with $C^* := \bar{C}_p \underline{C}_p (C_S + 1)$. (5c) gives rise to the recursive inequalities

$$\zeta_1 \leqq \zeta, \quad \zeta_l \leqq \zeta + C^* \zeta_{l-1}^{\gamma} \qquad \text{for all } l \geqq 2. \tag{7.1.6}$$

An elementary analysis shows

Lemma 7.1.6. Assume $C^* \gamma > 1$. If

$$\gamma \geqq 2, \quad \zeta \leqq \zeta_{\max} := \frac{\gamma - 1}{\gamma} (\gamma C^*)^{-1/(\gamma - 1)}, \tag{7.1.7a}$$

any solution of (6) is bounded by

$$\zeta_l \leqq \zeta^* < 1, \tag{7.1.7b}$$

where ζ^* and ζ are related by

$$\zeta = \zeta^* - C^* \zeta^{*\gamma}, \quad \zeta^* \leqq \frac{\gamma}{\gamma - 1} \zeta. \tag{7.1.7c}$$

In the case of the W-cycle (i.e. $\gamma = 2$), requirement (7a) becomes

$$\gamma = 2, \quad \zeta \leqq \zeta_{\max} := 1/(4C^*) \tag{7.1.7d}$$

and all ζ_l are bounded by

$$\zeta_l \leqq 2\zeta/(1 + \sqrt{1 - 4\zeta C^*}). \tag{7.1.7e}$$

Proof of Theorem 7.1.2. Without loss of generality we may assume that C^* from (5c) exceeds $1/\gamma$. Define $\zeta^* := \min\{\zeta', (\gamma C^*)^{-1/(\gamma-1)}\}$ with ζ' from Theorem 7.1.2 and set $\zeta := \zeta^* - C^* \zeta^{*\gamma}$. Note that ζ satisfies (7a): $\zeta \leqq \zeta_{\max}$. Theorem 6.1.7 shows

$$\| M_l(v, 0) \|_{\mathscr{U} \leftarrow \mathscr{U}} \leqq C_A \eta(v) \leqq \zeta \qquad (\underline{v} \leqq v < \bar{v}(h_1) \leqq \bar{v}(h_l), \ h_1 \leqq \bar{h})$$

for suitable $\underline{v}$ and $\bar{h}$. By (5b) the quantities $\zeta_l := \| M_l'(v, 0) \|_{\mathscr{U} \leftarrow \mathscr{U}}$ satisfy (6). Lemma 7.1.6 proves

$$\| M_l'(v, 0) \|_{\mathscr{U} \leftarrow \mathscr{U}} \leqq \zeta^* \leqq \zeta' \qquad \text{for all } v \in [\underline{v}, \bar{v}(h_1)), \ h_1 \leqq \bar{h}.$$

As explained in Theorem 6.1.7, $\bar{v}(h) = \infty$ implies $\bar{h} = \infty$; i.e., Inequality (3) holds for all $v \geqq \underline{v}$ without restriction on h_1. □

Since Lemma 7.1.6 requires $\gamma \geqq 2$, the V-cycle ($\gamma = 1$) is excluded. To prove $\zeta_l < 1$ from the inequalities (6), one needs $C^* < 1 - \zeta < 1$, which is unrealistic. Smaller values than $C^* \approx 1$ cannot be expected.

Theorem 7.1.2 applies to the case of no post-smoothing. A dual result can be proved in the case of no pre-smoothing (cf. Theorem 6.1.10):

Theorem 7.1.7 (multi-grid convergence of $MGM^{(0,v)}$). Suppose $\gamma \geqq 2$, the smoothing property (6.1.7a–c), the approximation property (6.1.8), and

$$\| L_l S_l^{(v)} L_l^{-1} \|_{\mathscr{F} \leftarrow \mathscr{F}} \leqq \hat{C}_S \qquad \text{for all } l \geqq 1, \ 0 < v < \bar{v}(h_1), \tag{7.1.8a}$$

$$\| r \|_{\mathscr{F} \leftarrow \mathscr{F}} \leqq \bar{C}_r, \quad \inf\{\| f_l \|_{\mathscr{F}} : f_l \in \mathscr{F}_l, \ r f_l = f_{l-1}\} \leqq \underline{C}_r \, \| f_{l-1} \|_{\mathscr{F}}$$
$$\text{for all } f_{l-1} \in \mathscr{F}_{l-1}. \tag{7.1.8b}$$

Then the assertions of Theorem 7.1.2 hold for $\| L_l M_l'(0, v) L_l^{-1} \|_{\mathscr{F} \leftarrow \mathscr{F}}$.

The analogue of Remark 7.1.1 is

Remark 7.1.8. Let $\mathscr{F}_l$ be a Hilbert space. An equivalent formulation of (8b) is the following assumption: There exists some prolongation $\hat{p}$: $\mathscr{F}_{l-1} \to \mathscr{F}_l$ with

$$\| \hat{p} \|_{\mathscr{F} \leftarrow \mathscr{F}} \leqq \underline{C}_r, \quad r\hat{p} = I, \quad \| r \|_{\mathscr{F} \leftarrow \mathscr{F}} \leqq \bar{C}_r. \tag{7.1.8b'}$$

The multi-grid iteration $MGM^{(\nu_1,\nu_2)}$ with $\nu_1 \neq 0$, $\nu_2 \neq 0$ will be discussed in the next chapter. A very rough analysis of $MGM^{(\nu_1,\nu_2)}$ can also be based on (6.1.17), which also holds for $M_l'(\nu_1,\nu_2)$. Note that Lemma 6.1.12 (or Lemma 7.2.4) does not carry over to $M_l'(\nu_1,\nu_2)$, if $\gamma \geqq 2$.

7.2 The Symmetric Case

The symmetric case is characterised by the assumptions (6.4.1 a–d):

$$L_l = L_l^* > O, \quad r = p^*, \quad L_{l-1} = r L_l p, \quad S_l = I - W_l^{-1} L_l, \quad W_l = W_l^* \geqq L_l.$$

To obtain a sharper estimate than in § 7.1, we have to analyse $M_l'(\nu_1,\nu_2)$ directly instead of using estimates of the two-grid matrix $M_l(\nu_1,\nu_2)$. We recall (1.4 b):

$$M_l'(\nu_1,\nu_2) = S_l^{(\nu_2)}\{I - p[I - (M_{l-1}'(\nu_1,\nu_2))^\gamma]\, L_{l-1}^{-1} r L_l\}\; S_l^{(\nu_1)}.$$

$\check{S}_l$, X_l, and Q_l have already been introduced in (6.4.4 a, b); in addition, we define

$$\check{M}_l'(\nu_1,\nu_2) := L_l^{1/2} M_l'(\nu_1,\nu_2)\, L_l^{-1/2}, \check{p} = L_l^{1/2} p L_{l-1}^{-1/2}, \check{r} = L_{l-1}^{-1/2} r L_l^{1/2}. \tag{7.2.1 a}$$

Note that

$$Q_l = I - \check{p}\check{r}, \quad \check{r} = \check{p}^*. \tag{7.2.1 b}$$

The representation of $M_l'(\nu_1,\nu_2)$ implies

$$\check{M}_l'(\nu_1,\nu_2) = \check{S}_l^{(\nu_2)}\{I - \check{p}[I - (\check{M}_{l-1}'(\nu_1,\nu_2))^\gamma]\,\check{r}\}\; \check{S}_l^{(\nu_1)}. \tag{7.2.2 a}$$

First we consider the case $\nu_1 = \nu_2 =: \nu/2$ and abbreviate $\check{M}_l'(\nu/2, \nu/2)$ by

$$\check{M}_l' := \check{M}_l'\left(\frac{\nu}{2}, \frac{\nu}{2}\right). \tag{7.2.2 b}$$

If $S_l^{(\mu)} = S_l^\mu$ and therefore

$$\check{S}_l^{(\mu)} = \check{S}_l^\mu = (I - X_l)^\mu, \qquad (X_l \text{ from } (6.4.4\,\text{a})) \tag{7.2.3}$$

formally, the definition (2 b) makes sense for any real value $\nu \geqq 0$. In particular, odd ν are permitted.

The main step of the following convergence proofs is

Lemma 7.2.1. Suppose (6.4.1 a–d), (2 b), (3), and the approximation property (6.4.8) for some $s > -1$. Then

$$O \leqq \check{M}_{l-1}' \leqq \zeta_{l-1} I, \quad 0 \leqq \zeta_{l-1} \leqq 1 \tag{7.2.4 a}$$

implies

$$O \leqq \check{M}_l' \leqq \zeta_l I, \quad \zeta_l := \min_{\zeta_{l-1}^\gamma \leqq \beta \leqq 1}\ \max_{0 \leqq \xi \leqq 1} f(\xi;\beta), \tag{7.2.4 b}$$

where

$$f(\xi;\beta) := (1-\xi)^{\nu}[\beta + (1-\beta)(C_A\xi)^{1+s}]. \tag{7.2.4 c}$$

Proof. $\check{M}'_{l-1} \geqq O$ proves $\check{M}'_{l-1} \geqq \check{M}_{l-1}$, which is positive semi-definite as stated in (6.4.5 d). Hence, $\check{M}'_l \geqq O$. Using (4 a), (1 b), (6.4.5 c), and (6.4.9), we obtain

$$\begin{aligned}\check{M}'_l &\leqq \check{S}_l^{\nu/2}[I - (1-\zeta_{l-1}^{\gamma})\check{p}\check{r}]\check{S}_l^{\nu/2} = \check{S}_l^{\nu/2}[(1-\zeta_{l-1}^{\gamma})Q_l + \zeta_{l-1}^{\gamma}I]\check{S}_l^{\nu/2} \\ &\leqq \check{S}_l^{\nu/2}\{(1-\zeta_{l-1}^{\gamma})[\alpha(C_A X_l)^{1+s} + (1-\alpha)I] + \zeta_{l-1}^{\gamma}I\}\check{S}_l^{\nu/2}\end{aligned}$$

for all $0 \leqq \alpha \leqq 1$. Define $\beta := (1-\zeta_{l-1}^{\gamma})(1-\alpha) + \zeta_{l-1}^{\gamma}$. For all $\beta \in [\zeta_{l-1}^{\gamma}, 1]$, we have

$$\check{M}'_l \leqq \check{S}_l^{\nu/2}\{(1-\beta)(C_A X_l)^{1+s} + \beta I\}\check{S}_l^{\nu/2} = f(X_l;\beta); \tag{7.2.5}$$

hence, by $O \leqq X_l \leqq I$, (4 b) follows. □

Theorem 7.2.2 (convergence of the *V*-cycle). Let $\|u_l\|_{\mathscr{U}} := |L_l^{1/2}u_l|_0$ be the energy norm and suppose (6.4.1 a–d), the approximation property (6.3.13) (i.e. (6.4.8) for $s = 0$), $S_l^{(\nu)} = S_l^{\nu}$, $\nu > 0$, $\gamma = 1$. Then the *V*-cycle converges:

$$\left\| M'_l\left(\frac{\nu}{2}, \frac{\nu}{2}\right)\right\|_{\mathscr{U}\leftarrow\mathscr{U}} \leqq \frac{C_A}{C_A + \nu} < 1. \tag{7.2.6}$$

Proof. $\|M'_l\|_{\mathscr{U}\leftarrow\mathscr{U}}$ equals $\|\check{M}'_l\|$. Formally, the proof by induction can be started with $\zeta_0 = \|\check{M}'_0\| = 0$ (cf. Lemma 7.1.4). Let the assertion hold for $l-1$, i.e. $\zeta_{l-1} \leqq C_A/(C_A+\nu)$. Choose $\beta^* := C_A/(C_A+\nu)$. $f(\xi;\beta^*)$ is maximum at $\xi = 0$ and one finds $\zeta_l = f(0;\beta^*) = \beta^* = C_A/(C_A+\nu)$. □

The contraction numbers improve with increasing γ. For $\gamma \to \infty$ they converge to the right side of (6.4.3). The result for $\gamma = 2$ reads as follows.

Theorem 7.2.3 (convergence of the *W*-cycle). Under the conditions of Theorem 7.2.2 (but $\gamma = 2$), the *W*-cycle converges. If $C_A > 1$ and $\nu \leqq (C_A - 1)[1 - (1 - 1/C_A)^{2\nu}]$, the contraction number is

$$\left\| M'_l\left(\frac{\nu}{2}, \frac{\nu}{2}\right)\right\|_{\mathscr{U}\leftarrow\mathscr{U}} \leqq (1 - 1/C_A)^{\nu} < 1. \tag{7.2.7 a}$$

Otherwise, the smallest root $\zeta = \zeta(\nu)$ of $\zeta = \eta_0(\nu)[\zeta^2 + (1-\zeta^2)C_A]^{\nu+1}$ satisfies

$$\left\| M'_l\left(\frac{\nu}{2}, \frac{\nu}{2}\right)\right\|_{\mathscr{U}\leftarrow\mathscr{U}} \leqq \zeta(\nu) \tag{7.2.7 b}$$

for all $l \geqq 0$.

The bounds in (6) and (7 a, b) are independent of the level number l. The contraction number of the *V*-cycle at a fixed level may be better, whereas the *W*-cycle contraction numbers are less dependent on l.

l	V-cycle	W-cycle
1	0.2963	0.2963
2	0.3699	0.3112
3	0.4041	0.3130
4	0.4239	0.313217
5	0.4369	0.313245
10	0.4659	0.313249
∞	0.5	0.313249

Table 7.2.1. l-dependent contraction numbers $\| M_l(\nu/2, \nu/2) \|_{\mathscr{U} \leftarrow \mathscr{U}}$ for $C_A = \nu = 2$ (computed by (4b))

For the analysis of $M_l'(\nu_1, \nu_2)$ with $\nu_1 \neq \nu_2$ we need a generalisation of Lemma 6.1.12:

Lemma 7.2.4. Suppose (6.4.1a–c) and $\gamma = 1$. Then

$$M_l'(\nu_1, \nu_2) = M_l'(0, \nu_2)\, M_l'(\nu_1, 0). \tag{7.2.8}$$

Proof by induction. (8) holds for $l = 1$ by Lemma 6.1.12. Suppose (8) for $l - 1$. If suffices to prove (8) for the transformed matrices $\check{M}_l'$ instead of M_l'. Since $\check{r} Q_l = O$, $Q_l \check{p} = O$, $Q_l^2 = Q_l$, and $\check{r}\check{p} = I$, one concludes that

$$\begin{aligned} \check{M}_l'(0, \nu_2)\, \check{M}_l'(\nu_1, 0) &= \check{S}_l^{(\nu_2)} [Q_l + \check{p} \check{M}_{l-1}'(0, \nu_2) \check{r}]\, [Q_l + \check{p} \check{M}_{l-1}'(\nu_1, 0) \check{r}]\, \check{S}_l^{(\nu_1)} \\ &= \check{S}_l^{(\nu_2)} [Q_l + \check{p} \check{M}_{l-1}'(0, \nu_2)\, \check{M}_{l-1}'(\nu_1, 0) \check{r}]\, \check{S}_l^{(\nu_1)} \\ &= \check{S}_l^{(\nu_2)} [Q_l + \check{p} \check{M}_{l-1}'(\nu_1, \nu_2) \check{r}]\, \check{S}_l^{(\nu_1)} = \check{M}_l'(\nu_1, \nu_2). \qquad \square \end{aligned}$$

An immediate result of Theorem 7.2.2 and Lemma 7.2.4 is

Theorem 7.2.5 (convergence of the V-cycle). Under the conditions of Theorem 7.2.2 and if $\nu_1 + \nu_2 > 0$, the V-cycle converges and its contraction number is

$$\| M_l'(\nu_1, \nu_2) \|_{\mathscr{U} \leftarrow \mathscr{U}} \leqq \frac{C_A}{\sqrt{C_A + \nu_1}\, \sqrt{C_A + \nu_2}} < 1. \tag{7.2.9}$$

Proof. Lemma 7.2.4 yields

$$\| M_l'(\nu_1, \nu_2) \|_{\mathscr{U} \leftarrow \mathscr{U}} \leqq \| M_l'(0, \nu_2) \|_{\mathscr{U} \leftarrow \mathscr{U}}\, \| M_l'(\nu_1, 0) \|_{\mathscr{U} \leftarrow \mathscr{U}}.$$

Since

$$\begin{aligned} \| M_l'(0, \nu_2) \|^2_{\mathscr{U} \leftarrow \mathscr{U}} &= \| \check{M}_l'(0, \nu_2) \|^2 = \| \check{M}_l'(0, \nu_2)^* \, \check{M}_l'(0, \nu_2) \| \\ &= \| \check{M}_l'(\nu_2, 0)\, \check{M}_l'(0, \nu_2) \| = \| \check{M}_l'(\nu_2, \nu_2) \| = \| \check{M}_l'(\nu_2, \nu_2) \|_{\mathscr{U} \leftarrow \mathscr{U}} \end{aligned}$$

and similarly $\| M_l'(\nu_1, 0) \|^2_{\mathscr{U} \leftarrow \mathscr{U}} = \| M_l'(\nu_1, \nu_1) \|_{\mathscr{U} \leftarrow \mathscr{U}}$, inequality (9) follows. $\square$

Although Lemma 7.2.4 does not extend to the W-cycle, Theorem 7.2.5 can be generalised to $\gamma \geqq 2$. The result is stated in Exercise 7.4.3.

Hitherto we assumed the approximation property (6.4.8) for $s = 0$. In the less regular case $-1 < s < 0$, one can prove convergence of the W-cycle, but not of the V-cycle.

7.3 Comments

As already hinted in an early paper by the author (cf. Hackbusch [1]), the multi-grid convergence can be derived from almost the same basic assumptions as the two-grid convergence. The additional suppositions (1.1) and (1.2) are very weak. In the symmetric case of § 7.2 these inequalities are not mentioned explicitly. However, it is immediate that under the conditions of § 7.2 the inequalities (1.1), (1.2), (1.8 a, b) hold with the optimal constants $C_S = \underline{C}_p = \bar{C}_p = \hat{C}_S = \underline{C}_r = \bar{C}_r = 1$ and $\bar{v}(h) = \infty$.

The proof of Theorem 7.1.2 is implicitly contained in the convergence proofs by Fedorenko [2], Bachvalov [1], Astrachancev [1], Nicolaides [2], etc. Usually, the theorem is proved slightly differently. First one can show that the two-grid iteration converges with a contraction number $\zeta' = \zeta'(\varepsilon)$ even if the coarse-grid equation is solved up to a relative error $\varepsilon > 0$. If $(\zeta'(\varepsilon))^\gamma \leqq \varepsilon$, one obtains the assertions of Theorem 7.1.2. This variant of the proof will be presented for the nonlinear multi-grid iteration in § 9.5.

The results of § 7.2, which have been described for the special case $W_l = \| L_l \| I$ by Hackbusch [29] and Braess-Hackbusch [1], proceeded from the proofs used by Braess [1], [2]. A V-cycle convergence proof requiring stronger assumptions on the finite element space $\mathscr{H}_l$ is presented by McCormick [3]. The V-cycle is also studied by Maitre-Musy [2], [3].

Bank-Douglas [1] gave the first V-cycle convergence proof, that applies to the smoothing iteration (6.4.1 d) with a general $W_l = W_l^* \geqq L_l$. Though the proof is formulated quite differently from our proof in § 7.2, it is completely equivalent. For instance, the inequality of their Theorem 4 (involving constants α, κ) is the approximation property (6.4.2) with the same α and $C_A := \kappa$.

7.4 Exercises

Exercise 7.4.1. Let $\mathscr{S}_l^{(v)}(u_l, f_l) = S_l^{(v)} u_l + T_l^{(v)} f_l$. Show that the matrix $N_l'(v_1, v_2)$ of the multi-grid iteration (cf. (1.3.2)) equals $N_0'(v_1, v_2) = L_0^{-1}$ for $l = 0$ and

$$N_l'(v_1, v_2) = T_l^{(v_2)} + S_l^{(v_2)} \left[T_l^{(v_1)} - p \sum_{k=0}^{\gamma-1} (M_{l-1}'(v_1, v_2))^k N_{l-1}'(v_1, v_2)\, r(L_l T_l^{(v_1)} - I) \right]$$

for $l \geqq 1$.

Exercise 7.4.2. Prove that Theorem 7.1.2 extends to the choice (4.1.5) of an alternating γ.

Exercise 7.4.3 (convergence of the *W*-cycle). Let $\|\cdot\|_{\mathcal{U}}$ be the energy norm $\|u_l\|_{\mathcal{U}} = |L_l^{1/2} u_l|_0$ and suppose (6.4.1 a–d), (6.4.8) for $s=0$, $S_l^{(\nu)} = S_l^{\nu}$, $\nu_1 + \nu_2 \geqq 1$, $\gamma = 2$. Then the multi-grid method $MGM^{(\nu_1,\nu_2)}$ (*W*-cycle) converges. Prove that its contraction number

$$\| M_l'(\nu_1, \nu_2) \|_{\mathcal{U} \leftarrow \mathcal{U}} \leqq \zeta(\nu_1, \nu_2) < 1$$

is defined as follows. Set

$$\varphi(\nu; z) := \min_{z \leqq \beta \leqq 1} \max_{0 \leqq \xi \leqq 1} f(\xi; \beta)$$

with f from (2.4c). Then $\zeta = \zeta(\nu_1, \nu_2)$ is the smallest root of

$$\zeta = \sqrt{\varphi(\nu_1; \zeta^2)\, \varphi(\nu_2; \zeta^2)}.$$

Note that the next result does not state uniform convergence (with respect to l).

Exercise 7.4.4. Assume the symmetry conditions (6.4.1 a–c) and

$$\| L_l^{1/2} S_l^{(\nu_1)} L_l^{-1/2} \| \; \| L_l^{1/2} S_l^{(\nu_2)} L_l^{-1/2} \| < 1.$$

Prove that the multi-grid iteration $MGM^{(\nu_1,\nu_2)}$ with $\gamma \geqq 1$ (e.g. the *V*-cycle) converges.

8. Fourier Analysis

For the analysis of difference discretisations of time-dependent problems the Fourier analysis has proved as a very useful tool (cf. Richtmyer-Morton [1]). In § 2.4 we applied the Fourier analysis to a one-dimensional model problem. In this chapter the analogous analysis is performed in two dimensions.

The analysis of the previous Chaps. 6, 7 yields insight into the behaviour of the two- and multi-grid contraction numbers depending on the smoothing iteration, on the number of smoothing steps, and on the involved norm. However, the bounds on the contraction numbers are upper bounds. Even the quantitative results of § 6.4 and § 7.2 cannot be used as predictions of the convergence rate, since the constant C_A from (6.4.2) is not easy to compute. The goal of this section is to compute the exact two-grid convergence rate or at least to give a rough approximation to this value.

Except for special model problems, the Fourier analysis can rarely be used as a precise basis for convergence proofs. In particular there is no application to irregular geometries which arise typically from finite element discretisation.

8.1 Model Examples

The model problems of this subsection admit an exact analysis by Fourier transformation.

8.1.1 Poisson's Equation in a Square

We consider the Poisson equation (3.1.1) in the unit square $\Omega = (0, 1) \times (0, 1)$ with Dirichlet conditions (3.1.2a). It is discretised by the five-point scheme (3.1.15) in the grid of size $h_l = 1/2^{1+l}$.

The eigenvectors of the matrix L_l are the grid functions $e_l^{\nu\mu}$ ($1 \leqq \nu$, $\mu \leqq N_l - 1$, $N_l := 1/h_l$) with

$$e_l^{\nu\mu}(x, y) = 2 \sin(\nu\pi x) \sin(\mu\pi y), \qquad (x, y) \in \Omega_l, \tag{8.1.1}$$

(cf. (2.2.4)). As in § 2.4 we can state

Remark 8.1.1. The Fourier transform

$$Q_l\colon \alpha = (\alpha_{\nu\mu})_{\nu,\mu=1,\dots,N_l-1} \mapsto \sum_{\nu,\mu=1}^{N_l-1} \alpha_{\nu\mu} e_l^{\nu\mu} \tag{8.1.2a}$$

is unitary (i.e. $Q_l^{-1} = Q_l^*$), since the vectors $e_l^{\nu\mu}$ form an orthonormal system with respect to $\langle\cdot,\cdot\rangle$ from (3.5.2). We prescribe a special ordering of the coefficients $\alpha_{\nu\mu}$ of α. Enumerate the pairs $\{(\nu,\mu)\colon 1 \leqq \nu,\mu \leqq N_{l-1}\}$ lexicographically and define

$$\alpha = (\alpha^{11}, \alpha^{12}, \dots, \alpha^{1,N_{l-1}}, \alpha^{21}, \dots, \alpha^{N_{l-1},N_{l-1}}), \tag{8.1.2b}$$

where the 4-tuples $\alpha^{\nu\mu}$ are

$$\begin{aligned} \alpha^{\nu\mu} &= (\alpha_{\nu\mu}, \alpha_{\nu'\mu}, \alpha_{\nu\mu'}, \alpha_{\nu'\mu'}), \qquad 1 \leqq \nu,\mu \leqq N_{l-1} \\ \nu' &= N_l - \nu, \quad \mu' = N_l - \mu. \end{aligned} \tag{8.1.2c}$$

If $\nu = N_{l-1} = \nu' \neq \mu$, $\alpha^{\nu\mu}$ degenerates to $(\alpha_{\nu\mu}, \alpha_{\nu\mu'})$; if $\mu = N_{l-1} = \mu' \neq \nu$, one has $\alpha^{\nu\mu} = (\alpha_{\nu\mu}, \alpha_{\nu'\mu})$, while $\alpha^{\nu\mu} = \alpha_{\nu\mu}$ for $\nu = \mu = N_{l-1}$. At level $l-1$ the vector $\alpha = (\alpha_{\nu\mu})_{\nu,\mu=1,\dots,N_{l-1}-1}$ is ordered lexicographically.[1]

Since $L_l e_l^{\nu\mu} = 4h_l^{-2}[\sin^2(\nu\pi h_l/2) + \sin^2(\mu\pi h_l/2)]\, e_l^{\nu\mu}$, one obtains

Lemma 8.1.2. The transformed matrices $\hat{L}_l := Q_l^{-1} L_l Q_l$ are

$$\hat{L}_l = \operatorname{blockdiag}\{\dots, L_l^{(\nu\mu)}, \dots\} \tag{8.1.3a}$$

with 4×4 blocks

$$L_l^{(\nu\mu)} = 4h_l^{-2} \operatorname{diag}\{s_\nu^2 + s_\mu^2,\ c_\nu^2 + s_\mu^2,\ s_\nu^2 + c_\mu^2,\ c_\nu^2 + c_\mu^2\}, \tag{8.1.3b}$$

$$s_\nu = \sin(\nu\pi h_l/2), \quad c_\nu = \cos(\nu\pi h_l/2). \tag{8.1.3c}$$

At level $l-1$ one has

$$\hat{L}_{l-1} = \operatorname{diag}\{\dots, L_{l-1}^{(\nu\mu)}, \dots\}, \quad L_{l-1}^{(\nu\mu)} = 4h_l^{-2}(s_\nu^2 c_\nu^2 + s_\mu^2 c_\mu^2). \tag{8.1.3d}$$

The definition of the degenerate 2×2 or 1×1 matrices $L_l^{(\nu\mu)}$ when $\nu = N_{l-1}$ and/or $\mu = N_{l-1}$ is obvious.

Lemma 8.1.3. The nine-point prolongation (3.4.2) yields $\hat{p} = Q_l^{-1} p Q_{l-1}$ with a structure as in Lemma 2.4.3. The blocks of $\hat{p}$ are

$$p^{(\nu\mu)} = \begin{bmatrix} c_\nu^2 c_\mu^2 \\ -s_\nu^2 c_\mu^2 \\ -c_\nu^2 s_\mu^2 \\ s_\nu^2 s_\mu^2 \end{bmatrix}, \qquad 1 \leqq \nu,\mu \leqq N_{l-1} - 1. \tag{8.1.4a}$$

---

1 This is possible since we shall study the two-grid situation with fixed l. It would be more consistent to define the ordering recursively. Instead of (2b), the ordering from level $l-1$ should be used

The blocks of the transformed nine-point restriction (3.5.4), $\hat{r} = Q_{l-1}^{-1} r Q_l$, are

$$r^{(\nu\mu)} = [c_\nu^2 c_\mu^2, \; -s_\nu^2 c_\mu^2, \; -c_\nu^2 s_\mu^2, \; s_\nu^2 s_\mu^2], \qquad 1 \leqq \nu, \mu \leqq N_{l-1} - 1. \tag{8.1.4b}$$

Proof. (4b) is proved as (2.4.16) in the one-dimensional case. Since $p = r^*$, (4a) is immediate. □

From Lemmata 8.1.2–3 we obtain

Lemma 8.1.4. The transformed matrix of the coarse-grid correction $C_l = I - p L_{l-1}^{-1} r L_l$ equals

$$\hat{C}_l = Q_l^{-1} C_l Q_l = \text{blockdiag}\{\ldots, C_l^{(\nu\mu)}, \ldots\}, \tag{8.1.5a}$$

$$C_l^{(\nu\mu)} = I - c c^T L_l^{(\nu\mu)}/(4 h_l^{-2}), \tag{8.1.5b}$$

$$c = \frac{1}{\sqrt{s_\nu^2 c_\nu^2 + s_\mu^2 c_\mu^2}} \begin{bmatrix} c_\nu^2 c_\mu^2 \\ -s_\nu^2 c_\mu^2 \\ -c_\nu^2 s_\mu^2 \\ s_\nu^2 s_\mu^2 \end{bmatrix}. \tag{8.1.5c}$$

Although the damped Jacobi iteration is the simplest to analyse, we shall study the Gauß-Seidel iteration (3.3.1) with red-black ordering.

In § 3.3.1 (Fig. 3.3.5) we defined a four-colour ordering and defined four subgrids $\Omega_l^1, \ldots, \Omega_l^4$. The points of $\Omega_l^r := \Omega_l^1 \cup \Omega_l^2$ are the 'red' ones and those of $\Omega_l^b := \Omega_l^3 \cup \Omega_l^4$ the 'black' ones of the red-black ordering. The Gauß-Seidel quarter step on Ω_l^j ($1 \leqq j \leqq 4$) is defined by

$$u_l' = \mathscr{S}_{lj}(u_l, f_l), \qquad j \in \{1, 2, 3, 4\},$$
$$u_l'(\mathbf{x}) = u_l(\mathbf{x}) \quad \text{for } \mathbf{x} \in \Omega_l^i (i \neq j), \quad (L_l u_l - f_l)(\mathbf{x}) = 0 \quad \text{for } \mathbf{x} \in \Omega_l^j.$$

Similarly, $\mathscr{S}_{lr}$ and $\mathscr{S}_{lb}$ denote Gauß-Seidel half steps on Ω_l^r and Ω_l^b, respectively. The corresponding iteration matrices are S_{lj}, S_{lr}, S_{lb}.

Note 8.1.5. Even for general nine-point schemes (3.1.16), the matrices S_{lj}, $1 \leqq j \leqq 4$, are projections. For a five-point formula one has in addition that $S_{lr} = S_{l1} S_{l2} = S_{l2} S_{l1}$, $S_{lb} = S_{l3} S_{l4} = S_{l4} S_{l3}$ and that S_{lr} and S_{lb} are projections.

A straightforward analysis yields:

Lemma 8.1.6. Let L_l be the five-point scheme (3.1.15) and $\mathscr{S}_{lj}$ as defined above. The Fourier transformed matrices $\hat{S}_{lj} = Q_l^{-1} S_{lj} Q_l$ are

$$\hat{S}_{lj} = \text{blockdiag}\{\ldots, S_{lj}^{(\nu\mu)}, \ldots\}, \tag{8.1.6a}$$

$$S_{lj}^{(\nu\mu)} = I - s_j s_j^T L_l^{(\nu\mu)}/(4 h_l^{-2}), \qquad 1 \leqq \nu, \mu \leqq N_{l-1} - 1, \tag{8.1.6b}$$

where

$$s_1 = \frac{1}{2}\begin{bmatrix} 1 \\ -1 \\ -1 \\ 1 \end{bmatrix}, \quad s_2 = \frac{1}{2}\begin{bmatrix} 1 \\ 1 \\ 1 \\ 1 \end{bmatrix}, \quad s_3 = \frac{1}{2}\begin{bmatrix} 1 \\ 1 \\ -1 \\ -1 \end{bmatrix}, \quad s_4 = \frac{1}{2}\begin{bmatrix} 1 \\ -1 \\ 1 \\ -1 \end{bmatrix}. \tag{8.1.6c}$$

If $\nu = N_{l-1} = \nu'$, $\mu \neq \mu'$, the vectors s_j become $s_1 = s_4 = \begin{bmatrix} 0 \\ 0 \end{bmatrix}$, $s_2 = \begin{bmatrix} 1 \\ 1 \end{bmatrix} \Big/ \sqrt{2}$, $s_3 = \begin{bmatrix} 1 \\ -1 \end{bmatrix} \Big/ \sqrt{2}$. When $\mu = \mu'$ and $\nu \neq \nu'$, one has $s_1 = s_3 = \begin{bmatrix} 0 \\ 0 \end{bmatrix}$, $s_2 = \begin{bmatrix} 1 \\ 1 \end{bmatrix} \Big/ \sqrt{2}$, $s_4 = \begin{bmatrix} 1 \\ -1 \end{bmatrix} \Big/ \sqrt{2}$, whereas $s_1 = s_3 = s_4 = 0$ and $s_2 = 1$ holds for $\nu = \mu = N_{l-1}$.

Lemma 2.4.1 applies again and proves

Theorem 8.1.7. Let L_l and L_{l-1} be the five-point discretisations (3.1.15), while p is the nine-point prolongation (3.4.2) and r is the nine-point restriction (3.5.4). Let $\mathscr{S}_l$ be the Gauß-Seidel iteration with red-black ordering: $\mathscr{S}_l = \mathscr{S}_{lb} \mathscr{S}_{lr}$. Then the contraction number of the two-grid iteration $TGM^{(\nu_1, \nu_2)}$ with respect to $\|\cdot\|$ is exactly

$$\| M_l(\nu_1, \nu_2) \| = \max \{ \| M_l^{(\nu\mu)} \| : 1 \leqq \nu, \mu \leqq N_{l-1} \}, \tag{8.1.7 a}$$

where the 4×4 matrices (or 2×2 or 1×1 matrices in the degenerate cases) are

$$M_l^{(\nu\mu)} = (S_l^{(\nu\mu)})^{\nu_2} C_l^{(\nu\mu)} (S_l^{(\nu\mu)})^{\nu_1}, \qquad S_l^{(\nu\mu)} := S_{l4}^{(\nu\mu)} S_{l3}^{(\nu\mu)} S_{l2}^{(\nu\mu)} S_{l1}^{(\nu\mu)}. \tag{8.1.7 b}$$

The spectral radius equals

$$\varrho(M_l(\nu_1, \nu_2)) = \max \{ \varrho(M_l^{(\nu\mu)}) : 1 \leqq \nu, \mu \leqq N_{l-1} \}. \tag{8.1.7 c}$$

By this theorem we reduce the analysis of $M_l(\nu_1, \nu_2)$ to that of at most 4×4 matrices. The right sides of (7 a, c) can be evaluated numerically.

In what follows we reduce the 4×4 matrices once more to 2×2 matrices. Fix $\nu, \mu \in \{1, \ldots, N_{l-1} - 1\}$ and introduce

$$L := L_l^{(\nu\mu)}, \qquad M := L^{1/2} M_l^{(\nu\mu)} L^{-1/2}, \qquad S_j := L^{1/2} S_{lj}^{(\nu\mu)} L^{-1/2},$$
$$C := L^{1/2} C_l^{(\nu\mu)} L^{-1/2}, \qquad S_r = S_2 S_1, \qquad S_b = S_4 S_3.$$

Since $S_r^2 = S_r$ (cf. Note 8.1.5), the spectral radius equals

$$\varrho(M_l^{(\nu\mu)}) = \varrho(C(S_b S_r)^{\nu_1 + \nu_2}) = \varrho(S_r C(S_b S_r)^{\nu_1 + \nu_2}).$$

An eigenvector of $S_r C(S_b S_r)^{\nu_1 + \nu_2}$ must lie in the range of S_r, which is spanned by the orthonormal vectors

$$e_1 = (\sqrt{(c_\nu^2 + c_\mu^2)/2}, 0, \qquad 0, \qquad -\sqrt{(s_\nu^2 + s_\mu^2)/2})^T,$$
$$e_2 = (0, \qquad \sqrt{(s_\nu^2 + c_\mu^2)/2}, \quad -\sqrt{(c_\nu^2 + s_\mu^2)/2}, 0)^T.$$

These vectors are invariant under S_r and eigenvectors of $S_r S_b$. One checks that $M(\alpha e_1 + \beta e_2) = \gamma e_1 + \delta e_2$ with $\begin{pmatrix} \gamma \\ \delta \end{pmatrix} = A \begin{pmatrix} \alpha \\ \beta \end{pmatrix}$, where

$$A = \left(D - \frac{1}{s_\nu^2 c_\nu^2 + s_\mu^2 c_\mu^2} \begin{bmatrix} d_1 & 0 \\ 0 & d_2 \end{bmatrix} \begin{bmatrix} 1 & 1 \\ 1 & 1 \end{bmatrix} \begin{bmatrix} d_3 & 0 \\ 0 & d_4 \end{bmatrix} \right) D^{\nu_1 + \nu_2 - 1}, \tag{8.1.8 a}$$

$$\begin{aligned} d_1 &= a(c_\nu^2 c_\mu^2 - s_\nu^2 s_\mu^2), & d_2 &= b(c_\nu^2 s_\mu^2 - s_\nu^2 c_\mu^2), \\ d_3 &= a(c_\nu^2 c_\mu^2 + s_\nu^2 s_\mu^2)(c_\nu^2 - s_\mu^2), & d_4 &= b(c_\nu^2 s_\mu^2 + s_\nu^2 c_\mu^2)(c_\nu^2 - c_\mu^2), \qquad \text{(8.1.8 b)} \\ a &= \sqrt{(s_\nu^2 + s_\mu^2)(c_\nu^2 + c_\mu^2)/2}, & b &= \sqrt{(c_\nu^2 + s_\mu^2)(s_\nu^2 + c_\mu^2)/2}, \\ D &= \begin{bmatrix} 1 - 2a^2 & 0 \\ 0 & 1 - 2b^2 \end{bmatrix}. \end{aligned}$$

Since $\varrho(M_l^{(\nu\mu)}) = \varrho(A)$, it suffices to calculate the eigenvalues of the 2×2 matrices A. A depends on ν, μ and ν_1, ν_2. Let $A(\xi, \eta; \nu_1 + \nu_2)$ denote the matrix from (8 a) for $s_\nu^2 = \xi, c_\nu^2 = 1 - \xi, s_\mu^2 = \eta, c_\mu^2 = 1 - \eta$. One can show that in spite of the degenerate cases $\nu = \nu'$ or $\mu = \mu'$, one obtains

$$\varrho(M_l(\nu_1, \nu_2)) \leqq \varrho_{\nu_1 + \nu_2} := \max\{\varrho(A(\xi, \eta; \nu_1, \nu_2)) : 0 \leqq \xi, \eta \leqq \tfrac{1}{2}\} \qquad \text{(8.1.9)}$$

from Theorem 8.1.7. The numerical evaluation of ϱ_ν yields the following convergence rates:

$\nu_1 + \nu_2$	1	2	3	4	5
$\varrho_{\nu_1 + \nu_2}$	0.25	0.07407	0.05273	0.04096	0.03349

It turns out that for $\nu_1 + \nu_2 \geqq 2$ the function $\varrho(A(\xi, \eta; \nu_1 + \nu_2))$ attains its maximum at $\xi = \eta = (1 - \sqrt{\nu/(1 + \nu)})/2$. Therefore the value of the two-grid rate equals

$$\varrho_1 = \tfrac{1}{4}, \quad \varrho_\nu = \tfrac{1}{2}\eta_0(\nu) \quad \text{for } \nu \geqq 2 \quad \text{with} \quad \eta_0(\nu) \quad \text{from (6.2.1 b)}.$$

Generalisations of the previous analysis are considered in

Remark 8.1.8.

a) The contraction numbers (7 a) can also be calculated from 2×2 matrices.
b) The previous analysis applies to all symmetric nine-point schemes. The 4×4 matrices can then be reduced to 3×3 matrices.
c) The generalisation to more than two dimensions is evident.
d) To extend the analysis to Neumann conditions one has to replace the sine-functions in (1) by cosine-functions.

The Fourier analysis in the case of Poisson's equation is described in detail by Stüben-Trottenberg-Witsch [2].

8.1.2 Periodic Boundary Conditions

As pointed out in Remark 8.1.8, the approach of § 8.1.1 can be extended to more general difference schemes with constant coefficients, but it fails when $L_l \neq L_l^*$. Unsymmetric L_l arise e.g. from all differential operators involving derivatives of odd order.

General constant difference schemes can still be analysed by Fourier transform technique if we pose periodic boundary conditions in

$\Omega = (-1,1)\times(-1,1)$:

$$\frac{\partial^\nu}{\partial x^\nu} u(-1, y) = \frac{\partial^\nu}{\partial x^\nu} u(1, y) \qquad (-1 < y < 1),$$

$$\frac{\partial^\nu}{\partial y^\nu} u(x, -1) = \frac{\partial^\nu}{\partial y^\nu} u(x, 1) \qquad (-1 < x < 1) \tag{8.1.10}$$

for all $\nu = 0, 1, \ldots, m$,

where $2m$ is the order of the differential operator. Instead of the unit square we could choose any rectangle Ω. The grid is

$$\Omega_l = \{(x, y) \in \Omega : x = \nu h_l,\ y = \mu h_l,\ 1 - N_l \leqq \nu,\ \mu \leqq N_l\}$$

with $N_l = 1/h_l$, $h_l = h_0/2^l$. A grid function u_l on Ω_l can be extended to a doubly 2-periodic grid function on the infinite grid Q_l from (3.1.13). Then $(L_l u_l)(\mathbf{x}) = f_l(\mathbf{x})$ has to hold at all $\mathbf{x} \in Q_l$.

The $4 N_l^2$ eigenvectors are $e_l^{\nu\mu}$, $1 - N_l \leqq \nu, \mu \leqq N_l$, with

$$e_l^{\nu\mu}(x, y) = \tfrac{1}{2} e^{\pi i(\nu x + \mu y)}, \quad (x, y) \in \Omega_l. \tag{8.1.11 a}$$

Each low frequency (ν, μ), $1 - N_{l-1} \leqq \nu, \mu \leqq N_{l-1}$, is associated with the corresponding high frequencies (ν', μ), (ν, μ'), (ν', μ'), where ν' and μ' are now defined by

$$\nu' = \begin{cases} \nu + N_l & \text{if } \nu \leqq 0, \\ \nu - N_l & \text{if } \nu > 0. \end{cases}$$

The Fourier transform

$$Q_l : \alpha = (\alpha_{\nu\mu})_{1 - N_l \leqq \nu, \mu \leqq N_l} \mapsto \sum_{\nu, \mu = 1 - N_l}^{N_l} \alpha_{\nu\mu} e_l^{\nu\mu} \tag{8.1.11 b}$$

is unitary. Again the analysis of the two-grid matrix M_l can be replaced by an analysis of the blocks of the block-diagonal matrix $\hat{M}_l := Q_l^{-1} M_l Q_l$, which is composed of $\hat{L}_l$, $\hat{r}$, $\hat{p}$, $\hat{L}_{l-1}$, $\hat{S}_l$.

In what follows we briefly describe $\hat{L}_l, \hat{r}, \hat{p}$ for general L_l, r, p. The difference operator

$$L_l = \sum_{\boldsymbol{\alpha} \in \mathbb{Z}^2} c_{\boldsymbol{\alpha}}(h_l) T^{\boldsymbol{\alpha}} \tag{8.1.12 a}$$

(T from (6.3.25 a)) is transformed into the diagonal matrix

$$\hat{L}_l = Q_l^{-1} L_l Q_l = \operatorname{blockdiag}\{\ldots, L_l^{(\nu\mu)}, \ldots\}, \tag{8.1.12 b}$$

$$L_l^{(\nu\mu)} = \sum_{\boldsymbol{\alpha} \in \mathbb{Z}^2} c_{\boldsymbol{\alpha}}(h_l) e^{\pi i(\nu\alpha_1 + \mu\alpha_2)h_l} \operatorname{diag}\{1, (-1)^{\alpha_1}, (-1)^{\alpha_2}, (-1)^{\alpha_1 + \alpha_2}\}$$

$$(1 - N_{l-1} \leqq \nu, \mu \leqq N_{l-1}). \tag{8.1.12 c}$$

To study the restriction we start with the trivial injection r_{inj} from (3.5.1). $\hat{r}_{\text{inj}} = Q_{l-1}^{-1} r_{\text{inj}} Q_l$ equals $\operatorname{blockdiag}\{\ldots, r_{\text{inj}}^{\nu\mu}, \ldots\}$ with

$$r_{\text{inj}}^{\nu\mu} = [1, 1, 1, 1], \quad 1 - N_{l-1} \leqq \nu, \mu \leqq N_{l-1},$$

since $r_{\text{inj}}\, e_l^{\nu\mu} = r_{\text{inj}}\, e_l^{\nu'\mu} = r_{\text{inj}}\, e_l^{\nu\mu'} = r_{\text{inj}}\, e_l^{\nu'\mu'} = e_{l-1}^{\nu\mu}$. A general restriction

$$(r\, u_l)(\mathbf{x}) = (\sum_{\boldsymbol{\alpha}} \varrho_{\boldsymbol{\alpha}}\, T^{\boldsymbol{\alpha}}\, u_l)(\mathbf{x}) \quad \text{for } \mathbf{x} \in \Omega_{l-1} \tag{8.1.13 a}$$

can be written as the product $r = r_{\text{inj}}\, \Lambda_l$, where $\Lambda_l = \sum_{\boldsymbol{\alpha}} \varrho_{\boldsymbol{\alpha}}\, T^{\boldsymbol{\alpha}}$. Using the representation of r_{inj} and applying (12 a–c) to Λ_l instead of L_l, one obtains

$$\hat{r} := Q_{l-1}^{-1}\, r\, Q_l = \text{blockdiag}\,\{\ldots, r^{\nu\mu}, \ldots\}, \tag{8.1.13 b}$$

$$r^{\nu\mu} = \sum_{\boldsymbol{\alpha}} \varrho_{\boldsymbol{\alpha}}\, e^{\pi i (\alpha_1 \nu + \alpha_2 \mu) h_l}\, [1, (-1)^{\alpha_1}, (-1)^{\alpha_2}, (-1)^{\alpha_1+\alpha_2}]. \tag{8.1.13 c}$$

To calculate $\hat{p}$ one has to apply (13 a–c) to $\tilde{r} := p^*$ and to use $p^{\nu\mu} = (\tilde{r}^{\nu\mu})^*$.

If $\hat{S}_l$ is also block-diagonal, $\hat{M}_l$ proves to be block-diagonal. Once again the two-grid contraction number and the spectral radius are then given by (7 a, c) if we replace "$1 \leqq \nu, \mu \leqq N_{l-1}$" by "$1 - N_{l-1} \leqq \nu, \mu \leqq N_{l-1}$".

If a Dirichlet problem with an unsymmetric operator L_l is to be analysed, one may replace the Dirichlet data by periodic boundary conditions and apply the analysis of this section. The question arises, whether the periodic problem yields convergence numbers which approximate those of the Dirichlet problem. For the one-dimensional singular perturbation problem

$$-\varepsilon\, u''(x) + u'(x) = f(x), \qquad \varepsilon \geq 0, \tag{8.1.14}$$

a positive answer is proved by Hackbusch [34].

8.1.3 The Case of $\Omega = \mathbb{R}^d$

Instead of posing periodic boundary conditions, we can omit the boundary conditions by choosing $\Omega = \mathbb{R}^2$ (or $\Omega = \mathbb{R}^d$, respectively). The grid Ω_l coincides with the infinite grid Q_l. Grid functions with finite Euclidean norm can be represented by

$$u_l(x, y) = \tfrac{1}{2} \int_{-1}^{1} \int_{-1}^{1} \hat{u}_l(\xi, \eta)\, e^{i\pi(x\xi + y\eta)/h_l}\, d\xi\, d\eta\,. \tag{8.1.15 a}$$

That means that $\{u_l(x, y) \colon (x, y) \in \Omega_l\}$ can be regarded as Fourier coefficients of the 2-periodic function

$$\hat{u}_l(\xi, \eta) := \tfrac{1}{2} \sum_{(x,y)\in\Omega_l} u_l(x, y)\, e^{-i\pi(x\xi + y\eta)/h_l}. \tag{8.1.15 b}$$

Using $e_l^{\nu\mu}$ from (11 a) we can rewrite (15 a) as

$$u_l = \int_{-1}^{1} \int_{-1}^{1} \hat{u}_l(\xi, \eta)\, e_l^{\xi/h_l, \eta/h_l}\, d\xi\, d\eta\,. \tag{8.1.15 a*}$$

The transform Q_l is defined by $u_l = Q_l\, \hat{u}_l$. All results of § 8.1.2 remain valid if we replace the integral frequencies $\nu, \mu \in (-N_l, N_l]$ by the real variables ξ/h_l, η/h_l, where $\xi, \eta \in (-1, 1]$. The frequency ξ' associated with ξ is

$$\xi' = \begin{cases} \xi + 1 & \text{if } \xi \leq 0 \\ \xi - 1 & \text{if } \xi > 0. \end{cases}$$

A detailed description of the Fourier analysis in the case of $\Omega = \mathbb{R}^d$ is given by Hemker [2].

The Gauß-Seidel iteration with red-black ordering considered in § 8.1.1 leads again to a coupling of the four frequencies (ξ, η), (ξ', η), (ξ, η'), (ξ', η'). But for several other variants of the Gauß-Seidel process the coupling can be avoided. In this case the Fourier transform diagonalises the infinite matrix S_l:

$$S_l\, e^{\pi i(x\xi + y\eta)/h_l} = \sigma_l(\xi, \eta)\, e^{\pi i(x\xi + y\eta)/h_l}. \tag{8.1.16}$$

Let L_l be given by (12a). The lexicographical Gauß-Seidel iteration $u_l^{k+1} = \mathscr{S}_l(u_l^k, f_l)$ on the infinite grid $\Omega_l = Q_l$ is defined by

$$u_l^{k+1} = [f_l - \textstyle\sum_I c_{\boldsymbol\alpha}(h_l)\, T^{\boldsymbol\alpha} u_l^{k+1} - \sum_{II} c_{\boldsymbol\alpha}(h_l)\, T^{\boldsymbol\alpha} u_l^k]/c_0(h_l).$$

where $\sum_I$ is taken over all $\boldsymbol\alpha = (\alpha_1, \alpha_2) \in \mathbb{Z}^2$ with either $\alpha_2 < 0$ or $\alpha_1 < 0$, $\alpha_2 = 0$. $\sum_{II}$ is the sum over the remaining $\boldsymbol\alpha \neq (0, 0)$. Thus, one obtains

Note 8.1.9. Equation (16) holds for the lexicographical Gauß-Seidel iteration with

$$\sigma_l^{\text{lex}}(\xi, \eta) = -\textstyle\sum_I c_{\boldsymbol\alpha}(h_l)\, e^{\pi i(\alpha_1\xi + \alpha_2\eta)/h_l} / [c_0(h_l) + \sum_{II} c_{\boldsymbol\alpha}(h_l)\, e^{\pi i(\alpha_1\xi + \alpha_2\eta)/h_l}]. \tag{8.1.17}$$

The corresponding symmetric Gauß-Seidel process leads to

$$\sigma_l^{sym}(\xi, \eta) = |\sigma_l^{\text{lex}}(\xi, \eta)|^2. \tag{8.1.18}$$

The characteristic value of the lexicographical x-line Gauß-Seidel iteration is

$$\sigma_l^{\text{line}}(\xi, \eta) = -\sum_{\substack{\boldsymbol\alpha \\ \alpha_2 > 0}} c_{\boldsymbol\alpha}(h_l)\, e^{\pi i(\alpha_1\xi + \alpha_2\eta)/h_l} \Big/ \sum_{\substack{\boldsymbol\alpha \\ \alpha_2 \leq 0}} c_{\boldsymbol\alpha}(h_l)\, e^{\pi i(\alpha_1\xi + \alpha_2\eta)/h_l}. \tag{8.1.19}$$

Example 8.1.10. Let L_l be the five-point scheme (3.1.15). The previous values (17), (19) are

$$\sigma_l^{\text{lex}}(\xi, \eta) = (e^{\pi i\xi} + e^{\pi i\eta})/(4 - e^{-\pi i\xi} - e^{-\pi i\eta}), \tag{8.1.20 a}$$

$$\sigma_l^{\text{line}}(\xi, \eta) = e^{\pi i\eta}/(4 - 2\cos\pi\xi - e^{-\pi i\eta}). \tag{8.1.20 b}$$

The value corresponding to the damped Jacobi iteration is given by

$$\sigma_l^{\text{Jac}}(\xi, \eta) = 1 - \omega_l[\sin^2(\pi\xi/2) + \sin^2(\pi\eta/2)]. \tag{8.1.20 c}$$

8.2 Local Mode Analysis

The analysis of Poisson's equation by means of sine-expansions was given in the very first multi-grid paper of Fedorenko [2] (although in a complicated and non-optimal way). The analysis of § 8.1.1 can be found in the author's first report on multi-grid methods (cf. Hackbusch [2])[2]. The 'local mode analysis'

2 A similar approach has been attempted by Brandt [3, Appendix C]; however, the analysis contains mistakes

introduced by Brandt [3], [6], [12] uses a simplified Fourier analysis in which the 'smoothing rates' play an important rôle.

8.2.1 Neglect of the Boundary Conditions

The Fourier transform applies only to difference schemes in domains $\Omega = (a_1, b_1) \times \ldots \times (a_d, b_d)$, where $-\infty \leqq a_i < b_i \leqq \infty$. Therefore, the boundary of a general domain together with the boundary conditions will be neglected. Either $\Omega \subset \mathbb{R}^d$ is enlarged to $\mathbb{R}^d$ or replaced by a square (or rectangle) with Dirichlet or periodic boundary conditions. This alternative problem can be studied as in § 8.1 and one may hope that the results of the Fourier analysis are approximately valid for the original problem.

Concerning the justification of this approach, we recall that one can study local phenomena by means of such a Fourier analysis. An example is the investigation of the interior regularity of finite difference solutions (cf. Thomée-Westergren [1]). The smoothing process, the prolongation, and the restriction are largely local operations.

However, the coarse-grid correction involves the inverse L_{l-1}^{-1}, which can be described only globally. For instance, the important question, whether L_{l-1} is non-singular or not, can depend on the underlying domain Ω.

That boundary conditions should not be neglected without scruples is supported by the following remarks. In the stability theory of time-dependent problems it is known that the stability cannot be guaranteed alone by the behaviour of the difference scheme in the interior (cf. Richtmyer-Morton [1, p. 141/142], Kreiss [1]). Another example is the 'pollution effect' known in finite element methods which shows that the shape of the boundary can influence the solution far off in the interior of the domain.

8.2.2 Treatment of Variable Coefficients

Let $L_l = \sum_{\boldsymbol{\alpha}} c_{\boldsymbol{\alpha}}(\,\cdot\,; h_l)\, T^{\boldsymbol{\alpha}}$ be a difference operator in the infinite grid $Q_l = h_l \mathbb{Z}^d$ (cf. (3.1.13)) with variable coefficients $c_{\boldsymbol{\alpha}}(\mathbf{x}; h_l)$. Replacing the variable $\mathbf{x}$ by a constant ξ, we obtain the difference operator $L_l(\xi)$ with 'frozen' coefficients:

$$L_l(\xi) := \sum_{\boldsymbol{\alpha} \in \mathbb{Z}^d} c_{\boldsymbol{\alpha}}(\xi; h_l)\, T^{\boldsymbol{\alpha}}, \qquad \xi \in \Omega \text{ fixed}. \tag{8.2.1}$$

Since ξ is fixed, $L_l(\xi)$ has constant coefficients and can be analysed by means of Fourier transforms. The analysis of frozen difference operators together with the 'partition of unity' (cf. Yosida [1, p. 60–62]) is a tool which is often used to prove properties of the variable difference operator. For instance, under certain conditions the smoothing property $\| L_l(\xi) S_l(\xi)^\nu \| \leqq h_l^{-\alpha} \tilde{\eta}(\nu)$ of the frozen operators $L_l(\xi)$ is necessary and sufficient for the smoothing property $\| L_l S_l^\nu \| \leqq h_l^{-\alpha} \eta(\nu)$ (cf. Exercise 6.6.7).

One must be aware of the fact that the frozen operator is not uniquely defined.

Example 8.2.1. The differential operator $Lu(x) = (a(x)u'(x))'$ with $a \in C^1([0,1])$ can be written as $L = D\,a\,D$, $L = a\,D^2 + a'\,D$, or $L = D^2 a - D\,a'$, where $D = \partial/\partial x$. Replacing the function $a(\cdot)$ by the 'frozen' value $a(\xi)$, we obtain three different frozen operators $L(\xi) = a(\xi)D^2$, $L(\xi) = a(\xi)D^2 \pm a'(\xi)D$. The same conclusion holds for a difference operator $L_l = \partial\, a \partial$.

The non-singularity of $L_l(\xi)$ is neither necessary nor sufficient for the non-singularity of L_l as can be seen from the following counter-examples.

Example 8.2.2. Consider the differential operator

$$(Lu)(x) = -(a(x)u'(x))' - \pi^2 a(x)u(x), \qquad 0 < x < 1,$$

with boundary conditions $u(0) = u(1) = 0$. The frozen operators $L(\xi)u = -a(\xi)[u'' + \pi^2 u]$ are singular for all $\xi \in [0,1]$. Nevertheless, L is even positive definite when we choose, e.g., $a(x) = e^x$.

Example 8.2.3. The boundary value problem

$$(Lu)(x) \equiv -(a(x)u'(x))' - b(x)u(x) = f(x), \quad 0 < x < 1, \quad u(0) = u(1) = 0,$$

is singular for the choice $a(x) = \exp(-x^2/2)$, $b(x) = a(x)\,(\pi^2 + x^2/4 - \frac{1}{2})$, since $Lu = 0$ for $u(x) = \exp(x^2/4)\sin\pi x$, but all frozen operators $L(\xi)u = -[a(\xi)u'' + b(\xi)u]$, $0 \leqq \xi \leqq 1$, are still positive definite.

Similar counter-examples can be shown for difference operators. Therefore, it is not possible to prove Exercise 6.6.7 for the approximation property instead for the smoothing property.

8.2.3 Idealisation of the Coarse-Grid Correction and Smoothing Rates

Due to the previous techniques a general difference scheme with variable coefficients in a general region subject to certain boundary conditions can be turned into a problem (or a family of problems) with constant coefficients and with boundary conditions as considered in §§ 8.1.1–3. But the two-grid analysis is still complicated.

To simplify the analysis we change the coarse-grid correction. The weighted restriction $r\colon \mathscr{F}_l \to \mathscr{F}_{l-1}$ is changed into the orthogonal projection $\tilde{r}$ onto $\mathscr{F}_{l-1}$:

$$\tilde{r}\,e_l^{\nu\mu} = \begin{cases} e_{l-1}^{\nu\mu} & \text{if } \xi = \nu h_l \in (-\frac{1}{2}, \frac{1}{2}],\ \eta = \mu h_l \in (-\frac{1}{2}, \frac{1}{2}] \\ 0 & \text{otherwise}, \end{cases} \tag{8.2.2}$$

where $e_l^{\nu\mu}$ is defined in §§ 8.1.2–3. If $e_l^{\nu\mu}$ is taken from § 8.1.1, the upper case is characterised by νh_l, $\mu h_l \in (0, \frac{1}{2}]$. The kernel of $\tilde{r}$ consists of all 'high frequencies', that means of all $(\xi, \eta) = (\nu h_l, \mu h_l) \in I_{\text{high}}$ with

$$I_{\text{high}} := \{(\xi,\eta) \in (-1,1] \times (-1,1] : (\xi,\eta) \notin (-\tfrac{1}{2}, \tfrac{1}{2}] \times (-\tfrac{1}{2}, \tfrac{1}{2}]\}. \tag{8.2.3}$$

Analogously, the prolongation p is replaced by $\tilde{p}$:

$$\tilde{p}\, e_{l-1}^{\nu\mu} = e_l^{\nu\mu}. \tag{8.2.4}$$

$\tilde{p} = \tilde{r}^*$ can be regarded as trigonometric interpolation. Substituting L_{l-1} by $\tilde{L}_{l-1} = \tilde{r}\, L_l \tilde{p}$, we obtain the coarse-grid correction step

$$u_l \mapsto u_l - \tilde{p}\, \tilde{L}_{l-1}^{-1}\, \tilde{r}(L_l u_l - f_l) \tag{8.2.5 a}$$

with the associated matrix

$$\tilde{C}_l = I - \tilde{p}\, \tilde{L}_{l-1}^{-1}\, \tilde{r}\, L_l. \tag{8.2.5 b}$$

Note 8.2.4. $\tilde{C}_l$ is the orthogonal projection onto the subspace of high frequencies:

$$\tilde{C}_l\, e_l^{\nu\mu} = \begin{cases} e_l^{\nu\mu} & \text{if } (\xi, \eta) = (\nu h_l, \mu h_l) \in I_{\text{high}}, \\ 0 & \text{otherwise}. \end{cases} \tag{8.2.5 c}$$

Definition 8.2.5. Assume $e_l^{\xi/h_l, \eta/h_l}$ to be the eigenvectors of $S_l^{(\nu)}$:

$$S_l^{(\nu)}\, e_l^{\xi/h_l, \eta/h_l} = \sigma_l^{(\nu)}(\xi, \eta) e_l^{\xi/h_l, \eta/h_l}. \tag{8.2.6}$$

The quantity

$$\varrho_B(\nu) := \max\{|\sigma_l^{(\nu)}(\xi, \eta)| : (\xi, \eta) \in I_{\text{high}}\} \tag{8.2.7}$$

is called the 'smoothing rate' of $S_l^{(\nu)}$.

The smoothing rate is introduced by Brandt [3], [6], [12] to roughly approximate the convergence rate as suggested by

Proposition 8.2.6. Let $\| u_l \|_{\mathscr{U}} = h_l [\sum_{\mathbf{x} \in Q_l} |u_l(\mathbf{x})|^2]^{1/2}$ be the l_2-norm. If (i) L_l is a difference operator on the infinite grid Q_l with constant coefficients, (ii) $S_l^{(\nu)}$ satisfies (6), (iii) the coarse-grid correction uses $\tilde{r}$, $\tilde{p}$, $\tilde{L}_{l-1}$ introduced above, then the convergence rate and the contraction number of the two-grid iteration cqual

$$\| M_l(\nu) \|_{\mathscr{U} \leftarrow \mathscr{U}} = \varrho(M_l(\nu)) \leqq \varrho_B(\nu). \tag{8.2.8}$$

Proof. Let $u_l = \int_{-1}^{1} \int_{-1}^{1} \hat{u}_l(\xi, \eta)\, e_l^{\xi/h_l, \eta/h_l}\, d\xi\, d\eta$ be arbitrary. (8) follows from

$$M_l(\nu)\, u_l = \int_{-1}^{1} \int_{-1}^{1} \tilde{\sigma}_l^{(\nu)}(\xi, \eta)\, \hat{u}_l(\xi, \eta)\, e_l^{\xi/h_l, \eta/h_l}\, d\xi\, d\eta$$

with

$$\tilde{\sigma}_l^{(\nu)}(\xi, \eta) := \begin{cases} \sigma_l^{(\nu)}(\xi, \eta) & \text{for } (\xi, \eta) \in I_{\text{high}} \\ 0 & \text{otherwise} \end{cases}$$

and from

$$\| M_l(\nu)\, u_l \|_{\mathscr{U}} = \| \tilde{\sigma}_l^{(\nu)}\, \hat{u}_l \|_{L^2} \leqq \| \tilde{\sigma}_l^{(\nu)} \|_{L^\infty}\, \| \hat{u}_l \|_{L^2} = \varrho_B(\nu)\, \| u_l \|_{\mathscr{U}}. \qquad \square$$

Example 8.2.7 (cf. Brandt [2], [3]). Let L_l be the five-point scheme (3.1.15) of the Poisson equation. Equations (20 a–c) result in the following smoothing rates:

lexicographical Gauß-Seidel iteration:	$\varrho_B(v) = 2^{-v}$,
symmetric Gauß-Seidel iteration:	$\varrho_B(v) = 4^{-v}$,
lexicographical x-line Gauß-Seidel iteration:	$\varrho_B(v) = 5^{-v/2}$,
damped Jacobi iteration (3.3.6), $0 \leqq \theta \leqq \frac{4}{5}$:	$\varrho_B(v) = (1 - \theta/2)^v$.

We conclude with the following remark. If we replace $\tilde{C}_l$ from (5 c) by

$$\tilde{\tilde{C}}_l e_l^{\xi/h_l, \eta/h_l} = L_l e_l^{\xi/h_l, \eta/h_l} / \| L_l \|_{\mathscr{U} \leftarrow \mathscr{U}} \quad \text{for} \quad -1 < \xi, \eta \leqq 1,$$

Inequality (8) holds with $\varrho_B(v)$ replaced by the smoothing number $\varrho_L(v)$ defined in (2.6.6). Note that $\tilde{\tilde{C}}_l$ is not a projection but $\| \tilde{\tilde{C}}_l \|_{\mathscr{U} \leftarrow \mathscr{U}} = 1$ is valid.

The estimation of $\| M_l(v) \|_{\mathscr{U} \leftarrow \mathscr{U}}$ by means of the smoothing rate $\varrho_B(v)$ under the simplifying assumptions of §§ 8.2.1–3 is called the *local mode analysis*.

9. Nonlinear Multi-Grid Methods

With the intention of not overloading the presentation with the notational and conceptual difficulties of nonlinear problems, we have so far considered only linear equations. However, it will turn out that multi-grid algorithms are perfectly suited for nonlinear boundary value problems. There are two different approaches. The first one explained in § 9.2 is based on a linearisation, thus allowing the application of linear multi-grid solvers. The remaining subsections §§ 9.3–5 are devoted to an intrinsic multi-grid treatment of non-linear problems. A third approach will be described in § 16.8.

9.1 The Nonlinear Problem

Equation (3.1.10), $Lu = f$, may become nonlinear when either the differential equation $L_\Omega u = f_\Omega$ or the boundary conditions $L_\Gamma u = f_\Gamma$ or both are replaced by nonlinear equations. The nonlinear counterpart of Eq. (3.1.10) is denoted by

$$\mathscr{L}(u) = 0. \tag{9.1.1}$$

Note that the choice of the right side in (1) is a matter of definition of $\mathscr{L}$. For example, the original linear equation $Lu = f$ is equivalent to (1) with $\mathscr{L}(u) := Lu - f$.

The discretisation of Eq. (1) yields

$$\mathscr{L}_l(u_l) = 0 \tag{9.1.2}$$

at the levels $l = 0, 1, \ldots$. Since the discretisation should make sense, we suppose that Eq. (2) has at least one solution u_l^*. We shall embed Eq. (2) into a family of equations

$$\mathscr{L}_l(u_l) = f_l. \tag{9.1.3}$$

The derivative (the Jacobian) of $\mathscr{L}_l$ is denoted by

$$L_l(v_l) := \mathscr{L}_l'(v_l), \qquad L_l := L_l(u_l^*) \quad (u_l^* \text{ is solution of (1)}). \tag{9.1.4}$$

Problem (2) is well-posed only if L_l is non-singular.

Remark 9.1.1. Let $\mathscr{L}_l(\cdot)$ be continuous in a neighbourhood of u_l^* and suppose $L_l = L_l(u_l^*)$ to be non-singular. Then the implicit function theorem guarantees that there are connected neighbourhoods $\mathscr{U}_l$ of u_l^* and $\mathscr{F}_l$ of 0 such that

$$\mathscr{L}_l : \mathscr{U}_l \to \mathscr{F}_l \quad \text{is a homeomorphism}, \tag{9.1.5}$$

i.e. $\mathscr{L}_l$ is a bijection and $\mathscr{L}_l$ as well as $\mathscr{L}_l^{-1}$ are continuous.

Therefore it makes sense to denote the solution of Eq. (3) by $\mathscr{L}_l^{-1}(f_l)$ as long as f_l is small enough.

9.2 Newton-Multi-Grid Iteration

The Newton iteration transforms the nonlinear problem (2) into a sequence of linear ones. Since the multi-grid method has proved to be a fast linear solver it is obvious how to formulate the algorithm:

$$\text{start:} \qquad u_l^0 \tag{9.2.1 a}$$

iteration: $j := 0, 1, \ldots$:

(i) $d_l^j := \mathscr{L}_l(u_l^j); \quad L_l^j := L_l(u_l^j);$ (9.2.1 b)

(ii) apply one step of the (linear) multi-grid iteration to $L_l^j v_l = d_l^j$ starting with $v_l^0 = 0$ resulting in v_l^1; (9.2.1 c)

(iii) $u_l^{j+1} := u_l^j - v_l^1$; (9.2.1 d)

Algorithm (1) is the prototype of a Newton-multi-grid iteration. Several modifications are possible:

a) L_l^j may be replaced by an approximation $\tilde{L}_l^j$ of the derivative $L_l(u_l^j)$. If $\tilde{L}_l^j$ does not tend to $L_l = L_l(u_l^*)$ as $j \to \infty$ one loses the quadratic convergence of the original Newton process. However, this does not matter since the multi-grid step (1 c) causes only linear convergence.

b) As mentioned above there is a conflict between the quadratic convergence of the Newton process and the linear multi-grid convergence. The quadratic convergence of (1) can be accomplished if we perform μ_j multi-grid iterations in (1 c) instead of only one. However, it is not obvious how to guess μ_j.

c) It is well-known that an under-relaxed correction $u_l^{j+1} := u_l^j - \omega_j v_l^1$, $0 < \omega_j \leqq 1$, may enlarge the convergence domain $\{u_l^0 : u_l^j$ are well-defined for $j \geqq 0$ and $u_l^j \to u_l^*\}$ of admissible starting values (cf. Bank-Rose [1]).

d) Applying iteration (1 b–d) i-times at the levels $l = 1, 2, \ldots, l_{\max}$ with $u_l^0 := \tilde{p}\, u_{l-1}^i$ we obtain a nested iteration comparable with (5.1.3). The last matrix L_l^i may serve as coarse-grid matrix in the multi-grid algorithms at the further levels $l+1, \ldots, l_{\max}$.

Since quadratic Newton convergence and linear multi-grid convergence are superposed, the errors $e_l^j := u_l^j - u_l^*$ behave like

$$\| e_l^{j+1} \|_{\mathscr{U}} \leqq \zeta \| e_l^j \|_{\mathscr{U}} + C \| e_l^j \|_{\mathscr{U}}^2,$$

Table 9.2.1. Euclidean norm of the corrections in iteration (1) for problem (2 a, b) with $h_l = h_4 = \frac{1}{32}$

j	0	1	2	3	4	5	6	7
$\|u_l^{j+1} - u_l^j\|_{\mathscr{U}}$	$4.4_{10}{-2}$	$1.1_{10}{-3}$	$2.9_{10}{-5}$	$7.1_{10}{-7}$	$1.9_{10}{-8}$	$4.5_{10}{-10}$	$1.4_{10}{-11}$	$3.9_{10}{-13}$

where ζ is the multi-grid contraction number. Asymptotically the multi-grid process dominates:

$$\overline{\lim_{j\to\infty}}\ \|e_l^{j+1}\|_{\mathscr{U}}/\|e_l^j\|_{\mathscr{U}} \leqq \zeta$$

if the process converges to u_l^*. This behaviour is confirmed by the following results.

Consider the nonlinear equation

$$-\Delta u(x, y) = e^{u(x,y)} \tag{9.2.2 a}$$

with homogeneous Dirichlet data on the boundary of the unit square $(0,1)\times(0,1)$. By symmetry it can be reduced to $\Omega = (0, \frac{1}{2}) \times (0, \frac{1}{2})$ with

$$u(x, y) = 0 \quad \text{if } x = 0 \quad \text{or} \quad y = 0, \qquad u_x(\tfrac{1}{2}, y) = u_y(x, \tfrac{1}{2}) = 0. \tag{9.2.2 b}$$

The program from § 4.4.3 can be applied to the linearised equation $-\Delta v - e^u v = f$ arising in Step (1 c). Starting with $u_l^0 = 0$ at level $l = 4$ ($h_k = 1/2^{k+1}$) we obtain the corrections $\|u_l^{j+1} - u_l^j\|_{\mathscr{U}} = \|v_l^1\|_{\mathscr{U}}$ (cf. (1 d)) listed in Table 9.2.1. The complete computer code generating these numbers can be found in Hackbusch [6], since in this report problem (2 a, b) was chosen as a test example. Evidently, the numbers in Table 9.2.1 decrease by the same factor as the linear problems reported in § 4.4.3. The converged discrete value at $x = y = \frac{1}{2}$ is $u_4(\frac{1}{2}, \frac{1}{2}) = 0.078044$, while the continuous solution is $u(\frac{1}{2}, \frac{1}{2}) = 0.078101$.

Equation (2 a, b) has two solutions. The first one is computed above. The second solution is harder to compute since the linearised problems become indefinite (cf. § 10.2.2). It can be obtained by a careful nested iteration (cf. Hackbusch [4]). A systematical approach is the use of the continuation technique (cf. § 13).

9.3 The Nonlinear Multi-Grid Iteration

The linearisation in the Newton process can be avoided by the nonlinear multi-grid algorithm which is as efficient as the linear multi-grid iteration. In the following subsections we develop this algorithm step by step. Its convergence will be studied in § 9.5.

9.3.1 Nonlinear Smoothing Iterations

There exist nonlinear counterparts of most of the smoothing iterations mentioned in § 3.3. The smoothing process is again denoted by

$$u_l \mapsto \mathscr{S}_l^{(\nu)}(u_l, f_l). \tag{9.3.1 a}$$

It is reasonable to assume that $\mathscr{L}_l^{-1}(f_l)$ is a fixed point of $\mathscr{S}_l^{(\nu)}$:

$$u_l = \mathscr{S}_l^{(\nu)}(u_l, \mathscr{L}_l(u_l)) \quad \text{for all } u_l \in \mathscr{U}_l, \quad \nu \geqq 1, \tag{9.3.1 b}$$

where $\mathscr{U}_l$ is the neighbourhood of u_l^* from (1.5). Condition (1 b) is satisfied for all iterations mentioned below.

The nonlinear version of the Gauß-Seidel iteration (3.3.1) reads as

$$\textbf{for } i := 1\,(1)\,n_l \textbf{ do } u_{l,i} := u'_{l,i} \tag{9.3.2 a}$$

$\{$where $u'_{l,i}$ is the solution of

$$\mathscr{L}_{l,i}(u_{l,1}, \ldots, u_{l,i-1}, u'_{l,i}, u_{l,i+1}, \ldots) = f_{l,i} \tag{9.3.2 b}$$

with $\mathscr{L}_{l,i}$ and $f_{l,i}$ being the ith components of $\mathscr{L}_l$ and $f_l\}$;

Note 9.3.1. a) It often happens that $\mathscr{L}_{l,i}$ is linear with respect to $u_{l,i}$ since the nonlinear terms are discretised by means of the neighbours of $u_{l,i}$. Iteration (2) is then as easy to perform as in the linear case.

b) If $\mathscr{L}_{l,i}$ is nonlinear in $u_{l,i}$, one can replace the exact solution of (2 b) by one Newton step. Let $L_{l,ii}(u_l) := \partial \mathscr{L}_{l,i}(u_l)/\partial u_{l,i}$. The resulting iteration $u_l^{j+1} = \mathscr{S}_l(u_l^j, f_l) = \mathscr{S}_l^{(j+1)}(u_l^0, f_l)$ is

$$u := u_l^j;$$

$$\textbf{for } \; i := 1\,(1)\,n_l \textbf{ do } u_{l,i} := u_{l,i} - (\mathscr{L}_{l,i}(u) - f_{l,i})/L_{l,ii}(u); \tag{9.3.3}$$

$$u_l^{j+1} := u_l;$$

c) Blockwise Gauß-Seidel variants corresponding to (3.3.12) require a greater effort to solve the nonlinear block-counterpart of (2 b).

The nonlinear analogue of Jacobi's iteration (3.3.6) may be written as $\mathscr{S}_l^{(j)}(u_l^0, f_l) = u_l^j$, where

$$u_l^{j+1} := (1 - \theta)\, u_l^j + \theta\, u_l'; \tag{9.3.4 a}$$

with u_l' being the solution of

$$\mathscr{L}_{l,i}(u_{l,1}^j, \ldots, u_{l,i-1}^j, u'_{l,i}, u_{l,i+1}^j, \ldots) = f_{l,i}. \tag{9.3.4 b}$$

Again the exact solution $u'_{l,i}$ may be replaced by the result of one Newton step applied to the starting value $u_{l,i}^j$.

The solution of nonlinear equations is completely avoided when we use the nonlinear counterpart of iteration (3.3.7). Only the evaluation of $\mathscr{L}_l$ is needed:

$$\mathscr{S}_l(u_l, f_l) := u_l - \omega_l h_l^{2m} \mathscr{L}_l(u_l, f_l). \tag{9.3.5}$$

It is plain how to define the semi-iterative version $\mathscr{S}_l^{(\nu)}(u_l, f_l; \omega)$ of (5) corresponding to (3.3.15).

9.3.2 The Nonlinear Two-Grid Iteration

Although Eq. (1.2), $\mathscr{L}_l(u_l) = 0$, is to be solved, we shall see in § 9.3.3 that the more general problem (1.3), $\mathscr{L}_l(u_l) = f_l$, has to be considered. The two-grid iteration consists again of a smoothing step and a coarse-grid correction. The smoothing step is $u_l' := \mathscr{S}_l^{(\nu)}(u_l^j, f_l)$ with $\mathscr{S}_l^{(\nu)}$ from § 9.3.1. It remains to develop the coarse-grid correction $u_l' \mapsto u_l^{j+1}$.

The desired exact correction v_l is defined by

$$\mathscr{L}_l(u_l' - v_l) = f_l. \tag{9.3.6 a}$$

The defect

$$d_l := \mathscr{L}_l(u_l') - f_l \tag{9.3.6 b}$$

is computable. Taylor's expansion yields

$$v_l = u_l' - (u_l' - v_l) = \mathscr{L}_l^{-1}(f_l + d_l) - \mathscr{L}_l^{-1}(f_l) = L_l^{-1}(u_l'')\, d_l + O(\| d_l \|_{\mathscr{F}}^2), \tag{9.3.6 c}$$

where $u_l'' := \mathscr{L}_l^{-1}(f_l)$, since the derivative of $\mathscr{L}_l^{-1}(f_l)$ is $L_l^{-1}(\mathscr{L}_l^{-1}(f_l))$. To approximate v_l at level $l-1$ we need some coarse-grid function $\tilde{u}_{l-1}$. From $\tilde{u}_{l-1}$ one determines

$$\tilde{f}_{l-1} := \mathscr{L}_{l-1}(\tilde{u}_{l-1}). \tag{9.3.6 d}$$

We define the right-hand side

$$d_{l-1} := \tilde{f}_{l-1} - s\, r\, d_l, \tag{9.3.6 e}$$

where r is a restriction and $s \neq 0$ some real number. According to the two-grid philosophy we assume that equations at level $l-1$ can be solved exactly:

$$v_{l-1} := \mathscr{L}_{l-1}^{-1}(d_{l-1}). \tag{9.3.6 f}$$

Taylor's expansion applied to

$$\tilde{v}_l := \frac{1}{s} p(\tilde{u}_{l-1} - v_{l-1}) \quad \text{with } s \text{ from (6 e)} \tag{9.3.6 g}$$

results in

$$\tilde{v}_l = \frac{1}{s} p[\mathscr{L}_{l-1}^{-1}(\tilde{f}_{l-1}) - \mathscr{L}_{l-1}^{-1}(\tilde{f}_{l-1} - s\, r\, d_l)]$$
$$= p\, L_{l-1}^{-1}(\tilde{u}_{l-1})\, r\, d_l + O(s \| r\, d_l \|_{\mathscr{F}}^2).$$

The starting value $v := \tilde{u}_{l-1}$ is chosen since this is the exact coarse-grid solution if $d = \tilde{f}_{l-1}$ in (12f) (i.e. if $s\,d = 0$ for d from (12d)).

The choice (9) leads to Brandt's [3, p. 346] FAS ("full approximation storage") algorithm, which is the multi-grid version of (10):

procedure $NMGM\,(l, u, f)$; **integer** l; **array** u, f;
if $l = 0$ **then** $u := \Phi_0(u, f)$ **else** (9.3.13 a)
begin integer i; **array** $v, d, \tilde{u}$;
$u := \mathscr{S}_l^{(\nu_1)}(u, f)$; $\tilde{u} := \tilde{r} * u$; (9.3.13 b)
$d := \mathscr{L}_{l-1}(\tilde{u}) - r * (\mathscr{L}_l(u) - f)$; (9.3.13 c)
$v := \tilde{u}$; **for** $i := 1\,(1)\,\gamma$ **do** $NMGM\,(l-1, v, d)$; (9.3.13 d)
$u := u + p * (v - \tilde{u})$; (9.3.13 e)
$u := \mathscr{S}_l^{(\nu_2)}(u, f)$ (9.3.13 f)
end;

The FORTRAN formulation of the algorithms (12) and (13) follows the pattern of subroutine (4.1.2).

Using Remark 9.3.4 one proves by induction the following remark.

Remark 9.3.5. Let $\mathscr{L}_k(u_k) = L_k u_k - g_k$ for $0 \leqq k \leqq l$ and suppose that one step of iteration (11) results in the exact solution of $\mathscr{L}_0(u_0) = f_0$. Then for any choice of $\tilde{u}_k$, $0 \leqq k \leqq l$, and $s \neq 0$ the results of $NMGM\,(l, u, f)$ from (12) or (13) and $MGM\,(l, u, f + g_l)$ from (4.1.1) coincide.

9.3.4 The Nonlinear Nested Iteration

In § 9.3.3 we did not explain how to choose the auxiliary values $\tilde{u}_k$, $k = 0, 1, \ldots, l-1$, which are involved in the algorithm (12). However, these quantities already appeared in the nested iteration (5.1.3). The only change in the present nonlinear case is that we compute $\tilde{f}_k$ from $\tilde{u}_k$:

nonlinear nested iteration (9.3.14)

$\tilde{u}_0 :=$ given approximation to u_0^*; (9.3.14 a)
for $k := 1$ **step** 1 **until** l **do**
begin $\tilde{u}_k := \tilde{p}\,\tilde{u}_{k-1}$; (9.3.14 b)
$\tilde{f}_{k-1} := \mathscr{L}_{k-1}(\tilde{u}_{k-1})$; (9.3.14 c)
for $j := 1$ **step** 1 **until** i **do** $NMGM\,(k, \tilde{u}_k, 0)$ (9.3.14 d)
end;

If $NMGM$ from (13) is used, the statement (14c) can be omitted since $\tilde{f}_{k-1}$ is computed inside of $NMGM$.

The optimal starting value would be $\tilde{u}_0 := u_0^* = \mathscr{L}_0^{-1}(0)$ which is usually not available. Depending on the problem it may be easy to guess $\tilde{u}_0$ or it may be the most difficult part of the whole problem. If more than one solution of $\mathscr{L}(u) = 0$ exists, the choice in (14 a) might decide what discrete solutions are approximated in (14 b–d).

It is plain that the modifications mentioned in (5.1.4 b) and § 5.4 are still applicable.

9.4 Numerical Example: Natural Convection in an Enclosed Cavity

This chapter is based on the author's contribution to a report of Jones-Thompson [1], who chose the buoyancy driven flow in a rectangular cavity as a test problem for computer programs.

The cavity is represented by the unit square $\Omega = \{(x,z): 0 < x, z < 1\}$. The two-dimensional flow can be described by

$$v_h \frac{\partial \zeta}{\partial x} + v_v \frac{\partial \zeta}{\partial z} = Ra \cdot Pr \frac{\partial T}{\partial x} + Pr\, \Delta \zeta, \tag{9.4.1 a}$$

$$\zeta = -\Delta \psi, \tag{9.4.1 b}$$

$$v_h \frac{\partial T}{\partial x} + v_v \frac{\partial T}{\partial z} = \Delta T, \tag{9.4.1 c}$$

$$v_h = -\frac{\partial \psi}{\partial z}, \quad v_v = \frac{\partial \psi}{\partial x}. \tag{9.4.1 d}$$

The physical meaning of the quantities is as follows. v_h: horizontal velocity, v_v: vertical velocity, T: temperature, ζ: vorticity, ψ: stream function, *Ra*: Rayleigh number, *Pr*: Prandtl number.

The boundary conditions are

$$\psi = 0 \quad \text{and} \quad \frac{\partial \psi}{\partial n} = 0 \quad \text{on } \Gamma, \tag{9.4.2 a}$$

$$T(0,z) = 1, \quad T(1,z) = 0, \quad \frac{\partial T}{\partial n}(0,z) = \frac{\partial T}{\partial n}(1,z) = 0 \quad (0 \leqq z \leqq 1), \tag{9.4.2 b}$$

where $\frac{\partial}{\partial n}$ is the outer normal derivation. (2 b) indicates that $x = 0$ represents a hot wall and $x = 1$ a cold one.

By means of (1 b, d) the vorticity ζ and the velocities v_h, v_v can be eliminated:

$$\Delta^2 \psi - Ra \frac{\partial T}{\partial x} + \frac{1}{Pr} \frac{\partial \psi}{\partial z} \frac{\partial}{\partial x} \Delta \psi - \frac{1}{Pr} \frac{\partial \psi}{\partial x} \frac{\partial}{\partial z} \Delta \psi = 0 \quad \text{in } \Omega, \tag{9.4.3 a}$$

$$-\Delta T - \frac{\partial \psi}{\partial z} \frac{\partial T}{\partial x} + \frac{\partial \psi}{\partial x} \frac{\partial T}{\partial z} = 0 \quad \text{in } \Omega. \tag{9.4.3 b}$$

Although Eq. (3 a, b) represents a system of two equations, it behaves like a scalar elliptic equation because each of the Eqs. (3 a, b) are of elliptic type. ψ and T are coupled in (3 a, b) by lower order terms.

We replace $-\Delta$ by the five-point formula Δ_l from (3.1.15), Δ^2 by the square Δ_l^2. All first differential operators $\partial/\partial x$ and $\partial/\partial z$ are discretised by centred differences. The unknowns $\psi_l(\mathbf{x})$, $\mathbf{x} \in \Gamma$, can be eliminated using (2 a). Similarly, $T_l(0,z)$ and $T_l(1,z)$ are known from (2 b). The difference scheme for ψ_l e.g. at $(x,z) = (h_l, z)$ involves $\psi_l(-h_l, z)$. This value can be eliminated by means of $\psi_l(-h_l, z) = \psi_l(h_l, z)$ (cf. (2 a)). Analogously, $\partial T/\partial n = 0$ $(x = 0, 1)$ can be used to determine $T_l(-h_l, z)$ and $T_l(1 + h_l, z)$. The following system results:

$$\mathscr{L}_l(u_l) = 0, \quad u_l = \begin{bmatrix} \psi_l \\ T_l \end{bmatrix}, \tag{9.4.4}$$

with $(h_l^{-1} - 1)^2$ ψ_l-variables and $(h_l^{-1} - 1)\,(h_l^{-1} + 1)$ T_l-variables.

The choice of the prolongation $p u_l = (p_\psi \psi_l, p_T T_l)$ has to take into account that Eq. (3 a) for ψ is of fourth order, whereas Eq. (3 b) for T is of second order. Following § 3.4.3 we define p_ψ by cubic interpolation, while linear interpolation (3.4.2) suffices for p_T.

Split the defect $d_l = L_l(u_l) - f_l$ into the components $d_l = (d_l^\psi, d_l^T)$. The restriction $r\, d_l = (r_\psi d_l^\psi, r_T d_l^T)$ is chosen here as follows. r_ψ as well as r_T are the nine-point restriction (3.5.4). Thus, r_T is the adjoint of p_T (cf. 3.5.3)), whereas r_ψ differs from p_ψ^*. As pointed out in Example 3.5.4 this choice is sensible.

The smoothing iteration $\mathscr{S}_l$ is chosen as the pointwise Gauß-Seidel iteration (first applied to T_l then ψ_l). Eq. (3 a) is nonlinear with respect to ψ because of the terms $\psi_z \Delta \psi_x$ and $\psi_x \Delta \psi_z$. Since we discretise ψ_z and ψ_x by central differences, Note 9.3.1 a) applies: At each $\mathbf{x} \in \Omega_l^\psi$ the corresponding equation is linear in $\psi_l(\mathbf{x})$. Therefore the pointwise Gauß-Seidel process is easy to perform. The same conclusion would hold for an improved Gauß-Seidel iteration, which uses the blocks $(\psi_l(\mathbf{x}), T_l(\mathbf{x}))^T$ instead of the single variables.

The further parameters of the multi-grid iteration (3.12) are $\nu_1 = 2$, $\nu_2 = 0$, $\gamma = 2$. The parameter s from (3.12 e) is chosen as $s = 0.001$. The grid sizes start at $h_0 = \frac{1}{4}$, i.e. $h_l = 2^{-2-l}$. The physical quantities are chosen as

$$Pr = 0.71, \qquad Ra = 10\,000\,.$$

The results of Tables 9.4.1 a–d are obtained by means of the non-linear nested iteration (3.14). The approximation $\tilde{u}_0$ in the grid with step size $h_0 = \frac{1}{4}$ (24 unknowns) is the result of several (20 or 40) Gauß-Seidel iterations applied to the starting value $\psi_0(x, y) = 0$, $T_0(x, y) = 1 - x$. The observed rate of convergence is about $\varrho = 0.5$. According to Table 5.2.1 a reasonable value of the number i in (3.14 d) is $i = 3$. In Tables 9.4.1 a–d the numerical results are compared for different values $i = 3, 5, 7, 10$. Obviously, the choice $i = 3$ is sufficient. In the third column $i = 7$ is used only at the levels $k = 1, 2$ ($h_1 = \frac{1}{8}$, $h_2 = \frac{1}{16}$), whereas only $i = 2$ iterations are performed at the highest level with $h_3 = \frac{1}{32}$. In this case the computational work is comparable to that of $i = 3$.

Table 9.4.1 a. Values of the stream function ψ at $(x, y) = (\frac{1}{8}, \frac{1}{8})$

h_l \ i	3	5	7/7/2	10
$\frac{1}{8}$	−2.8491	−2.8567	−2.8507	−2.8718
$\frac{1}{16}$	−2.4955	−2.4493	−2.4404	−2.4367
$\frac{1}{32}$	−2.3205	−2.3197	−2.3189	−2.3199

Table 9.4.1 b. Values of the horizontal velocity v_h at $(x, y) = (\frac{1}{8}, \frac{1}{8})$

h_l \ i	3	5	7/7/2	10
$\frac{1}{8}$	10.668	11.135	11.225	11.554
$\frac{1}{16}$	11.001	10.869	10.834	10.817
$\frac{1}{32}$	10.616	10.615	10.605	10.616

Table 9.4.1 c. Values of the vertical velocity v_v at $(x, y) = (\frac{1}{8}, \frac{1}{8})$

h_l \ i	3	5	7/7/2	10
$\frac{1}{8}$	−10.185	−9.7178	−9.5050	−9.1902
$\frac{1}{16}$	− 9.1840	−9.0160	−8.9942	−8.9830
$\frac{1}{32}$	− 8.8680	−8.8749	−8.8648	−8.8757

Table 9.4.1 d. Values of the vorticity ζ at $(x, y) = (\frac{1}{8}, \frac{1}{8})$

h_l \ i	3	5	7/7/2	10
$\frac{1}{8}$	−62.981	−61.399	−62.559	−69.137
$\frac{1}{16}$	−81.073	−70.063	−67.782	−66.801
$\frac{1}{32}$	−66.143	−66.764	−68.012	−66.758

Table 9.4.2. Results of the vertical velocity v_v for $h_3 = \frac{1}{32}$

y \ x	$\frac{1}{8}$	$\frac{2}{8}$	$\frac{3}{8}$	$\frac{4}{8}$	$\frac{5}{8}$	$\frac{6}{8}$	$\frac{7}{8}$
$\frac{1}{8}$	− 4.507	− 3.496	−0.963	0.576	0.901	1.398	4.727
$\frac{2}{8}$	−11.486	− 8.876	−3.087	0.814	2.769	5.524	12.302
$\frac{3}{8}$	−17.127	−12.011	−4.351	0.560	4.081	9.574	17.833
$\frac{4}{8}$	−19.544	−12.029	−4.572	−0.001	4.571	12.030	19.544
$\frac{5}{8}$	−17.833	− 9.574	−4.082	−0.562	4.351	12.011	17.127
$\frac{6}{8}$	−12.301	− 5.524	−2.769	−0.814	3.087	8.877	11.486
$\frac{7}{8}$	− 4.727	− 1.399	−0.901	−0.576	0.963	3.496	4.507

The CPU times on a CDC Cyber 70/76 for the complete nested iteration (including the print out) are 1.18 s ($i = 3$), 1.76 s ($i = 5$), 1.18 s ($i = 7/7/2$), 3.18 s ($i = 10$).

Comparing the results of ψ, v_h, v_v, ζ for different step sizes h_l, we are able to guess the accuracy of the difference solution.

9.5 Convergence Analysis

We analyse the two- and multi-grid iterations of §§ 9.3.2–3. The main result will be the following: Asymptotically the nonlinear two- and multi-grid iterations behave as the linear ones when applied to the linearised equations. Other than in the linear case one has first to prove that the iterations are well-defined since it might happen that the iterates leave a certain domain of definition and go astray.

9.5.1 Convergence of the Nonlinear Two-Grid Iteration

According to Remark 9.1.1 the grid functions u_l and f_l must belong to the respective subsets $\mathscr{U}_l$ and $\mathscr{F}_l$. We have still to restrict these sets. For a fixed norm $\|\cdot\|_{\mathscr{U}}$ on $\mathscr{U}_l$ there is some $\varepsilon_l > 0$ such that

$$\mathscr{U}_l(\varrho) := \{u_l : \|u_l - u_l^*\|_{\mathscr{U}} \leqq \varrho\} \tag{9.5.1 a}$$

satisfies

$$\mathscr{U}_l(\varrho) \subset \mathscr{U}_l \quad \text{for all } 0 < \varrho \leqq \varepsilon_l, \qquad l \geqq 0. \tag{9.5.1 b}$$

Usually we cannot expect ε_l to be independent of l for the following reason. Assume that the coefficients of the differential equation depend on u (or even on u_x, u_y). Then reasonable neighbourhoods of u_l^* should be described by means of the supremum norm $\|\cdot\|_\infty$ (or $\|\cdot\|_{1,\infty}$ being the supremum of the grid function and its first divided differences). But $\|\cdot\|_\infty$ (and $\|\cdot\|_{1,\infty}$) is not uniformly equivalent, e.g., to the Euclidean norm $\|\cdot\|_{\mathscr{U}}$. In the two-dimensional case we have $\|\cdot\|_\infty \leqq C h_l^{-1} \|\cdot\|_{\mathscr{U}}$ (or $\|\cdot\|_{1,\infty} \leqq C h_l^{-2} \|\cdot\|_{\mathscr{U}}$). Hence, the radii ε_l behave like $\varepsilon_l = \varepsilon h_l$ (or $\varepsilon_l = \varepsilon h_l^2$, respectively). However under other circumstances $\varepsilon_l = \varepsilon$ may hold.

The image of $\mathscr{U}_l(\varrho)$, $0 < \varrho \leqq \varepsilon_l$, is

$$\mathscr{F}_l(\varrho) := \mathscr{L}_l(\mathscr{U}_l(\varrho)) = \{f_l = \mathscr{L}_l(u_l) : u_l \in \mathscr{U}_l(\varrho)\} \subset \mathscr{F}_l. \tag{9.5.1 c}$$

To analyse the two-grid iteration we shall use divided differences $D\mathscr{L}_l$ and $D\mathscr{S}_l^{(\nu)}$ satisfying

$$\mathscr{L}_l(u_l) - \mathscr{L}_l(u_l') = D\mathscr{L}_l(u_l, u_l')(u_l - u_l'), \qquad u_l, u_l' \in \mathscr{U}_l, \tag{9.5.2 a}$$

$$\mathscr{S}_l^{(\nu)}(u_l, f_l) - \mathscr{S}_l^{(\nu)}(u_l', f_l) = D\mathscr{S}_l^{(\nu)}(u_l, u_l', f_l)(u_l - u_l'), \quad u_l, u_l' \in \mathscr{U}_l, f_l \in \mathscr{F}_l. \tag{9.5.2 b}$$

Note 9.5.1. a) The differences $D\mathscr{L}_l$, $D\mathscr{S}_l$ are not uniquely defined by (2a, b).

b) If $\mathscr{L}_l$ and $\mathscr{S}_l^{(v)}$ are differentiable in the convex domain $\mathscr{U}_l$, one may define

$$D\mathscr{L}_l(u_l, u_l') = \int_0^1 L_l(u_l + t(u_l' - u_l))\,dt, \tag{9.5.2c}$$

$$D\mathscr{S}_l^{(v)}(u_l, u_l', f_l) = \int_0^1 \frac{\partial}{\partial u_l} \mathscr{S}_l^{(v)}(u_l + t(u_l' - u_l), f_l)\,dt. \tag{9.5.2d}$$

c) If $\mathscr{L}_l$ is differentiable at u_l and $D\mathscr{L}_l$ is continuous at (u_l, u_l), one has

$$L_l(u_l) = D\mathscr{L}_l(u_l, u_l). \tag{9.5.2e}$$

d) The differences $D\mathscr{L}_l^{-1}(\cdot,\cdot)$ of $\mathscr{L}_l^{-1}$ defined by

$$\mathscr{L}_l^{-1}(f_l) - \mathscr{L}_l^{-1}(f_l') = D\mathscr{L}_l^{-1}(\mathscr{L}_l^{-1}(f_l), \mathscr{L}_l^{-1}(f_l'))(f_l - f_l') \tag{9.5.2f}$$

satisfy

$$D\mathscr{L}_l^{-1}(u_l, u_l') = [D\mathscr{L}_l(u_l, u_l')]^{-1}. \tag{9.5.2g}$$

The two-grid iteration (3.7) is well-defined if the input u and the intermediate values u appearing in the right sides of (3.7b, c, h) lie in $\mathscr{U}_l$, so that $\mathscr{S}_l^{(v)}$ and $\mathscr{L}_l$ are defined, and if d from (3.7f) lies in $\mathscr{F}_{l-1}$. The former conditions can be satisfied if $u_l = \mathscr{L}_l^{-1}(f_l)$ is in the interior of $\mathscr{U}_l$ and if the input u is close enough to u_l. The latter condition is discussed in

Remark 9.5.2. Assume $\tilde{u}_{l-1} \in \mathscr{U}_{l-1}(\varepsilon_{l-1}/2)$ for $\tilde{u}_{l-1}$ from (3.7a) and

$$\| D\mathscr{L}_{l-1}^{-1}(u_{l-1}, u_{l-1}') \|_{\mathscr{U} \leftarrow \mathscr{F}} \leqq C_{DL} \quad \text{for all } u_{l-1}, u_{l-1}' \in \mathscr{U}_{l-1}(\varepsilon_{l-1}).$$

Then the choice $\sigma_{l-1} := \varepsilon_{l-1}/(2C_{DL})$ in (3.8) ensures $d \in \mathscr{F}_{l-1}(\varepsilon_{l-1})$ for d from the left-hand side of (3.7e).

In general, one cannot ensure $d \in \mathscr{F}_{l-1}$ for algorithm (3.10). It may happen that the set $\mathscr{L}_{l-1}\tilde{r}\mathscr{U}_l$ is disjoint from $\mathscr{F}_{l-1}$. Then $d \notin \mathscr{F}_{l-1}$ holds even for u_l arbitrarily close to the solution $\mathscr{L}_l^{-1}(f_l)$. We give a sufficient condition in

Remark 9.5.3. Let all $u_l \in \mathscr{U}_l$ satisfy the consistency condition

$$\| D\mathscr{L}_{l-1}^{-1}(u_{l-1}, u_{l-1}^*)[\mathscr{L}_{l-1}(\tilde{r}u_l) - r\mathscr{L}_l(u_l)] \|_{\mathscr{U}} \leqq C h_{l-1}^{\kappa}$$

for all $u_{l-1} \in \mathscr{U}_{l-1}(\varepsilon_{l-1})$ and assume $\|rf_l\|_{\mathscr{F}}/C_{DL} + C h_{l-1}^{\kappa} \leqq \varepsilon_{l-1}$ with C_{DL} from Remark 9.5.2. Then d from (3.10b) lies in $\mathscr{F}_{l-1}(\varepsilon_{l-1})$.

In the nonlinear case the contraction number ζ_l can still be defined by

$$\| u_l^{j+1} - \mathscr{L}_l^{-1}(f_l) \|_{\mathscr{U}} \leqq \zeta_l \| u_l^j - \mathscr{L}_l^{-1}(f_l) \|_{\mathscr{U}}, \quad \zeta_l < 1. \tag{9.5.3a}$$

One may also consider

$$\| u_l^{j+1} - \mathscr{L}_l^{-1}(f_l) \|_{\mathscr{U}} \leqq \zeta_{l,j} \| u_l^j - \mathscr{L}_l^{-1}(f_l) \|_{\mathscr{U}}, \quad \sup_j \zeta_{l,j} < 1. \tag{9.5.3b}$$

(3b) implies (3a) with $\zeta_l = \sup_j \zeta_{l,j}$, while the asymptotic behaviour is governed by the possibly smaller number ζ_l^*:

$$\zeta_l^* := \overline{\lim_{j\to\infty}}\, \zeta_{l,j} \leqq \zeta_l := \sup_j \zeta_{l,j} < 1\,. \tag{9.5.3 c}$$

Note 9.5.4. Suppose (3.1 b). Then $\mathcal{L}_l^{-1}(f_l)$ is a fixed point of the two-grid iteration (3.7).

The computation of the contraction number can be based on

Lemma 9.5.5. Suppose (3.1 b). If the two-grid iteration (3.7) is well-defined, the errors $\delta u_l^j := u_l^j - \mathcal{L}_l^{-1}(f_l)$ satisfy

$$\delta u_l^{j+1} = \tag{9.5.4}$$
$$D\mathcal{S}_l^{(\nu_2)}(u_l'', u_l, f_l)\,[I - p\,D\mathcal{L}_{l-1}^{-1}(v_{l-1}, \tilde{u}_{l-1})\,r\,D\mathcal{L}_l(u_l', u_l)]\,D\mathcal{S}_l^{(\nu_1)}(u_l^j, u_l, f_l)\,\delta u_l^j,$$

where $u_l = \mathcal{L}_l^{-1}(f_l)$, $u_l' = \mathcal{S}_l^{(\nu_1)}(u_l^j, f_l)$, $v_{l-1} = \mathcal{L}_{l-1}^{-1}(d_{l-1})$, $d_{l-1} = \tilde{f}_{l-1} - s\,r(\mathcal{L}_l(u_l') - f_l)$, $u_l'' = u_l' + p(v_{l-1} - \tilde{u}_{l-1})/s$.

Proof. By (3.1 b) the error of u_l' from the left-hand side of (3.7 b) is

$$\delta u_l' := u_l' - u_l = \mathcal{S}_l^{(\nu)}(u_l^j, f_l) - \mathcal{S}_l^{(\nu)}(u_l, f_l) = D\mathcal{S}_l^{(\nu)}(u_l^j, u_l, f_l)\,\delta u_l^j\,.$$

d from (3.7 c) equals $r(\mathcal{L}_l(u_l') - f_l) = r(\mathcal{L}_l(u_l') - \mathcal{L}_l(u_l)) = r\,D\mathcal{L}_l(u_l', u_l)\,\delta u_l'$. The left-hand side v_{l-1} of (3.7 f) satisfies

$$\begin{aligned} v_{l-1} - \tilde{u}_{l-1} &= \mathcal{L}_{l-1}^{-1}(\tilde{f}_{l-1} - s\,r(\mathcal{L}_l(u_l') - f_l)) - \mathcal{L}_{l-1}^{-1}(\tilde{f}_{l-1})) \\ &= -\,s\,D\mathcal{L}_{l-1}^{-1}(v_{l-1}, \tilde{u}_{l-1})\,r\,D\mathcal{L}_l(u_l', u_l)\,\delta u_l'. \end{aligned}$$

Hence the error of u_l'' defined in (3.7 g) is

$$\delta u_l'' := u_l'' - u_l = \delta u_l' + p(v_{l-1} - \tilde{u}_{l-1})/s = [I - p\,D\mathcal{L}_{l-1}^{-1}\,r\,D\mathcal{L}_l]\,\delta u_l'$$

and the post-smoothing (3.7 h) yields (4). □

If $\mathcal{L}_l$ and $\mathcal{S}_l^{(\nu)}$ are affine, the matrices $D\mathcal{L}_l$, $D\mathcal{S}_l^{(\nu)}$ become $D\mathcal{L}_l = L_l$, $D\mathcal{S}_l^{(\nu)} = S_l^{(\nu)}$ and (4) proves again Remark 9.3.4.

Proposition 9.5.6 (convergence of $NTGM^{(\nu,0)}$). Let (7.1.1), the smoothing property (6.1.7 a–c), and the approximation property (6.1.8) be satisfied uniformly for all $L_l := D\mathcal{L}_l(u_l', u_l)$, $L_{l-1} := D\mathcal{L}_{l-1}(u_{l-1}, u_{l-1}')$, $S_l^{(\nu)} := D\mathcal{S}_l^{(\nu)}(u_l'', u_l, f_l)$ with arguments $u_l, u_l', u_l'' \in \mathcal{U}_l(\varrho_l)$, $u_{l-1}, u_{l-1}' \in \mathcal{U}_{l-1}(\varrho_{l-1})$, where ϱ_l and ϱ_{l-1} satisfy

$$\varrho_l \leqq \varepsilon_l, \quad \varrho_{l-1} \leqq \varepsilon_{l-1}, \quad \tilde{u}_{l-1} \in \mathcal{U}_l(\varrho_l/2), \quad C_S\,\varrho_l \leqq \varepsilon_l \quad \text{with } C_S \text{ from (7.1.1).}$$

Let σ_{l-1} from (3.8) satisfy $\sigma_{l-1} \leqq \varepsilon_{l-1}/(2C_{DL})$ (cf. Remark 9.5.2) and assume (5 a) or (5 b):

$$f_l = 0 \quad \text{and} \quad u_l^0 \in \mathcal{U}_l(\varrho_l)\,, \tag{9.5.5 a}$$

$$f_l \in \mathcal{F}_l(\varrho_l/3) \quad \text{and} \quad u_l^0 \in \mathcal{U}_l(\varrho_l/3)\,. \tag{9.5.5 b}$$

Then there exist $\underline{\nu}$ and $\bar{h}$ such that (3 a) holds for the iterates of (3.7) with $\zeta_l \leqq C_A\,\eta(\nu) < 1$, provided that $\nu \in [\underline{\nu}, \bar{\nu}(h_l))$ and $h_l \leqq \bar{h}$. All iterates u_l^j lie in $\mathcal{U}_l(\varrho_l)$.

Proof by induction. a) For $j=0$ $u_l^0 \in \mathscr{U}_l(\varrho_l)$ holds. Assume $u_l^j \in \mathscr{U}_l(\varrho_l)$. $C_S\varrho_l \leqq \varepsilon_l$ ensures $u_l' = \mathscr{S}_l^{(\nu)}(u_l^j, f_l) \in \mathscr{U}_l(\varepsilon_l) \subset \mathscr{U}_l$. By Remark 9.5.2 the left-hand side d_{l-1} of (3.7e) lies in $\mathscr{F}_{l-1}(\varrho_{l-1})$. Hence the application of iteration (3.7) to u_l^j is well-defined. From Lemma 9.5.5 one concludes that $\|\delta u_l^{j+1}\|_{\mathscr{U}} \leqq \|M_l(\nu)\|_{\mathscr{U}\leftarrow\mathscr{U}} \|\delta u_l^j\|_{\mathscr{U}}$ with $M_l(\nu)$ from (6.1.4). As in Theorem 6.1.7 one proves $\|M_l(\nu)\|_{\mathscr{U}\leftarrow\mathscr{U}} \leqq C_A\eta(\nu) =: \zeta_l < 1$ for suitable ν.

b) It remains to prove $u_l^{j+1} \in \mathscr{U}_l(\varrho_l)$. (5a) and (5b) are special cases of

$$f_l \in \mathscr{F}_l(\varrho'), \qquad u_l^0 \in \mathscr{U}_l(\varrho_l - 2\varrho') \quad \text{and} \quad \varrho' \leqq \varrho_l/2. \tag{9.5.5c}$$

Let $u_l := \mathscr{L}_l^{-1}(f_l)$. $u_l^{j+1} \in \mathscr{U}_l(\varrho_l)$ is shown by

$$\begin{aligned}\|u_l^{j+1} - u_l^*\|_{\mathscr{U}} &\leqq \|u_l^{j+1} - u_l\|_{\mathscr{U}} + \|u_l - u_l^*\|_{\mathscr{U}}\\ &\leqq \zeta_l^{j+1}\|u_l^0 - u_l\|_{\mathscr{U}} + \|u_l - u_l^*\|_{\mathscr{U}}\\ &\leqq \zeta_l^{j+1}(\|u_l^0 - u_l^*\|_{\mathscr{U}} + \|u_l^* - u_l\|_{\mathscr{U}}) + \|u_l - u_l^*\|_{\mathscr{U}}\\ &\leqq \zeta_l^{j+1}(\varrho_l - 2\varrho' + \varrho') + \varrho' < \varrho_l. \qquad \square\end{aligned}$$

In Proposition 9.5.6 we postulate conditions for a family of matrices L_l and $S_l^{(\nu)}$. It is more convenient to require the conditions (e.g. the approximation and smoothing properties) for one $L_l = D\mathscr{L}_l$ and one $S_l^{(\nu)} = D\mathscr{S}_l^{(\nu)}$ with fixed arguments. Since we are interested in Eq. (1.2) with $f_l = 0$ the favourable arguments are $u_l = u_l^*$ and $f_l = 0$.

We recall (2e). If $\mathscr{L}_l$ and $\mathscr{S}_l^{(\nu)}$ are continuously differentiable, the derivatives

$$L_l := \frac{\partial}{\partial u_l}\mathscr{L}_l(u_l^*), \qquad S_l^{(\nu)} := \frac{\partial}{\partial u_l}\mathscr{S}_l^{(\nu)}(u_l^*, 0) \tag{9.5.6a}$$

are approached by $D\mathscr{L}_l(u_l, u_l')$, $D\mathscr{S}_l^{(\nu)}(u_l, u_l', f_l)$ as $u_l, u_l' \to u_l^*$ and $f_l \to 0$. More precisely, we require

$$\|D\mathscr{S}_l^{(\nu)}(u_l, u_l', f_l) - S_l^{(\nu)}\|_{\mathscr{U}\leftarrow\mathscr{U}} \leqq \Theta(\varrho_l/\varepsilon_l) \quad \text{for all} \quad u_l, u_l' \in \mathscr{U}_l(\varrho_l), \quad f_l \in \mathscr{F}_l(\varrho_l) \tag{9.5.6b}$$

$$\|L_l^{-1} D\mathscr{L}_l(u_l, u_l') - I\|_{\mathscr{U}\leftarrow\mathscr{U}} \leqq \Theta(\varrho_l/\varepsilon_l) \quad \text{for all} \quad u_l, u_l' \in \mathscr{U}_l(\varrho_l), \tag{9.5.6c}$$

where $\Theta(\varepsilon)$ decreases monotonely to $\Theta(0) = 0$:

$$\Theta(\varepsilon) \searrow 0 \quad \text{as } \varepsilon \searrow 0. \tag{9.5.6d}$$

In Exercise 9.6.1 Inequality (6c) is stated for a particular example.

Theorem 9.5.7 (convergence of $NTGM^{(\nu,0)}$). Let the conditions of Proposition 9.5.6 hold except that (6.1.7), (6.1.8), (7.1.1) are to be satisfied only for L_l, L_{l-1}, $S_l^{(\nu)}$ from (6a). In addition assume (6b–d), (7.1.2). Then for sufficiently small ϱ_l, ϱ_{l-1} the two-grid iteration (3.7) satisfies (3a) with a contraction number

$$\zeta_l = \zeta(\varrho_l/\varepsilon_l, \varrho_{l-1}/\varepsilon_{l-1}, \nu) < 1 \tag{9.5.7a}$$

with

$$\zeta(\alpha, \beta, \nu) \to \|M_l(\nu)\|_{\mathscr{U}\leftarrow\mathscr{U}} \qquad \text{as } \alpha, \beta \to 0. \tag{9.5.7b}$$

Proof. Abbreviate the left-hand sides of (6 b, c) by $\| Y_l(u_l, u_l', f_l) \|_{\mathscr{U} \leftarrow \mathscr{U}}$, $\| X_l(u_l, u_l') \|_{\mathscr{U} \leftarrow \mathscr{U}}$ and note that $D\mathscr{S}_l^{(v)} = S_l^{(v)} + Y_l$, $D\mathscr{L}_k = L_k(I + X_k)$ for $k = l - 1, l$. By Lemma 9.5.5 one has to estimate

$$[I - p(I + X_{l-1})^{-1} L_{l-1}^{-1} r L_l(I + X_l)](S_l^{(v)} + Y_l).$$

Using $\|p\|_{\mathscr{U} \leftarrow \mathscr{U}} \leqq \bar{C}_p$, $\| L_{l-1}^{-1} r L_l \|_{\mathscr{U} \leftarrow \mathscr{U}} \leqq$ const (cf. Lemma 7.1.5),

$$\| I - (I + X_{l-1})^{-1} \|_{\mathscr{U} \leftarrow \mathscr{U}} \leqq O(\Theta), \qquad \| X_l \|_{\mathscr{U} \leftarrow \mathscr{U}} \leqq \Theta, \qquad \| Y_l \|_{\mathscr{U} \leftarrow \mathscr{U}} \leqq \Theta$$

with $\Theta = \Theta(\text{const}(\varrho_l/\varepsilon_l + \varrho_{l-1}/\varepsilon_{l-1})) \to 0$ as $\varrho_l, \varrho_{l-1} \to 0$, one proves (7 a). □

The asymptotic contraction number ζ_l^* from (3 c) can be characterised in the following cases:

Corollary 9.5.8. a) Instead of (3.8) define

$$\sigma(k, d_k) := \min\{\text{const}, \sigma_k / \| d_k \|_{\mathscr{F}}\}.$$

Then the two-grid iteration (3.7) has the asymptotic contraction number

$$\zeta_l^* = \left\| \frac{\partial}{\partial u_l} \mathscr{S}_l^{(v_2)}(u_l, f_l) [I - p L_{l-1}^{-1}(\tilde{u}_{l-1}) r L_l(u_l)] \frac{\partial}{\partial u_l} \mathscr{S}_l^{(v_1)}(u_l, f_l) \right\|_{\mathscr{U} \leftarrow \mathscr{U}} \tag{9.5.7 c}$$

with $u_l := \mathscr{L}_l^{-1}(f_l)$.

b) If the modification (3.10) is well-defined, its asymptotic contraction number is given by

$$\zeta_l^* = \left\| \frac{\partial}{\partial u_l} \mathscr{S}_l^{(v_2)}(u_l, f_l) [I - p L_{l-1}^{-1}(\tilde{r} u_l) r L_l(u_l)] \frac{\partial}{\partial u_l} \mathscr{S}_l^{(v_1)}(u_l, f_l) \right\|_{\mathscr{U} \leftarrow \mathscr{U}}. \tag{9.5.7 d}$$

9.5.2 Convergence of the Nonlinear Multi-Grid Iteration

First we consider a modified two-grid iteration. Let $u_{l-1} \mapsto \Psi_{l-1}(u_{l-1}, f_{l-1})$ be an iteration converging to $\mathscr{L}_{l-1}^{-1}(f_{l-1})$ with contraction number $\psi_{l-1} < 1$:

$$\| \Psi_{l-1}(u_{l-1}, f_{l-1}) - \mathscr{L}_{l-1}^{-1}(f_{l-1}) \|_{\mathscr{U}} \leqq \psi_{l-1} \| u_{l-1} - \mathscr{L}_{l-1}^{-1}(f_{l-1}) \|_{\mathscr{U}}$$
$$\text{for all } u_{l-1} \in \mathscr{U}_{l-1}(\varepsilon_{l-1}), f_{l-1} \in \mathscr{F}_{l-1}(\varepsilon_{l-1}). \tag{9.5.8 a}$$

The modification of iteration (3.7) is described by

$$\text{replace (3.7 f) by "} v := \Psi_{l-1}(\tilde{u}_{l-1}, d);\text{"} \tag{9.5.8 b}$$

Denote the result of (8 b) by $\tilde{v}_{l-1} = v_{l-1} - \delta v_{l-1}$, where $v_{l-1} := \mathscr{L}_{l-1}^{-1}(d_{l-1})$ is the exact solution. (8 a) implies

$$\| \delta v_{l-1} \|_{\mathscr{U}} \leqq \psi_{l-1} \| v_{l-1} - \tilde{u}_{l-1} \|_{\mathscr{U}}.$$

$v_{l-1} - \tilde{u}_{l-1} = - s D\mathscr{L}_{l-1}^{-1} r D\mathscr{L}_l \delta u_l'$ is already mentioned in the proof of Lemma 9.5.5. Hence the modification (8 b) causes the following change. Instead of an error $\delta u_l''$ we obtain $\delta \tilde{u}_l''$ with

$$\delta \tilde{u}_l'' = \delta u_l' + p(\tilde{v}_{l-1} - \tilde{u}_{l-1})/s = \delta u_l'' - p\,\delta v_{l-1}/s;$$

thus

$$\|\delta \tilde{u}_l''\|_{\mathscr{U}} \leqq \|\delta u_l''\|_{\mathscr{U}} + \psi_{l-1}\,\|p\|_{\mathscr{U}\leftarrow\mathscr{U}}\,\|D\mathscr{L}_{l-1}^{-1}\,r\,D\mathscr{L}_l\,D\mathscr{S}_l^{(\nu)}\|_{\mathscr{U}\leftarrow\mathscr{U}}\,\|\delta u_l^j\|_{\mathscr{U}}$$

shows that Theorem 9.5.7 has to be changed as follows.

Lemma 9.5.9. Under the assumptions of Theorem 9.5.7 the two-grid iteration (3.7) modified according to (8 b) with Ψ_{l-1} satisfying (8 a) has a contraction number

$$\zeta_l = \zeta(\varrho_l/\varepsilon_l, \varrho_{l-1}/\varepsilon_{l-1}, \nu) + C^*\,\psi_{l-1} \tag{9.5.8 c}$$

with ζ from (7 a). Convergence holds for sufficiently small ψ_{l-1}.

Let the iteration Φ_0 from (3.11) satisfy

$$\|\Phi_0(u_0, f_0) - \mathscr{L}_0^{-1}(f_0)\|_{\mathscr{U}} \leqq \varphi_0\,\|u_0 - \mathscr{L}_0^{-1}(f_0)\|_{\mathscr{U}}$$
$$\text{for all } u_0 \in \mathscr{U}_0(\varepsilon_0), \quad f_0 \in \mathscr{F}_0(\varepsilon_0). \tag{9.5.9}$$

Note 9.5.10. a) At level $l = 1$ the iteration $NMGM^{(\nu,0)}$ from (3.12) is identical with the two-grid iteration considered in Lemma 9.5.9 when we define $\Psi_0 := \Phi_0^\gamma$ (γ-fold application of Φ_0).

b) The multi-grid iteration (3.12) can be described by

$$\Phi_l(u_l, f_l) := \text{output of } NMGM(l, u_l, f_l),$$

where $NMGM$ is the two-grid iteration of Lemma 9.5.9 with $\Psi_{l-1} = \Phi_{l-1}^\gamma$.

Denote the contraction number of Φ_l by ζ_l. From (8 c) and Note 9.5.10 one concludes that

$$\zeta_l = \zeta + C^*\,\zeta_{l-1}^\gamma \quad \text{for } l \geqq 1, \qquad \zeta_0 = \varphi_0, \tag{9.5.10}$$

with $\zeta = \zeta(\varrho_l/\varepsilon_l, \varrho_{l-1}/\varepsilon_{l-1}, \nu) < 1$. This is the inequality (7.1.6), only the starting value is different.

Remark 9.5.11. Assume $C^*\gamma > 1$ and (7.1.7 a) and let ζ^* be the value defined in Lemma 7.1.6. The estimate $\zeta_l \leqq \zeta^*$ ($l \geqq 1$) holds if $\varphi_0 \leqq \zeta^*$.

Theorem 9.5.12 (convergence of $NMGM^{(\nu,0)}$). (i) Let $NMGM^{(\nu,0)}$ be the iteration (3.12) with $\gamma \geqq 2$,

$$\tilde{u}_k \in \mathscr{U}_k(\varrho_k/6) \quad \text{for } k = 0, 1, \ldots, l-1 \tag{9.5.11 a}$$

and with $\sigma(k, d_k)$ (cf. (3.12e)) defined by (3.8) involving

$$\sigma_k \leqq \varrho_k/(6\,C_{DL}), \tag{9.5.11 b}$$

where C_{DL} is defined in Note 9.5.2.

(ii) Let ϱ_k/ε_k ($0 \leqq k < l$) and the contraction number φ_0 of Φ_0 be sufficiently small.

(iii) L_l and $S_l^{(\nu)}$ defined by (6 a) have to satisfy the smoothing property (6.1.7), the approximation property (6.1.8), and (7.1.1). In addition assume (7.1.2) and (6 b–d).

Then for any $\zeta^* < 1$ there are $\bar{h} > 0$ and $\underline{\nu}$ such that the following assertion holds for $NMGM^{(\nu,0)}$ with $h_1 \leqq \bar{h}$, $\underline{\nu} \leqq \nu < \bar{\nu}(h_1)$. The multi-grid iteration (3.12) is well-defined for all starting values u_l^0 and right-hand sides f_l satisfying (5 a), (5 b), or (5 c). The iterates remain in $\mathscr{U}_l(\varrho_l)$. The contraction number is bounded by $\zeta^* \approx \text{const} \cdot \eta(\nu)$ with $\eta(\nu)$ from (6.1.7 a). If $\bar{\nu}(h) = \infty$ (cf. (6.1.7 c)) there is no restriction on h_1 and $\nu \geqq \underline{\nu}$ is the only condition on ν.

Proof. a) Given any $\zeta^* < 1$ one finds ζ such that Remark 9.5.11 implies $\zeta_l \leqq \zeta^*$. According to assumption (ii) $\varphi_0 \leqq \zeta^*$ must hold. There exist $\bar{h}$ and $\underline{\nu}$ such that $\| M_l(\nu) \|_{\mathscr{U} \leftarrow \mathscr{U}} \leqq C_A \eta(\nu) \leqq \zeta/2$ for $M_l(\nu) := (I - p L_{l-1}^{-1} r L_l) S_l^{(\nu)}$ provided that $h_1 \leqq \bar{h}, \underline{\nu} \leqq \nu < \bar{\nu}(h_1)$ (cf. Theorem 6.1.7). Hence the two-grid contraction number $\zeta(\varrho_l/\varepsilon_l, \varrho_{l-1}/\varepsilon_{l-1}, \nu)$ from Theorem 9.5.7 is bounded by ζ if ϱ_l and ϱ_{l-1} are sufficiently small (cf. (7 b) and assumption (ii)).

b) The assertion of the theorem is obvious for $l = 0$ since $\varphi_0 \leqq \zeta^*$.

c) Let the assertion hold for the level $l-1$. By Note 9.5.10 the multi-grid iteration can be regarded as the two-grid method (3.7) modified by (8 b) with $\Psi_{l-1} = \Phi_{l-1}^\gamma$, where Φ_{l-1} is the multi-grid iteration (3.12) at level $l-1$. We have to check whether Φ_{l-1}^γ is well-defined. The arguments of Φ_{l-1}^γ are $u_{l-1}^0 = \tilde{u}_{l-1}$ and $f_{l-1} = \tilde{f}_{l-1} - sd$ (d from (3.12d)). By the choice of s (cf. (11 b)) and because of $\tilde{f}_{l-1} \in \mathscr{F}_{l-1}(\varrho_{l-1}/6)$ it follows that $f_{l-1} \in \mathscr{F}_{l-1}(\varrho_{l-1}/3)$. Since $\tilde{u}_{l-1} \in \mathscr{U}_{l-1}(\varrho_{l-1}/3)$ (cf. (11 a)) the requirement (5 b) is satisfied. For the given arguments Φ_{l-1}^γ is well-defined and has the contraction number ζ_{l-1}. The contraction number ζ_l at level l is estimated in (10) by $\zeta_l = \zeta(\varrho_l/\varepsilon_l, \varrho_{l-1}/\varepsilon_{l-1}, \nu) + C^* \zeta_{l-1}^\gamma \leqq \zeta + C^* \zeta_{l-1}^\gamma$.

d) Remark 9.5.11 proves that $\zeta_l \leqq \zeta^*$ for all l. Because of (7 b) the contraction number behaves like const $\cdot\, \eta(\nu)$ for sufficiently small ϱ_k. □

9.5.3 Analysis of the Nonlinear Nested Iteration

As in the linear case we assume (5.2.3–4):

$$\|\tilde{p}\, u_{k-1}^* - u_k^*\|_{\mathscr{U}} \leqq C_1 h_k^\kappa, \qquad 1 \leqq k \leqq l, \tag{9.5.12}$$

$$\|\tilde{p}\|_{\mathscr{U} \leftarrow \mathscr{U}} \leqq C_{20}, \qquad h_{k-1}^\kappa \leqq C_{21} h_k^\kappa, \qquad C_2 := C_{20} C_{21}. \tag{9.5.13}$$

The exponent κ from the relative discretisation error estimate (12) is the consistency order. Although the nonlinear nested iteration (3.14) behaves almost as the linear one, there is a characteristic distinction. Even if the approximation $\tilde{u}_{k-1}$ is the exact solution u_{k-1}^*, the starting value $\tilde{p}\,\tilde{u}_{k-1}$ may be outside of $\mathscr{U}_k(\varepsilon_k)$ so that the iteration at level k may fail to converge to u_k^*. $\tilde{p}\, u_{k-1}^* \in \mathscr{U}_k(\varepsilon_k)$ can be ensured by

$$C_1 h_k^\kappa < \varepsilon_k, \qquad 1 \leqq k \leqq l. \tag{9.5.14}$$

We recall that ε_k may depend on h_k as $\varepsilon_k = \varepsilon h_k^\beta$. If $\beta \leqq \kappa$ Inequality (14) becomes $C_1 h_1^{\kappa-\beta} < \varepsilon$, otherwise $C_1 h_l^{\kappa-\beta} < \varepsilon$.

Theorem 9.5.13. Assume that the iteration *NMGM* involved in (3.14 d) is well-defined for $f_k - 0$, $u_k \in \mathscr{U}_k(\varepsilon_k)$ (cf. (5 a)) and has a contraction number $\zeta < 1$ for all $1 \leqq k \leqq l$. Suppose (12), (13), (14). Using (14) the number i involved in (3.14 d) can be chosen so that

$$C_2 \zeta^i \leqq 1 - C_1 h_k^\kappa/\varepsilon_k, \qquad 1 \leqq k \leqq l. \tag{9.5.15}$$

Then the nested iteration (3.14) is well-defined and produces results $\tilde{u}_k$ with

$$\|\tilde{u}_k - u_k^*\|_{\mathscr{U}} \leqq C_3(\zeta, i)\, C_1 h_k^\kappa \quad \text{with} \quad C_3(\zeta, i) = \zeta^i/(1 - C_2 \zeta^i)$$
$$\text{for } 0 \leqq k \leqq 1, \tag{9.5.16}$$

provided that the starting value $\tilde{u}_0$ from (3.14 a) satisfies (16).

The proof is exactly the same as that for Theorem 5.2.3. In addition one has to check that (15) implies $\tilde{p}\,\tilde{u}_{k-1} \in \mathscr{U}_k(\varepsilon_k)$. It is obvious, that we may replace *NMGM* in (3.14d) by any iterative solver defined on $\mathscr{U}_k(\varepsilon_k)$ with a contraction number ζ. For instance, *NMGM* may be replaced by the Newton-multi-grid iteration from § 9.2. If we use the multi-grid iteration $NMGM^{(\nu_1,\nu_2)}$ defined in (3.12), we need the previously computed approximations $\tilde{u}_0, \tilde{u}_1, \ldots, \tilde{u}_{l-1}$ in the iteration at level l. In Theorem 9.5.12 we required $\tilde{u}_k \in \mathscr{U}_k(\varrho_k/6)$. This can be achieved by the nested iteration as stated in

Corollary 9.5.14. Theorem 9.5.12 requires $\varrho_k = \xi\,\varepsilon_k$ with sufficiently small ξ. If i from (3.14d) satisfies $C_3(\zeta, i)\, C_1 h_k^\kappa/\varepsilon_k \leqq \xi/6$ and if $\tilde{u}_0 \in \mathscr{U}_0(\varrho_0/6)$, the condition $\tilde{u}_k \in \mathscr{U}_k(\varrho_k/6)$, $0 \leqq k \leqq l$, is met. If we replace ε_k in (14) and (15) by ϱ_k, a sufficient condition is

$$\zeta^i \leqq \tfrac{1}{6}. \tag{9.5.17}$$

9.6 Exercises

Exercise 9.6.1. Discretise the equation

$$-\Delta u + u\, u_x = f \quad \text{in } \Omega = (0,1) \times (0,1), \qquad u = 0 \quad \text{on } \Gamma,$$

by $\mathscr{L}_l(u_l) := -\Delta_l u_l + u_l \hat{\partial}_x u_l - f_l$, where Δ_l is the standard five-point scheme and $\hat{\partial}_x u_l(x, y) = [u_l(x + h_l, y) - u_l(x - h_l, y)]/2h_l$.

(i) Prove that $D\mathscr{L}_l$ can be represented such that

$$(D\mathscr{L}_l(u_l, u_l') - L_l)\, v_l = v_l \hat{\partial}_x (u_l - u_l^*) + (u_l' - u_l^*)\, \hat{\partial}_x v_l .$$

(ii) Choose the energy norm $\|\cdot\|_{\mathscr{U}} = |\cdot|_1$ and prove inequality (5.6 c) with the right-hand side $\Theta(\varrho_l/\varepsilon_l) = \text{const} \cdot \varrho_l$. Therefore, the choice $\varepsilon_0 = \varepsilon_1 = \ldots = \varepsilon_l = \varepsilon$ is possible. Hint: Use the discrete analogue of $\|\cdot\|_{L^4(\Omega)} \leqq \|\cdot\|_{H^1(\Omega)}$ (cf. Adams [1, p. 97]).

(iii) Choose $\|\cdot\|_{\mathscr{U}} = |\cdot|_0$ and prove (5.6c) with $\varepsilon_l = \varepsilon\, h_l^{1+\delta}\,(\delta > 0)$ and $\Theta(\eta) = \text{const} \cdot \eta$.

Exercise 9.6.2. In Theorem 9.5.12 we required $\tilde{u}_k \in \mathscr{U}_k(\varrho_k/6)$. Prove that $\mathscr{U}_k(\varrho_k/6)$ can be replaced by $\mathscr{U}_k(\Theta\,\varrho_k)$ with $\Theta = (5-\zeta)/(6\,(2\zeta+1))$, where ζ is the multi-grid contraction number. As a consequence, Inequality (5.17) may be weakened (cf. Hackbusch [33, p. 28]).

10. Singular Perturbation Problems

An operator $L = L(\varepsilon)$ depending on a parameter ε is called singularly perturbed if the limiting operator $L(0) = \lim_{\varepsilon \to 0} L(\varepsilon)$ is of a type other than $L(\varepsilon)$ for $\varepsilon > 0$. For instance, an elliptic operator $L(\varepsilon) = \varepsilon L^I + L^{II}$ $(\varepsilon > 0)$ is singularly perturbed if L^{II} is non-elliptic or elliptic of a lower order. Different types of L^{II} give rise to the subsections 10.1–3.

When L^{II} is not elliptic (of order $2m$) the given boundary conditions can be satisfied for $\varepsilon = 0$ only partially. For $\varepsilon > 0$ the remaining parts of the boundary conditions generate a boundary layer. Under certain conditions such layers may also lie in the interior of the domain.

Instead of $L(\varepsilon) = \varepsilon L^I + L^{II}$ one may also consider $L^I + cL^{II}$ for $c \to \infty$. For ease of exposition we consider linear boundary value problems. They may serve as model problems approaching a more complicated nonlinear situation.

The discretisation of singular perturbation problems contains characteristic difficulties. This fact influences the multi-grid iteration, since the discretisation process is hidden in the hierarchy of operators L_k $(0 \leqq k \leqq l)$. A related difficulty concerns the smoothing process. When the discrete problem is also singularly perturbed, one has to construct special smoothing iterations which are well-suited to the discrete analogue of $L(0)$.

10.1 Anisotropic Elliptic Equations

10.1.1 A Model Problem

A particular model problem of an anisotropic problem is

$$-\varepsilon u_{xx}(x, y) - u_{yy}(x, y) = f(x, y) \quad \text{in } \Omega, \qquad u = 0 \quad \text{on } \Gamma, \tag{10.1.1}$$

where $\varepsilon \ll 1$ (or $\varepsilon \gg 1$) and $\Omega = (0,1) \times (0,1)$. The general anisotropic problem is given in Eq. (14) below. However, most of the difficulties can be illustrated even for Eq. (1). In a square grid Ω_l of size h_l Eq. (1) is discretised by the

10.1.2 Smoothing with Respect to Alternating Directions

Equation (1) is a model problem for the more general equation

$$-a(x, y)\,u_{xx}(x, y) - b(x, y)\,u_{yy}(x, y) = f(x, y), \tag{10.1.5}$$

where $0 < a(x, y) \ll b(x, y)$ (perhaps only locally). The smoothers mentioned in § 10.1.1 work for Eq. (5) if $a \leqq b$. If $a \geqq b$ one has to replace the y-lines by x-lines. Therefore, alternating versions have been proposed for the general situation, where $a \ll b$ and $a \gg b$ occur in different parts of Ω. Denote the x-line (y-line) Gauß-Seidel iteration by $\mathscr{S}_l^x$ ($\mathscr{S}_l^y$). The alternating linewise Gauß-Seidel process is defined by

$$\mathscr{S}_l = \mathscr{S}_l^y \circ \mathscr{S}_l^x. \tag{10.1.6}$$

According to the ordering of the lines one distinguishes the lexicographical version from the alternating zebra-line version. Obviously, the computational work of (6) is doubled. Except when $a \approx b$ half of the work spent for smoothing is wasted, since $\mathscr{S}_l^x$ ($\mathscr{S}_l^y$) has almost no effect if $a \ll b$ ($a \gg b$).

Another smoothing process suited for anisotropic problems is the well-known ADI (alternating direction implicit iterative) method (3.3.14). Its second half-step has the form (3 a), while the first one corresponds to the x-direction.

Since the mixed term u_{xy} is missing in Eq. (5), the only dominating directions can be x or y. As soon as the mixed term is added, every direction $x \sin\varphi = y\cos\varphi$, $0 \leqq \varphi < 2\pi$, may be dominant. This situation can even arise for the isotropic Poisson equation, if for some reason the discretisation uses a curvi-linear grid. The dominant direction of the operator $L(\varepsilon) = -[(\partial/\partial x + \partial/\partial y)^2 + \varepsilon(\partial/\partial x - \partial/\partial y)^2]/2$ corresponds to the angle $\varphi = \frac{\pi}{2}$. $L(\varepsilon)$ may be discretised by the star

$$L_l = h_l^{-2}\begin{bmatrix} 0 & -\varepsilon & -(1+\varepsilon)/2 \\ -\varepsilon & 1+3\varepsilon & -\varepsilon \\ -(1+\varepsilon)/2 & -\varepsilon & 0 \end{bmatrix}. \tag{10.1.7}$$

Note 10.1.4. Neither the alternating line-Gauß-Seidel process nor the ADI method is suited for Eq. (7) when $\varepsilon \ll 1$.

Obviously, Eq. (7) requires a diagonal line method. In order to construct a smoother fitted for all angles φ, one should replace $\mathscr{S}_l$ from (6) by

$$\mathscr{S}_l = \mathscr{S}_l^{y+x} \circ \mathscr{S}_l^y \circ \mathscr{S}_l^{y-x} \circ \mathscr{S}_l^x \tag{10.1.8}$$

where $\mathscr{S}_l^{y\pm x}$ are linewise Gauß-Seidel iterations with respect to the two diagonal directions $y \pm x = 0$. This method is safer than (6) but it costs twice as much as (6) and four-times the work of the usual Gauß-Seidel iteration.

Methods safer than alternating line-Gauß-Seidel iterations are the ILU and ILLU smoothers described below. In particular, ILLU works for all dominant directions.

10.1.3 The Incomplete LU Decomposition as a Smoother

The exact LU decomposition $L_l = A\,B$ (where A and B are lower and upper triangular matrices respectively) cannot be recommended except for one-dimensional problems. The reason is the well-known fill-in: The factors A and B are not as sparse as L_l. However, a prescribed sparsity pattern can be achieved if we replace $L_l = A\,B$ by

$$L_l = AB - C \quad (A \text{ is a lower, } B \text{ an upper triangular matrix}). \qquad (10.1.9\,\text{a})$$

In the two-dimensional case $\Omega \subset \mathbb{R}^2$ the sparsity pattern P can be described by a subset of all pairs $\{(i,j): 1 \leqq i, j \leqq n_l\}$ of indices. A $n_l \times n_l$-matrix M has the pattern P_M if $(i,j) \notin P_M$ implies $M_{ij} = 0$. On the other hand $M_{ij} = 0$ is permitted for $(i,j) \in P_M$. For instance, L_l is a five-point star if P_L is the five-point pattern which is defined by "$(i,j) \in P_L$ if and only if i and j are indices of horizontally or vertically neighboured grid points". This five-point pattern will be symbolised by

$$P_L = \begin{bmatrix} & * & \\ * & * & * \\ & * & \end{bmatrix}.$$

The notation of other patterns by

$$P = \begin{bmatrix} & * & * \\ * & * & * \\ * & * & \end{bmatrix}, \qquad P = \begin{bmatrix} & * & \\ \cdot & * & * \\ & \cdot & \end{bmatrix}, \qquad P = \begin{bmatrix} & \cdot & \\ * & * & \cdot \\ & * & \end{bmatrix}$$

needs no explanation.

In the case of (9 a) we have to fix the patterns P_A, P_B, P_C of A, B, and C. The usual assumption is

$$P_A \cap P_C = P_B \cap P_C = \emptyset, \qquad (10.1.9\,\text{b})$$

i.e. A and C have no common non-zero entries (similarly B and C). If

$$P_L \subset P_A \cup P_B \qquad (10.1.9\,\text{c})$$

did not hold, the decomposition (9 a) would be too poor. In the following we assume that the factors A, B can be normalised by $A_{ii} = B_{ii}$. Then (9 a) may be rewritten as

$$L_l = (A' + D)\,D^{-1}(B' + D) - C, \qquad (10.1.9\,\text{a*})$$

where A' (B') is a strictly lower (upper) triangular matrix, while in the first instance D is diagonal. A', B', and D can be computed by the following algorithm for ILU factorisation (10.1.10)

$F := L_l;$ (10.1.10 a)

for $i := 1\,(1)\, n_l$ **do for** $j := 1 - i\ (1)\ n_l - i$ **do**

  if $(i, j+i) \in P_A \cup P_B$ **then**

$$F_{i,j+i} := F_{i,j+i} - \sum_{\substack{m<0 \\ m<j \\ (i,i+m)\in P_A \\ (i+m,i+j)\in P_B}} F_{i,i+m} F^{-1}_{i+m,i+m} F_{i+m,i+j}; \tag{10.1.10 b}$$

comment $A'_{ik} = F_{ik}$ for $i > k$, $D_{ii} = F_{ii}$, $B'_{ik} = F_{ik}$ for $i < k$;

The matrix C can be computed by

$$C = A' D^{-1} B' + A' + B' + D - L_l. \tag{10.1.11}$$

Note that $C_{ik} = 0$ for $(i, k) \in P_A \cup P_B$, while $C_{ik} = (A' D^{-1} B')_{ik}$ otherwise (cf. (9c)).

Remark 10.1.5. a) Factorisation (9 a*) with non-singular D exists if L_l is a symmetric M-matrix (cf. Meijerink-van der Vorst [1]).

b) The ILU decomposition depends on the ordering of the components of u_l.

c) If $(i, j) \in P_A$ is equivalent to $(j, i) \in P_B$ and if $L_l = L_l^* > 0$, algorithm (10) yields A', B', D with $A' = B'^*$, $D = D^*$. The factorisation (9 a) with $A = (A' + D)D^{-1/2}$ and $B = A^*$ is called an *incomplete Cholesky decomposition.*

By Eq. (11) the sparsity pattern P_C of C can be determined. Below we list two examples of P_A, P_B and the resulting pattern of C. The underlying ordering of the grid points is the lexicographical one (left to right, bottom to top).

$$P_L = \begin{bmatrix} & * & \\ * & * & * \\ & * & \end{bmatrix}, \quad P_A = \begin{bmatrix} & \cdot & \\ * & * & \cdot \\ & * & \end{bmatrix}, \quad P_B = \begin{bmatrix} & * & \\ \cdot & * & * \\ & \cdot & \end{bmatrix}, \quad P_C = \begin{bmatrix} * & \cdot & \\ \cdot & \cdot & \cdot \\ & \cdot & * \end{bmatrix}, \tag{10.1.12 a}$$

$$P_L = \begin{bmatrix} & * & * \\ * & * & * \\ * & * & \end{bmatrix}, \quad P_A = \begin{bmatrix} \cdot & \cdot & \cdot \\ * & * & \cdot \\ * & * & \cdot \end{bmatrix}, \quad P_B = \begin{bmatrix} & * & * \\ \cdot & * & * \\ & \cdot & \end{bmatrix}, \quad P_C = \begin{bmatrix} * & \cdot & \cdot \\ \cdot & \cdot & \cdot \\ \cdot & \cdot & * \end{bmatrix}, \tag{10.1.12 b}$$

$$P_L = \begin{bmatrix} * & * & \cdot \\ * & * & * \\ \cdot & * & * \end{bmatrix}, \quad P_A = \begin{bmatrix} & \cdot & \\ * & * & \cdot \\ & * & * \end{bmatrix}, \quad P_B = \begin{bmatrix} * & * & \\ \cdot & * & * \\ & \cdot & \end{bmatrix},$$

$$P_C = \begin{bmatrix} * & \cdot & \cdot & \cdot & \\ & \cdot & \cdot & \cdot & \\ & \cdot & \cdot & \cdot & * \end{bmatrix}, \tag{10.1.12 c}$$

$$P_L = \begin{bmatrix} * & * & * \\ * & * & * \\ * & * & * \end{bmatrix}, \quad P_A = \begin{bmatrix} \cdot & \cdot & \cdot \\ * & * & \cdot \\ * & * & * \end{bmatrix}, \quad P_B = \begin{bmatrix} * & * & * \\ \cdot & * & * \\ \cdot & \cdot & \cdot \end{bmatrix},$$

$$P_C = \begin{bmatrix} * & \cdot & \cdot & \cdot & \\ * & \cdot & \cdot & \cdot & * \\ & \cdot & \cdot & \cdot & * \end{bmatrix}. \tag{10.1.12 d}$$

(12 a), (12 c), (12 d) are called five-, seven-, and nine-point ILU, respectively. The computational work of algorithm (10) is comparatively small. In the case of (12 a) the sum in (10 b) is empty for all $j \neq 0$. When $j = 0$, the sum is taken over two indices m (such that $i + m$ is the index of the left or lower neighbour of i). Therefore, algorithm (10) requires 6 operations per grid point.

An algorithm computing A, B from (9 a) with $A_{ii} = B_{ii}$ (instead of A', B', D) is described by Wesseling [3]. Operation counts of the five- and seven-point ILU can also be found there.

The representations (9 a) and (9 a*) of L_l give rise to the following iteration:

$$\begin{aligned} \mathscr{S}_l(u_l, f_l) &:= u_l - B^{-1} A^{-1} (L_l u_l - f_l) = B^{-1} A^{-1} (f_l + C u_l) \\ &= (B' + D)^{-1} D (A' + D)^{-1} (f_l + C u_l). \end{aligned} \tag{10.1.13}$$

An ILU smoothing process was first introduced by Wesseling-Sonneveld [1].

Remark 10.1.6. If $\mathscr{S}_l$ from (13) is used as smoothing iteration one need not store the matrix L_l, but only A', B', C, D. If $P_A \cup P_B = P_L$ the matrices A', B', D require the same storage as L_l. The matrix C is obviously very sparse. The coarse-grid correction requires the defect $d_l = L_l u_l' - f_l$ of $u_l' = \mathscr{S}_l(u_l, f_l)$. This grid function can be computed cheaply from $d_l = C(u_l - u_l')$.

Remark 10.1.6 shows that the ILU smoothing (13) is easier to perform than it might seem at first glance. The next result indicates that (13) is a good candidate for a smoother of anisotropic problems.

Remark 10.1.7. The ILU smoothing process (13) with A, B, C from (9 a, b) satisfies Criterion 10.1.1 for $L_l(\varepsilon)$ from (2), for the analogous discretisation of $-u_{xx} - \varepsilon u_{yy} = f$, and for $L_l(\varepsilon)$ from (7).

Proof. In all cases $L_l(0) u_l = f_l$ represents a family of decoupled systems for vertical, horizontal, or diagonal lines. Since each system is tridiagonal, it admits an exact factorisation without fill-in. Hence, $L_l = AB - C$ holds with $C = O$ and one step of $\mathscr{S}_l$ yields the true solution. □

Wesseling [3, Table 5.5] applied a multi-grid algorithm with seven-point ILU smoothing to $L_l(\varepsilon) u_l = f_l$ with L_l from (2). The observed averaged reduction ratios $\tilde{\varrho}$ are repeated in Table 10.1.2. The further details of the multi-grid program are: $\nu_1 = 0$, $\nu_2 = 1$, $\gamma = 1$, p: seven-point prolongation, r: seven-point restriction, $L_{l-1} = r L_l p$.

Table 10.1.2. Multi-grid convergence for the anisotropic problem (2) with seven-point ILU smoother

ε	0.5	0.1	0.01	0.0001	2	10	100	1000
Averaged rate	0.036	0.023	0.0016	$4_{10}-8$	0.047	0.10	0.052	$31_{10}-5$

The convergence behaviour shown in Table 10.1.2 differs from the prediction by means of the smoothing rates ϱ_B (cf. § 8.2.3). Kettler [1, Theorem 3.2] proved $\varrho_B(\nu) \to (3+2\sqrt{2})^{-\nu}$ as $\varepsilon \to 0$ and $\varrho_B(\nu) \to 1$ as $\varepsilon \to \infty$. The smoothing rates can be computed since for $\Omega = \mathbb{R}^2$ the matrices A, B become Toeplitz operators which are diagonalised by the Fourier transform (cf. also Hemker [3], [9], Wesseling [3]).

Rotating Eq. (1) by an angle α we obtain

$$-(\varepsilon c^2 + s^2)\, u_{xx} - 2(\varepsilon - 1)\, s\, c\, u_{xy} - (\varepsilon s^2 + c^2)\, u_{yy} = f, \tag{10.1.14}$$

where $c = \cos\alpha$, $s = \sin\alpha$, $0 \leqq \alpha \leqq \pi$. As pointed out by Hemker [11] ILU is an excellent smoother for $\frac{\pi}{4} \leqq \alpha \leqq \pi$, but may cause divergence for values $\alpha \in (0, \frac{\pi}{4})$. Nevertheless, rapid convergence can be achieved by using the line-version of (9 a*), where A' (B') is a strictly lower (upper) block-triangular matrix and D is tridiagonal. The algorithm for the incomplete line LU decomposition (ILLU) is explained below.

Define the lexicographically ordered x (or y) lines as blocks. Any nine-point scheme involves three lines; hence its block pattern is $P_L = (*\ \ *\ \ *)$. Interpret the indices of F_{ik} in algorithm (10) as indices of the blocks. Then the sum in (10 b) is empty for $j \neq 0$ and is taken over $m = -1$ for $j = 0$. With this modification $F_{ii} = D_{ii}$ would become a full block of the block matrix F. To ensure the tridiagonal structure of the matrix D the algorithm is changed into the

algorithm for the ILLU factorisation (10.1.15)

$F := L_l$; **comment** L_l is at most a nine-point scheme; (10.1.15 a)

comment $L_{l,ik}$ are the blocks of L_l; (10.1.15 b)

for $i := 2(1)\, n_l^{\text{line}}$ **do**

$F_{ii} := F_{ii} - \operatorname{tridiag}\{F_{i,i-1} F_{i-1,i-1}^{-1} F_{i-1,i}\}$; (10.1.15 c)

comment meaning of the blocks F_{ik}: A'_{ik} if $i > k$, D_{ii} if $i = k$, B'_{ik} if $i < k$;

The symbol tridiag$\{C\}$ denotes the tridiagonal matrix $\tilde{C}$ with $\tilde{C}_{ik} = C_{ik}$ for $|i-k| \leqq 1$. n_l^{line} is the number of lines (blocks).

The numerical results reported by Sonneveld-Wesseling-de Zeeuw [1] confirm that the multi-grid iteration with ILLU smoother works perfectly for Eq. (14) with arbitrary angle α.

10.2 Indefinite Problems

Another singular perturbation problem arises for operators $L(\varepsilon)$ like

$$L(\varepsilon) = \varepsilon L^P + L^S, \qquad L^S = (L^S)^* \geqq 0, \tag{10.2.1}$$

where order $(L^S) <$ order $(L^P) = 2m$. L^P may be assumed to be positive definite. The two cases of $\varepsilon > 0$ and $\varepsilon < 0$ are to be distinguished. In the first case $L(\varepsilon)$ is always positive definite and relatively easy to solve (cf. § 10.2.1), whereas the second case leads to indefinite problems (cf. § 10.2.2), i. e. $L(\varepsilon)$ has positive and negative eigenvalues.

10.2.1 Strongly Positive Definite Problems

The discrete analogue of (1) is

$$L_l(\varepsilon) = \varepsilon L_l^P + L_l^S, \qquad L_l^S = (L_l^S)^* \geqq O, \qquad \varepsilon > 0. \tag{10.2.2}$$

Usually, the perturbed problem $L_l(\varepsilon) u_l = f_l$ is easier to solve than $L_l^P u_l = f_l$, provided that Criterion 10.1.1 is satisfied: The multigrid iteration must be fast for the equation $L_l^S u_l = f_l$.

In the case of the standard five-point difference scheme of Helmholtz' equation $-\Delta u + cu = f$, $c = 1/\varepsilon > 0$, the matrix L_l^S is diagonal. Then for instance any variant of the Gauß-Seidel iteration yields the exact solution of $L_l^S u_l = f_l$ in one step. Therefore, a multigrid iteration with such a smoother satisfies Criterion 10.1.1.

The perturbation may also be induced by the boundary condition as shown in the next example. The finite element method with penalty replaces the Poisson equation $-\Delta u = f_\Omega$ in Ω, $u = f_\Gamma$ on Γ by

$$-\Delta u = f_\Omega \quad \text{in } \Omega, \qquad u + \varepsilon \frac{\partial u}{\partial n} = f_\Gamma \quad \text{on } \Gamma \tag{10.2.3}$$

(cf. Babuška [1], Hackbusch [31]). The variational formulation of Eq. (3) is

$$a_\varepsilon(u, v) = \varepsilon \int_\Omega f_\Omega v \, d\mathbf{x} + \int_\Gamma f_\Gamma v \, d\Gamma \quad \text{for all } v \in H^1(\Omega)$$

with

$$a_\varepsilon(u, v) := \varepsilon \int_\Omega (\operatorname{grad} u) \cdot (\operatorname{grad} v) \, d\mathbf{x} + \int_\Gamma u v \, d\Gamma. \tag{10.2.3'}$$

The operator $L(\varepsilon)$ associated with $a_\varepsilon(.,.)$ can be split into $\varepsilon L^P + L^S$ as in Eq. (1). L^P corresponds to the Poisson equation with Neumann conditions, while L^S defined by $(L^S u, v) = \int_\Gamma u v \, d\Gamma$ is an operator of order 1, which is obviously positive semi-definite. The finite element discretisation yields a stiffness matrix (2), where L_l^S has non-zero entries $(L_l^S)_{ij}$ only if i, j are indices of elements $\varphi_{l,i}$, $\varphi_{l,j}$ (cf. (3.1.29 a)) with support $(\varphi_{l,i}) \cap$ support $(\varphi_{l,j}) \neq \emptyset$. To satisfy

Criterion 10.1.1 one may apply a multi-grid iteration with the following Gauß-Seidel process: Let the coefficients of all $\varphi_{l,i}$ with support $(\varphi_{l,i}) \cap \Gamma \neq \emptyset$ form the first block, while the further blocks may consist of only one variable.

10.2.2 Weakly Indefinite Problems

For negative ε the non-singularity of the operator $L(\varepsilon)$ from (1) cannot be guaranteed. Under usual conditions $L(\varepsilon)$ is singular for $\varepsilon^{(\mu)}$ with

$$0 > \varepsilon^{(1)} < \varepsilon^{(2)} < \ldots < \varepsilon^{(\mu)} < \ldots \nearrow 0.$$

As soon as $\varepsilon^{(1)} < \varepsilon < 0$, $L(\varepsilon)$ is indefinite. The strength of indefiniteness is given by the largest ν such that $\varepsilon^{(\nu)} < \varepsilon < 0$. Here we consider "weakly" indefinite Problems, i.e. ν is not too large.

The discretisation by $L_l(\varepsilon)$ represents a non-singular system if and only if ε is not equal to one of the eigenvalues

$$\varepsilon_l^{(1)} < \varepsilon_l^{(2)} < \ldots < \varepsilon_l^{(n_l)} < 0.$$

Since $\varepsilon_l^{(\nu)} \to \varepsilon^{(\nu)}$ $(l \to \infty)$ should be assumed, we obtain the following result: Let $L(\varepsilon)\,u = f$ be solvable (i.e. $\varepsilon \neq \varepsilon^{(\mu)}$ for all μ); then for h_l sufficiently small (e.g. for l large enough) the discrete problem $L_l(\varepsilon)\,u_l = f_l$ is non-singular.

First we consider a numerical example, where $\varepsilon^{(1)} < \varepsilon < \varepsilon^{(2)}$, $\varepsilon_l^{(1)} < \varepsilon < \varepsilon_l^{(2)}$. The problem

$$-e^x u_{xx} - e^{-y} u_{yy} + \sin(\pi x)\,u_y - [\lambda + \pi^2(e^x + e^{-y}) + \pi\cos(\pi x)]\,u = 1$$
$$\text{in } \Omega, \qquad u = 0 \quad \text{on } \Gamma \tag{10.2.4}$$

with $\Omega = (0,1) \times (0,1)$ has $\lambda = 0$ as first eigenvalue. Hence it is indefinite for $\lambda > 0$. Multiplying (4) by $\varepsilon := -1/(\lambda + 1)$ we can cast (4) into the form $\varepsilon L^P u + L^S u = \varepsilon$ with $L^S = I$. Decreasing values $\lambda \to \infty$ correspond to $\varepsilon \nearrow 0$. The observed convergence factors are listed in Table 10.2.1. The applied multi-grid program is taken from Hackbusch [6]. Since the discrete systems are indefinite, almost all classical iterations are divergent. Nevertheless they can be used as smoothing iteration (cf. Lemma 6.1.5). But the smoothing iteration cannot be applied at the lowest level to solve approximately for $u_0 = L_0^{-1} f_0$. Therefore one has to follow

Remark 10.2.1. The equations $L_0 u_0 = f_0$ appearing at the lowest level $l = 0$ have to be solved exactly as formulated in (4.1.1 a). An iterative solution as in (9.3.12 b) does not suffice unless the iteration applies to indefinite systems as e.g. Kaczmarz' algorithm (3.3.20).

The theoretical considerations of §§ 6–7 proved multi-grid convergence provided that the coarsest grid size h_0 is chosen small enough. The dependence on h_0 is illustrated by the numbers of Table 10.2.1. $h_0 = \frac{1}{4}$ yields worse results than $h_0 = \frac{1}{8}$. In particular for $\lambda = +1$ the multi-grid iteration turns from convergence ($h_0 = \frac{1}{8}$) to divergence ($h_0 = \frac{1}{4}$).

Table 10.2.1. Convergence rates of problem (4) for $h_l = \frac{1}{64}$

λ	-5	-2	-1	1	2	3	5	8	10	15
$h_0 = \frac{1}{8}$	0.042	0.14	0.32	0.23	0.12	0.083	0.049	0.078	0.11	0.34
$h_0 = \frac{1}{4}$	0.12	0.37	0.64	3.7	0.95	0.55	0.29	0.15	0.21	0.55

To understand this effect one has to study the behaviour of an error $\tilde{u}_l - u_l = v_l^{(1)}$, where $v_l^{(1)}$ is the eigenvector defined by $(\varepsilon_l^{(1)} L_l^P + L_l^S)\, v_l^{(1)} = 0$. The error will be slightly increased by the (divergent) smoothing iteration. Convergence of the multi-grid process can only be achieved by the coarse-grid correction. For that reason $v_l^{(1)}$ must be sufficiently well approximated by $v_{l-1}^{(1)}$. In particular, $v_0^{(1)}$ has to approximate well the vector $v_1^{(1)}$. If h_0 is too coarse, $v_0^{(1)}$ is only a rough approximation and the multi-grid convergence deteriorates. Moreover, additional difficulties arise even if the approximation is perfect as pointed out below.

Table 10.2.1 shows also that the rates depend on the distance of λ to the next eigenvalue. Since $\lambda = +1$ is close to the first eigenvalue $\lambda^{(1)} = 0$ and since $\lambda = +15$ is close to the second eigenvalue $\lambda^{(2)}$, the rates are worse than in between (e.g. for $\lambda = +5$). A brief explanation of this effect follows.

Let $v_l^{(1)}$ be the eigenvector satisfying $(\varepsilon_l^{(1)} L_l^P + L_l^S)\, v_l^{(1)} = 0$, where $L_l^S = I$ holds as for Eq. (4). Assume $\varepsilon_l^{(1)} < \varepsilon < \varepsilon_l^{(2)}$ for $l = 0, 1$ and study the behaviour of the error $\tilde{u}_1 - u_1 = v_1^{(1)}$ under the optimal assumption that $v_1^{(1)}$ and $v_0^{(1)}$ are connected by $r\, v_1^{(1)} = v_0^{(1)}$ and $p\, v_0^{(1)} = v_1^{(1)}$. The defect of $\tilde{u}_1$ equals $(1 - \varepsilon/\varepsilon_1^{(1)})\, v_1^{(1)}$ and the coarse-grid correction results in the new error $\tilde{\tilde{u}}_1 - u_1 = \alpha\, v_1^{(1)}$ with $\alpha = [\varepsilon/\varepsilon_1^{(1)} - \varepsilon_0^{(1)}]/[1 - \varepsilon/\varepsilon_0^{(1)}]$. Replacing ε by $-1/(\lambda - \text{const})$ as in Eq. (4) we obtain

$$\alpha = [\lambda_1^{(1)} - \lambda_0^{(1)}]/[\lambda - \lambda_0^{(1)}]. \tag{10.2.5}$$

The coarse-grid correction would be perfect if $\lambda_1^{(1)} = \lambda_0^{(1)}$. However, $\lambda_1^{(1)}$ and $\lambda_0^{(1)}$ differ by some discretisation error $\delta\lambda_{01}^{(1)}$. We conclude from Eq. (5) that the coarse-grid corrections becomes poor if $|\lambda - \lambda_0^{(1)}| \lessapprox |\delta\lambda_{01}^{(1)}|$. In particular $\min\limits_{l=0,1} \lambda_l^{(1)} < \lambda < \max\limits_{l=0,1} \lambda_l^{(1)}$ is disastrous since $|\alpha| > 1$.

Following this advice in the case of $\varepsilon^{(1)} < \ldots < \varepsilon^{(\nu)} < \varepsilon < \varepsilon^{(\nu+1)}$ we have to ensure

$$\lambda_l^{(\nu)} < \lambda < \lambda_l^{(\nu+1)} \tag{10.2.6}$$

for all $l = 0, 1, \ldots, l_{\max}$. For larger ν three severe problems arise:

- Possibly the νth eigenvector $v_1^{(\nu)}$ is hardly representable on the coarsest grid Ω_0. Then h_0 must be refined.
- Usually, inequality (6) cannot be guaranteed. In the case of Fig. 10.2.1 (6) does not hold for $\nu = 2$ since $\lambda_0^{(3)} < \lambda_3^{(2)}$. In general, not even $\lambda_0^{(\nu)} < \lambda_0^{(\nu+1)}$ can be ensured which is necessary for (6).
- If $\nu > n_0$ there is no eigenvector $v_0^{(\nu)}$ and the coarse-grid correction cannot work.

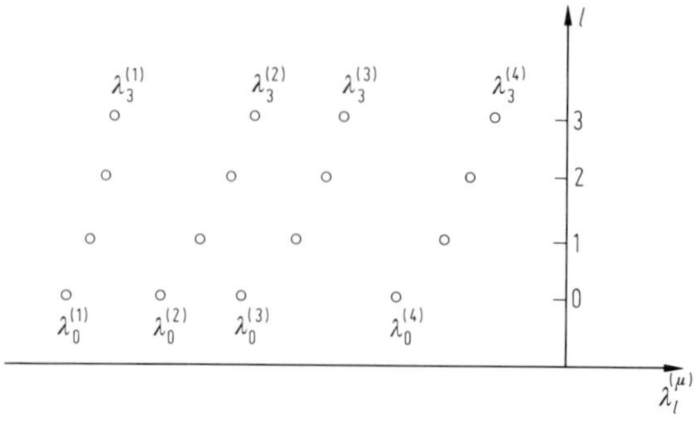

Fig. 10.2.1.
Distribution of eigenvalues

In order to overcome these difficulties, one has to use smoothing iterations working for indefinite problems or one has to have some knowledge about the eigenvalues and possibly the eigenvectors. In the situation of Fig. 10.2.1 one can replace the coarse-grid matrix L_l, $l<3$, by $\tilde{L}_l := L_l + (\lambda_3^{(2)} - \lambda_l^{(2)}) I$ provided that $\lambda_0^{(2)}, \dots, \lambda_3^{(2)}$ are known. Since the new eigenvalues are $\tilde{\lambda}_0^{(2)} = \tilde{\lambda}_1^{(2)} = \tilde{\lambda}_2^{(2)} = \lambda_3^{(2)}$, there is a better chance for the modified multi-grid process.

If also the eigenvectors $v_l^{(\mu)}$ of the first eigenvalues are known, they can be used within the smoothing process: The projection

$$\tilde{u}_l \mapsto \tilde{u}_l - \alpha\, v_l^{(\mu)}, \qquad \alpha = \langle v_l^{(\mu)}, d_l \rangle / \| v_l^{(\mu)} \|^2, \qquad d_l = L_l u_l - f_l,$$

annihilates the $v_l^{(\mu)}$-error component provided that $L_l^P = (L_l^P)^*$, $L_l^S = I$. As soon as the smoothing step deals well with these components, it does not matter that the coarse-grid correction is poor or even divergent.

We have restricted our interest to weakly indefinite problems since severe difficulties arise for very small ε. For instance, one has to choose an unrealistically small h_l to obtain a difference scheme that approximates the strongly indefinite differential operator (but compare Ta'asan [1]).

10.3. Interface Problems

10.3.1 A One-Dimension Model Example

The equation

$$-(a(x)\,u'(x))' = f(x) \quad \text{in } \Omega = (0,1); \qquad u(0) = 0, \qquad u(1) = 1 \tag{10.3.1 a}$$

describes the stationary temperature u in a wire which is heated according to the source term f. The heat conductivity $a(x)$ is a material constant. In the classical sense the derivative of $a(x)\,u'(x)$ is reasonable only if $a \in C^1(\Omega)$. However, when the pieces $[0, \xi]$ and $[\xi, 1]$ of the wire consist of different materials,

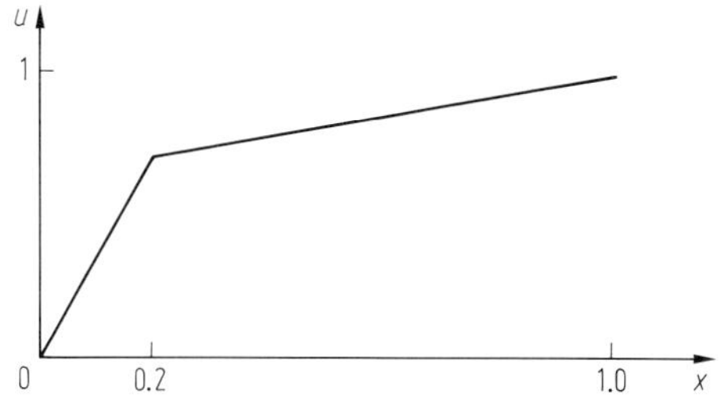

Fig. 10.3.1. Solution of Eq. (1 a, b) with $f=0,\ a^-=1,\ a^+=10,\ \xi=0.2$

$a(x)$ becomes a piecewise constant function, which is discontinuous at ξ:

$$a(x)=\begin{cases} a^- & 0<x<\xi, \\ a^+ & \xi<x<1. \end{cases} \tag{10.3.1 b}$$

To treat the arising problem we have to write Eq. (1 a) in the variational formulation (cf. (3.1.11)):

$$\int_0^1 a(x)\,u'(x)\,v'(x)\,dx=\int_0^1 f(x)\,v(x)\,dx \quad \text{for all } v\in H_0^1(\Omega). \tag{10.3.2}$$

Equation (2) makes sense even for the discontinuous $a(x)$ from (1 b). Separate integration by parts over $[0,\xi]$ and $[\xi,1]$ yields a new classical formulation:

$$\begin{aligned} &-(a^-u')'=f \quad \text{in } (0,\xi), \qquad -(a^+u')'=f \quad \text{in } (\xi,1), \\ &u(0)=0, \qquad u(1)=1, \qquad u(\xi-0)=u(\xi+0), \\ &a^-u'(\xi-0)=a^+u'(\xi+0), \end{aligned} \tag{10.3.3}$$

where the argument $\xi-0$ $(\xi+0)$ denotes the limit from the left (right) side. Because of the additional condition at ξ, Eq. (3) is called an interface problem. The solution $u(x)$ of the equivalent problems (2) and (3) is no longer smooth in Ω. As depicted in Fig. 10.3.1 the solution is continuous, but has a discontinuous derivative at ξ.

Formulation (2) can be used to define a finite element discretisation. However, the discretisation error will be relatively large unless the finite element space fits to the behaviour of the solution. This is the case only when ξ coincides with a grid point of Ω_l.

Also carefully defined difference schemes are suited. The scheme (3.1.18) reduces in the present case to $L_l u_l=f_l$, where L_l is the stencil

$$L_l=h_l^{-2}\left[-a\left(x-\frac{h_l}{2}\right) \quad a\left(x-\frac{h_l}{2}\right)+a\left(x+\frac{h_l}{2}\right) \quad -a\left(x+\frac{h_l}{2}\right)\right].$$

If ξ is a grid point, the choice (1 b) implies

$$L_l=\begin{cases} h_l^{-2}a^-[-1 \quad 2 \quad -1] & \text{if } \xi>x\in\Omega_l, \quad (10.3.4\text{ a}) \\ h_l^{-2}[-a^- \quad a^-+a^+ \quad -a^+] & \text{if } \xi=x\in\Omega_l, \quad (10.3.4\text{ b}) \\ h_l^{-2}a^+[-1 \quad 2 \quad -1] & \text{if } \xi<x\in\Omega_l. \quad (10.3.4\text{ c}) \end{cases}$$

$$L_l^y = \begin{bmatrix} L_{l,1}^y(x,y) \\ L_{l,0}^y(x,y) \\ L_{l,-1}^y(x,y) \end{bmatrix} \quad \text{with } L_{l,j}^y := \sum_{i=-1}^{1} L_{l,ij}, \tag{10.3.11 b}$$

where $L_{l,ij}(x, y)$ are the entries of L_l:

$$(L_l u_l)(x,y) = \sum_{i,j=-1}^{1} L_{l,ij}(x,y)\, u_l(x + i h_l, y + j h_l). \tag{10.3.11 c}$$

Given any three-point formulae L_l^x, L_l^y, e.g. those from (11 a, b), we may define

$$\begin{aligned}(p\, v_{l-1})(x,y) \\ = [L_{l,-1}^x(x,y)\, v_{l-1}(x-h_l, y) + L_{l,1}^x(x,y)\, v_{l-1}(x+h_l,y)]/L_{l,0}^x(x,y) \\ \text{for } (x,y) \in \Omega_l^3,\end{aligned} \tag{10.3.10 c}$$

$$\begin{aligned}(p\, v_{l-1})(x,y) \\ = [L_{l,-1}^y(x,y)\, v_{l-1}(x, y-h_l) + L_{l,1}^y(x,y)\, v_{l-1}(x,y+h_l)]/L_{l,0}^y(x,y) \\ \text{for } (x,y) \in \Omega_l^4.\end{aligned} \tag{10.3.10 d}$$

The choice (11 a, b) leads us to the

Example 10.3.2 (cf. Dendy [2] 1983). Choose L_l^x, L_l^y by (11 a, b) and define p by (10 a–d), r by (8), L_{l-1} by (7).

Further approaches using other L_l^x, L_l^y have been described.

Example 10.3.3 (cf. Hackbusch [2] 1976). Define L_l^x and L_l^y by $h_l^{-2}[-1 \;\; 2 \;\; -1]$ and p, r, L_{l-1} as in Example 10.3.2.

Example 10.3.4 (cf. Kettler-Meijerink [1], Kettler [1] 1981). Set

$$L_l^x = [L_{l,-1,0}(x,y) \quad -L_{l,-1,0}(x,y) \quad -L_{l,1,0}(x,y) \;\; L_{l,1,0}], \tag{10.3.12 a}$$

$$L_l^y = \begin{bmatrix} L_{l,0,1}(x,y) \\ -L_{l,0,-1}(x,y) - L_{l,0,1}(x,y) \\ L_{l,0,-1}(x,y) \end{bmatrix}, \tag{10.3.12 b}$$

and define p, r, L_{l-1} as in Example 10.3.2. Further, the coarse-grid correction of each multi-grid iteration has to be followed by a quarter Gauß-Seidel step with respect to Ω_l^2. According to Note 4.3.4 (ii) $p\, v_{l-1}$ may remain undefined on Ω_l^2, since the Gauß-Seidel step defines $u_l(x,y)$ by $(L_l u_l)(x,y) = f_l(x,y)$, $(x,y) \in \Omega_l^2$.

Example 10.3.5 (cf. Alcouffe-Brandt-Dendy-Painter [1] 1981). Choose the quantities as in Example 10.3.4 but omit the post-smoothing step on Ω_l^2 and replace (10 a) by

$$(p\, v_{l-1})(x,y) = \sum_{(i,j) \neq (0,0)} L_{l,ij}(x,y)\, u_l(x + i h_l, y + j h_l) / \sum_{(i,j) \neq (0,0)} L_{l,ij}(x,y) \quad \text{for } (x,y) \in \Omega_l^2, \tag{10.3.13}$$

where the values of $u_l = p\, v_{l-1}$ are already defined by (10 b–d).

Discretise Eq. (9) by (3.1.18). Then the definitions of L_l^x, L_l^y by (11 a, b) and (12 a, b) coincide. Also (10 a) and (13) are equivalent. But (12 a) and (13) are less suited for general equations as can be seen from the following example. Consider the harmless selfadjoint equation $-\Delta u + u_{xy} = f$. Adding the discretisation $h_l^{-2}\begin{bmatrix} -1 & 1 & 0 \\ 1 & -1 & 0 \\ 0 & 0 & 0 \end{bmatrix}$ of u_{xy} to the five-point Laplacian operator, one obtains a reasonable matrix L_l (even an M-matrix). Then (12 a) yields $L_l^x = [0 \quad 1 \quad -1]/h_l^2$, which describes an extremely unbalanced interpolation, whereas definition (11 a, b) describes the usual linear interpolation.

10.4 Convection Diffusion Equation

10.4.1 The Continuous Problem

The model problem of the convection diffusion equation is

$$-\varepsilon \Delta u + \mathbf{c} \operatorname{grad} u = f_\Omega \quad \text{in } \Omega, \quad u = f_\Gamma \quad \text{on } \Gamma. \tag{10.4.1}$$

When $0 < \varepsilon \ll 1$, the convection term $\mathbf{c} \cdot \operatorname{grad} u = \sum_i c_i u_{x_i}$ dominates and $\mathbf{c} \cdot \operatorname{grad} u \approx f_\Omega$ holds in the interior. The curves $\mathbf{x}(s)$, where s is a parameter, defined by $\dot{\mathbf{x}} = \mathbf{c}$ are called characteristics when $\varepsilon = 0$ and subcharacteristics when $\varepsilon \neq 0$. The prescription of one boundary value per characteristic determines the solution of the reduced equation $\mathbf{c} \cdot \operatorname{grad} u = f_\Omega$ uniquely. In general, $u = f_\Gamma$ cannot be satisfied for $\varepsilon = 0$. Therefore boundary layers are generated for $0 < \varepsilon \ll 1$.

Equation (1) is a non-selfadjoint problem whenever $\mathbf{c} \neq 0$. However a suitable transformation $u \mapsto v$ changes Eq. (1) into a selfadjoint problem. For constant $\mathbf{c}$ the choice $v(\mathbf{x}) = u(\mathbf{x}) \exp(-\mathbf{c} \cdot \mathbf{x}/2\varepsilon)$ with $\mathbf{c} \cdot \mathbf{x} = \sum c_i x_i$ leads us to the equation

$$-\varepsilon \Delta v + \left(\frac{1}{4\varepsilon} \sum_i c_i^2\right) v = f \cdot \exp(-\mathbf{c} \cdot \mathbf{x}/2\varepsilon) \quad \text{in } \Omega, \tag{10.4.1*}$$

which is again a singular perturbation problem, but now a strongly positive definite Helmholtz equation (cf. § 10.2.1). The equivalence to Eq. (1*) proves the solvability of Eq. (1) for all $\mathbf{c}$ and $\varepsilon \neq 0$.

Some difficulties of the multi-grid solution of Eq. (1) originate from the discretisation process. Therefore we shall first discuss the discretisation of Eq. (1). For simplicity we shall study in particular the case of $\mathbf{c} = (1,0)^T$:

$$-\varepsilon \Delta u + u_x = f_\Omega \quad \text{in } \Omega = (0,1) \times (0,1), \qquad u = f_\Gamma \quad \text{on } \Gamma. \tag{10.4.2}$$

10.4.2 Stable Discretisations

Difference formula (3.1.18) applied to Eq. (2) yields the star

$$L_l = h_l^{-2} \begin{bmatrix} & -\varepsilon & \\ -\varepsilon - h_l/2 & 4\varepsilon & -\varepsilon + h_l/2 \\ & -\varepsilon & \end{bmatrix}. \tag{10.4.3}$$

Remark 10.4.1. Scheme (3) has the consistency order 2. If $h_l < 2\varepsilon$, L_l is an M-matrix. Therefore standard arguments prove a discretisation error $\| u_l - u \|_{\mathscr{U}} = 0(h_l^2)$.

We recall that a matrix A is called an *M-matrix* if $A_{ii} > 0$, $A_{ik} \leqq 0\,(i \neq k)$, $(A^{-1})_{ik} > 0$ (cf. Varga [1]). If $h_l \leqq 2\varepsilon$ is violated, the discretisation becomes unstable. The larger h_l/ε the more oscillatory is u_l. In Table 10.4.1 the solution $u_l(x, \frac{1}{2})$ of scheme (3) with $f_\Omega = 0$, $f_\Gamma(0, y) = \sin(\pi y)$, $f_\Gamma(x, y) = 0$ elsewhere, is shown for $h_l = \frac{1}{32}$. The upper part corresponds to ε with $h_l \leqq 2\varepsilon$. Since L_l is an M-matrix, the maximum principle holds: $\min f_\Gamma = 0 \leqq u_l \leqq 1 = \max f_\Gamma$. For ε slightly smaller than $h_l/2 = \frac{1}{64} = 0.015625$ the maximum principle is obviously violated. For decreasing ε the solution $u_l(x, \frac{1}{2})$, x/h_l even, tends to $1 - x$, whereas 'wiggles' of size $O(h_l^2/\varepsilon)$ form in between. Concerning a stability analysis we refer e.g. to de Zeeuw-van Asselt [1].

If ε is sufficiently small, a grid size h_l with $h_l/\varepsilon \leqq 2$ may be too fine. In particular, the coarsest grid sizes $h_0, h_1, \ldots$ can hardly satisfy $h_l/\varepsilon \leqq 2$. Therefore, one might try to change ε into ε_l such that h_l/ε_l is reasonable. This approach yields the star

$$L_l = h_l^{-2} \begin{bmatrix} & -\varepsilon_l & \\ -\varepsilon_l - h_l/2 & 4\varepsilon_l & -\varepsilon_l + h_l/2 \\ & -\varepsilon_l & \end{bmatrix} \tag{10.4.4 a}$$

e.g. with $\varepsilon_l = \max(\varepsilon, h_l/2)$ or $\varepsilon_l = \varepsilon + h_l/2$.

Table 10.4.1. Solution $u_l(x, \frac{1}{2})$ of Eqs. (2), (3) with $f_\Omega = 0$, $f_\Gamma(x, y) = \sin(\pi y)(1 - x)$, $h_l = \frac{1}{32}$

ε \ x	0	$\frac{1}{32}$	$\frac{2}{32}$	$\frac{3}{32}$	$\frac{4}{32}$	…	$\frac{30}{32}$	$\frac{31}{32}$	1
0.5	1	0.93	0.87	0.81	0.75	…	0.04	0.02	0
0.1	1	0.97	0.95	0.92	0.89	…	0.22	0.13	0
0.05	1	0.99	0.97	0.96	0.94	…	0.47	0.31	0
0.02	1	0.99	0.99	0.98	0.98	…	0.82	0.73	0
0.01	1	1.00	0.99	0.99	0.99	…	0.87	1.11	0
$1_{10} - 3$	1	1.03	0.99	1.04	0.99	…	0.23	1.89	0
$1_{10} - 4$	1	4.95	0.95	5.00	0.90	…	0.08	5.88	0
$1_{10} - 5$	1	48.6	0.94	48.7	0.88	…	0.06	49.6	0
$1_{10} - 7$	1	4859.5	0.94	4859.6	0.88	…	0.06	4860.4	0

On the other hand the stability of the discretisation can be ensured if we replace u_x by the one-sided (backward) difference $h_l^{-1}[-1 \;\; 1 \;\; 0]$. The resulting scheme is

$$L_l = h_l^{-2}\begin{bmatrix} & -\varepsilon & \\ -\varepsilon - h_l & 4\varepsilon + h_l & -\varepsilon \\ & -\varepsilon & \end{bmatrix}. \tag{10.4.4 b}$$

Remark 10.4.2. Scheme (4 b) is stable for all $\varepsilon > 0$, $h_l > 0$, since L_l is an M-matrix. The consistency order of (4 b) is only 1.

When the formulae (4 a, b) are applied to a smooth function u, Taylor's expansion shows that

$$L_l u = -\varepsilon \Delta u + u_x - \kappa h_l \Delta u + O(h_l^2) \quad \text{for (4 a)} \quad \text{with} \quad \varepsilon_l = \varepsilon + \kappa h_l,$$
$$L_l u = -\varepsilon \Delta u + u_x - \tfrac{1}{2} h_l u_{xx} + O(h_l^2) \quad \text{for (4 b)}.$$

In the former case one says that an artificial viscosity (or artificial ellipticity) $-\kappa h_l \Delta u$ is added, while in the latter case $-\frac{1}{2} h_l u_{xx}$ is called the numerical viscosity.

The extension of the difference schemes (4 a, b) to Eq. (1) with $\mathbf{c} = (c_1, c_2)^T$ yields

$$L_l = h_l^{-2}\begin{bmatrix} & -\varepsilon_l + c_2 h_l/2 & \\ -\varepsilon_l - c_1 h_l/2 & 4\varepsilon_l & -\varepsilon_l + c_1 h_l/2 \\ & -\varepsilon_l - c_2 h_l/2 & \end{bmatrix}, \tag{10.4.5 a}$$

$$L_l = h_l^{-2}\begin{bmatrix} & -\varepsilon + c_2^- h_l & \\ -\varepsilon - c_1^+ h_l & 4\varepsilon + |c_1| h_l + |c_2| h_l & -\varepsilon + c_1^- h_l \\ & -\varepsilon - c_2^+ h_l & \end{bmatrix}, \tag{10.4.5 b}$$

where $c_i^+ = \max(0, c_i)$, $c_i^- = \min(0, c_i)$. In contrast to scheme (5 a), formula (5 b) can also make use of diagonal differences. If e.g. $\mathbf{c} = (1, 1)^T$, the convection term $\mathbf{c} \cdot \operatorname{grad} u$ may be discretised by $h_l^{-1}\begin{bmatrix} 0 & 0 & -1 \\ 0 & 1 & 0 \\ 0 & 0 & 0 \end{bmatrix}$ yielding a six-point formula L_l.

Remark 10.4.3. Scheme (5 a) yields an M-matrix if $\varepsilon_l = \max\{|c_1|, |c_2|\} h_l/2$, whereas scheme (5 b) is an M-matrix for all $\varepsilon > 0$, $h_l > 0$. The consistency errors are $|\varepsilon - \varepsilon_l| + O(h_l^2)$ for (5 a) and $O(h_l)$ for (5 b).

The instability of difference scheme (3) occurs also for the finite element discretisation since the usual approach yields centred formulae for $\mathbf{c} \cdot \operatorname{grad} u$. To produce one-sided differences one has to generalise the Galerkin method by the Petrov-Galerkin method: Seek $u \in \mathscr{H}_l$ such that $a(u, v) = (f, v)$ for all $v \in \mathscr{K}_l$, where $\mathscr{K}_l$ is different from $\mathscr{H}_l$ but $\dim(\mathscr{H}_l) = \dim(\mathscr{K}_l)$. For the choice of $\mathscr{H}_l$ and $\mathscr{K}_l$ we refer e.g. to Brooks-Hughes [1], Johnson-Nävert [1], Fletcher [1], Thomasset [1].

10.4.3 Multi-Grid Algorithms

The choice of the smoothing process depends essentially on the discretisation. Scheme (5 a) ("artificial viscosity approach") is considered in § 10.4.3.3, while §§ 10.4.3.1–2 refer to the "upwind differencing scheme" (5 b).

10.4.3.1 The Case of the Stable Discretisation (5 b)

Consider again the special scheme $L_l = L_l(\varepsilon)$ from (4 b). According to Criterion 10.1.1 one should study the multi-grid behaviour for $L_l(0) = h_l^{-1}\begin{bmatrix} & 0 & \\ -1 & 1 & 0 \\ & 0 & \end{bmatrix}$. In particular, one has to check whether the smoothing iteration annihilates high-frequency errors which cannot be reduced by the coarse-grid correction.

The Gauß-Seidel process can easily be analysed. The lexicographical version (left-to-right ordering in each x-line, bottom-to-top ordering of the x-lines) results in $S_l u_l = 0$ for all u_l, i.e. it is an exact solver of $L_l(0) u_l = f_l$. The inverted lexicographical iteration (top-to-bottom, right-to-left) yields $u_l^{j+1} = S_l u_l^j$ with $u_l^{j+1}(x, y) = u_l^j(x - h_l, y)$. Hence, even oscillating errors are only shifted but not reduced. A similar result holds for the red-black ordering, since then $u_l^{j+1} = S_l u_l^j$ with $u_l^{j+1}(x, y) = u_l^{j+1}(x + h_l, y) = u_l^j(x - h_l, y)$ for $(x + y)/h_l$ even. Obviously, the x-line Gauß-Seidel iteration is an exact solver. The y-line version (left-to-right, right-to-left, or zebra) yields the same results as the corresponding pointwise iteration. We summarise:

Note 10.4.4. Let L_l be defined by (4 b). The only Gauß-Seidel versions working as smoothing iterations are the x-line Gauß-Seidel iteration, the pointwise version with left-to-right ordering in each x-line, and the y-line Gauß-Seidel iteration with left-to-right ordering of the blocks. The respective inverted orderings or the chequer-board version are not suited for smoothing.

The treatment of $L_l u_l = f_l$ with L_l from (4 b) is very similar to the iterative solution of the one-dimensional problem $-\varepsilon u'' + u' = f$ with

$$L_l = h_l^{-2}[-\varepsilon - h_l \quad 2\varepsilon + h_l \quad -\varepsilon] \tag{10.4.6}$$

in $\Omega = \mathbb{R}$. Table 10.4.2 exhibits the contraction numbers ζ_ν of the following two-grid iteration: $\nu_1 = \nu$, $\nu_2 = 0$, L_{l-1} defined by (6), r from (2.3.7), p from (2.3.9), $\mathscr{S}_l$: lexicographical Gauß-Seidel iteration. The two-grid contraction number ζ_ν is determined according to § 8.1.3 by Fourier analysis (for details see Hackbusch [34]). ζ_ν depends on the ratio $\kappa_l := h_l/\varepsilon$. $\kappa_l \to \infty$ indicates increasing perturbation ($\varepsilon \to 0$), while $\kappa_l = 0$ corresponds to the unperturbed stencil $L_l = \text{const} \cdot h_l^{-2}[-1 \;\; 2 \;\; -1]$. Table 10.4.2 shows that $\kappa_l = 0$ is the worst case. This reflects the fact that the smoother becomes an exact solver as $\varepsilon \to 0$ ($\kappa_l \to \infty$).

Table 10.4.2. The two-grid contraction numbers ζ_ν, $\nu = 1, 2$, for scheme (6)

h_l/ε	0	0.1	0.5	1.0	2.0	10.0	$\to\infty$
ζ_1	$\sqrt{2}/3$	0.425	0.348	0.286	0.212	0.069	$\to 0$
ζ_2	0.2	0.178	0.122	0.099	0.072	0.022	$\to 0$

Scheme (4 b) corresponds to the perturbation $+u_x$. If we replace $-\varepsilon\Delta u + u_x = f$ by $-\varepsilon\Delta u - u_x = f$, the left-to-right ordering mentioned in Note 10.4.4 has to be replaced by the right-to-left ordering. If $\pm u_x$ is turned into $\pm u_y$, the x-line and y-line versions change their rôles. Roughly speaking, the ordering has to follow the characteristics.

In the general case of scheme (5 b) the situation becomes more complicated. Assume e.g. $c_1 > 0$ and $c_2 > 0$. Then the Gauß-Seidel iteration with one of the following orderings is an exact solver: pointwise lexicographical, x-line (bottom-to-top), y-line (left-to-right). If $c_1 < 0$ ($c_2 < 0$) the ordering has to be reversed with respect to the x-direction (y-direction). Obviously, there is no ordering that works for all directions $\mathbf{c} = (c_1, c_2)^T \in \mathbb{R}^2$, but combinations of different iterations can be used as pointed out in

Remark 10.4.5. The symmetric x-line (bottom-top-bottom) or y-line (left-right-left) Gauß-Seidel iterations are possible smoothers for the upwind discretisation (5 b).

An excellent smoother is given by the ILU or ILLU decomposition defined in § 10.1.3. Set $\mathbf{c} = (\cos\alpha, \sin\alpha)^T$. For $0 \leqq \alpha \leqq \frac{\pi}{2}$ and small ε ILU is almost exact (cf. Exercise 10.6.3) as confirmed by the left part of Table 10.4.3, which gives the observed average reduction factor (approximating the convergence rate) of a multi-grid iteration with the following components: p: seven-point prolongation, $r = p^*$, $L_{l-1} = r L_l p$, $\nu_1 = 0$, $\nu_2 = \gamma = 1$, $\mathscr{S}_l$: seven-point ILU iteration (cf. (1.12c)), $h_l = 2^{-1-l}$. ILU is no longer an exact solver of $L_l(0)\, u_l = f_l$ if $\frac{\pi}{2} < \alpha < \pi$ (unless the lexicographical ordering is changed into top-to-bottom and left-to-right numbering). This fact partly explains the poorer rates $\tilde{\varrho}$ from the right part of Table 10.4.3.

Another reason is the use of the standard Galerkin approximation $L_{l-1} = r L_l p$, as explained in Wesseling [4], Hemker-Kettler-Wesseling-de Zeeuw [1]. With each step $L_l \mapsto L_{l-1} = r L_l p$ the star becomes closer to the finite element discretisation (cf. Notes 3.4.1–2, Exercise 3.9.4), which involves centred differences. Hence the one-sided difference formulae for $\mathbf{c} \cdot \operatorname{grad} u$ in L_l change during the repeated Galerkin approximations $L_l \mapsto L_{l-1} = r L_l p$ into more balanced differences. Consequently, the instability of L_k grows as $k \to 0$ and poisons the multi-grid iteration at least if the maximum level l is large enough. For instance, if $\varepsilon \leqq 10^{-5}$ and $\alpha = 11\pi/12$, the multi-grid algorithm described above diverges as soon as $l \geqq 6$. A simple remedy is the use of the prolongations and restrictions defined in § 10.3.2.

In the case of matrix (5 a) the condition for L_l being an M-matrix is $\kappa_l \leqq 2$ (cf. Remark 10.4.1). The choice

$$\varepsilon_l = \max\{\varepsilon, |c_1| h_l/\kappa, |c_2| h_l/\kappa\} \tag{10.4.9}$$

implies that the numbers κ_l are kept constant ($\kappa_l = \kappa$) as long as $h_l \geqq \varepsilon/\|\mathbf{c}\|_\infty$. If $\varepsilon \ll 1$, the latter inequality holds for all realistic grid sizes. The constancy of $\kappa_l = \kappa$ implies that the parameters $\varepsilon_{l-1} = 2\varepsilon_l$ must increase for decreasing l. This fact reveals the drawback of the artificial viscosity approach: The fine-grid equation $L_l(\varepsilon_l)u_l = f_l$ and the coarse-grid analogue $L_{l-1}(\varepsilon_{l-1})u_{l-1} = f_{l-1}$ are not consistent with respect to their principal terms. As pointed out below the resulting multi-grid convergence is comparatively slow.

As an illustration we consider the one-dimensional problem $-\varepsilon u'' + u' = f$ with periodic boundary conditions in $[-1, 1]$. The discrete problem with

$$L_l = h_l^{-2}[-\varepsilon_l - h_l/2 \quad 2\varepsilon_l \quad -\varepsilon_l + h_l/2], \qquad \varepsilon_l = h_l/\kappa_l,$$

is solved by a two-grid iteration characterised by $v_1 = v \in \{1, 2\}$, $v_2 = 0$, p and r as in § 2, $\mathcal{S}_l$: Gauß-Seidel iteration with red-black ordering (first odd indices). The two-grid contraction numbers $\zeta_v = \zeta_v(\kappa_l)$ depending on $\kappa_l = h_l/\varepsilon_l$ and v are computed according to § 8.1.2 (cf. also Hackbusch [34]). Tables 10.4.5–6 exhibit the contraction numbers ζ_1 and ζ_2 in the case of $\kappa_l = \kappa_{l-1} = \kappa$ for different values of κ. Although $\kappa = 2$ yields a reasonable difference scheme (L_l is an M-matrix), not only the contraction number but also the spectral radius (convergence rate) is $\geqq 1$ for $\kappa \geqq 2$. The divergence is not a result of the 'wrong' choice $\varepsilon_{l-1} = 2\varepsilon_l$. The values in Table 10.4.5 marked by an asterisk cannot be improved by any choice of ε_{l-1}. These values $\zeta_l(\kappa_l)$ exceed $\frac{1}{2}\kappa_l\sqrt{1 + \kappa_l^2/4}$, which is a lower bound independent of κ_{l-1} or ε_{l-1}.

Table 10.4.5. The two-grid contraction numbers for $v = 1$, $\kappa_l = \kappa_{l-1} = \kappa$ (i.e. $\varepsilon_{l-1} = 2\varepsilon_l$)

κ	2.0	1.5	1.148	1.0	0.8	0.7	0.6	0.5	0.1	$\to 0$
$\zeta_1(\kappa)$	$\sqrt{2}$*	15/16*	0.707	0.662	0.630	0.623	0.622	0.626	0.684	$\to 1/\sqrt{2}$

Table 10.4.6. The two-grid contraction numbers for $v = 2$, $\kappa_1 = \kappa_{l-1} = \kappa$

κ	2.0	1.5	1.0	0.7	0.5	0.2	$\to 0$
$\zeta_2(\kappa)$	$\sqrt{2}$	0.590	0.460	0.476	0.512	0.610	$\to 1/\sqrt{2}$

Table 10.4.7. The two-grid contraction numbers for $v = 1$, $\kappa_l = \kappa_{l-1}/2$ (i.e. $\varepsilon_{l-1} = \varepsilon_l$)

$\kappa_l = \kappa_{l-1}/2$	2.0	1.5	1.148	1.0	0.8	0.7	0.6	0.5	0.1	$\to 0$
$\zeta_1(\kappa_l)$	$\sqrt{2}$	15/16	0.662	0.565	0.484	0.456	0.435	0.418	0.386	$\to 0.385$

The choice $\varepsilon_{l-1} = \varepsilon_l$ instead of $\varepsilon_{l-1} = 2\varepsilon_l$ does improve the other contraction numbers of Table 10.4.5 as can be seen in Table 10.4.7. The entries of Table 10.4.5 show the uniform estimate $\zeta_1(\kappa) \leqq 1/\sqrt{2}$ of the two-grid contraction number for all $\kappa = h_l/\varepsilon_l \leqq 1.148$. A good choice of κ is 0.6.

Similar results can be found for multi-dimensional problems. Several test results for Eq. (1) with constant and variable $\mathbf{c}$ are reported by Börgers [1]. He considers the following multi-grid algorithm: $\nu_1 = 1$, $\nu_2 \in \{1, 2\}$, $\gamma \in \{1, 2\}$, $\mathscr{S}_l$: chequer-board Gauß-Seidel iteration, p: (3.4.2), r: (3.5.4) [but also (3.8.8), alternatively]. For $L_l(\varepsilon_l)\,u_l = f_l$ with L_l from (5a), ε_l from (9), $\kappa = 1$, $h_l = \frac{1}{64}$, the following reduction numbers (approximating the convergence rate) are observed (cf. Börgers [1, p. 89 and p. 111]):

Table 10.4.8 Convergence rates $\varrho(\nu_1, \nu_2)$ and smoothing rate $\varrho_B(\nu_1 + \nu_2)$

	V-cycle rate ($\gamma = 1$)				W-cycle rate ($\gamma = 2$)				Two-grid rate ($\nu_1 = 1$)		Smoothing rate $\varrho_B(\nu)$	
	$\nu_1 = 1$		$\nu_1 = 2$		$\nu_1 = 1$		$\nu_1 = 2$					
	$\nu_2=0$	$\nu_2=1$	$\nu_2=0$	$\nu_2=1$	$\nu_2=0$	$\nu_2=1$	$\nu_2=0$	$\nu_2=1$	$\nu_2=0$	$\nu_2=1$	$\nu=1$	$\nu=2$
$\mathbf{c}=\binom{0}{1}$	0.80	0.71	0.66	0.59	0.46	0.33	0.34	0.27	0.39	0.32	0.31	0.096
$\mathbf{c}=\binom{1}{1}$	0.78	0.63	0.55	0.46	0.44	0.29	0.30	0.23	0.38	0.30	0.34	0.116

It can be shown that the smoothing rates $\varrho_B(\nu)$, $\nu = \nu_1 + \nu_2$, satisfy

$$\max_{\mathbf{c}\in\mathbb{R}^2} \varrho_B(\nu) = 1 \quad \text{if } \kappa = 2, \qquad \max_{\mathbf{c}\in\mathbb{R}^2} \varrho_B(\nu) \geqq 1 \quad \text{if } \kappa > 2$$

for all ν. The divergence for $\kappa = 2$ has already been observed in the case of the one-dimensional example described above. The analysis of the two-grid iteration in $\Omega = \mathbb{R}^2$ (cf. § 8.1.3) yields the following result, which in particular confirms that the choice of a large ν is needless.

Lemma 10.4.7 (Börgers [1, p. 113]). Let Eq. (1) in $\Omega = \mathbb{R}^2$ with a constant vector $\mathbf{c} \in \mathbb{R}^2$ be discretised by matrix (5a) with $\varepsilon_l = \max_i h_l |c_i|/\kappa$. The two-grid iteration is characterised by p: nine-point prolongation, r: nine-point restriction, $\mathscr{S}_l$: chequer-board Gauß-Seidel iteration. Then for all $\mathbf{c} \in \mathbb{R}^2$, $h_l > 0$, $\kappa > 0$, and all $\nu = \nu_1 + \nu_2 > 0$ the contraction number and the convergence rate exceed $1 - \varepsilon_l/\varepsilon_{l-1} = 0.5$:

$$\| M_l(\nu_1, \nu_2) \|_{\mathscr{U} \leftarrow \mathscr{U}} \geqq \varrho(M_l(\nu_1, \nu_2)) \geqq 0.5. \tag{10.4.10}$$

For the proof one has to study $M_l u_l$ for $u_l = \exp\{i\pi(x\xi + y\eta)/h_l\}$ with ξ, η such that $c_1\xi + c_2\eta = 0$, $|\xi| + |\eta| \to 0$, A similar result is stated by van Asselt [1], who analysed the coarse-grid correction step. Inequality (10) contradicts inequality (6.1.9), where ϱ may be arbitrarily small. Theorem 6.1.7 does not apply since the approximation property does not hold. Note that the different values $\varepsilon_l \neq \varepsilon_{l-1}$, involved in the coarse-grid correction process, determine the lower bound in (10). We summarise the results in

Remark 10.4.8. The artificial viscosity approach (5 a) for L_l and L_{l-1} permits the use of simple pointwise Gauß-Seidel smoothing iterations (chosen independently of the characteristic direction **c**). On the other hand the two-grid and multi-grid convergence is very slow in comparison with the rates reported in § 4.4 or §§ 10.4.3.1–2. In spite of the centred differences for $\mathbf{c} \cdot \operatorname{grad} u$, the discretisation error is of first order: $O(\varepsilon_l - \varepsilon) = O(h_l)$, as long as $\varepsilon < \varepsilon_l$.

The latter disadvantages can partly be removed by the defect correction technique mentioned in § 14.3.4.

10.5 Comments

Anisotropic problems. One has to be aware of the fact that isotropic differential operators can also lead to anisotropic discrete problems. The discretisation of Poisson's equation by means of a curvi-linear grid corresponds to a difference method in a square grid applied to the transformed equation, which usually contains also the mixed term. The more the transformation differs from a conformal mapping the more anisotropic is the arising discrete problem. A similar effect may happen in the finite element case. If for some reason Poisson's equation is approximated by finite elements consisting of triangles distorted in one or another direction, the discrete problem is anisotropic. Vice versa, anisotropic problems produce isotropic discrete equations, if one succeeds in constructing a suitable grid. However, this task turns out to be very difficult.

Semi-coarsening. The anisotropic problem (1.1) admits another very simple multi-grid approach. Instead of the standard coarsening we may use a coarse grid Ω_{l-1} with anisotropic grid widths $h^x_{l-1} = h_l$, $h^y_{l-1} = 2h_l$ (cf. Fig. 3.4.2). Then pointwise Gauß-Seidel iteration is still applicable. Since there is no coarsening in the x-direction, errors highly oscillating in the x-direction are annihilated by the coarse-grid correction. The smoothing process has to damp only the high frequencies in the y-direction. This is done perfectly by a pointwise Gauß-Seidel iteration as long as $\varepsilon \lessapprox 1$. The five-point scheme of Eq. (1.1) in the rectangular grid with $h_l = h^x_{l-1} = h^y_{l-1}/2$ is the star

$$L_{l-1} = h_l^{-2} \begin{bmatrix} & -\frac{1}{4} & \\ -\varepsilon & 2\varepsilon - \frac{1}{2} & -\varepsilon \\ & -\frac{1}{4} & \end{bmatrix}.$$

The operator is still anisotropic when $\varepsilon < \frac{1}{4}$. Then the grid sizes at level $l-2$ must be $h^x_{l-2} = h_l$, $h^y_{l-2} = 4h_l$. An alternating coarsening ($h^x_{l-2} = h^y_{l-2} = 2h_l$) is forbidden if we do not apply linewise smoothing processes as in § 10.1.1. The semi-coarsening in the y-direction has to be repeated up to level $l-k$ until $\varepsilon \approx 4^{-k}$, i.e. until L_{l-k} is no longer anisotropic.

Unfortunately, this approach cannot be extended to problem (1.5) with varying coefficients, unless we allow an irregular coarsening as in the algebraic multi-grid (AMG) method (cf. Stüben [2]).

ILU and ILLU. The ILU factorisation was first used as a smoother for multi-grid methods by Wesseling-Sonneveld [1] in 1979 and Hemker [3] in 1980. Previously, the ILU iteration (1.13) has proved to be an efficient pre-conditioner for cg methods (cf. Meijerink-van der Vorst [1], Kershaw [1]). The first ILLU smoother was introduced by Kettler [1].

Discontinuous grid sizes. In § 10.3.1 we discretised Eq. (3.1 a) with discontinuous coefficient $a(x)$ in a grid with equidistant width h_l. We obtain a discrete problem of the same kind if we discretise the equation $-u''(x) = f(x)$ in a grid Ω_l with step size Δx^I in $\Omega_l \cap [0, \xi]$ and step size Δx^{II} in $\Omega_l \cap [\xi, 1]$ for some $\xi \in \Omega_l$. The prolongation $p = [\frac{1}{2}\ 1\ \frac{1}{2}]$ would yield $u_l(\xi) = [u_l(x - \Delta x^I) + u_l(x + \Delta x^{II})]/2$, which is obviously different from the linear interpolation. The linear interpolation is again given by formula (3.6). Therefore, multi-grid algorithms should use the prolongations and restrictions constructed in § 10.3, if the difference or finite element equation is based on a grid with discontinuously varying sizes.

Nonlinear singular perturbation problems. In §§ 10.1–4 we considered only linear equations. According to § 9 nonlinear problems can be treated by two multi-grid approaches: by the Newton-multi-grid iteration (9.2.1) and by the nonlinear multi-grid algorithm (cf. § 9.3.3). The choice between these algorithms is closely connected with the choice of the smoothing process. For the artificial viscosity approach of § 10.4.3.3 it suffices to use a pointwise Gauß-Seidel iteration. This smoother can easily be extended to a nonlinear equation (cf. (9.3.2)). Hence it is possible to perform the nonlinear multi-grid iteration (9.3.12). To apply the same algorithm to the nonlinear analogue of scheme (4.5 b) one would need appropriate smoothers, e. g., nonlinear counterparts of the ILU or ILLU iteration, which are not available. Nevertheless, the Newton-multi-grid method is still applicable, since the powerful linear smoothers of § 10.4.3.2 can be used for the linearised problems (cf. Wesseling-Sonneveld [1]). Choosing the Newton-multi-grid algorithm one has to ensure that L_l is differentiable. Usually, discretisations of a differentiable operator L yield a differentiable L_l. However, scheme (4.5 b) is a counter-example. If $\mathbf{c} = \mathbf{c}(\mathbf{x}) \in C^1(\Omega, \mathbb{R}^2)$ has components $c_i(\mathbf{x})$ with changing signs, the coefficients of L_l as defined in (4.5 b) vary non-differentiably. An easy remedy follows.

The Il'in scheme for the convection diffusion equation. Il'in [1] proposes the discretisation of $-\varepsilon u_{xx} + c_1 u_x$ by $L_l^x u_l$ with

$$L_l^x = \gamma h_l^{-2}[-1\ \ 2\ \ -1] + c_1 h_l^{-1}[-\tfrac{1}{2}\ \ 0\ \ -\tfrac{1}{2}], \qquad \gamma = \tfrac{1}{2} c_1 h_l \coth(c_1 h_l / 2\varepsilon),$$

and a similar scheme L_l^y for $-\varepsilon u_{yy} + c_2 u_y$. $L_l := L_l^x + L_l^y$ is an M-matrix depending smoothly on $\mathbf{c}$. Moreover, γ is chosen such that in the one-dimensional case the Dirichlet problem $-\varepsilon u'' + c u' = 0, u(a) = u_a, u(b) = u_b$, is solved exactly.

Non-elliptic problems. The reduced equation $\mathbf{c} \cdot \operatorname{grad} u = f$, which results in the limit $\varepsilon \to 0$ from (4.1), is of hyperbolic type. Usually, such hyperbolic initial-boundary value problems are solved by other numerical methods; but, in principle, it would be possible to apply the multi-grid algorithms developed in §§ 10.4.3.1–2 (because they satisfy Criterion 10.1.1) or add a further boundary condition and an elliptic part $-\varepsilon \Delta u$ in order to solve the arising elliptic problem $-\varepsilon \Delta u + \mathbf{c} \cdot \operatorname{grad} u = f$ as described in § 10.4.3.3.

It is more interesting that non-elliptic problems can be solved without introducing artificial or numerical ellipticity. In § 3.1.1 we defined the ellipticity of the continuous problem. A difference operator L_l may be called uniformly elliptic of order $2m$ if the characteristic function $\lambda(\mathbf{x}, \xi; h)$ of L_l satisfies

$$|\lambda(\mathbf{x}, \xi; 0)| \geqq \varepsilon |\xi|^{2m} \quad \text{for all } \xi \in [-\pi, \pi]^d, \quad \mathbf{x} \in \Omega \quad \text{with fixed } \varepsilon > 0 \tag{10.5.1}$$

(cf. Thomée-Westergren [1]). Re λ instead of $|\lambda|$ is used in Theorem 6.3.33 (cf. Hackbusch [23, p. 74]). The characteristic function, for instance of the nine-point scheme (3.1.16) with coefficients $c_{\alpha\beta} = c_{\alpha\beta}(\mathbf{x}; h_l)$ is given by

$$\lambda(\mathbf{x}, \xi; h) := \sum_{-1 \leqq \alpha, \beta \leqq 1} c_{\alpha\beta}(\mathbf{x}; h)\, e^{i(\alpha\xi_1 + \beta\xi_2)}. \tag{10.5.2}$$

The discrete ellipticity is essential for the discrete regularity needed for proving the approximation property (cf. § 6.3.2.1). As pointed out by Brandt-Dinar [1, p. 89] the 'h-ellipticity measure'

$$E(h) = \min \left\{ \frac{|\lambda(\mathbf{x}, \xi; h)|}{|\lambda(\mathbf{x}, \xi'; h)|} : \tfrac{\pi}{2} \leqq |\xi|_\infty, |\xi'|_\infty \leqq \pi, \mathbf{x} \in \Omega \right\} \tag{10.5.3}$$

must not be too small to promise a success of the smoothing process. It is possible to satisfy $E(h) \geqq \text{const} > 0$ by a difference operator approximating a non-elliptic equation, since definition (3) does not involve low frequencies $|\xi|_\infty < \frac{\pi}{2}$. For instance, the hyperbolic wave equation $u_{xx} - u_{yy} = f$ can be discretised by the scheme

$$L_l = h_l^{-2} \begin{bmatrix} & & -1 & & \\ 0 & 0 & 3 & -2 & 1 \\ & & -1 & & \end{bmatrix}.$$

From $|\lambda(\mathbf{x}, \xi; h)|^2 = 4[(c_2 - c_1)^2 + 2(1 - c_2)(1 - c_1)^2]$, $c_i = \cos \xi_i$, one concludes that $E(h_l) \geqq \text{const} > 0$, although inequality (1) is violated because of $\lambda(\mathbf{x}, \xi; h) = O(|\xi|^3)$ as $\xi_1 = \xi_2 \to 0$. It still remains to analyse the complete two- or multi-grid iteration.

These considerations are important for the multi-grid solution of the transonic potential equation

$$(\varrho\,\Phi_x)_x + (\varrho\,\Phi_y)_y = 0, \qquad \varrho\,(x,y) := \left[1 + \frac{\gamma-1}{2}\,M_\infty^2\,(1 - |\mathrm{grad}\,\Phi|^2)\right]^{1/(\gamma-1)}, \tag{10.5.4}$$

(cf. Glowinski [1], T. J. Baker [1]), which is elliptic in the subsonic region $\{\varrho\,(x,y)\colon |\mathrm{grad}\,\Phi|^2 = \Phi_x^2 + \Phi_y^2 < 1\}$ and hyperbolic when $|\mathrm{grad}\,\Phi| > 1$. A multi-grid approach to the transonic small perturbation equation $((\alpha - \beta\,\Phi_x)\,\Phi_x)_x + \Phi_{yy} = 0$ is described by South-Brandt [1] using Murman's conservative difference scheme and by Fuchs [2]. Jameson [1] treated Eq. (4) by a multi-grid method with a special alternating direction smoothing process (cf. also Schmidt-Jameson [1]). Deconinck-Hirsch [1], [2] discretised Eq. (4) by a finite element method. A successful application of program (4.7) to the transonic potential flow is reported by Nowak-Wesseling [1].

10.6 Exercises

Exercise 10.6.1. Let $\mathscr{H}_l$ be the finite element space of piecewise linear functions over triangles as in Exercise 3.9.3 b. a) Show that the finite element method applied to the Dirichlet problem (1.1) in $\Omega = (0,1)\times(0,1)$ produces a stiffness matrix L_l, which is identical with (1.2). b) Let p (r) be the seven-point prolongation (restriction). Prove the approximation property (6.1.8) with $C_A \equiv C_A(\varepsilon) = C'_A/\varepsilon$, where C'_A is independent of ε.

Exercise 10.6.2. Let L_l be a nine-point scheme and define the patterns of the ILU factors by $P_B = \begin{bmatrix} \cdot & \cdot & \cdot \\ \cdot & * & \cdot \\ \cdot & \cdot & \cdot \end{bmatrix}$, $P_A = \begin{bmatrix} \cdot & \cdot & \cdot \\ * & * & \cdot \\ * & * & \cdot \end{bmatrix}$. Prove that the ILU iteration (1.13) coincides with the lexicographical Gauß-Seidel iteration. Note that (1.9 c) is violated. Formulate the analogous statement for ILLU. Does the five-point ILU (1.12 a) coincide with the symmetric Gauß-Seidel iteration (6.2.41)?

Exercise 10.6.3. Let $L_l(\varepsilon)$ be the scheme defined in (4.5 b) with $\mathbf{c} = (\cos\alpha, \sin\alpha)^T$. Prove that the seven-point factorisation (1.12 c) of $L_l(0)$ is exact (i.e. $L_l(0) = AB$) for $0 \leqq \alpha \leqq \frac{\pi}{2}$ and $\pi \leqq \alpha \leqq \frac{3\pi}{2}$. The range of α can be extended to $[0, \frac{3\pi}{4}] \cup [\pi, \frac{7\pi}{4}]$ if $u_x - u_y$ is discretised by a diagonal difference.

Exercise 10.6.4. Consider the ILLU factorisation with the patterns $P_{A'} = \begin{bmatrix} \cdot & \cdot & \cdot \\ \cdot & \cdot & \cdot \\ * & * & * \end{bmatrix}$, $P_D = \begin{bmatrix} \cdot & \cdot & \cdot \\ * & * & * \\ \cdot & \cdot & \cdot \end{bmatrix}$, $P_{B'} = \begin{bmatrix} * & * & * \\ \cdot & \cdot & \cdot \\ \cdot & \cdot & \cdot \end{bmatrix}$ of A', D, B'. Prove that

a) the ILLU factorisation is exact if $P_L = \begin{bmatrix} * & * & * \\ * & * & * \\ \cdot & \cdot & \cdot \end{bmatrix}$ or $P_L = \begin{bmatrix} \cdot & \cdot & \cdot \\ * & * & * \\ * & * & * \end{bmatrix}$;

b) the ILLU iteration (1.13) is a direct solver of $L_l(0)\,u_l = f_l$ with L_l as in Exercise 10.6.3, whatever the direction $\mathbf{c} \in \mathbb{R}^2$ is.

Exercise 10.6.5. Define p, r, L_{l-1} according to § 10.3.1. Prove that

a) the three-point scheme $L_l = h_l^{-2}[c_{-1}\ c_0\ c_1]$ yields the coarse-grid matrix $\Phi(L_l) := L_{l-1} = h_{l-1}^{-2}[c'_{-1}\ c'_0\ c'_1]$ with $c'_{\pm 1} = -2c_{\pm 1}^2/c_0$, $c'_0 = 2c_0 - 4c_1 c_{-1}/c_0$;
b) (weak) diagonal dominance is preserved under Φ: $L_l \mapsto L_{l-1}$;
c) the schemes $L_l = \text{const} \cdot h_l^{-2}[-1\ 2\ -1]$, $\text{const} \cdot h_l^{-1}[-1\ 1\ 0]$, and $\text{const} \cdot h_l^{-1}[0\ 1\ -1]$ are fixed points of Φ;
d) unbalanced schemes (i.e. $c_{-1} \neq c_1$) tend to one-sided schemes under the repeated application of Φ.

11. Elliptic Systems

In the preceding chapters only scalar differential equations $Lu = f$ have been considered. In the present chapter it will be pointed out that the multi-grid approach applies to elliptic systems as well. In § 11.1 we introduce examples of elliptic systems. The main difficulty of the multi-grid solution is the choice of an appropriate smoother. In § 11.2 we present a smoother which is related to the Jacobi iteration for scalar problems. A counterpart of the Gauß-Seidel iteration is the distributed relaxation described in § 11.3.

11.1 Examples

11.1.1 Systems of Coupled Elliptic Equations

Two kinds of elliptic systems will be distinguished: (i) systems of n equations

$$L_\alpha u_\alpha = \varphi_\alpha(u_1, \ldots, u_m) \qquad (\alpha = 1, \ldots, n), \tag{11.1.1}$$

where each equation can be regarded as an elliptic equation for the respective unknown u_α, which is coupled via the right-hand side φ_α with low order derivatives of u_β $(\beta = 1, \ldots, n)$ – and (ii) systems which do not have the property described in (i). In this subsection we study case (i).

A nonlinear system of the first category is already described in § 9.4. The principal terms in Eq. (9.4.3 a, b) are $L_1 \Phi = \Delta^2 \Phi$ and $L_2 T = -\Delta T$. Both equations are elliptic, but of different orders. Another example with $L_1 = L_2 = \Delta^2$ is the von Kármán equation describing large deflections of plates:

$$\Delta^2 \Phi + \alpha[w, w] = 0, \qquad \Delta^2 w - \beta[\Phi, w] = f_2,$$

where $[\varphi, \psi] := \varphi_{xx}\psi_{yy} + \varphi_{yy}\psi_{xx} - 2\varphi_{xy}\psi_{xy}$ (cf. Nečas-Hlaváček [1]). Three coupled second order equations arise in the physics of semiconductor devices (cf. Bank-Jerome-Rose [1]). In § 16.10.1 we shall mention a system of elliptic equations which are coupled by means of the boundary conditions.

The discretisation of the nonlinear version of Eq. (1) yields

$$\mathscr{L}_l(u_l) = f_l, \tag{11.1.2 a}$$

where $u_l \in \mathscr{U}_l$ and $f_l \in \mathscr{F}_l$ are vector-valued grid functions:

$$u_l = (u_{l,1}, \ldots, u_{l,n})^T, \quad f_l = (f_{l,1}, \ldots, f_{l,n})^T. \tag{11.1.2 b}$$

Equation (2 a) can be rewritten blockwise:

$$\mathscr{L}_{l,\alpha}(u_l) = f_{l,\alpha} \quad (\alpha = 1, \ldots, n). \tag{11.1.2 a*}$$

The Jacobian $L_l(u_l) := \partial \mathscr{L}_l(u_l)/\partial u_l$ may be split into the blocks $L_{l,\alpha\beta} = \partial \mathscr{L}_{l,\alpha}/\partial u_{l,\beta}$. By assumption the diagonal blocks $L_{l,\alpha\alpha}$ contain the principal terms.

The multi-grid treatment of Eq. (2 a) is hardly different from the scalar case. Usually, the prolongation $p: \mathscr{U}_{l-1} \to \mathscr{U}_l$ is blockdiagonal, i.e.

$$p = \operatorname{blockdiag}\{p_1, \ldots, p_n\}, \quad (p\, u_{l-1})_\alpha = p_\alpha u_{l-1,\alpha} \quad (1 \leqq \alpha \leqq n). \tag{11.1.3}$$

One has to define p_α so that p_α is suitable for $L_{l,\alpha} u_{l,\alpha} = f_{l,\alpha}$. In particular, the degree of the interpolation p_α has to be chosen according to the order of the αth differential equation (cf. Note 3.5.1). This is important when the orders of $L_1, \ldots, L_n$ do not coincide as in the example of § 9.4. The blockdiagonal structure (3) is proposed to obtain a sparse matrix p. A prolongation of the general form $u_{l,\alpha} = \sum_{\beta=1}^{n} p_{\alpha\beta} u_{l-1,\beta}$ would result if we extend the construction of § 10.3.2 to the present case.

Similar comments hold for the restriction r. Usually, we assume

$$r = \operatorname{blockdiag}\{r_1, \ldots, r_n\}, \quad (r\, u_l)_\alpha = r_\alpha u_{l,\alpha} \quad (1 \leqq \alpha \leqq n). \tag{11.1.4}$$

The choice $r := p^*$ with p from (3) yields r satisfying (4) with $r_\alpha = p_\alpha^*$.

The Gauß-Seidel smoothing iteration allows various orderings of the unknowns. In the case of a vector-valued grid function u_l, new variants arise. Below we describe appropriate versions for system (2) depending on the strength of the coupling.

(i) Assume that the coupling of the $u_{l,\alpha}$-components in Eq. (2 a) is sufficiently weak, i.e. the off-diagonal blocks $L_{\alpha\beta}$ $(\alpha \neq \beta)$ are small enough. Then the pointwise Gauß-Seidel iteration with the following ordering of unknown variables can be applied:

$$u_{l,1}(\mathbf{x}^1), \quad u_{l,1}(\mathbf{x}^2), \ldots, u_{l,1}(\mathbf{x}^{n_l}), \quad u_{l,2}(\mathbf{x}^1), \ldots, u_{l,n}(\mathbf{x}^{n_l}), \tag{11.1.5 a}$$

where $\mathbf{x}^1, \ldots, \mathbf{x}^{n_l}$ is some ordering (lexicographical, chequer-board, etc.) of the grid points of Ω_l. Here we assume for simplicity that all components $u_{l,\alpha}$ are defined on the same grid Ω_l. Otherwise, one has to describe the set of grids by $\Omega_l = (\Omega_{l,1}, \ldots, \Omega_{l,n})$.

(ii) Let $n = 2$. Assume that $u_{l,1}$ is strongly coupled with $u_{l,2}$ by means of $\mathscr{L}_{l,1}(u_l) = f_{l,1}$, whereas $u_{l,2}$ in $\mathscr{L}_{l,2}(u_l) = f_{l,2}$ is weakly coupled with $u_{l,1}$. Then the pointwise Gauß-Seidel process has to start with $u_{l,2}$. Hence, the ordering is

$$u_{l,2}(\mathbf{x}^1), \ldots, u_{l,2}(\mathbf{x}^{n_l}), \qquad u_{l,1}(\mathbf{x}^1), \ldots, u_{l,1}(\mathbf{x}^{n_l}). \tag{11.1.5 b}$$

(iii) Assume that all components $u_{l,\alpha}(1 \leqq \alpha \leqq n)$ are strongly coupled by $\mathscr{L}_{l,\alpha}(u_l) = f_{l,\alpha}$ with the other ones. Then a blockwise Gauß-Scidel iteration is required, where each block consists of the n components of $u_l(\mathbf{x}^\nu)$. The ordering of the blocks that arises,

$$u_l(\mathbf{x}^1), \ldots, u_l(\mathbf{x}^{n_l}), \tag{11.1.5 c}$$

is geometrically pointwise.

(iv) If the diagonal equations $L_{l,\alpha\alpha} u_{l,\alpha} = f_{l,\alpha}$ require a linewise Gauß-Seidel iteration, one has to construct the linewise counterparts of (5 a–c). The blocks that arise consist of $\{u_{l,\alpha}(\mathbf{x}): \mathbf{x} \text{ in a certain line}\}$ in the case (5 a, b) and of $\{u_l(\mathbf{x}): \mathbf{x} \text{ in a certain line}\}$ in the case of (5 c).

11.1.2 Examples: Systems of Cauchy-Riemann, Stokes, and Navier-Stokes

Let u be the solution of the Neumann problem $\Delta u = f_\Omega$ in $\Omega \subset \mathbb{R}^2$, $\partial u/\partial n = f_\Gamma$ on Γ, and set $v := u_x$, $w := u_y$. Then the pair (v, w) is a solution of the *Cauchy-Riemann system*

$$v_x + w_y = f \quad \text{in } \Omega, \tag{11.1.6 a}$$

$$v_y - w_x = 0 \quad \text{in } \Omega, \tag{11.1.6 b}$$

$$(v, w) \cdot \mathbf{n} = f \quad \text{on } \Gamma. \tag{11.1.6 c}$$

Neither Eq. (6 a) nor Eq. (6 b) can be considered as an elliptic equation for v or w.

Similarly, the biharmonic equation $\Delta^2 u = f_\Omega$ with boundary conditions $u = f_{1\Gamma}$, $\partial u/\partial n = f_{2\Gamma}$ can be split into two equations of second order for the functions u and $v := \Delta u$:

$$v - \Delta u = 0 \qquad \text{in } \Omega, \tag{11.1.7 a}$$

$$-\Delta v = -f_\Omega \qquad \text{in } \Omega, \tag{11.1.7 b}$$

$$u = f_{1\Gamma}, \quad \partial u/\partial n = f_{2\Gamma} \quad \text{on } \Gamma. \tag{11.1.7 c}$$

Because of the boundary condition $\partial u/\partial n = \ldots$ the system (7 a, b) cannot be regarded as a system of two coupled elliptic equations.

Stokes' equations for the velocities $\mathbf{u} = (u_1, \ldots, u_d)$ and for the pressure p reads as

$$-\Delta \mathbf{u} + \operatorname{grad} p = f_\Omega \quad \text{in } \Omega \subset \mathbb{R}^d, \tag{11.1.8 a}$$

$$-\operatorname{div} \mathbf{u} = 0 \quad \text{in } \Omega, \tag{11.1.8 b}$$

$$\mathbf{u} = 0 \quad \text{on } \Gamma, \tag{11.1.8 c}$$

$\sum_{i,j=1}^{d} \int_\Omega \partial u_{1,i}/\partial x_j \overline{\partial v_{1,i}/\partial x_j}\, d\mathbf{x}$. u_1 in (13) equals $\mathbf{u} = (u_1, \ldots, u_d)$ from (8), while u_2 from (13) equals p in (8).

The variational form of system (7) reads as (13) with

$$\begin{aligned} &\mathscr{W}_1 = H^1(\Omega), \quad \mathscr{W}_2 = H_0^1(\Omega), \\ &a(u_1, v_1) = (w_1, v_1), \quad b(u_2, v_1) = c(u_2, v_1) = (\operatorname{grad} u_2, \operatorname{grad} v_1), \\ &f_1 = 0, \quad f_2 = f_\Omega, \end{aligned} \tag{11.1.14 b}$$

if $f_{i\Gamma} = 0$ (cf. Ciarlet [1]). u_1 from (13) equals v in (7), u_2 from (13) equals u in (7).

Defining $u = \begin{pmatrix} u_1 \\ u_2 \end{pmatrix}$, $v = \begin{pmatrix} v_1 \\ v_2 \end{pmatrix}$, $f = \begin{pmatrix} f_1 \\ f_2 \end{pmatrix}$, we may rewrite (13) as

$$A(u, v) = (f, v) \quad \text{for all } u \in \mathscr{W}_1 \times \mathscr{W}_2 \tag{11.1.13'}$$

with $A(u, v) := a(u_1, v_1) + b(u_2, v_1) + \overline{c(v_2, u_1)}$. (13) is called a 'saddle point problem' because of

Note 11.1.2. The form $A(\cdot\,,\cdot)$ is indefinite.

Remark 11.1.3. The saddle point problem (13) is equivalent to Eq. (10) with

$$L = \begin{bmatrix} L_{11} & L_{12} \\ L_{21} & 0 \end{bmatrix}, \tag{11.1.15}$$

where L_{11}, L_{12}, L_{21}^* are the operators associated with the respective forms a, b, c.

Though it is often possible to eliminate one of the variables and to obtain a variational formulation with a positive definite form $\hat{a}$, one prefers formulation (13). E.g. the biharmonic problem (7) is described by a $H_0^2(\Omega)$-elliptic form $\hat{a}$; but finite elements in $H_0^2(\Omega)$ are more complicated than elements in $H_0^1(\Omega)$ and $H^1(\Omega)$ as required in (14 b). Using the $\mathscr{H}$-elliptic formulation of Stokes' problem (8): $(\operatorname{grad} \mathbf{u}, \operatorname{grad} \mathbf{v}) = (f_\Omega, \mathbf{v})$ for all $\mathbf{v} \in \mathscr{H}$, would require the need for finite elements in the space $\mathscr{H} = \{\mathbf{u} \in H_0^1(\Omega) : \operatorname{div} \mathbf{u} = 0\}$, which would not be practical. There are even numerical justifications for splitting a convenient equation such as Poisson's equation $\Delta u = f$ into the system $\operatorname{div} w = f$, $\operatorname{rot} w = 0$ (cf. Neittaanmäki-Křížek [1]).

11.1.5 Discretisation

Replacing the differential operators L_{ij} in (10 b) by suitable difference schemes or restricting the variational formulation to finite element subspaces, one obtains

$$L_l u_l = f_l. \tag{11.1.16 a}$$

Numbering first the grid function $u_{l,1}$ (corresponding to the first component u_1 of u from (10 a)), then $u_{l,2}$ etc., one may write L_l as

$$L_l = \begin{bmatrix} L_{l,11} & \dots & L_{l,1n} \\ \vdots & & \vdots \\ L_{l,n1} & \dots & L_{l,nn} \end{bmatrix}. \tag{11.1.16 b}$$

according to (10b). $L_{l,ij}\,(i \neq j)$ may be non-square blocks.

The difficulties of the discretisation process can be illustrated by the saddle point problem (13). Because of Note 11.1.2 the discretisation of problem (13) may be singular or ill-posed. L_l is singular, if $L_{l,21}\,L_{l,11}^{-1}\,L_{l,12}$ is singular. There are well-known simple ill-conditioned finite element discretisations of Stokes' problem. In the finite element case, Eq. (16 a) is well-posed if Brezzi's [1] condition is met, which reads

$$\sup\{|(\operatorname{div}\mathbf{v}, q)| : \mathbf{v} \in \mathscr{H}_{l,1} \subset H_0^1(\Omega)^d, |\mathbf{v}|_1 = 1\} \geqq \frac{1}{C}|q|_0$$

$$\text{for all } q \in \mathscr{H}_{l,2} \subset L^2(\Omega)/\mathbb{C}$$

for Stokes' problem. As described by Brezzi-Pitkäranta [1] well-posedness can be ensured by adding special finite elements or by modifying the discrete incompressibility condition.

11.2 A Multi-Grid Approach to a General System

11.2.1 Appropriate Norms and Scales

11.2.1.1 The Continuous Case

For a single differential equation we have introduced a scale of spaces $\mathscr{H}^s$ such that L is an isomorphism of $\mathscr{H}^{s+2}$ onto $\mathscr{H}^s$. The standard choice in § 6 is $\mathscr{H}^s \subset H^{sm}(\Omega)$ depending on the order $2m$ of L. In the present case there are different orders $m_i + m_j'$ of the operators L_{ij} in (1.10 b). For each component u_i, f_i of the vector functions u, f we define spaces $\mathscr{X}_i^s$, $\mathscr{Y}_i^s$ $(i = 1, \dots, n;\ s \in \mathbb{R})$, where s indicates the order of differentiability. u and f belong to the respective product spaces

$$\mathscr{X}^{s_1,\dots,s_n} := \mathscr{X}_1^{s_1} \times \dots \times \mathscr{X}_n^{s_n}, \tag{11.2.1 a}$$

$$\mathscr{Y}^{s_1,\dots,s_n} := \mathscr{Y}_1^{s_1} \times \dots \times \mathscr{Y}_n^{s_n}. \tag{11.2.1 b}$$

$\mathscr{X}_i^s$ and $\mathscr{Y}_i^s$ have to be chosen such that (at least for suitable values of t) the mappings

$$L\colon \quad \mathscr{X}^{t+m_1,\dots,t+m_n} \to \mathscr{Y}^{t-m_1',\dots,t-m_n'}, \tag{11.2.2 a}$$

$$L^{-1}\colon \mathscr{Y}^{t-m_1',\dots,t-m_n'} \to \mathscr{X}^{t+m_1,\dots,t+m_n}, \tag{11.2.2 b}$$

are bounded. m_i and m'_j are the numbers given by Definition 11.1.1. In contrast to L_Ω, the operator L includes the boundary conditions, which are specified only for the examples in § 11.1.2.

Example 11.2.1 (Stokes' equations). Split u from (1.10 a) into $(u_1, u_2)^T$ with $u_1 = \mathbf{u}$, $u_2 = p$, where the right-hand quantities are those from system (1.8). Define

$$\mathscr{X}_1^s = \begin{cases} [H^s(\Omega) \cap H_0^{\min(1,s)}(\Omega)]^d & \text{if } s \geqq 0, \quad s \neq \frac{1}{2}, \\ [H^{-s}(\Omega)^d]' = \text{dual space of } H^{-s}(\Omega)^d & \text{if } s < 0, \end{cases}$$

$$\mathscr{X}_2^s = \begin{cases} H^s(\Omega)/\mathbb{C} & \text{if } s \geqq 0, \\ [H^{-s}(\Omega)]'/\mathbb{C} & \text{if } s < 0, \end{cases}$$

$$\mathscr{Y}_1^s = (\mathscr{X}_1^{-s})', \qquad \mathscr{Y}_2^s = (\mathscr{X}_2^{-s})'.$$

For a saddle point problem, conditions (2 a, b) can easily be described by means of the operators L_{ij}.

Proposition 11.2.2. Let L be given by (1.15). Assume that the following mappings are bounded:

$$L_{11}: \mathscr{X}_1^{t+m_1} \to \mathscr{Y}_1^{t-m'_1}, \qquad L_{11}^{-1}: \mathscr{Y}_1^{t-m'_1} \to \mathscr{X}_1^{t+m_1}, \tag{11.2.3 a}$$

$$L_{12}: \mathscr{X}_2^{t+m_2} \to \mathscr{Y}_1^{t-m'_1}, \qquad L_{21}: \mathscr{X}_1^{t+m_1} \to \mathscr{Y}_2^{t-m'_2}. \tag{11.2.3 b}$$

Then

$$S := L_{21} L_{11}^{-1} L_{12}: \mathscr{X}_2^{t+m_2} \to \mathscr{Y}_2^{t-m'_2} \tag{11.2.3 c}$$

is well-defined. L and L^{-1} satisfy (2 a, b) if

$$S^{-1}: \mathscr{Y}_2^{t-m'_2} \to \mathscr{X}_2^{t+m_2} \text{ is bounded.} \tag{11.2.3 d}$$

For Stokes' problem $S = -\operatorname{div} \Delta^{-1} \operatorname{grad}$ is an isomorphism of $X_2^0 = Y_2^0 = L^2(\Omega)/\mathbb{C}$ onto itself and (3 d) [hence (2 a, b)] can be shown also for $t > 0$ (cf. Kellogg-Osborn [1]).

11.2.1.2 Discrete Spaces

In the discrete case, L_{ij} is replaced by the blocks $L_{l,ij}$ from (1.16 b). The Hilbert spaces

$$\mathscr{X}_l^{s_1, \ldots, s_n} = \mathscr{X}_{l,1}^{s_1} \times \ldots \times \mathscr{X}_{l,n}^{s_n}, \qquad \mathscr{Y}_l^{s_1, \ldots, s_n} = \mathscr{Y}_{l,1}^{s_1} \times \ldots \times \mathscr{Y}_{l,n}^{s_n} \tag{11.2.4}$$

are the discrete analogues of (1 a, b) (cf. (6.2.4 b) and § 6.3.2.1). The counterparts of the conditions (2 a, b) are

$$\|L_l\|_{\mathscr{Y}_l^{t-m'_1, \ldots, t-m'_n} \leftarrow \mathscr{X}_l^{t+m_1, \ldots, t+m_n}} \leqq C \quad \text{for all } l \geqq 0, \tag{11.2.5 a}$$

$$\|L_l^{-1}\|_{\mathscr{X}_l^{t+m_1, \ldots, t+m_n} \leftarrow \mathscr{Y}_l^{t-m'_1, \ldots, t-m'_n}} \leqq C \quad \text{for all } l \geqq 0, \tag{11.2.5 b}$$

The latter inequality replaces (6.3.26) in the scalar case.

Note 11.2.3. Since the spaces $\mathscr{X}_i^0$, $\mathscr{Y}_i^0$ equal $L^2(\Omega)$ (or $L^2(\Omega)^d$, $L^2(\Omega)/\mathbb{C}$ etc) the discrete spaces $\mathscr{X}_{l,i}^0$ and $\mathscr{Y}_{l,i}^0$ are endowed with the usual Euclidean norm, which is again denoted by $|\cdot|_0$. As in (6.2.6a) the matrix norm $\|\cdot\|$ means the corresponding spectral norm. Note that a common symbol is used for the different cases

$$\|\cdot\| = \|\cdot\|_{\mathscr{X}_{l,i}^0 \leftarrow \mathscr{Y}_{l,j}^0}, \quad \|\cdot\| = \|\cdot\|_{\mathscr{X}_{l,i}^0 \leftarrow \mathscr{X}_{l,j}^0}, \text{ etc.} \quad (0 \leqq i, j \leqq n) \tag{11.2.6}$$

Proposition 11.2.2 carries over to the discrete case. To simplify the notation, we shall write

$$|\cdot|_{s \leftarrow t},$$

where "$s \leftarrow t$" means one of the subscripts "$\mathscr{X}_{l,i}^s \leftarrow \mathscr{X}_{l,j}^t$", "$\mathscr{X}_{l,i}^s \leftarrow \mathscr{Y}_{l,i}^t$", ... For example, in (8 a) $|L_{l,11}|_{t-m_1' \leftarrow t+m_1}$ means $\|L_{l,11}\|_{\mathscr{Y}_{l,1}^{t-m_1'} \leftarrow \mathscr{Y}_{l,1}^{t+m_1}}$. The full notation is used, when the meaning is not quite obvious as e.g. in (8 e).

Lemma 11.2.4. Consider the saddle point problem (1.16 a) with

$$L_l = \begin{bmatrix} L_{l,11} & L_{l,12} \\ L_{l,21} & 0 \end{bmatrix}, \tag{11.2.7}$$

and define $S_l := L_{l,21} L_{l,11}^{-1} L_{l,12}$. Inequalities (5 a, b) hold if

$$|L_{l,11}|_{t-m_1' \leftarrow t+m_1} \leqq C, \tag{11.2.8 a}$$

$$|L_{l,12}|_{t-m_1' \leftarrow t+m_2} \leqq C, \tag{11.2.8 b}$$

$$|L_{l,21}|_{t-m_2' \leftarrow t+m_1} \leqq C, \tag{11.2.8 c}$$

$$|L_{l,11}^{-1}|_{t+m_1 \leftarrow t-m_1'} \leqq C, \tag{11.2.8 d}$$

$$|S_l^{-1}|_{t+m_2 \leftarrow t-m_2'} = \|S_l^{-1}\|_{\mathscr{X}_{l,2}^{t+m_2} \leftarrow \mathscr{Y}_{l,2}^{t-m_2'}} \leqq C. \tag{11.2.8 e}$$

Corollary 11.2.5. In the case of a finite element discretisation, Brezzi's [1] condition is equivalent to inequality (8 e) with $t = 0$, $m_2 = m_2'$.

Criteria of condition (8 e) are mentioned by Hackbusch [19].

11.2.1.3 Approximation Property

The approximation property (6.1.8) requires an estimation of $L_l^{-1} - p L_{l-1}^{-1} r$ with respect to suitable norms $\|\cdot\|_{\mathscr{U}}$ and $\|\cdot\|_{\mathscr{F}}$. In § 6 we chose e.g. $\|\cdot\|_{\mathscr{U}} = \|\cdot\|_{\mathscr{F}} = |\cdot|_0$ and $\alpha = 2m =$ order of L. Since now the different blocks L_{ij} have different orders, $\|\cdot\|_{\mathscr{U}}$ cannot be the norm of $\mathscr{X}_l^{0,\ldots,0}$, which would be the simplest choice. Instead we may define

$$\|u_l\|_{\mathscr{U}} := \sqrt{\sum_{i=1}^{n} h_l^{-2m_i'} \|u_{l,i}\|_{\mathscr{X}_{l,i}^0}^2}, \quad \|f_l\|_{\mathscr{F}} := \sqrt{\sum_{i=1}^{n} h_l^{2m_i} \|f_{l,i}\|_{\mathscr{Y}_{l,i}^0}^2}, \tag{11.2.9}$$

which are weighted Euclidean norms (cf. Note 11.2.3).

Remark 11.2.6. Denote the blocks of L_l^{-1} by $Z_{l,ij}$. Assume that p and r are block-diagonal (cf. (1.3), (1.4)) and define the norms of $\mathscr{U}_l$ and $\mathscr{F}_l$ by (9). Then the approximation property

$$\| L_l^{-1} - p L_{l-1}^{-1} r \|_{\mathscr{U} \leftarrow \mathscr{F}} \leqq C_A \tag{11.2.10}$$

is equivalent to

$$\| Z_{l,ij} - p_i Z_{l-1,ij} r_j \| \leqq c_{ij} h_l^{m_i' + m_j} \quad \text{for all } 1 \leqq i,\ j \leqq n, \quad l \geqq 1, \tag{11.2.11}$$

where $\| \cdot \|$ is the spectral norm from (6).

Hence, each block of $L_l^{-1} - p L_{l-1}^{-1} r$ satisfies an estimate of the form (6.1.8), where the exponent $\alpha = m_i + m_j'$ depends on the block location (i, j). The right-hand side in (10) is independent of h_l, since the norms are suitably scaled (cf. Note 6.1.4).

For the saddle point problem (7) the following result can be proved by the techniques presented in § 6.3.2.2.

Proposition 11.2.7. Consider the saddle point problem with L_l from (7). Assume $m_1 = m_1'$, $m_2 = m_2'$, and define the norms by (9). Suppose (8 a–e) for $t = 0$, $\pm m_1$, $\pm m_2$. Assume that there exists a block-diagonal restriction $r': \mathscr{U}_l \to \mathscr{U}_{l-1}$ such that

$$\begin{aligned}
|I - p_1 r_1'|_{s \leftarrow t} &\leqq C h_l^{t-s} && \text{for } t = 2m_1,\ m_1 + m_2 \text{ and } s = 0,\ m_1 - m_2, \\
|I - p_2 r_2'|_{s \leftarrow t} &\leqq C h_l^{t-s} && \text{for } t = 2m_2,\ m_1 + m_2 \text{ and } s = 0,\ m_2 - m_1, \\
|r_1|_{s \leftarrow s} &\leqq C && \text{for } s = 0,\ m_2 - m_1,\ -m_1 - m_2,\ -2m_1, \\
|r_2|_{s \leftarrow s} &\leqq C && \text{for } s = -2m_2,\ -m_1 - m_2, \\
|p_1|_{s \leftarrow s} &\leqq C && \text{for } s = 0,\ m_1 - m_2,\ 2m_1, \\
|p_2|_{s \leftarrow s} &\leqq C && \text{for } s = 0,\ m_2 - m_1,\ m_2 + m_1,\ 2m_2, \\
|r_1'|_{s \leftarrow s} &\leqq C && \text{for } s = 2m_1,\ m_1 + m_2, \\
|r_2'|_{s \leftarrow s} &\leqq C && \text{for } s = 2m_2,\ m_1 + m_2.
\end{aligned}$$

The consistency assumptions are

$$|r_i L_{l,ij} p_j - L_{l-1,ij}|_{s \leftarrow t} \leqq C h_l^{t-s-m_i-m_j}$$
$$\text{for } t = 2m_j,\ m_1 + m_2,\ s = -2m_i,\ -m_1 - m_2$$

for $(i, j) = (1,1), (1,2), (2,1)$. Then the approximation property (10) is valid.

Proof (cf. Hackbusch [19]). We introduce the following abbreviations:

$$\begin{aligned}
&A = L_{l,11}, && B = L_{l,12}, && C = L_{l,21}, && S = S_l = C A^{-1} B, \\
& && && && Z = Z_{l,21}, \\
&A' = L_{l-1,11}, && B' = L_{l-1,12}, && C' = L_{l-1,21}, && S' = S_{l-1}, \\
& && && && Z' = Z_{l-1,21}.
\end{aligned}$$

The blocks $Z_{l,ij}$ of the inverse matrix L_l^{-1} are

$$Z_{l,11} = A^{-1}(I - BZ), \quad Z_{l,12} = A^{-1} B S^{-1}, \quad Z_{l,21} = Z = S^{-1} C A^{-1},$$
$$Z_{l,22} = -S^{-1}.$$

As in the proof of Criterion 6.3.35 we split $A^{-1} - p_1 A'^{-1} r_1$ into

$$\begin{aligned} &A^{-1} - p_1 A'^{-1} r_1 \\ &= [I - p_1 A'^{-1} r_1 A][I - p_1 r_1'] A^{-1} - p_1 A'^{-1} [r_1 A p_1 - A'] r_1' A^{-1} \end{aligned}$$

and obtain

$$\begin{aligned} &|A^{-1} - p_1 A'^{-1} r_1|_{s \leftarrow t} \leqq C h_l^{t-s+2m_1} \\ &\text{for } t = 0, \quad m_2 - m_1 \quad \text{and} \quad s = 0, \quad m_1 - m_2 \end{aligned} \tag{11.2.12 a}$$

as in Criterion 6.3.35. From

$$\begin{aligned} r_2 S p_2 - S' &= r_2 C A^{-1} B p_2 - C' A'^{-1} B' \\ &= r_2 C [A^{-1} - p_1 A'^{-1} r_1] B p_2 + [r_2 C p_1 - C'] A'^{-1} r_1 B p_2 \\ &\quad + C' A'^{-1} [r_1 B p_2 - B'] \end{aligned}$$

we conclude that

$$\begin{aligned} &|r_2 S p_2 - S'|_{t \leftarrow s} \leqq C h_l^{t-s-2m_2} \\ &\text{for } s = 2m_2, \ m_1 + m_2 \quad \text{and} \quad t = -2m_2, \ -m_1 - m_2 . \end{aligned}$$

Hence

$$\begin{aligned} p_2 S'^{-1} r_2 - S^{-1} &= [p_2 S'^{-1} r_2 S - I][I - p_2 r_2'] S^{-1} \\ &\quad + p_2 S'^{-1} [r_2 S p_2 - S'] r_2' S^{-1} \end{aligned}$$

can be estimated by

$$\begin{aligned} &|p_2 S'^{-1} r_2 - S^{-1}|_{s \leftarrow t} \leqq C h_l^{t-s+2m_2} \\ &\text{for } t = 0, \ m_1 - m_2 \quad \text{and} \quad s = 0, \ m_2 - m_1 \end{aligned} \tag{11.2.12 b}$$

proving (11) for $i = j = 2$ since $Z_{l,22} = -S^{-1}$. The splitting

$$\begin{aligned} p_2 Z' r_2 - Z &= p_2 S'^{-1} C' A'^{-1} r_2 - S^{-1} C A^{-1} \\ &= p_2 S'^{-1} [C' - r_2 C p_1] A'^{-1} r_1 \\ &\quad + [p_2 S'^{-1} r_2 - S^{-1}] C p_1 A'^{-1} r_1 \\ &\quad + S^{-1} C [p_1 A'^{-1} r_1 - A^{-1}] \end{aligned}$$

implies

$$|p_2 Z' r_2 - Z|_{s \leftarrow 0} \leqq C h_l^{-s+m_1+m_2} \quad \text{for } s = 0, \ m_2 - m_1 . \tag{11.2.12 c}$$

The particular case $s = 0$ demonstrates the validity of (11) for $i = 2$, $j = 1$. Similarly, the representation

$$\begin{aligned} Z_{l,12} - p_1 Z_{l-1,12} r_2 &= A^{-1} B S^{-1} - p_1 A'^{-1} B' S'^{-1} r_2 \\ &= [A^{-1} - p_1 A'^{-1} r_1] B S^{-1} \\ &\quad + p_1 A'^{-1} r_1 B [S^{-1} - p_2 S'^{-1} r_2] \\ &\quad + p_1 A'^{-1} [r_1 B p_2 - B'] S'^{-1} r_2 \end{aligned}$$

proves (11) for $i = 1, j = 2$. The remaining block $Z_{l,11}$ is the difference of A^{-1}

and $A^{-1}BZ$. The estimation of the first term is already given by (12 a), $s=t=0$. The estimation of the second term can be obtained from (12 c) and

$$\begin{aligned} A^{-1}BZ - p_1 A'^{-1}B'Z'r_1 = {} & [A^{-1} - p_1 A' r_1]BZ \\ & + p_1 A'^{-1} r_1 B[Z - p_2 Z' r_1] \\ & + p_1 A'^{-1}[r_1 B p_2 - B']Z' r_1 . \end{aligned}$$ □

This Proposition is especially suited to difference schemes L_l. But it can also be applied to finite element discretisations.

Example 11.2.8. Discretise the Stokes equations (1.8) in a convex domain by linear finite elements for **u** and p, where the triangles of the p-elements have twice the size of the **u**-elements (cf. Bercovier-Pironneau [1]). Then the assumptions of Proposition 11.2.7 are satisfied.

Another approach suitable for finite element methods is mentioned below in Proposition 11.2.13.

11.2.2 A Smoothing Iteration for a General System

The smoothing iterations developed in § 3 for the scalar case do not work for the elliptic system considered here. For instance, the Gauß-Seidel iteration is not applicable to any saddle point problem (7), since the diagonal of the matrix L_l is singular. Even more sophisticated iterative methods like the Kaczmarz iteration (3.3.20), which is a convergent method when applied e.g. to the discrete Stokes problem, are not suited for smoothing, since the multi-grid convergence rate would be h_l-dependent.

In this chapter we describe the simplest smoother that applies to a general linear system. In the scalar case of § 6, the smoother simplest to analyse was the modified Jacobi iteration $\mathscr{S}_l(u_l, f_l) = u_l - \omega_l h_l^{2m}(L_l u_l - f_l)$ (cf. (3.3.7)) or the squared version $\mathscr{S}_l(u_l, f_l) = u_l - \omega_l^2 h_l^{4m} L_l^*(L_l u_l - f_l)$ in the non-symmetric case (cf. (3.3.10)). First we shall extend the latter variant to the present problem.

Let m_i, m'_j be the numbers from Definition 11.1.1. Assume that L_l has the block structure (1.16 b) and define the block-diagonal matrices

$$H_l := \operatorname{blockdiag}\{h_l^{m_1} I, \ldots, h_l^{m_n} I\}, \quad H'_l := \operatorname{blockdiag}\{h_l^{m'_1} I, \ldots, h_l^{m'_n} I\}, \tag{11.2.13}$$

which will be used for scaling.

Note 11.2.9. The norms defined in (9) can be written

$$\|u_l\|_{\mathscr{U}} = |H_l'^{-1} u_l|_0, \qquad \|f_l\|_{\mathscr{F}} = |H_l f_l|_0,$$

where $|u_l|_0 = \left(\sum_{i=1}^{n} |u_{l,i}|_0^2\right)^{1/2}$ is the Euclidean norm.

Consider

$$\hat{L}_l := H_l L_l H'_l. \tag{11.2.14}$$

The blocks of $\hat{L}_l$ are $\hat{L}_{l,ij} = h_l^{m_i+m'_j} L_{l,ij}$, $1 \leqq i, j \leqq n$. Because of order $(L_{ij}) \leqq m_i + m'_j$, we may assume the boundedness of the spectral norms, $\|\hat{L}_{l,ij}\| \leqq c_{ij}$, independent of l, which corresponds to inequality (6.3.23) in the scalar case. Hence, we obtain

$$1/\omega_l := \|\hat{L}_l\| \leqq C_L < \infty \quad \text{for all } l \geqq 0 \quad (\|\cdot\|\text{: spectral norm}). \tag{11.2.15}$$

The generalisation of iteration (3.3.10) is the smoothing process

$$\mathscr{S}_l(u_l, f_l) := u_l - \omega_l^2 H_l'^2 L_l^* H_l^2 (L_l u_l - f_l) \tag{11.2.16}$$

with ω_l from (15). Its iteration matrix is

$$S_l = I - \omega_l^2 H_l'^2 L_l^* H_l^2 L_l. \tag{11.2.16'}$$

The smoothing property of (16) is stated in

Proposition 11.2.10. Assume (15) and define the norms $\|\cdot\|_{\mathscr{U}}, \|\cdot\|_{\mathscr{F}}$ by (9). The smoothing iteration (16) satisfies the smoothing property (6.1.7):

$$\|L_l S_l^\nu\|_{\mathscr{F}\leftarrow\mathscr{U}} \leqq \eta(\nu) := C_L \sqrt{\eta_0(2\nu)} \quad \text{for all } \nu \geqq 0, \quad l \geqq 0 \tag{11.2.17}$$

with C_L from (15) and η_0 from (6.2.1 b).

Proof. Note that $S_l = I - \omega_l^2 H_l' \hat{L}_l^* \hat{L}_l H_l'^{-1}$ and $\hat{S}_l := H_l'^{-1} S_l H_l' = I - \omega_l^2 \hat{L}_l^* \hat{L}_l = I - X_l$, where $X_l := \omega_l^2 \hat{L}_l^* \hat{L}_l$ satisfies $O \leqq X_l \leqq I$. The smoothing property (17) follows from Note 11.2.9 and Lemma 6.2.1:

$$\begin{aligned}\|L_l S_l^\nu\|^2_{\mathscr{F}\leftarrow\mathscr{U}} &= \|H_l L_l S_l^\nu H_l'\|^2 = \|H_l L_l H_l' H_l'^{-1} S_l^\nu H_l'\|^2 = \|\hat{L}_l \hat{S}_l^\nu\|^2 \\ &= \|\hat{S}_l^{*\nu} \hat{L}_l^* \hat{L}_l \hat{S}_l^\nu\| = \omega_l^{-2} \|X_l (I - X_l)^{2\nu}\|.\end{aligned}$$ □

Since the norms (9) are also suited for the approximation property (10), the conditions (6.1.7/8) can be satisfied simultaneously and Theorem 7.1.2 ensures multi-grid convergence.

Proposition 11.2.10 remains valid if we replace definition (13) by

$$H_l = \operatorname{blockdiag}\{\ldots, c_i h_l^{m_i} I, \ldots\}, \quad H_l' = \operatorname{blockdiag}\{\ldots, c_i' h_l^{m_i'} I, \ldots\},$$
$$c_i, c_i' > 0.$$

The constants c_i, c_i' can be used to balance the blocks of $\hat{L}_l$.

Example 11.2.11. Consider Stokes' problem $-\nu\Delta u + \operatorname{grad} p = f$, $\operatorname{div} u = 0$ with very small or large ν. The blocks of L_Ω are $L_{11} = -\nu\Delta$, $L_{12} = \operatorname{grad}$, $L_{21} = -\operatorname{div}$, $L_{22} = 0$. The weighting factors should be $c_1 = c_1' = \text{const}/\sqrt{\nu}$ and $c_2 = c_2' = \text{const} \cdot \sqrt{\nu}$. Then the multi-grid convergence rate is independent of ν.

A further improvement can be achieved by the semi-iterative approach discussed in § 6.2.5.

Remark 11.2.12. Let $\mathscr{S}_l^{(\nu)}(u_l, f_l; \boldsymbol{\omega})$ be the semi-iterative process defined by (16) with ω_l^2 replaced by $\omega_\mu/\|\hat{L}_l\|^2$, $\mu = 1, \ldots, \nu$, where ω_μ is taken from (6.2.50 b).

Then the following estimate holds:

$$\| L_l S_l^{(\nu)}(\omega) \|_{\mathscr{F} \leftarrow \mathscr{U}} \leqq C_L/(2\nu + 1) \quad \text{for all } \nu \geqq 0, \quad l \geqq 0.$$

In the symmetric case of $L_l = L_l^*$ the preceding iteration is equivalent to the semi-iterative process with 2ν steps of the form

$$\mathscr{S}_l(u_l, f_l; \omega_\mu) := u_l - (\omega_\mu / \| \hat{L}_l \|) H_l^2 (L_l u_l - f_l), \tag{11.2.18}$$

where $\omega_1 = -\omega_2$, $\omega_3 = -\omega_4$, etc. (cf. Exercise 3.9.1). This choice is optimal if only the bounds $-C_L I \leqq \hat{L}_l \leqq C_L I$ are known. If $-C_1 I \leqq \hat{L}_l \leqq C_2 I$ with $C_1 \neq C_2$, other parameters ω_μ might be optimal for (18).

In Example 11.2.8 we stated the approximation property of a finite element discretisation of Stokes' equations. The latter equation is also studied by Verfürth [2]. He chooses the same multi-grid method, in particular, the same smoother (16). In his proof the smoothing property takes the form $|||S_l^\nu u_l|||_2 \leqq \eta(\nu) |||u_l|||_0$, with specially defined norms $|||\cdot|||_0$, $|||\cdot|||_2$. However, since $|||v_l|||_2 = \| L_l v_l \|_{\mathscr{F}} / h_l$ and $|||v_l|||_0 = h_l \| v_l \|_{\mathscr{U}}$, with $\|\cdot\|_{\mathscr{U}}$ and $\|\cdot\|_{\mathscr{F}}$ from (9), his result coincides with (17). His conclusion is stated in

Proposition 11.2.13 (Verfürth [2]). Consider Stokes' equations (1.8) in a domain Ω smooth enough to satisfy the regularity assumption $|u|_2 + |p|_1 \leqq C |f|_0$ (cf. Kellogg-Osborn [1]). Let $\mathscr{H}_{l,1} \subset H_0^1(\Omega)^d$ and $\mathscr{H}_{l,2} \subset L^2(\Omega)/\mathbb{C}$ be finite element subspaces with

$$\inf_{v \in \mathscr{H}_{l,1}} |u - v|_s \leqq C h_l^{t-s} |u|_t$$

$$\text{for all } u \in H^t(\Omega)^d \cap H_0^1(\Omega)^d, \qquad 0 \leqq s \leqq 1 \leqq t \leqq 2,$$

$$\inf_{q \in \mathscr{H}_{l,2}} |p - q|_s \leqq C h_l^{t-s} |p|_t \quad \text{for all } p \in H^t(\Omega)/\mathbb{C}, \qquad 0 \leqq s \leqq t \leqq 1,$$

$$|u|_1 \leqq C h_l^{-1} |u|_0 \quad \text{for all } u \in \mathscr{H}_{l,1},$$

$$\sup\{|b(p,u)| : u \in \mathscr{H}_{l,1}, |u|_1 = 1\} \geqq \frac{1}{C} |p|_0 \quad \text{for all } p \in \mathscr{H}_{l,2}, \quad l \geqq 0$$

with $b(\cdot,\cdot)$ from (1.14 a) and Sobolev norms $|\cdot|_s$ of order s. Define the multi-grid iteration by (4.1.1) with $\nu_1 = \nu$, $\nu_2 = 0$, and $\mathscr{S}_l$ from (16). Then the smoothing and approximation properties hold with respect to the norms (9) and imply the multi-grid convergence for suitable $\nu \geqq \underline{\nu}$.

The assumptions of Proposition 11.2.13 are, e.g., satisfied for Example 11.2.8. Unfortunately, Brezzi's condition $\sup\{|b(p,u)|: u \in \mathscr{H}_{l,1}, |u|_1 = 1\} \geqq \frac{1}{C} |p|_0$ is violated if $\mathscr{H}_{l,1}$ ($\mathscr{H}_{l,2}$) consists of piecewise linear (constant) functions over triangles of a common triangulation $\mathscr{T}_l$ (cf. Johnson-Pitkäranta [1]). Pitkäranta-Saarinen [1] confirm by numerical examples that in this case the multi-grid algorithm yields poor convergence. Moreover, they prove that this choice of finite element spaces still leads to multi-grid convergence, when the second equation of the variational formulation (1.13), (1.14 a) is changed to

$$b(u_1, v_2) + d(u_2, v_2) = (f_2, v_2),$$

where the form $d(\cdot\,,\cdot)$ is specially defined.

Concerning the choice of the Uzawa method as a smoothing iteration we refer to a recent paper of Maitre-Musy-Nigon [1].

11.2.3 Nonlinear Systems

Let $\mathscr{L}_l(u_l) = f_l$ be a discretisation of a nonlinear elliptic system and define $L_l(u_l) := \mathscr{L}_l'(u_l)$ as in (9.1.4). The nonlinear variant of iteration (16) is

$$\mathscr{S}_l(u_l, f_l) = u_l - \omega_l^2 H_l'^2 L_l^*(u_l) H_l^2 (\mathscr{L}_l(u_l) - f_l), \tag{11.2.19}$$

since the derivative $S_l = \partial \mathscr{S}_l(u_l^*, 0)/\partial u_l$ (cf. (9.5.6 a)) equals S_l from (16′). If $H_l' = H_l$ and if $L_l := L_l(u_l)$ and L_l^* have the same principal terms or even $L_l = L_l^*$, the computation of L_l can be avoided by the choice

$$\mathscr{S}_l(u_l, f_l; \omega_\mu) = u_l - \omega_\mu H_l^2 (\mathscr{L}_l(u_l) - f_l), \tag{11.2.20}$$

which is the nonlinear analogue of iteration (18).

In the case of the Navier-Stokes equations (1.9) the linearisation of $\mathscr{L}_\Omega$ equals

$$L_\Omega(u) = \begin{bmatrix} -\nu\Delta + \delta L_{11}(u) & \operatorname{grad} \\ -\operatorname{div} & 0 \end{bmatrix}, \tag{11.2.21}$$

where $(\delta L_{11} v)_i = \sum_{j=1}^{d} (u_j \partial v_i/\partial x_j + v_j \partial u_i/\partial x_j)$ is a lower order term. The principal term is the Stokes operator. The analysis of § 9.5.2 would yield multi-grid convergence provided h_0 is small enough. However, a difficulty arises for small ν. From Example 11.2.11 the iterations (19) and (20) are well-suited for the principal part of operator (21), but δL_{11} leads to a singular perturbation problem. As pointed out in § 8, the modifications of the Jacobi iteration do not work in the singular perturbation case.

11.3 The Distributive Relaxation Smoother

In § 11.1.2 we split the Poisson equation into the Cauchy-Riemann system with $L_\Omega = \begin{bmatrix} \partial/\partial x & \partial/\partial y \\ \partial/\partial y & -\partial/\partial x \end{bmatrix}$. On the other hand we regain Poisson's equations by squaring L_Ω: $L_\Omega^2 = \begin{bmatrix} \Delta & 0 \\ 0 & \Delta \end{bmatrix}$. For system (1.7) with $L_\Omega = \begin{bmatrix} I & -\Delta \\ -\Delta & 0 \end{bmatrix}$ the multiplication by $\bar{L}_\Omega := \begin{bmatrix} 0 & -\Delta \\ -\Delta & -I \end{bmatrix}$ reproduces the biharmonic operator:

$L_\Omega \bar{L}_\Omega = \begin{bmatrix} \Delta^2 & 0 \\ 0 & \Delta^2 \end{bmatrix}$. Multiplying the Stokes operator $L_\Omega = \begin{bmatrix} -\Delta & \text{grad} \\ -\text{div} & 0 \end{bmatrix}$ by

$$\bar{L}_\Omega = \begin{bmatrix} I & \text{grad} \\ 0 & \Delta \end{bmatrix}, \tag{11.3.1 a}$$

one obtains

$$L_\Omega \bar{L}_\Omega = \begin{bmatrix} -\Delta & 0 \\ -\text{div} & -\Delta \end{bmatrix}, \tag{11.3.1 b}$$

which differs from a diagonal operator only by a lower order term. Instead of solving the system $L_\Omega u_\Omega = f_\Omega$, one can look for a solution of $(L_\Omega \bar{L}_\Omega) w = f_\Omega$, where $L_\Omega \bar{L}_\Omega$ is (almost) diagonal.

Similarly, one can try to replace the discrete problem $L_l u_l = f_l$ by $L_l \bar{L}_l w_l = f_l$, where $L_l \bar{L}_l$ has a diagonal block structure. According to § 11.1.1 such systems are easier to handle. In particular, one can formulate a Gauß-Seidel iteration with respect to the new unknown grid function w_l. This method defines a new iteration with respect to $u_l = \bar{L}_l w_l$, which is called the 'distributive relaxation method' of Brandt-Dinar [1].

First we shall discuss the Stokes problem. Let u_l be partitioned into $u_l = \begin{pmatrix} \mathbf{u}_l \\ p_l \end{pmatrix}$, where $\mathbf{u}_l = (u_{l,1}, u_{l,2})$ and p_l correspond to $\mathbf{u}$ and p from Eq. (1.8). The induced block structure of L_l is $L_l = \begin{bmatrix} -\Delta_l^{\mathbf{u}} & \partial_l \\ \partial_l^* & 0 \end{bmatrix}$, where $\Delta_l^{\mathbf{u}} \mathbf{u}_l = (\Delta_{l,1} u_{l,1}, \Delta_{l,2} u_{l,2})$, $\partial_l p_l = (\partial_{l,1} p_l, \partial_{l,2} p_l)$, $\partial_l^* \mathbf{u}_l = \sum_i \partial_{l,i}^* u_{l,i}$ are discrete counterparts of the entries in $L_\Omega = \begin{bmatrix} -\Delta & \text{grad} \\ -\text{div} & 0 \end{bmatrix}$. The simplest difference scheme can be formulated when the staggered grids of Fig. 11.3.1 are used. $u_{l,i}$ $(i = 1, 2)$ is a grid function on $\Omega_l^i = \{\mathbf{x} \in \Omega : x_j/h \in \mathbb{Z} \text{ for } j \neq i,\ x_i/h - \frac{1}{2} \in \mathbb{Z}\}$, while p_l is defined on $\Omega_l^p = \{\mathbf{x} \in \Omega : \mathbf{x}/h \in \mathbb{Z}^d\}$. In the interior of the grids we may define $\Delta_{l,i} :=$ standard five-point scheme on Ω_l^i (cf. (3.1.15)),

$$(\partial_{l,i} p_l)(\mathbf{x}) = [p_l(\mathbf{x} + h_l \mathbf{e}^i/2) - p_l(\mathbf{x} - h_l \mathbf{e}^i/2)]/h_l \qquad \text{for interior } \mathbf{x} \in \Omega_l^i,$$

$$(\partial_{l,i}^* u_{l,i})(\mathbf{x}) = [u_{l,i}(\mathbf{x} - h_l \mathbf{e}^i/2) - u_{l,i}(\mathbf{x} + h_l \mathbf{e}^i/2)]/h_l \quad \text{for interior } \mathbf{x} \in \Omega_l^p,$$

where $\mathbf{e}^i \in \mathbb{Z}^d$ is the ith unit vector. Near the boundary other discretisations are needed. According to (1 a) we may define

$$\bar{L}_l := \begin{bmatrix} I & \partial_l \\ 0 & \Delta_l^p \end{bmatrix}, \qquad \Delta_l^p\text{: standard five-point scheme on } \Omega_l^p.$$

Set

$$T_l := L_l \bar{L}_l = \begin{bmatrix} -\Delta_l^{\mathbf{u}} & W_{12} \\ \partial_l^* & W_{22} \end{bmatrix}$$

$$\text{with } W_{12} = -\Delta_l^{\mathbf{u}} \partial_l + \partial_l \Delta_l^p, \qquad W_{22} = \partial_l^* \partial_l. \tag{11.3.2}$$

```
2       2       2       2

p   1   p   1   p   1   p   1
2       2       2       2

p   1   p   1   p   1   p   1
```

Fig. 11.3.1. Two-dimensional staggered grid. 1, grid points for $u_{l,1}$; 2, grid points for $u_{l,2}$; p, grid points for p_l

Obviously, one has $W_{12} = 0$ and $W_{22} = \Delta_l^p$ in the interior of Ω (i.e. when applied to p_l vanishing in a certain neighbourhood of Γ). Hence, according to (1 b), T_l coincides with $\begin{bmatrix} -\Delta_l^{\mathbf{u}} & 0 \\ \partial_l^* & -\Delta_l^p \end{bmatrix}$ in the interior.

We consider the well-known Gauß-Seidel iteration for $T_l w_l = f_l$, where $w_l = (w_{l,1}, w_{l,2}, q_l)$ is defined by $\bar{L}_l w_l = u_l$. All components of w_l can be indexed by $\mathbf{x} \in \Omega_l := \Omega_l^1 \cup \Omega_l^2 \cup \Omega_l^p$: $w_l(\mathbf{x}) = w_{l,1}(\mathbf{x})$ if $\mathbf{x} \in \Omega_l^1$ etc. Denote the diagonal entries of T_l by $t_{\mathbf{x}}$. Then the Gauß-Seidel correction at $\mathbf{x} \in \Omega_l$ reads

$$w_l(\mathbf{x}) \mapsto w_l(\mathbf{x}) + \delta w_l(\mathbf{x}) \quad \text{with} \quad \delta w_l(\mathbf{x}) := t_{\mathbf{x}}^{-1}(f_l - T_l w_l)(\mathbf{x}) \tag{11.3.3}$$

and all other components of w_l remain unchanged. By (2), $t_{\mathbf{x}} = 4h_l^{-2}$ except perhaps near the boundary.

The 'distributive relaxation' results, when we write (3) in terms of $u_l = \bar{L}_l w_l$:

(i) Let $\mathbf{x} \in \Omega_l^1$. Define the grid function δw_l by $\delta w_l(\mathbf{y}) = 0$ if $\mathbf{y} \neq \mathbf{x}$, while $\delta w_l(\mathbf{x})$ is given by (3). Set $\delta u_l := \bar{L}_l \delta w_l$ and note that $\delta u_l = \delta w_l$ by definition of $\bar{L}_l$. Since $f_l - T_l w_l$ equals $f_l - L_l u_l$, the Gauß-Seidel step at $\mathbf{x} \in \Omega_l^1$ effects the change

$$u_l(\mathbf{x}) \mapsto u_l(\mathbf{x}) + \delta u_l(\mathbf{x}) \quad \text{with} \quad \delta u_l(\mathbf{x}) = t_{\mathbf{x}}^{-1}(f_l - L_l u_l)(\mathbf{x}), \tag{11.3.4 a}$$

while all other components $u_l(\mathbf{y})$, $\mathbf{y} \neq \mathbf{x}$, are unchanged. Since $t_{\mathbf{x}}$ is also the diagonal entry of L_l, (4 a) represents a usual Gauß-Seidel step applied to $(L_l u_l - f_l)(\mathbf{x}) = 0$.

(ii) The analogous result holds for the correction at $\mathbf{x} \in \Omega_l^2$.

(iii) Let $\mathbf{x} \in \Omega_l^p$. Define $\delta w_l := (0, 0, \delta q_l)$ with $\delta q_l(\mathbf{y}) = 0$ for $\mathbf{y} \neq \mathbf{x}$, $\delta q_l(\mathbf{x}) = \delta w_l(\mathbf{x})$ as in (3). $\delta q_l(\mathbf{x})$ equals

$$\begin{aligned} \delta q_l(\mathbf{x}) &= t_{\mathbf{x}}^{-1}(f_l - T_l w_l)(\mathbf{x}) = t_{\mathbf{x}}^{-1}(f_l - L_l u_l)(\mathbf{x}) \\ &= t_{\mathbf{x}}^{-1}(f_{l,p} - \partial_{l,1} u_{l,1} - \partial_{l,2} u_{l,2})(\mathbf{x}), \end{aligned} \tag{11.3.4 b}$$

where $t_{\mathbf{x}} = 4h_l^{-2}$ for interior $\mathbf{x}$ as mentioned above. δw_l induces the correction $\delta u_l = \bar{L}_l \delta w_l$:

$$\begin{aligned} &u_l \mapsto u_l + \delta u_l \quad \text{with} \\ &\delta u_l = (\delta u_{l,1}, \delta u_{l,2}, \delta p_l), \quad \delta u_{l,i} = \partial_{l,i} \delta q_l, \quad \delta p_l = \Delta_l^p \delta q_l. \end{aligned} \tag{11.3.4 c}$$

Therefore, the following nine components are changed:

$$\begin{aligned} &\delta u_{l,1}(\mathbf{x} \pm h_l \mathbf{e}^1/2) = h_l^{-1} \delta q_l(\mathbf{x}), \quad \delta u_{l,2}(\mathbf{x} \pm h_l \mathbf{e}^2/2) = h_l^{-1} \delta q_l(\mathbf{x}), \\ &\delta p_l(\mathbf{x}) = 4h_l^{-2} \delta q_l(\mathbf{x}), \quad \delta p_l(\mathbf{x} \pm \mathbf{e}^i h_l) = -h_l^{-2} \delta q_l(\mathbf{x}), \quad i = 1, 2. \end{aligned} \tag{11.3.4 c'}$$

In the general case one has to determine a matrix $\bar{L}_l = (\bar{L}_{ij})_{i,j=1,\ldots,n}$ such that $L_l \bar{L}_l$ is suitable for the Gauß-Seidel method. Then the Gauß-Seidel step related to $u_{l,i}(\mathbf{x})$ reads as

- determine $\delta := t_{\mathbf{x}}^{-1}(f_{l,i} - L_{i1} u_{l,1} - \ldots - L_{in} u_{l,n}(\mathbf{x})$, where $t_{\mathbf{x}}$ is the diagonal entry of $\sum_j L_{ij} \bar{L}_{ji}$ at $\mathbf{x}$;
- set $\delta w_i(\mathbf{x}) := \delta$, $\delta w_i(\mathbf{y}) = 0$ for $\mathbf{y} \neq \mathbf{x}$;
- correct u_l by $u_l \mapsto u_l + \delta u_l$ with $\delta u_{l,j} := \bar{L}_{ji} \delta w_i$ $(j = 1, \ldots, n)$.

We conclude with the following comments:

(i) The favourable form of $L_l \bar{L}_l$ is a block-diagonal matrix. At least no principal terms should appear above the diagonal. However, the following difficulties might arise:

(ii) Consider the off-diagonal block W_{12} defined in (2). If Δ_l^u, ∂_l, Δ_l^p are constant difference operators on regular grids, they commute (i.e. $\partial_{l,1} \Delta_l^p = \Delta_{l,1} \partial_{l,1}$, etc.) and prove $W_{12} = 0$. This result does not hold for difference operators with variable coefficients. Replace, e.g., Δ in Stokes' problem by $a(\mathbf{x})\Delta$. Then even in the continuous case $\bar{L}_\Omega = \begin{bmatrix} I & \text{grad} \\ 0 & a\Delta \end{bmatrix}$ yields $L_\Omega \bar{L}_\Omega = \begin{bmatrix} -a\Delta & R \\ -\text{div} & -\Delta \end{bmatrix}$ with $R = -a\Delta\,\text{grad} + \text{grad}(a\Delta) = \text{grad}(a)\Delta$, which is also of second order. Although R is not a principal term of $L_\Omega \bar{L}_\Omega$, the remainder can impair the Gauß-Seidel process, especially if a is not smooth enough.

(iii) If L_l represents a system of constant difference schemes in the interior of Ω, the blocks do not commute because of the irregularity near the boundary. Concerning this defect, some advices are given by Brandt-Dinar [1, p. 129].

(iv) Even for regular discretisations problems may arise. Consider the finite element discretisation of Stokes' equations by a regular triangulation. Approximate $\mathbf{u}$ by linear elements on triangles of size h_l, whereas p is approximated by linear elements on triangles of size $2h_l$ (cf. Example 11.2.8). As a result, Ω_l^p has only about a quarter of the points of Ω_l^1 or Ω_l^2. The operator Δ_l^p in $\bar{L}_l$ has to be defined with respect to the step size $2h_l$. Therefore, W_{12} from (2) cannot vanish.

(v) If L_l is obtained from a finite element discretisation with some irregular triangulation (cf. Fig. 3.8.6), there will be no chance to find a suitable $\bar{L}_l$.

(vi) Concerning the prolongation p for the staggered grids of Fig. 11.3.1 we refer to § 3.8.4.

11.4 An Elliptic Problem Augmented by an Algebraic Equation

11.4.1 The Linear Case

Often a function v and a scalar quantity λ is to be determined by the system

$$L v + \lambda w = g, \tag{11.4.1 a}$$

$$\Phi(v) + \alpha \lambda = \mu, \tag{11.4.1 b}$$

where w, Φ, α, g, μ are given quantities. w is a function, $\Phi(\cdot)$ represents a linear functional, and α, μ are scalar numbers.

The discretisation of (1 a, b) yields the system

$$L_l u_l = f_l \quad \text{with} \quad L_l = \begin{bmatrix} \bar{L}_l & w_l \\ \varphi_l^T & \alpha_l \end{bmatrix}, \quad f_l = \begin{bmatrix} \bar{f}_l \\ \mu_l \end{bmatrix}, \quad u_l = \begin{bmatrix} \bar{u}_l \\ \lambda_l \end{bmatrix}. \tag{11.4.2}$$

The multi-grid solution of the augmented system (2) is as simple as the solution of the reduced equation $\bar{L}_l \bar{u}_l = \bar{f}_l$. Since $\lambda_l \in \mathbb{R}$ or $\mathbb{C}$ for all levels l, this component requires no prolongation, no restriction, and in particular, no smoothing process. Let $\bar{p}$ ($\bar{r}$) be the prolongation (restriction) for the reduced problem $\bar{L}_l \bar{u}_l = \bar{f}_l$. We define

$$p = \begin{bmatrix} \bar{p} & 0 \\ 0 & 1 \end{bmatrix}, \quad r = \begin{bmatrix} \bar{r} & 0 \\ 0 & 1 \end{bmatrix}, \tag{11.4.3}$$

and assume

$$\alpha_l = \alpha_{l-1} = \ldots = \alpha_0, \tag{11.4.4 a}$$

$$\varphi_{l-1}^T = \varphi_l^T \bar{p} \quad \text{for } l = 1, 2, \ldots \tag{11.4.4 b}$$

Hence the coarse-grid vector φ_{l-1} should be defined by $\bar{p}^T \varphi_l$. We recall that the transposed $\bar{p}^T$ and the adjoint $\bar{p}^*$ differ by a constant factor (cf. Exercise 2.7.2).

Let $\bar{u}_l \mapsto \bar{\mathscr{S}}_l(\bar{u}_l, \bar{f}_l)$ be a suitable smoothing iteration for the reduced problem. Define $\mathscr{S}_l$ by

$$u_l = \begin{bmatrix} \bar{u}_l \\ \lambda_l \end{bmatrix} \mapsto \mathscr{S}_l(u_l, f_l) = \begin{bmatrix} \bar{u}_l' \\ \lambda_l \end{bmatrix} \quad \text{with} \quad \bar{u}_l' := \bar{\mathscr{S}}_l(\bar{u}, \bar{f}_l - \lambda_l w_l), \quad f_l = \begin{bmatrix} \bar{f}_l \\ \mu_l \end{bmatrix}. \tag{11.4.5}$$

$\mathscr{S}_l$ is the simplest extension of $\bar{\mathscr{S}}_l$ to the augmented system and leaves λ_l unchanged.

Proposition 11.4.1. Assume (3), (4 a, b) and define the multi-grid algorithm by (4.1.1) with $\nu_2 = 0$, $\gamma \geqq 1$. Then all iterates u_l^i satisfy the additional equation $[\varphi_l^T, \alpha_l] u_l = \mu_l$ exactly.

Proof. The assertion holds for $l = 0$ because of (4.1.1 a). Let the assertion be valid for $l - 1$. The pre-smoothing at level l yields the defect

$$d_{l-1} = r\, d_l = \begin{bmatrix} \bar{d}_{l-1} \\ \delta_{l-1} \end{bmatrix}, \quad \delta_{l-1} = \varphi_l^T \tilde{u}_l + \alpha_l \tilde{\lambda}_l - \mu_l.$$

By induction, the coarse-grid correction produces $\delta u_l = \begin{bmatrix} \delta w_l \\ \delta \lambda_l \end{bmatrix} := p\, \delta u_{l-1} = \begin{bmatrix} \bar{p}\, \delta w_{l-1} \\ \delta \lambda_{l-1} \end{bmatrix}$, where

$$[\varphi_{l-1}^T, \alpha_{l-1}] \delta u_{l-1} = \varphi_{l-1}^T \delta w_{l-1} + \alpha_{l-1} \delta \lambda_{l-1} = \delta_{l-1}.$$

Therefore, $u_l^{i+1} := \tilde{u}_l - \delta u_l$ satisfies

$$[\varphi_l^T, \alpha_l] u_l^{i+1} = [\varphi_l^T, \alpha_l]\tilde{u}_l - [\varphi_l^T, \alpha_l]\delta u_l = \mu_l + \delta_{l-1} - [\varphi_l^T, \alpha_l]\delta u_l .$$

Since

$$\delta_{l-1} = [\varphi_{l-1}^T, \alpha_{l-1}]\delta u_{l-1} = [\varphi_l^T \bar{p}, \alpha_l]\delta u_{l-1} = [\varphi_l^T, \alpha_l] p\,\delta u_{l-1} = [\varphi_l^T, \alpha_l]\delta u_l$$

by (3) and (4 a, b), Proposition 11.4.1 is proved. □

Let $\mathcal{S}_l^I$ be the smoother described above. If also the λ_l-components should be corrected, one can apply the combination $\mathcal{S}_l := \mathcal{S}_l^{II} \circ \mathcal{S}_l^I$ (i.e. $\mathcal{S}_l(u_l, f_l) := \mathcal{S}_l^{II}(\mathcal{S}_l^I(u_l, f_l), f_l)$), where $\mathcal{S}_l^{II}$ is defined by

$$\mathcal{S}_l^{II}(u_l, f_l) := u_l + \frac{\mu_l - \varphi_l^T \bar{u}_l - \alpha_l \lambda_l}{\varphi_l^T \bar{v}_l + \alpha_l} \begin{bmatrix} \bar{v}_l \\ 1 \end{bmatrix} \quad \text{for } u_l = \begin{bmatrix} \bar{u}_l \\ \lambda_l \end{bmatrix}, \quad f_l = \begin{bmatrix} \bar{f}_l \\ \mu_l \end{bmatrix}. \tag{11.4.6 a}$$

The grid function $\bar{v}_l$ has to satisfy (approximately)

$$\bar{L}_l \bar{v}_l = \bar{w}_l . \tag{11.4.6 b}$$

Note 11.4.2. Denote the defects of u_l and $u_l' = \mathcal{S}_l^{II}(u_l, f_l)$ by $d_l = \begin{bmatrix} \bar{d}_l \\ \delta_l \end{bmatrix}$ and $d_l' = \begin{bmatrix} \bar{d}_l' \\ \delta_l' \end{bmatrix}$, respectively. $\mathcal{S}_l^{II}$ is defined such that $\delta_l' = 0$, while $\bar{d}_l' = \bar{d}_l$. The denominator $\varphi_l^T \bar{v}_l + \alpha_l$ in (6 a) does not vanish if L_l is regular and if $\bar{v}_l$ satisfies (6 b).

11.4.2 The Nonlinear Case

The nonlinear counterpart of system (1 a, b) is

$$\mathcal{L}(v, \lambda) = g , \tag{11.4.7 a}$$

$$\Phi(v, \lambda) = \mu . \tag{11.4.7 b}$$

For instance, eigenvalue problems can be formulated by (7 a, b) (cf. § 12.2.1). A further example will be mentioned in § 13.

The discrete problem is denoted by $\mathcal{L}_l(u_l) = f_l$ with

$$u_l = \begin{bmatrix} \bar{u}_l \\ \lambda_l \end{bmatrix}, \quad f_l = \begin{bmatrix} \bar{f}_l \\ \mu_l \end{bmatrix}, \quad \bar{\mathcal{L}}_l(\bar{u}_l, \lambda_l) = \bar{f}_l , \quad \Phi_l(\bar{u}_l, \lambda_l) = \mu_l .$$

The construction of φ_{l-1} and α_{l-1} by (4 a, b) can be translated into the nonlinear situation. Let $(\tilde{u}_l, \tilde{\lambda}_l)$ and $(\tilde{u}_{l-1}, \tilde{\lambda}_{l-1})$ be given approximations.

The given functional Φ_{l-1} can be adjusted to Φ_l by the modification

$$\hat{\Phi}_{l-1}(\bar{u}_{l-1}, \lambda_{l-1}) := \Phi_{l-1}(\tilde{u}_{l-1}, \tilde{\lambda}_{l-1}) + \frac{\partial \Phi_l(\tilde{u}_l, \tilde{\lambda}_l)}{\partial \bar{u}_l} p(\bar{u}_{l-1} - \tilde{u}_{l-1}) + \frac{\partial \Phi_l(\tilde{u}_l, \tilde{\lambda}_l)}{\partial \lambda_l}(\lambda_{l-1} - \tilde{\lambda}_{l-1}). \tag{11.4.8}$$

Since $\hat{\Phi}_{l-1}$ is affine, it is very easy to define $\hat{\Phi}_{l-2}, \ldots, \hat{\Phi}_0$ in the same manner. Proposition 11.4.1 remains almost true: A multi-grid step without post-smoothing using $\hat{\Phi}_k$ for $k < l$, yields u_l' so that $\Phi_l(\bar{u}_l', \lambda_l')$ differs from the desired right-hand side μ_l only by $O(\|\bar{u}_l - \tilde{u}_l\|_{\mathscr{U}}^2 + |\lambda_l - \tilde{\lambda}_l|^2)$.

It is also possible to extend definition (6) by

$$\mathscr{S}_l^{II}(u_l, f_l) := u_l + \frac{\mu_l - \Phi_l(\bar{u}_l, f_l)}{\varphi_l^T \bar{v}_l + \alpha_l} \begin{bmatrix} \bar{v}_l \\ 1 \end{bmatrix}, \tag{11.4.9}$$

if $\varphi_l \approx \partial \Phi_l / \partial \bar{u}_l$, $\alpha_l \approx \partial \Phi_l / \partial \lambda_l$ are available and $(\partial \bar{\mathscr{L}}_l / \partial \bar{u}_l) \bar{v}_l \approx \partial \bar{\mathscr{L}}_l / \partial \lambda_l$.

12. Eigenvalue Problems and Singular Equations

12.1 Discussion of the Problems

We shall study two closely related problems: the eigenvalue problem $L_l u_l = \lambda_l u_l$, and the system $L_l u_l = f_l$ with a singular matrix L_l.

12.1.1 Eigenvalue Problem

Under usual assumptions the inverse operator L^{-1} of the elliptic equation $Lu = f$ is compact and the Riesz-Schauder theory (cf. Yosida [1, p. 283]) ensures the existence of isolated eigenvalues λ and eigenvectors e with

$$Le = \lambda e. \tag{12.1.1}$$

The discretisation of Eq. (1) is written

$$L_l e_l = \lambda_l e_l. \tag{12.1.2}$$

Concerning the approximation of λ and e by λ_l and e_l we refer to Chatelin [1].

We recall that the left eigenvector e_l^* is the non-trivial solution of

$$L_l^* e_l^* = \overline{\lambda_l} e_l^*. \tag{12.1.3}$$

The *eigenspaces* are denoted by

$$\mathcal{N}_l(\lambda_l) = \{e_l \text{: solution of (2)}\}, \qquad \mathcal{N}_l^*(\lambda_l) := \{e_l^* \text{: solution of (3) with } \overline{\lambda_l}\}.$$

If λ_l is fixed, we write $\mathcal{N}_l$, $\mathcal{N}_l^*$. The geometrical multiplicity of an eigenvalue λ_l is defined by $\dim(\mathcal{N}_l(\lambda_l)) = \dim(\mathcal{N}_l^*(\lambda_l))$, whereas the algebraic multiplicity is the multiplicity of the root λ_l in the characteristic polynomial $\det(L_l - \lambda I)$. Both coincide for symmetric $L_l = L_l^*$. λ_l is called a single eigenvalue if its algebraic (and hence also the geometrical) multiplicity is 1.

Note 12.1.1. (i) Suppose $\lambda_l \neq \mu_l$. Then $\langle e_l, e_l^* \rangle = 0$ for all $e_l \in \mathcal{N}_l(\lambda_l)$, $e_l^* \in \mathcal{N}_l^*(\mu_l)$.

(ii) Let k be the geometric and algebraic multiplicity of λ_l. There are eigenvectors e_{jl}, e_{jl}^*, $1 \leqq j \leqq k$, spanning $\mathcal{N}_l(\lambda_l)$ and $\mathcal{N}_l^*(\lambda_l)$, respectively, such that $\langle e_{il}, e_{jl}^* \rangle = \delta_{ij}$ (δ_{ij}: Kronecker's symbol).

We recall the computation of an eigenvalue approximation from an eigenvector approximation $\tilde{e}_l$ by means of the *Rayleigh quotient*

$$\Lambda_l(\tilde{e}_l) := \langle L_l \tilde{e}_l, \tilde{e}_l \rangle / |\tilde{e}_l|_0^2 \quad \text{where } |\tilde{e}_l|_0^2 = \langle \tilde{e}_l, \tilde{e}_l \rangle. \tag{12.1.4 a}$$

Lemma 12.1.2. Assume $e_l \in \mathcal{N}_l(\lambda_l)$ and $\delta_e := \min\{|e_l - \alpha \tilde{e}_l|_0 : \alpha \in \mathbb{C}\}$. The eigenvalue error estimation

$$|\lambda_l - \Lambda_l(\tilde{e}_l)| \leqq C\,\delta_e \tag{12.1.5 a}$$

holds in the general case, while

$$|\lambda_l - \Lambda_l(\tilde{e}_l)| \leqq C\,\delta_e^2 \tag{12.1.5 b}$$

holds for symmetric $L_l = L_l^*$ or, more generally, for normal L_l (i.e. $L_l^* L_l = L_l L_l^*$).

The quadratic estimate (5 b) can also be achieved in the general case, provided that an approximation $\tilde{e}_l^*$ of the corresponding left eigenvector e_l^* is available.

Lemma 12.1.3. Assume $e_l = e_l^{(j)}$, $e_l^* = e_l^{(j)*}$ for some $j \in \{1, \dots, k\}$, where $e_l^{(j)}$, $e_l^{(j)*}$, and k are defined in Note 12.1.1 (ii). Define the errors δ_e and δ_e^* of $\tilde{e}_l$ and $\tilde{e}_l^*$ as in Lemma 12.1.2. Then the generalised Rayleigh quotient

$$\Lambda_l(\tilde{e}_l, \tilde{e}_l^*) := \langle L_l \tilde{e}_l, \tilde{e}_l^* \rangle / (|e_l|_0\,|e_l^*|_0) \tag{12.1.4 b}$$

yields an eigenvalue approximation satisfying

$$|\lambda_l - \Lambda_l(\tilde{e}_l, \tilde{e}_l^*)| \leqq C\,\delta_e\,\delta_e^*. \tag{12.1.5 c}$$

12.1.2 Singular Equations

Let L_l be a singular matrix and consider the problem $L_l u_l = f_l$.

Lemma 12.1.4. Let $\mathcal{N}_l = \mathcal{N}_l(0)$, $\mathcal{N}_l^* = \mathcal{N}_l^*(0)$ and suppose $f_l \perp \mathcal{N}_l^*$, i.e. $\langle f_l, v_l \rangle = 0$ for all $v_l \in \mathcal{N}_l^*$. Then the equation $L_l u_l = f_l$ is well-posed and has a unique solution $u_l \perp \mathcal{N}_l^*$:

$$L_l u_l = f_l, \quad f_l \perp \mathcal{N}_l^*, \quad u_l \perp \mathcal{N}_l^*. \tag{12.1.6}$$

Such singular equations have already appeared in previous chapters. Often, the solution is determined up to a constant; i.e.

$$\mathcal{N}_l = \mathcal{N}_l^* = \{u_l\text{: constant grid function}\} \tag{12.1.7 a}$$

in Eq. (6). Examples are the Poisson equation $\Delta u = f$ with Neumann or periodic boundary conditions (cf. § 4.4.3). In Stokes' problem (11.1.8) the pressure p is determined up to a constant. Consequently, the integrability condition $f_l \perp \mathcal{N}_l^*$ must be met. In the former case, this condition reads $0 = \langle \mathbb{1}_l, f_l \rangle = h_l^d \sum_{\mathbf{x} \in \Omega_l} f_l(\mathbf{x})$, where $\mathbb{1}_l$ is the grid function with constant value 1.

In the case of (7 a), the multi-grid computation of Eq. (6) yields no complication because of

Remark 12.1.5. Suppose (7 a), $f_l \perp \mathbb{1}_l$ and

$$L_l \mathbb{1}_l = 0, \quad p \mathbb{1}_{l-1} = \mathbb{1}_l, \quad r^* \mathbb{1}_{l-1} = \mathbb{1}_l, \quad S_l \mathbb{1}_l = \sigma_l \mathbb{1}_l \text{ for some } \sigma_l. \tag{12.1.7 b}$$

Then the usual multi-grid algorithm (4.1.1) does not deteriorate in spite of the singularity of L_l. The convergence $u_l^j \to u_l$ holds modulo $\mathcal{N}_l$, i.e. with respect to the norm of the quotient space $\mathcal{U}_l/\mathcal{N}_l$. In particular, $u_l^0 \perp \mathbb{1}_l$ implies $u_l^i \perp \mathbb{1}_l$, provided that the solution of $L_0 u_0 = f_0$ in (4.1.1. a) satisfies $u_0 \perp \mathbb{1}_0$ according to (6).

Corollary 12.1.6. Suppose (7 a, b) and, in addition,

$$r \mathbb{1}_l = \mathbb{1}_{l-1}, \qquad T_l \mathbb{1}_l = \tau_l \mathbb{1}_l \quad (T_l \text{ from (2.2.6 a)}, \tau_l \in \mathbb{C}). \tag{12.1.7 c}$$

Even the condition $f_l \perp \mathbb{1}_l$ can be omitted, since for any $\alpha \in \mathbb{C}$ the right-hand side $f_l + \alpha \mathbb{1}_l$ yields the same multi-grid iterates u_l^i modulo $\mathcal{N}_l$.

In the case of (7 a) it is sometimes recommended to remove the singularity of $L_l u_l = f_l$ by fixing u_l at some $\mathbf{x}_0 \in \Omega_l$ and to solve the reduced equation $\bar{L}_l \bar{u}_l = \bar{f}_l$, where $\bar{u}_l$ is defined on $\Omega_l \setminus \{\mathbf{x}_0\}$. This approach is not suited for multi-grid methods. The artificial equation $u_l(\mathbf{x}_0) = c$ is a condition on a boundary, which degenerates to the isolated point $\mathbf{x}_0$. Such a condition yields bad regularity and causes the convergence speed to deteriorate considerably.

If $\mathcal{N}_l$ and $\mathcal{N}_l^*$ are not of the special form (7 a), conditions (7 b, c) cannot be expected for $e_l \in \mathcal{N}_l$, $e_l^* \in \mathcal{N}_l^*$ instead of $\mathbb{1}_l$. Therefore, the following problems arise:

(i) Iteration errors in $\mathcal{N}_l$ will not be reduced during the multi-grid process, but
(ii) error components in $\mathcal{N}_l$ induce new errors orthogonal to $\mathcal{N}_l^*$.
(iii) The coarse-grid equation reads $L_{l-1} u_{l-1} = f_{l-1} := r(L_l u_l - f_l)$. Even if $u_l \perp \mathcal{N}_l^*$, $f_l \perp \mathcal{N}_l^*$, the condition $f_{l-1} \perp \mathcal{N}_{l-1}^*$ of (6) is violated unless $r^* \mathcal{N}_{l-1}^* = \mathcal{N}_l^*$.

In practice, the matrix L_l will not be singular but almost singular; L_l is to be replaced by $L_l + \varepsilon I$ with small ε:

$$(L_l + \varepsilon I) u_l = f_l, \quad u_l \perp \mathcal{N}_l^*, \quad f_l \perp \mathcal{N}_l^*. \tag{12.1.8}$$

Even this problem is well-posed as stated in

Lemma 12.1.7. For all $|\varepsilon| \leqq \varepsilon_0$, with ε_0 sufficiently small, Eq. (8) admits a unique solution u_l with $|u_l|_0 \leqq C |f_l|_0$, where C is independent of ε.

Proof. The geometrical sum $u_l = \sum_{i=0}^{\infty} (-\varepsilon L_l^{-1})^i L_l^{-1} f_l$ is well-defined. □

Appropriate multi-grid solutions of Eq. (6) or Eq. (8) are described in § 12.2.2 and § 12.3.5.

12.2 Reformulation of the Problems

12.2.1 Reformulation of the Eigenvalue Problem as a Nonlinear Equation

Assume that λ_l is a single eigenvalue with eigenvector e_l. To obtain a unique solution (e_l, λ_l) one can choose various normalisations of e_l, e.g. (1 a) or (1 b):

$$|e_l|_0^2 = 1\,, \tag{12.2.1 a}$$

$$\langle e_l, w_l \rangle = 1 \tag{12.2.1 b}$$

for some $w_l \notin \mathcal{N}_l$. In the latter case, e_l is uniquely determined, whereas (1 a) has two solutions $\pm e_l$. In general, the normalisation condition is of the form

$$\varphi_l(e_l) = 0\,, \tag{12.2.1 c}$$

where φ_l is an affine or nonlinear, differentiable mapping from $\mathcal{U}_l$ into $\mathbb{C}$. Hence, the eigenvalue problem (1.2) can be written as

$$\mathcal{L}_l(u_l) := \begin{bmatrix} L_l e_l - \lambda_l e_l \\ \varphi_l(e_l) \end{bmatrix} = 0, \qquad u_l = \begin{bmatrix} e_l \\ \lambda_l \end{bmatrix}. \tag{12.2.2}$$

The nonlinear system (2) is well-posed because of

Remark 12.2.1. Let $e_l \neq 0$ and λ_l be a solution of Eq. (2) and assume that λ_l is a single eigenvalue. Suppose $w_l^T e_l \neq 0$ for $w_l^T := \partial \varphi_l(e_l)/\partial e_l$. Then $u_l = (e_l, \lambda_l)$ is an isolated solution of Eq. (2). The derivative $\mathcal{L}_l'(u_l)$ is non-singular.

Equation (2) can be solved as discussed in § 11.4.2. The functional $\hat{\Phi}_{l-1}$ from (11.4.8) equals

$$\hat{\Phi}_{l-1}(e_{l-1}, \lambda_{l-1}) = |\tilde{e}_{l-1}|_0^2 - 1 + 2\langle p^* e_l, e_{l-1} - \tilde{e}_{l-1} \rangle \tag{12.2.3 a}$$

in the case of (1 a), and is given by (3 b) in the case of (1 b):

$$\hat{\Phi}_{l-1}(e_{l-1}, \lambda_{l-1}) = \langle \tilde{e}_{l-1}, w_{l-1} \rangle - 1 + \langle e_{l-1} - \tilde{e}_{l-1}, p^* w_l \rangle. \tag{12.2.3 b}$$

A good choice of φ_l is (1 b) where w_l is some approximation to the desired eigenvector e_l (cf. Ruge [1], Brandt-McCormick-Ruge [1]). On the coarser grids we may define $w_{l-1} = p^* w_l$, $w_{l-2} = p^* w_{l-1}$, etc. according to (3 b).

The additional scalar equation in system (2) can be avoided. Let $\varphi_l(e_l)$ be a linear or nonlinear functional and normalise e_l by

$$\varphi_l(e_l) = \lambda_l, \tag{12.2.4}$$

where again $\varphi_l(e_l) := |e_l|_0^2 - \text{const}$, $\varphi_l(e_l) := \langle e_l, w_l \rangle - \text{const}$, but also

$$\varphi_l(e_l) := e_l(\mathbf{x}) - \text{const} \quad \text{for some fixed } \mathbf{x} \in \Omega_l$$

are possible choices. φ_l should be differentiable. Eliminating λ_l in $L_l e_l = \lambda_l e_l$ by (4), one obtains the new nonlinear equation

$$\mathcal{L}_l(e_l) := L_l e_l - \varphi_l(e_l)\, e_l = 0\,. \tag{12.2.5}$$

The equivalence of Eq. (5) to eigenvalue problem (1.2) is stated in

Note 12.2.2. (i) Let $e_l \neq 0$ be an eigenvector of L_l corresponding to the eigenvalue λ_l. Assume that λ_l is in the range $\{\varphi_l(\alpha e_l): \alpha \in \mathbb{R}$ or $\alpha \in \mathbb{C}$, respectively$\}$. Then, after a suitable normalisation $\hat{e}_l := \alpha e_l$, $\hat{e}_l$ is a solution of Eq. (5) with $\varphi_l(e_l) = \lambda_l$. One should choose φ_l such that the scaling factor α does not vanish: $\varphi_l(0) \neq \lambda_l$.

(ii) Let $e_l \neq 0$ be a solution of Eq. (5). Then e_l is an eigenvector corresponding to the eigenvalue $\lambda_l := \varphi_l(e_l)$.

The well-posedness of Eq. (5) is ensured by

Remark 12.2.3. Let e_l with $\varphi_l'(e_l)\, e_l \neq 0$ be a solution of Eq. (5) and suppose that $\lambda_l = \varphi_l(e_l)$ is a single eigenvalue. Then e_l is an isolated solution and $\mathscr{L}_l'(e_l)$ is non-singular.

Equation (5) is a nonlinear problem which can be solved as described in § 9. Note that the linearised problem is weakly indefinite.

A recommendable modification of the prolongation $\tilde{p}$ in the nested iteration (9.3.14) is

$$\tilde{u}_k := \tilde{\tilde{p}}\,(\tilde{u}_{k-1}) := \sigma\, \tilde{p}\, \tilde{u}_{k-1} \quad \text{with } \sigma \text{ such that } \varphi_k(\tilde{u}_k) = \varphi_{k-1}(\tilde{u}_{k-1})$$

in (9.3.14 b), since then the eigenvalue approximation $\tilde{\lambda}_k = \varphi_k(\tilde{u}_k)$ coincides with $\tilde{\lambda}_{k-1}$ at level $k-1$. $\tilde{\lambda}_k$ can also be obtained by extrapolation from $\tilde{\lambda}_{k-1}$ and $\tilde{\lambda}_{k-2}$ (cf. Hackbusch [18]).

Note 12.2.4. If λ_l is a double eigenvalue, one can use the formulation (2) or (5) with an additional scalar equation $\langle e_l, v_l\rangle = 0$, where $v_l \notin \mathcal{N}_l^{\perp}$. However, often higher multiplicities occur due to geometrical reasons and for the same reasons one-dimensional subspaces of $\mathcal{N}_l$ can be characterised by a symmetry condition $\Sigma_l e_l = e_l$, where e.g. $\Sigma_l u_l(x, y) := -u_l(1-x, y)$ for $0 \leqq x,\ y \leqq 1$, or $\Sigma_l u_l(x, y) := u_l(y, x)$. If these symmetries are preserved by the discretisation, the restriction, the prolongation, and the smoothing iteration (i.e. $L_l \Sigma_l = \Sigma_l L_l$, $r\,\Sigma_l = \Sigma_{l-1} r$, etc.), a starting value e_l^0 with $\Sigma_l e_l^0 = e_l^0$ yields e_l^i satisfying the same symmetry condition and the iteration is not influenced by the existence of further eigenvectors.

12.2.2 Reformulation of the Singular Equation

Assume that $\lambda_l = 0$ is a single eigenvalue of the matrix L_l. The condition $u_l \perp \mathcal{N}_l^*$, which is added in Eq. (1.6) to ensure uniqueness of u_l, can be replaced by $u_l \perp v_l$ for any v_l with $\langle v_l, e_l\rangle \neq 0$, $e_l \in \mathcal{N}_l$. The resulting equations are

$$L_l u_l = f_l, \quad f_l \perp \mathcal{N}_l^*, \quad u_l \perp v_l. \tag{12.2.6}$$

For $v_l := e_l^* \in \mathcal{N}_l^*$ we regain Eq. (1.6).

Lemma 12.2.5. Assume that $\mathcal{N}_l = \text{span}\{e_l\}$ and $\langle v_l, e_l\rangle \neq 0$. Then problem (6) has a unique solution u_l, which differs from the solution u_l' of Eq. (1.6) only by $u_l' - u_l \in \mathcal{N}_l$.

The augmented system

$$\hat{L}_l \hat{u}_l = \hat{f}_l, \quad \hat{L}_l = \begin{bmatrix} L_l & w_l \\ v_l^T & 0 \end{bmatrix}, \quad \hat{u}_l = \begin{bmatrix} u_l \\ \mu_l \end{bmatrix}, \quad \hat{f}_l = \begin{bmatrix} f_l \\ 0 \end{bmatrix} \tag{12.2.7}$$

is equivalent to problem (6) in the following sense:

Lemma 12.2.6. Assume that $\mathcal{N}_l = \text{span}\{e_l\}$, $\mathcal{N}_l^* = \text{span}\{e_l^*\}$, $\langle v_l, e_l\rangle \neq 0$, $\langle w_l, e_l^*\rangle \neq 0$. Then the matrix $\hat{L}_l$ from (7) is non-singular. If $f_l \perp \mathcal{N}_l^*$, the solutions of Eq. (6) and Eq. (7) are related by $\hat{u}_l = \begin{bmatrix} u_l \\ 0 \end{bmatrix}$.

Since $\hat{L}_l$ is non-singular, Eq. (7) makes sense even if $f_l \perp \mathcal{N}_l^*$ is violated, or if the singular submatrix L_l is replaced by the almost singular matrix $L_l + \varepsilon I$ with $|\varepsilon|$ sufficiently small.

The multi-grid solution of $\hat{L}_l \hat{u}_l = \hat{f}_l$ is described in § 11.4.1.

12.3 Direct Multi-Grid Approach to Eigenvalue Problems

12.3.1 Derivation of a Two-Grid Iteration

A direct multi-grid solution of $L_l e_l = \lambda_l e_l$ is described by Astrachancev [2], Frederickson [3], Strachovskaja [1], and Hackbusch [11]. Before we explain this method, we recall the inverse iteration or Wielandt's iteration, which is quite different from ours:

$$\begin{aligned} &\lambda_l^{(j)} = \Lambda_l(u_l^{j-1}); \quad \hat{u}_l^j := L_l^{(j)^{-1}} u_l^{j-1} \quad \text{with} \quad L_l^{(j)} := L_l - \lambda_l^{(j)} I; \\ &u_l^j := \hat{u}_l^j / |\hat{u}_l^j|_0, \end{aligned} \tag{12.3.1}$$

where Λ_l denotes the Rayleigh quotient (1.4 a). It seems obvious to apply the multi-grid method to the linear problems $L_l^{(j)} u_l = f_l$ with the almost singular matrices $L_l^{(j)}$ (cf. Wachspress [1]). However, note that these equations are completely different from Eq. (1.6). In Eq. (1.6) the right-hand side is orthogonal to $\mathcal{N}_l^*$, whereas now $f_l = u_l^{j-1}$ is almost in $\mathcal{N}_l$. The convergence of Wielandt's iteration is based upon the fact that the $\mathcal{N}_l$-component in u_l^{j-1} is greatly increased by the iteration step $u_l^{j-1} \mapsto \hat{u}_l^j = L_l^{(j)^{-1}} u_l^{j-1}$. In contrast to problem (1.6), a multi-grid iteration applied to $L_l^{(j)} \hat{u}_l^j = u_l^{j-1}$ has to treat the $\mathcal{N}_l$-component correctly, which is a very hard task. Bank [3] analyses the case of a finite element discretisation with symmetric $L_l = L_l^* > 0$, where the inequalities $\lambda < \lambda_0 < \lambda_1 < \ldots < \lambda_l$ can be ensured for the first continuous and discrete eigenvalues. He proposes a fixed shift μ_l instead of $\lambda_l^{(j)}$ such that

$\lambda_l - \mu_l > 0$ is large enough. Then multi-grid convergence can be proved, provided that λ is the smallest eigenvalue, λ is single, and h_0 is sufficiently small.

The following method avoids the difficulties sketched above, since the linear problems involved are of the form (1.6). The two-grid version again consists of a smoothing step and a coarse-grid correction. Let $\mathscr{S}_l(u_l, f_l; \lambda)$ be a smoothing iteration for the linear problem $(L_l - \lambda I) u_l = f_l$. The linear mapping $S_l(\lambda) u_l := \mathscr{S}_l(u_l, 0; \lambda)$ serves as smoothing process for the eigenvalue problem $L_l e_l = \lambda_l e_l$. For instance, the damped Jacobi iteration (3.3.7′) yields

$$S_l(\lambda) := I - \omega_l h_l^{2m} (L_l - \lambda I). \tag{12.3.2 a}$$

If $L_l = D_l - A_l - B_l$ according to (3.3.2), the Gauß-Seidel iteration is given by

$$S_l(\lambda) := (D_l - A_l - \lambda I)^{-1} B_l. \tag{12.3.2 b}$$

When λ coincides with the discrete eigenvalue λ_l, one concludes from Eq. (1.3.4) that

$$S_l(\lambda_l) e_l = e_l \quad \text{for } e_l \in \mathscr{N}_l(\lambda_l). \tag{12.3.3}$$

Let $\tilde\lambda$ be an approximation to λ_l. The application of smoothing steps $u_l \mapsto \tilde u_l := S_l^\nu(\tilde\lambda) u_l$ has the same effect as for linear equations: High frequency errors of the eigenvector approximation u_l are reduced, while, in particular, the desired eigenvector component is nearly unchanged (cf. (3) in the case of $\tilde\lambda = \lambda_l$).

For the coarse-grid correction we need the defect

$$d_l := (L_l - \tilde\lambda I) \tilde u_l. \tag{12.3.4 a}$$

Assuming $\mathscr{N}_l = \operatorname{span}\{e_l\}$, $\mathscr{N}_l^* = \operatorname{span}\{e_l^*\}$ with $\langle e_l, e_l^* \rangle = 1$ according to Note 12.1.1 (ii), we may split the approximation $\tilde u_l$ into the sum $\alpha e_l + v_l$ with $v_l \perp \mathscr{N}_l^*$. Hence, d_l can be decomposed into

$$d_l = (\lambda_l - \tilde\lambda) \alpha e_l + (L_l - \tilde\lambda I) v_l. \tag{12.3.4 b}$$

The interesting part of d_l is $(L_l - \tilde\lambda I) v_l$. We introduce the projection onto the orthogonal space $\mathscr{N}_l^{*\perp} := \{v_l : v_l \perp \mathscr{N}_l^*\}$:

$$Q_l u_l := u_l - \langle u_l, e_l^* \rangle e_l. \tag{12.3.5}$$

Since $v_l \perp \mathscr{N}_l^*$ implies $(L_l - \tilde\lambda I) v_l \in \mathscr{N}_l^{*\perp}$, the projection of d_l results in

$$d_l^\perp := Q_l d_l = (L_l - \tilde\lambda I) v_l.$$

The exact correction of $\tilde u_l$ would be $\tilde u_l \mapsto \tilde u_l - v_l = \alpha e_l \in \mathscr{N}_l$, where v_l is the solution of the equation

$$(L_l - \tilde\lambda I) v_l = d_l^\perp \quad \text{with} \quad d_l^\perp \perp \mathscr{N}_l^*, \qquad v_l \perp \mathscr{N}_l^*. \tag{12.3.6}$$

By Lemma 12.1.7, Eq. (6) is well-posed, although the matrix $L_l - \tilde\lambda I$ is almost singular as $\tilde\lambda$ approaches λ_l. Again, we emphasize this difference between the linear equation in (1) and Eq. (6). The solution of Eq. (6) is denoted by

$v_l = (L_l - \tilde{\lambda} I)^{-1} d_l^{\perp}$. To indicate the domain $\mathcal{N}_l^{*\perp}$ and the range $\mathcal{N}_l^{*\perp}$ of the inverse operator, we shall write $Q_l (L_l - \tilde{\lambda} I)^{-1} Q_l$ instead of $(L_l - \tilde{\lambda} I)^{-1}$:

$$v_l = Q_l (L_l - \tilde{\lambda} I)^{-1} Q_l d_l^{\perp} = Q_l (L_l - \tilde{\lambda} I)^{-1} Q_l d_l .$$

The exact correction (fine-grid correction) reads $\tilde{u}_l \mapsto \tilde{u}_l - v_l$ or

$$\tilde{u}_l \mapsto \tilde{u}_l - Q_l (L_l - \tilde{\lambda} I)^{-1} Q_l (L_l - \tilde{\lambda} I) \tilde{u}_l = (I - Q_l) \tilde{u}_l : \mathcal{U}_l \to \mathcal{N}_l .$$

Transferring the correction process to the coarse grid, we obtain

$$\tilde{u}_l \mapsto \tilde{u}_l - p Q_{l-1} (L_{l-1} - \tilde{\lambda} I)^{-1} Q_{l-1} r (L_l - \tilde{\lambda} I) \tilde{u}_l . \tag{12.3.7}$$

The resulting two-grid method reads as follows:

two-grid iteration for the eigenvalue problem $L_l e_l = \lambda_l e_l$ (12.3.8)

$\tilde{\lambda} := \Lambda_l (u_l^i);$	(Λ_l from (1.4 a))	(12.3.8 a)
$\tilde{u}_l := S_l^{\nu} (\tilde{\lambda}) u_l^i;$	(smoothing step)	(12.3.8 b)
$d_{l-1} := r (L_l - \tilde{\lambda} I) \tilde{u}_l;$	(coarse-grid correction)	(12.3.8 c$_1$)
$d_{l-1}^{\perp} := Q_{l-1} d_{l-1};$		(12.3.8 c$_2$)
$v_{l-1} := (L_{l-1} - \tilde{\lambda} I)^{-1} d_{l-1}^{\perp} \in \mathcal{N}_{l-1}^{*\perp};$		(12.3.8 d)
$u_l^{i+1} := \tilde{u}_l - p Q_{l-1} v_{l-1}$		(12.3.8 e)

which can be summarised as:

$$\begin{aligned} &\tilde{\lambda} := \Lambda_l (u_l^i); \quad u_l^{i+1} := M_l (\tilde{\lambda}) u_l^i \\ &\text{with } M_l (\lambda) := [I - p Q_{l-1} (L_{l-1} - \lambda I)^{-1} Q_{l-1} r (L_l - \lambda I)] S_l^{\nu} (\lambda) \end{aligned} \tag{12.3.8'}$$

The exact solution in (8 d) will be removed in the multi-grid version. The projection in (8 c$_2$) requires the knowledge of the eigenvectors e_{l-1}, e_{l-1}^*. We shall approximate Q_{l-1} by means of the results obtained from the nested iteration described below.

The behaviour of the two-grid iteration can easily be understood, when $\tilde{\lambda}$ coincides with the exact eigenvalue λ_l. Let $u_l^i = \alpha e_l + v_l$, $v_l \in \mathcal{N}_l^{*\perp}$. Equation (3) implies $M_l (\lambda_l) e_l = e_l$; therefore, the eigenvector component remains unchanged during the iteration. The result $M_l (\lambda_l) v_l$ is similar to the result $M_l v_l$ with the usual two-grid iteration matrix M_l from (6.1.1), since $L_{l-1} - \lambda_{l-1} I$ is non-singular in $\mathcal{N}_l^{*\perp}$. Again, the inequality $\| M_l (\lambda_l) \|_{\mathcal{U} \leftarrow \mathcal{U}} \leqq C(\nu)$, $\lim_{\nu \to \infty} C(\nu) = 0$, can be expected. The convergence analysis in the case of $\tilde{\lambda} \neq \lambda_l$ is described by Hackbusch [11] for the situation where: (i) L_l is obtained from a finite element discretisation, (ii) r and p are the canonical restrictions and prolongation of § 3.6, (iii) $S_l (\lambda)$ is given by (2 a).

12.3.2 Modifications and Generalisations

If $\mathcal{N}_l \neq \mathcal{N}_l^*$, the quadratic eigenvalue approximation requires the Rayleigh quotient (1.4 b). Therefore approximations $\hat{u}_l^i$ of the left eigenvector e_l^* are also to be computed:

Simultaneous two-grid solution of $L_l e_l = \lambda_l e_l, \quad L_l^* e_l^* = \overline{\lambda_l} e_l^*$ (12.3.9)

$\tilde{\lambda} := \Lambda_l(u_l^i, \hat{u}_l^i);$	(Λ_l from (1.4 b))	(12.3.9 a)
$u_l^{i+1} := M_l(\tilde{\lambda})\, u_l^i;$	(M_l from (8′))	(12.3.9 b)
$\hat{u}_l^{i+1} := \hat{M}_l(\tilde{\lambda})\, \hat{u}_l^i;$		(12.3.9 c)

where

$$\hat{M}_l(\lambda) := [I - p\, Q_{l-1}^* (L_{l-1}^* - \bar{\lambda} I)^{-1} Q_{l-1}^*\, r (L_l^* - \bar{\lambda} I)]\, \hat{S}_l^\nu(\bar{\lambda}). \tag{12.3.9 d}$$

The adjoint operator Q_{l-1}^* equals

$$Q_l^* u_l = u_l - \langle u_l, e_l \rangle e_l^* \tag{12.3.10}$$

and is the projection onto $\mathcal{N}_l^\perp$. $\hat{S}_l(\bar{\lambda})$ is a smoothing iteration for $(L_l^* - \bar{\lambda} I) u_l = 0$. The analogue of $S_l(\lambda)$ from (2 a) is $\hat{S}_l(\bar{\lambda}) = S_l(\lambda)^*$, whereas $\hat{S}_l(\bar{\lambda}) \neq S_l(\lambda)^*$ in the case of the Gauß-Seidel smoother (2 b).

Another modification is required for multiple eigenvalues. Assume $\mathcal{N}_l = \mathcal{N}_l(\lambda_l) = \operatorname{span}\{e_{1l}, \ldots, e_{kl}\}$, $\mathcal{N}_l^* = \operatorname{span}\{e_{1l}^*, \ldots, e_{kl}^*\}$, and $\langle e_{il}, e_{jl}^* \rangle = \delta_{ij}$ according to Note 12.1.1 (ii). Then iteration (8) or (9) has to be applied simultaneously to the starting values $\{u_{1l}^0, \ldots, u_{kl}^0\}$, which span $\mathcal{N}_l$. The projection Q_k is defined by

$$Q_k u_k := u_k - \sum_{i=1}^{k} \langle u_k, e_{ik}^* \rangle e_{ik}.$$

In algorithm (8) we chose equal shifts $\tilde{\lambda}$ at the levels l and $l-1$ (see (8 c$_1$) and (8 d)). Since $L_l - \lambda_l I$ is related to $L_{l-1} - \lambda_{l-1} I$ rather than to $L_{l-1} - \lambda_l I$, the optimal shift at level $l-1$ would be $L_{l-1} - \lambda' I$ with $\lambda' - \lambda_{l-1} = \tilde{\lambda} - \lambda_l$, where λ_k denotes the eigenvalue at level k. Since $\tilde{\lambda} \to \lambda_l$ is expected, a good choice is $\lambda' = \lambda_{l-1}$ in (8 d):

$$v_{l-1} := (L_{l-1} - \lambda_{l-1} I)^{-1} d_{l-1}^\perp \in \mathcal{N}_{l-1}^{*\perp}. \tag{12.3.8 d*}$$

The exact solution involved in (8 d) or (8 d*) can be replaced by an approximation $\tilde{v}_{l-1}$ satisfying $\| \tilde{v}_{l-1} - v_{l-1} \|_{\mathcal{U}} \leq \kappa \| v_{l-1} \|_{\mathcal{U}}$ with sufficiently small κ.

As soon as the exact projection Q_{l-1} is replaced by an approximation $\tilde{Q}_{l-1}$ defined by

$$\tilde{Q}_{l-1} u_{l-1} = u_{l-1} - \langle u_{l-1}, \tilde{e}_{l-1}^* \rangle \tilde{e}_{l-1}, \tag{12.3.11}$$

the vector $\tilde{d}_{l-1}^\perp := \tilde{Q}_{l-1} d_{l-1}$ does not necessarily belong to $\mathcal{N}_{l-1}^{*\perp}$.

12.3.3 Multi-Grid Iteration for the Eigenvalue Problem and the Nested Iteration

Since the coarse-grid problem in (8 d) is a singular equation and not an eigenvalue problem, algorithm (8) cannot be applied recursively. In § 12.3.5 we shall describe a multi-grid method *SMGM* for the approximate solution of (8 d). Together with the modifications (8 d*) and (11) we obtain the following

multi-grid iteration *EMGM* for the eigenvalue problem $L_l e_l = \lambda_l e_l$ (12.3.12)

procedure *EMGM* (l, u); **integer** l; **array** u;
comment l: level number,
u: u_l^i as input, u_l^{i+1} as output;
begin real λ; **array** d, v; **integer** j;
$\lambda := \Lambda_l(u);$ (12.3.12 a)
$u := S_l^\nu(\lambda)\, u;$ (12.3.12 b)
$d := \tilde{Q}_{l-1}\, r\,(L_l - \lambda I)\, u;$ (12.3.12 c)
$v := 0$; **for** $j := 1$ **step** 1 **until** γ **do** *SMGM* $(l-1, v, d)$; (12.3.12 d)
$u := u - p\, v$
end;

The implementation of algorithm (12) requires

(i) the knowledge of approximations $\tilde{e}_{l-1}, \tilde{e}^*_{l-1}$, which are involved in $\tilde{Q}_{l-1}$ (cf. (11)).
(ii) the knowledge of an eigenvalue approximation $\tilde{\lambda}_{l-1}$, since *SMGM* has to solve $(L_{l-1} - \tilde{\lambda}_{l-1} I)\, v_{l-1} = d^\perp_{l-1}$.

These quantities can be obtained by means of the nested iteration. Two cases are to be distinguished. If $\mathcal{N}_l = \mathcal{N}_l^*$ since $L_l = L_l^*$ or since L_l is normal, and if $\dim(\mathcal{N}_l) = 1$, one can apply algorithm (13), while algorithm (14) is required in the case $\mathcal{N}_l \neq \mathcal{N}_l^*$:

nested iteration for solving $L_l e_l = \lambda_l e_l, \quad 0 \leqq l \leqq l_{\max}$ (12.3.13)

comment $\tilde{e}_0$ is a given approximation to e_0; (12.3.13 a)
for $l := 1$ **step** 1 **until** $l_{\max}$ **do**
begin $\tilde{e}_{l-1} := \tilde{e}_{l-1}/|\tilde{e}_{l-1}|_0$; $\tilde{\lambda}_{l-1} := \Lambda_{l-1}(\tilde{e}_{l-1})$; (12.3.13 b)
$\tilde{e}_l := \tilde{p}\, \tilde{e}_{l-1}$; (12.3.13 c)
for $j := 1$ **step** 1 **until** i **do** *EMGM* $(l, \tilde{e}_l)$ (12.3.13 d)
end;

nested iteration for $L_l e_l = \lambda_l e_l$, $L_l^* e_l^* = \overline{\lambda_l} e_l^*$, $0 \leqq l \leqq l_{\max}$ (12.3.14)

comment $\tilde{e}_0$, $\tilde{e}_0^*$ are given approximations to e_0, e_0^*; (12.3.14 a)
for $l := 1$ **step** 1 **until** $l_{\max}$ **do**
begin $\tilde{e}_{l-1} := \tilde{e}_{l-1}/\langle \tilde{e}_{l-1}, \tilde{e}_{l-1}^* \rangle$; $\tilde{\lambda}_{l-1} := \Lambda_{l-1}(\tilde{e}_{l-1}, \tilde{e}_{l-1}^*)$; (12.3.14 b)
$\tilde{e}_l := \tilde{p}\,\tilde{e}_{l-1}$; $\tilde{e}_l^* := \tilde{p}\, e_{l-1}^*$; (12.3.14 c)
for $j := 1$ **step** 1 **until** i **do**
begin $EMGM(l, \tilde{e}_l)$; $EMGM^*(l, \tilde{e}_l^*)$ **end** (12.3.14 d)
end;

$EMGM^*$ is the counterpart of $EMGM$ for the adjoint problem $L_l^* e_l^* = \overline{\lambda_l} e_l^*$ (cf. (9 d)). In particular, $EMGM^*$ involves the procedure $SMGM^*$, which is obtained from $SMGM$ by interchanging L_l and L_l^*, $\tilde{\lambda}_l$ and λ_l.

If $\dim(\mathcal{N}_l) = k > 1$, approximations $\tilde{e}_{1l}, \ldots, \tilde{e}_{kl}$ and possibly $\tilde{e}_{1l}^*, \ldots, \tilde{e}_{kl}^*$ have to be computed simultaneously by a similar nested iteration.

Note 12.3.1. Because of the approximation $\tilde{Q}_{l-1}$ in (12 c), the vector d from (12 c) does not necessarily belong to $\mathcal{N}_{l-1}^{*\perp}$. Therefore, the condition $f_{l-1} \perp \mathcal{N}_{l-1}^*$ of Eq. (1.8) is violated. Split d into $d = d_{l-1}^{\perp} + \beta e_{l-1}$. We must require that the iteration $SMGM$ produces a solution $v = v_{l-1} + w_{l-1}$, where v_{l-1} is the result of a right-hand side $d_{l-1}^{\perp}$, while $|w_{l-1}|_0 \leqq C|\beta|$ with C independent of $\tilde{\lambda}_{l-1} - \lambda_{l-1}$.

Let $\lambda_1 < \lambda_2 < \ldots$ be the eigenvalues of the continuous problem: $Le_j = \lambda_j e_j$. Choosing an approximation $\tilde{e}_0 = \tilde{e}_{j0}$ to e_j in (13 a) one decides that the jth eigenvalues λ_{jl}, $1 \leqq l \leqq l_{\max}$, and the corresponding eigenvectors e_{jl} will be computed. One might expect difficulties if j is large, or if λ_j is close to one of the neighbouring eigenvalues $\lambda_{j\pm 1}$. However, such complications do not arise for the following numerical example reported by Hackbusch-Hofmann [1]. Let the biharmonic eigenvalue problem

$$\Delta^2 u = \lambda u \quad \text{in } \Omega = (0,1) \times (0,1), \qquad u = \partial u/\partial n = 0 \quad \text{on } \Gamma,$$

be discretised by a finite element method with grid size $h_0 = \frac{1}{16}, \ldots, h_3 = \frac{1}{128}$. For the first 46 eigenvalues (24 single and 11 double ones) the multi-grid iteration succeeds, although several eigenvalues are in close proximity. For example λ_{38} and λ_{39}:

$$\lambda_{37} \approx 3.41_{10^5} < \lambda_{38} \approx 4.19_{10^5} < \lambda_{39} \approx 4.20_{10^5} < \lambda_{40} = \lambda_{41} \approx 4.59_{10^{+5}}.$$

The corresponding eigenvectors are depicted in Fig. 12.3.1. The nodal lines (i.e. the curves defined by $e_l(x, y) = 0$) are shown in Fig. 12.3.2. The figures are reprinted from ZAMP 31 (1980) p. 733–737.

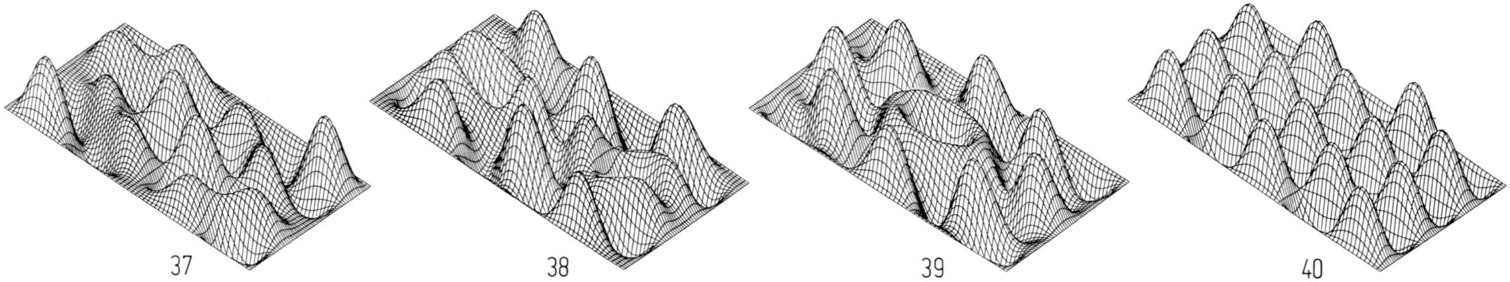

Fig. 12.3.1. Eigenvectors corresponding to λ_{37}, λ_{38}, λ_{39}, λ_{40}

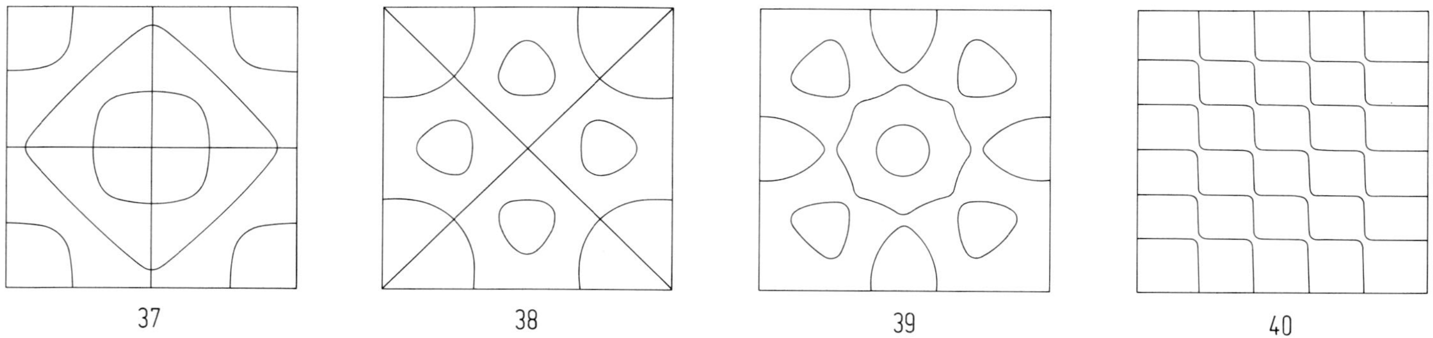

Fig. 12.3.2. Nodal lines of the eigenvectors

12.3.4 Ritz' Projection

As mentioned above, the multi-grid method (13) permits the computation of a selected eigenvalue and eigenvector. However, if approximations to the first n eigenvalues and eigenvectors are desired, the eigenvector iteration can be improved by means of the Ritz projection as proposed by Ruge [1]. According to Note 12.1.1, the exact eigenvectors can be normalised such that $\langle e_{il}^*, e_{jl}\rangle = \delta_{ij}$ $(1 \leqq i, j \leqq n)$. Given approximations $\tilde{e}_{jl}$ and $\tilde{e}_{jl}^*$, we can enforce the analogous equations by the bi-orthonormalisation process

$$\begin{aligned} \hat{e}_{jl} &:= \tilde{e}_{jl} - \sum_{i=1}^{j-1} \langle \tilde{e}_{jl}, \hat{e}_{il}^*\rangle \, \hat{e}_{il}, \\ \hat{e}_{jl}^* &:= \left(\tilde{e}_{jl}^* - \sum_{i=1}^{j-1} \langle \tilde{e}_{jl}^*, \hat{e}_{il}\rangle \, \hat{e}_{il}^* \right) \Big/ \langle \tilde{e}_{jl}^*, \hat{e}_{jl}\rangle \end{aligned} \qquad (j = 1, \ldots, n). \tag{12.3.15 a}$$

If $e_{jl} = e_{jl}^*$ because of $L_l = L_l^*$, iteration (15 a) is the usual Gram-Schmidt orthonormalisation where we scale $\hat{e}_{jl}$ and $\hat{e}_{jl}^*$ such that $\hat{e}_{jl} = \hat{e}_{jl}^*$, $|\hat{e}_{jl}|_0 = 1$. Define an $n \times n$ matrix A by

$$A = (a_{ij})_{i,j=1,\ldots,n}, \qquad a_{ij} := \langle L_l \hat{e}_{jl}, \hat{e}_{il}\rangle . \tag{12.3.15 b}$$

Were $\hat{e}_{il}, \hat{e}_{jl}^*$ exact eigenvectors, A would be the diagonal matrix $\operatorname{diag}\{\lambda_{1l}, \ldots, \lambda_{nl}\}$. Therefore, we look for vectors $\check{e}_{il} \in \operatorname{span}\{\hat{e}_{jl}: 1 \leqq j \leqq n\} = \operatorname{span}\{\tilde{e}_{jl}: 1 \leqq j \leqq n\}$, $\check{e}_{il}^* \in \operatorname{span}\{\hat{e}_{jl}^*: 1 \leqq j \leqq n\} = \operatorname{span}\{\tilde{e}_{jl}^*: 1 \leqq j \leqq n\}$, i.e.

$$\check{e}_{il} = \sum_{j=1}^{n} t_{ij} \hat{e}_{jl}, \qquad \check{e}_{il}^* = \sum_{j=1}^{n} t_{ij}^* \hat{e}_{jl}^*, \tag{12.3.15 c}$$

such that $\langle L_l \check{e}_{jl}, \check{e}_{il}^*\rangle$ is diagonal. One has to solve the eigenvalue problems

$$A t_i = \mu_i t_i, \qquad A^H t_i^* = \overline{\mu_i} t_i^*, \qquad t_i^{*H} t_j = \delta_{ij} \quad (1 \leqq i, j \leqq n), \tag{12.3.15 d}$$

which consume relatively little computational effort as long as n is not too large. The coefficients t_{ij} and t_{ij}^* in (15 c) are the components of the eigenvectors $t_i = (t_{i1}, \ldots, t_{in})^T$, $t_i^* = (t_{i1}^*, \ldots, t_{in}^*)^T$. Since the vectors $\check{e}_{il}$ and $\check{e}_{il}^*$ from (15 c) satisfy $\langle \check{e}_{il}, \check{e}_{jl}^*\rangle = \delta_{ij}$ and $\langle L_l \check{e}_{il}, \check{e}_{jl}^*\rangle = \mu_i \delta_{ij}$, they are better approximations than $\tilde{e}_{il}$, $\tilde{e}_{il}^*$, and $\tilde{\lambda}_{il} := \mu_i$ are the new eigenvalue approximations.

The combination of the multi-grid iteration with Ritz' projection reads as follows:

(i) Compute approximations $\tilde{e}_{1l}, \ldots, \tilde{e}_{nl}$ and $\tilde{e}_{1l}^*, \ldots, \tilde{e}_{nl}^*$ (unless $e_{jl} = e_{jl}^*$) by one iteration of algorithm (12) or by some other multi-grid iteration (e.g. applied to the reformulated problems (2.2) or (2.5)).

(ii) Perform (15 a), calculate the matrix A from (15 b), and diagonalise A:

$$T^{-1} A T = \operatorname{diag}\{\mu_1, \ldots, \mu_n\} . \tag{12.3.16}$$

The latter task is equivalent to (15 d) since $T = (t_1, \ldots, t_n)$, $T^{-1} = (t_1^*, t_2^*, \ldots)^H$.

(iii) Define $\check{e}_{1l}, \ldots, \check{e}_{nl}, \check{e}^*_{1l}, \ldots, \check{e}^*_{nl}$ by (15 c). Repeat the process (i) to (iii) with $\tilde{e}_{jl} := \check{e}_{jl}$, $\tilde{e}^*_{jl} := \check{e}^*_{jl}$. The eigenvalue approximations $\tilde{\lambda}_1, \ldots, \tilde{\lambda}_n$ are given by $\mu_1, \ldots, \mu_n$ from (15 d), since (15 b) can be regarded as a generalisation of the Rayleigh quotient (1.4 b).

It is not necessary to apply the Ritz projection with e_{jl}, $1 \leqq j \leqq n$, corresponding to all first n eigenvalues $\lambda_{1l} \leqq \lambda_{2l} \leqq \ldots \leqq \lambda_{nl}$. It suffices to treat simultaneously a cluster of eigenvalues $\lambda_{i+1,l} \leqq \lambda_{i+2,l} \leqq \ldots \leqq \lambda_{i+n,l}$, which are well separated from λ_{il} and $\lambda_{i+n+1,l}$.

Another multi-grid algorithm for the simultaneous eigenvalue computation is recently described by Hackbusch [38].

12.3.5 Direct Multi-Grid Solution of the Singular Equation

The equation arising in (8 d) is

$$(L_l - \tilde{\lambda}_l I) u_l = f_l, \quad u_l \perp \mathcal{N}_l^*, \quad f_l \perp \mathcal{N}_l^*, \tag{12.3.17}$$

where for ease of notation $l-1$ is replaced by l. Equation (17) coincides with Eq. (1.8), where we rename $L_l - \lambda_l I$ by L_l and set $\varepsilon = \lambda_l - \tilde{\lambda}_l$. $\tilde{\lambda}_l$ is assumed to be an approximation to the exact eigenvalue λ_l. $\tilde{\lambda}_l$ as well as $\tilde{\lambda}_{l-1}, \tilde{\lambda}_{l-2}, \ldots$ are known from the nested iteration (13) or (14).

The two-grid iteration for Eq. (17) is

$$\begin{aligned} &\tilde{u}_l := \mathcal{S}_l^{(\nu)}(u_l^i, f_l; \tilde{\lambda}_l); \\ &u_l^{i+1} := Q_l[\tilde{u}_l - p(L_{l-1} - \tilde{\lambda}_{l-1} I)^{-1} Q_{l-1} r(L_l \tilde{u}_l - \tilde{\lambda}_l \tilde{u}_l - f_l)], \end{aligned} \tag{12.3.18}$$

where $\mathcal{S}_l^{(\nu)}(u_l, f_l; \tilde{\lambda}_l)$ is the smoothing iteration for Eq. (17), possibly different from the choice in (8 b) and (12 b). The coarse-grid correction process is almost the same as in (8 c_1, c_2, d, e). The only difference is the additional projection Q_l applied to the result of (8 e). The justification is related to Note 12.3.1. Assume that $f_l \perp \mathcal{N}_l^*$ is violated since an approximate projection (11) is used in (8 c_2). The perturbation βe_l of $f_l = f_l^\perp + \beta e_l, f_l^\perp \in \mathcal{N}_l^{*\perp}$, must not induce a perturbation of the size of $\beta (L_l - \tilde{\lambda}_l I)^{-1} e_l = (\beta/(\lambda_l - \tilde{\lambda}_l)) e_l$, which would be increasingly large as $\tilde{\lambda}_l \to \lambda_l$. Iteration (18) can be written as

$$\begin{aligned} &u_l^{i+1} = M_l u_l^i + N_l f_l \quad \text{with} \\ &N_l = Q_l N_l^I + Q_l N_l^{II} T_l^{(\nu)}, \qquad N_l^I := p(L_{l-1} - \tilde{\lambda}_{l-1} I)^{-1} Q_{l-1} r, \\ &N_l^{II} := I - p(L_{l-1} - \tilde{\lambda}_{l-1} I)^{-1} Q_{l-1} r(L_l - \tilde{\lambda}_l I), \end{aligned}$$

where $T_l^{(\nu)}$ is defined by $\mathcal{S}_l^{(\nu)}(u_l, f_l; \tilde{\lambda}_l) = S_l^{(\nu)} u_l + T_l^{(\nu)} f_l$. The influence of a perturbation βe_l in f_l is characterised by $N_l e_l$. Let $T_l^{(\nu)} e_l = v_l + \alpha e_l$, $v_l \in \mathcal{N}_l^{*\perp}$. One observes that $N_l^{II} e_l = e_l + O(|\tilde{\lambda}_l - \lambda_l|)$, hence $Q_l N_l^{II} e_l = O(|\tilde{\lambda}_l - \lambda_l|)$ is small. The remaining term $N_l^{II} v_l$ is small, provided that $v_l = Q_l T_l^{(\nu)} e_l$ is smooth. Thus $Q_l N_l^{II} T_l^{(\nu)} e_l$ is small. The first term $Q_l N_l^I e_l$ is small since $Q_{l-1} r e_l = O(h_l^\kappa)$, where κ is the consistency order (cf. Hackbusch [11; Lemma 1]). We obtain

Note 12.3.2. If $f_l = f_l^{\perp} + \beta e_l, f_l^{\perp} \in \mathcal{N}_l^{*\perp}$, in spite of the condition in (17), iteration (18) results in $u_l^{i+1} = \bar{u}_l^{i+1} + \beta w_l$, where $\bar{u}_l^{i+1}$ is the result for the unperturbed right-hand side $f_l^{\perp}$, and $|w_l|_{\mathcal{U}} \leqq C$ with small $C = O(h_l + \| N_l^{II} Q_l \|)$ even if $\tilde{\lambda}_l$ approaches λ_l.

The foregoing considerations reveal that the $\mathcal{N}_l^{*\perp}$-part of $T_l^{(\nu)} e_l$ yields the major contribution to the perturbation of u_l^{i+1}. If $T_l^{(\nu)} e_l \in \mathcal{N}_l$, this term vanishes and even $[N_l^I + N_l^{II} T_l^{(\nu)}] e_l$ is small:

Note 12.3.3. Assume $T_l^{(\nu)} e_l = \tau_{l,\nu} e_l$, which holds, e.g., for the damped Jacobi smoother. Then the projection Q_l can be omitted in (18).

The coarse-grid equation arising in (18) is of the same form as Eq. (17). Hence, the construction of the multi-grid iteration follows algorithm (4.1.1):

multi-grid solution of the singular Eq. (17) (12.3.19)

$\tilde{\lambda}_k$, $0 \leqq k \leqq l$: given eigenvalue approximations;
$\tilde{Q}_k$, $0 \leqq k \leqq l$: approximate projections onto $\mathcal{N}_l^{*\perp}$; e.g. defined by Eq. (11).

procedure *SMGM* (l, u, f); **integer** l; **array** u, f;
if $l = 0$ **then** $u := Q_0 (L_0 - \tilde{\lambda}_0 I)^{-1} Q_0 f$ **else** (12.3.19 a)
begin integer j; **array** v, d;
  $u := \mathcal{S}_l^{(\nu)}(u, f; \tilde{\lambda}_l)$; (12.3.19 b)
  $d := \tilde{Q}_{l-1} * r * (L_l u - \tilde{\lambda}_l u - f)$; (12.3.19 c)
  $v := 0$; **for** $j := 1$ **step** 1 **until** γ **do** *SMGM* $(l-1, v, d)$; (12.3.19 d)
  $u := \tilde{Q}_l * (u - p * v)$ (12.3.19 e)
end;

A convergence analysis of algorithm (19) is given by Hackbusch [11]. According to Note 12.3.3 the projection in (19 e), or even in (19 c), can be omitted. These modifications of *SMGM* and *EMGM* are discussed by Hofmann [1].

12.3.6 Simultaneous Solution of *n* Singular Equations

The Ritz projection of § 12.3.4 yields approximations $\tilde{E}_l$ and $\tilde{E}_l^*$ of

$$E_l := (e_{1l}, \ldots, e_{nl}), \qquad E_l^* := (e_{1l}^*, \ldots, e_{nl}^*),$$

i.e. of n right and n left eigenvectors, such that $\tilde{E}_l^{*H} \tilde{E}_l = I$ and $\tilde{E}_l^{*H} L_l \tilde{E}_l = \tilde{D}_l = \operatorname{diag}\{\tilde{\lambda}_{1l}, \ldots, \tilde{\lambda}_{nl}\}$. Hence the defect F_l defined by

$$F_l := L_l \tilde{E}_l - \tilde{E}_l \tilde{D}_l$$

satisfies $\tilde{E}_l^{*H} F_l = 0$, i.e all n column vectors of F_l are almost orthogonal to

$$\mathcal{N}_l^* := \operatorname{span}\{e_{1l}^*, \ldots, e_{nl}^*\},$$

whereas previously the ith eigenvector defect $f_l := L_l e_{il} - \tilde{\lambda}_i e_{il}$ was assumed to be orthogonal only to $\mathcal{N}_{il}^* := \operatorname{span}\{e_{il}^*\}$. Instead of Eq. (17) one now has to solve the problem

$$L_l U_l - U_l \tilde{D}_l = F_l, \qquad U_l \in (\mathcal{N}_l^{*\perp})^n, \qquad F_l \in (\mathcal{N}_l^{*\perp})^n, \tag{12.3.20}$$

which consists of n equations

$$L_l u_{il} - \tilde{\lambda}_i u_{il} = f_{il}, \qquad u_{il}, f_{il} \in \mathcal{N}_l^{*\perp}, \qquad i = 1, \ldots, n. \tag{12.3.20'}$$

The simultaneous solution of these equations is more stable than the solution of Eq. (17), since the ith equation in (20′) is not impaired by the existence of possibly closely neighbouring eigenvalues λ_j, $1 \leqq j \leqq n$, $j \neq i$.

The multi-grid solution of problem (20) is also given by algorithm (12), if u, d, f, v denote n-tuples $u = (u_1, \ldots, u_n)$ etc. The meaning of $p\,v$, $r\,f$, $Q_l u$ is $(p\,v_1, \ldots, p\,v_n)$ etc. Q_k approximates the projection onto $\mathcal{N}_k^{*\perp}$:

$$Q_k u_k := u_k - \sum_{i=1}^{n} \langle u_k, e_{ik}^* \rangle e_{ik}.$$

Assignment (19 b) becomes $u := \mathcal{S}_l^{(\nu)}(u, f; \tilde{D}_l)$; $\tilde{\lambda}_l u$ in (19 c) has to be replaced by $u \tilde{D}_l$.

Analogously, *EMGM* from (12) can be applied to the simultaneous computation of n eigenvectors E_l. Note that the Rayleigh quotient in (12 a) has to be replaced by Ritz' projection from § 12.3.4.

12.4 The Generalised Eigenvalue Problem

Usual finite element discretisations of $Le = \lambda e$ do not yield $L_l e_l = \lambda_l e_l$ but

$$L_l e_l = \lambda_l M_l e_l, \tag{12.4.1}$$

where $M_l = R_l P_l$ is positive definite (P_l, R_l are defined in (3.1.30–31)). Equation (1) is called a generalised eigenvalue problem. In particular, this equation is obtained if the continuous problem is a generalised eigenvalue problem $Le = \lambda Me$. The left eigenvectors e_l^* are characterised by

$$L_l^* e_l^* = \overline{\lambda_l} M_l^* e_l^*. \tag{12.4.2}$$

Lemma 12.4.1. If $L_l = L_l^*$ and $M_l = M_l^* > O$, all eigenvalues λ_l are finite and real; furthermore, $\mathcal{N}_l(\lambda_l) = \mathcal{N}_l^*(\lambda_l)$.

The condition $M_l = M_l^* > O$ excludes the case $\lambda_l = \infty$, which corresponds to $\mathcal{N}_l(\infty) :=$ kernel of M_l. However, there are reasonable eigenvalue problems with singular M_l:

Example 12.4.2 (Steklov eigenvalue problem). The discretisation of

$$-\Delta u = 0 \quad \text{in } \Omega, \qquad \partial u / \partial n = \lambda u \quad \text{on } \Gamma \tag{12.4.3}$$

yields Eq. (1), where M_l is singular, since the rank of M_l equals the number of grid points on Γ. $M_l = M_l^* \geqq O$ holds in the finite element case. See also § 16.10.3.

The equivalent of Note 12.1.1 is

Note 12.4.3. (i) Suppose $\lambda_l \neq \mu_l$. Then $\langle M_l e_l, e_l^* \rangle = 0$ for all $e_l \in \mathcal{N}_l(\lambda_l)$, $e_l^* \in \mathcal{N}_l^*(\mu_l)$.

(ii) Let $k := \dim(\mathcal{N}_l(\lambda_l))$. Then we presuppose that there exist $e_{il} \in \mathcal{N}_l(\lambda_l)$, $e_{il}^* \in \mathcal{N}_l^*(\lambda_l)$, $1 \leqq i \leqq k$, such that $\langle M_l e_{il}, e_{jl}^* \rangle = \delta_{ij}$.

The Rayleigh quotients (1.4 a, b) are to be replaced by

$$\Lambda_l(\tilde{e}_l) := \langle L_l \tilde{e}_l, \tilde{e}_l \rangle / \langle M_l \tilde{e}_l, \tilde{e}_l \rangle, \tag{12.4.4 a}$$

$$\Lambda_l(\tilde{e}_l, \tilde{e}_l^*) := \langle L_l \tilde{e}_l, \tilde{e}_l^* \rangle / \langle M_l \tilde{e}_l, \tilde{e}_l^* \rangle. \tag{12.4.4 b}$$

$(M_l \mathcal{N}_l)^\perp$ and $(M_l^* \mathcal{N}_l^*)^\perp$ are the orthogonal spaces of

$$M_l \mathcal{N}_l := \{M_l e_l : e_l \in \mathcal{N}_l\}, \qquad M_l^* \mathcal{N}_l^* := \{M_l^* e_l^* : e_l^* \in \mathcal{N}_l^*\}.$$

In Eq. (1.6) and Eq. (1.8) the condition $u_l \perp \mathcal{N}_l^*$ is to be changed to $u_l \perp M_l^* \mathcal{N}_l^*$:

$$(L_l + \varepsilon M_l) u_l = f_l, \quad f_l \perp \mathcal{N}_l^*, \quad u_l \perp M_l^* \mathcal{N}_l^*. \tag{12.4.5}$$

Then Lemmata 12.1.4 and 12.1.7 remain valid.

The necessary modifications of the multi-grid methods are described below.

Remark 12.4.4 (modified *EMGM*). Λ_l in (3.12 a) is given by (4 a). In (3.12 c) I is to be replaced by M_l. $\tilde{Q}_{l-1}$ from (3.12 c) has to be an approximation to the following projection onto $\mathcal{N}_{l-1}^{*\perp}$:

$$\hat{Q}_k^* u_k = u_k - \langle u_k, e_k^* \rangle M_k e_k \quad (\text{if } \dim(\mathcal{N}_k) = 1, \langle M_k e_k, e_k^* \rangle = 1), \tag{12.4.6 a}$$

$\hat{Q}_l^*$ is the adjoint of the projection $\hat{Q}_l$ onto $(M_l \mathcal{N}_l)^\perp$:

$$\hat{Q}_l u_l = u_l - \langle u_l, M_l e_l \rangle e_l^*. \tag{12.4.6 b}$$

Similar comments hold for the lines (3.8 a), (3.8 c$_2$) of the two-grid iteration. In line (3.8 d) $\mathcal{N}_{l-1}^{*\perp}$ has to be replaced by $(M_{l-1}^* \mathcal{N}_{l-1}^*)^\perp$.

Remark 12.4.5 (modified *EMGM**). The iteration *EMGM** involved in (3.14 d) requires the (approximate) projection Q_{l-1}^*, in the statement corresponding to (3.12 c). Q_l^* with

$$Q_l^* u_l = u_l - \langle u_l, e_l \rangle M_l^* e_l^* \tag{12.4.6 c}$$

is the adjoint of

$$Q_l u_l = u_l - \langle u_l, M_l^* e_l^* \rangle e_l. \tag{12.4.6 d}$$

Remark 12.4.6 (modified nested iteration). Λ_l from (3.13 b/14 b) is given by (4 a/b). The normalisation reads $\tilde{e}_{l-1} := \tilde{e}_{l-1} / \langle M_{l-1} \tilde{e}_{l-1}, \tilde{e}_{l-1} \rangle^{1/2}$ in (3.13 b) and $\tilde{e}_{l-1} := \tilde{e}_{l-1} / \langle M_{l-1} \tilde{e}_{l-1}, \tilde{e}_{l-1}^* \rangle$ in (3.14 b).

Remark 12.4.7 (modified Ritz projection). (3.15 a) becomes

$$\hat{e}_{jl} := \tilde{e}_{jl} - \sum_{i=1}^{l-1} \langle M_l \tilde{e}_{jl}, \hat{e}^*_{il} \rangle \hat{e}_{il},$$

$$\hat{e}^*_{jl} := \left(\tilde{e}_{jl} - \sum_{i=1}^{l-1} \langle \tilde{e}^*_{jl}, M_l \hat{e}_{il} \rangle \hat{e}^*_{il} \right) \Big/ \langle \hat{e}^*_{jl}, M_l \hat{e}_{jl} \rangle .$$

Let $B = (b_{ij})$, $b_{ij} := \langle M_l \hat{e}_{jl}, \hat{e}^*_{jl} \rangle$. (3.15 d) has to be replaced by

$$A\, t_i = \mu_i B\, t_i, \qquad A^H t_i^* = \overline{\mu_i} B^H t_i^*, \qquad t_i^{*H} t_j = \delta_{ij}.$$

Step (ii) requires the computation of T with $T^{-1} A T = \operatorname{diag}\{\mu_1, \ldots, \mu_n\} B$.

Remark 12.4.8 (modified *SMGM*). The assignment in (3.19 a) becomes $u := Q_0 (L_0 - \tilde{\lambda}_0 M_0)^{-1} \hat{Q}_0^* f$, (3.19 c) becomes $d := \hat{Q}^*_{l-1} r (L_l u - \tilde{\lambda}_l M_l u - f)$, while Q_l from (3.19 e) is now defined by (6 d).

12.5 The Nonlinear Eigenvalue Problem

Equation (4.1) is a special example of the nonlinear eigenvalue problem

$$L_l e_l = M_l(\lambda_l)\, e_l \tag{12.5.1}$$

arising from the continuous problem

$$L e = M(\lambda)\, e. \tag{12.5.2}$$

For instance, bifurcation problems lead to Eq. (2). Consider the nonlinear equation

$$\mathscr{L}(u; \lambda) = 0 \quad \text{with} \quad \mathscr{L}(u; \lambda) = L u - \mathscr{M}(u; \lambda).$$

If $\mathscr{M}(0; \lambda) = 0$ for all λ, $u(\lambda) = 0$ is the trivial solution. Other branches bifurcate at $\lambda = \lambda^*$, where λ^* is characterised by Eq. (2) with $M(\lambda) := \partial \mathscr{M}(0; \lambda)/\partial u$.

We presuppose that there exists a generalised Rayleigh quotient $\Lambda_l(\tilde{e}_l)$ or $\Lambda_l(\tilde{e}_l, \tilde{e}_l^*)$, respectively, such that the inequalities (1.5 a–c) hold. If, for example, a quadratic eigenvalue problem with $M_l(\lambda_l) = \lambda_l A_l + \lambda_l^2 B_l$ is given, $\Lambda_l(\tilde{e}_l)$ can be defined to be one of the roots of $\langle L_l \tilde{e}_l, \tilde{e}_l \rangle = \langle A_l \tilde{e}_l, \tilde{e}_l \rangle \lambda + \langle B_l \tilde{e}_l, \tilde{e}_l \rangle \lambda^2$.

Remark 12.5.1. Define Λ_l as mentioned above. The modified algorithms of § 12.4 apply to the nonlinear eigenvalue problem (1), where the matrices M_l, M_l^* in (4.6 a–d) and in Remark 12.4.6 are replaced by $M_l(\lambda_l)$, $M_l^*(\lambda_l)$. Evidently, all expressions $\lambda_l M_l$, for example in $L_l u_l - \lambda_l M_l u_l - f_l$, are to be changed to $M_l(\lambda_l)$.

The convergence analysis presented by Hackbusch [11] applies to the multi-grid algorithms for the nonlinear eigenvalue problem (1).

13. Continuation Techniques

13.1 Continuous Continuation Problem

Instead of the elliptic equation $\mathscr{L}(u) = 0$ we consider the family of nonlinear equations

$$\mathscr{L}(u(\lambda), \lambda) = 0, \tag{13.1.1}$$

depending on the scalar parameter λ. For ease of notation we assume that Eq. (1) is defined for $\lambda \geqq 0$. For each $\lambda \geqq 0$ there may exist several solutions $u(\lambda)$ of Eq. (1). A smooth curve $(u(\lambda), \lambda) \in \mathscr{U} \times \mathbb{R}$ is called a solution branch. Intersection points of branches are called bifurcation points. At such points $\partial \mathscr{L}(u^*, \lambda^*)/\partial u$ is singular (see λ_A in Fig. 13.1.1). For the computation of bifurcation points we refer to § 12.5 and the approaches of Weber [1] and Mittelmann-Weber [1].

In this section we shall study branches without bifurcation points. $u(\lambda)$ is not necessarily a function of λ. Branch B in Fig. 13.1.1 is multi-valued in the interval (λ_B, λ_C). λ_B and λ_C are called turning points (or limit points). To describe the branch as a function of a parameter one has to choose a new parameter. A suitable curve parameter of all branches in Fig. 13.1.1 would be

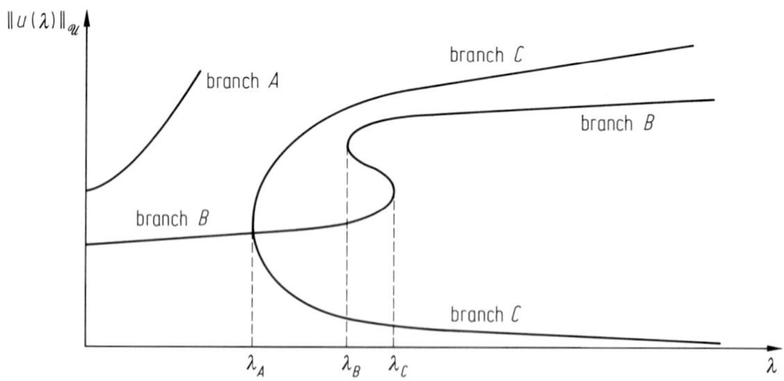

Fig. 13.1.1. Solution branches, bifurcation point, turning points

the norm $\mu = \|u\|_{\mathscr{U}}$. A general approach is the parametrisation of u and λ by the arc length μ:

$$\|du(\mu)/d\mu\|_{\mathscr{U}}^2 + |d\lambda(\mu)/d\mu|^2 = 1 \tag{13.1.2}$$

(cf. Menzel-Schwetlick [1], Keller [1]). A modified approach is described by Bank-Chan [1]. The problem that arises takes the form of Eq. (11.4.7), where the additional equation $\Phi(u, \lambda) = \mu$ is needed to determine λ as a function of μ.

13.2 Modified Nested Iteration

The solution of the discrete continuation problem

$$\mathscr{L}_l(u_l(\lambda), \lambda) = 0 \tag{13.2.1}$$

requires the solution of Eq. (1) for several values $\lambda_0 < \lambda_1 < \ldots < \lambda_\nu < \ldots$. One could use the nested iteration (9.3.14) to solve each equation $\mathscr{L}_l(u_l(\lambda_\nu), \lambda_\nu) = 0$ separately, without exploiting the already computed values $u_l(\lambda_{\nu-1})$. On the other hand, the nested iteration could be omitted, since at all levels l the foregoing values $u_l(\lambda_{\nu-1})$ could be used as starting iterates $u_l^0(\lambda_\nu)$, provided that $\lambda_\nu \approx \lambda_{\nu-1}$. In § 13.2.2 we shall study a modified nested iteration, which makes use of the known approximations at the previous value $\lambda = \lambda_{\nu-1}$. We shall restrict our considerations to branches without turning points, since by virtue of § 13.2.1 one may assume without loss of generality that $u_l(\lambda)$ is a function of λ.

13.2.1 Treatment of Turning Points

Given a new parameter μ of $u(\mu)$ and $\lambda(\mu)$, Eq. (1) has to be replaced by Eq. (11.4.7), for which the multi-grid solution is almost identical with the multi-grid solution of Eq. (1). As a particular example we consider the parametrisation (1.2) by means of the arc length μ. Initially, for $\lambda = 0$ set $\mu_0 := 0$. At level $l = 0$ the differential equation (1.2) can be discretised by

$$\mu_\nu := \mu_{\nu-1} + \sqrt{\|u_0(\mu_\nu) - u_0(\mu_{\nu-1})\|_{\mathscr{U}}^2 + |\lambda_0(\mu_\nu) - \lambda_0(\mu_{\nu-1})|^2}, \tag{13.2.2}$$

which together with $\mathscr{L}_0(u_0(\mu_\nu), \lambda_0(\mu_\nu)) = 0$ forms the augmented system for $u_0(\mu_\nu)$ and $\lambda_0(\mu_\nu)$. The analogous parametrisation could be used also for the levels $l \geqq 1$. However, a better coincidence of $u_0(\mu_\nu), u_1(\mu_\nu), \ldots, u_l(\mu_\nu)$ can be achieved, if the difference of $(u_l(\mu_\nu), \lambda_l(\mu_\nu))$ and $(u_{l-1}(\mu_\nu), \lambda_{l-1}(\mu_\nu))$ is perpendicular to one of the branches $u_l(\mu)$ or $u_{l-1}(\mu)$ (cf. Fig. 13.2.1). More precisely, for given $u_{l-1}^* = u_{l-1}(\mu_\nu)$ and $\lambda_{l-1}^* = \lambda_{l-1}(\mu_\nu)$ the values u_l^* and λ_l^* are defined as solutions of

$$\begin{aligned} &\mathscr{L}_l(u_l, \lambda_l) = 0, \\ &\Phi_l(u_l, \lambda_l) := \langle r' u_l - u_{l-1}^*, v_{l-1}^* \rangle + m_{l-1}^*(\lambda_l - \lambda_{l-1}^*) = 0, \end{aligned} \tag{13.2.3}$$

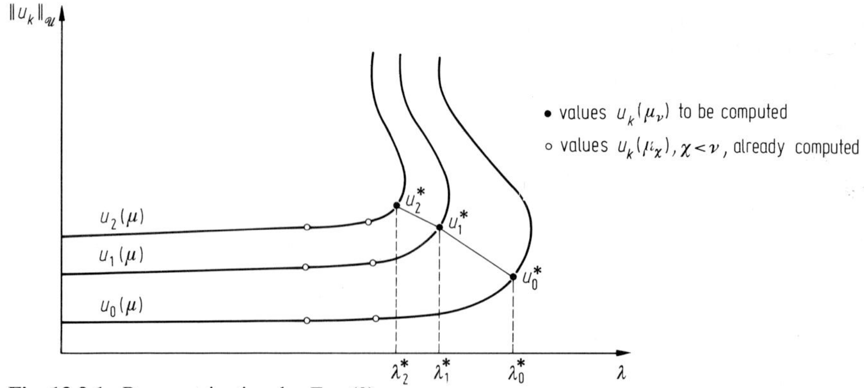

Fig. 13.2.1. Parametrisation by Eq. (3)

where the ratio $v_{l-1}^*:m_{l-1}^*$ approximates $du_{l-1}/d\mu:d\lambda_{l-1}/d\mu$; e.g. $v_{l-1}^*:= u_{l-1}(\mu_\nu)-u_{l-1}(\mu_{\nu-1})$, $m_{l-1}^*:=\lambda_{l-1}(\mu_\nu)-\lambda_{l-1}(\mu_{\nu-1})$ can be chosen.

According to Proposition 11.4.1, the scalar equations $\Phi_k(u_k,\lambda_k)=0$ at the coarser grids $k<l$ should satisfy (11.4.4 a, b). Definition (11.4.8) yields

$$\Phi_k(u_k,\lambda_k):=\langle r' p_{lk}(u_k-\tilde{u}_k), v_{l-1}^*\rangle + m_{l-1}^*(\lambda_k-\tilde{\lambda}_k),\quad k\leqq l-1, \tag{13.2.4}$$

where p_{lk} is the product $p_{l,l-1}p_{l-1,l-2}\cdots p_{k+1,k}$ of the prolongations $p=p_{m,m-1}:\mathscr{U}_{m-1}\to\mathscr{U}_m$. $\tilde{u}_k$ and $\tilde{\lambda}_k$ are the previously computed approximations (see nested iteration (5)). A possible choice of the restriction $r':\mathscr{U}_l\to\mathscr{U}_{l-1}$ is $r'=r_{inj}$ from (3.5.1). Then, since $r'p=I$ for any interpolation $p=p_{l,l-1}$, one has $r'p_{lk}=p_{l-1,k}$ and Eq. (4) can be rewritten

$$\Phi_k(u_k,\lambda_k):=\langle u_k-\tilde{u}_k, v_k^{**}\rangle + m_{l-1}^*(\lambda_k-\tilde{\lambda}_k),\quad k\leqq l-1, \tag{13.2.4'}$$

where $v_k^{**}:=p_{l-1,k}^* v_{l-1}^* (=r_{k,l-1}v_{l-1}^*$ if $r=p^*)$.

13.2.2 Modified Nested Iteration

In the sequel we assume that λ is a reasonable parameter (otherwise introduce instead of λ a new parameter μ as described in § 13.2.1).

Starting with given values $\tilde{u}_k(\lambda_0)$, one can compute $\tilde{u}_k(\lambda_\nu)$ by applying algorithm (5) for $\nu=1,2,\ldots$ (cf. Fig. 13.2.2):

. . . .

modified nested iteration for solving $\mathscr{L}_k(u_k(\lambda_\nu),\lambda_\nu)=0,\quad 0\leqq k\leqq l$ (13.2.5)

. . . .

comment compute $\tilde{u}_0(\lambda_\nu)$ from $\tilde{u}_0(\lambda_{\nu-1})$ by some continuation method; (13.2.5 a)

for $k:=1$ **step** 1 **until** l **do**

begin $\tilde{u}_k(\lambda_\nu):=\tilde{u}_k(\lambda_{\nu-1})+\tilde{p}[\tilde{u}_{k-1}(\lambda_\nu)-\tilde{u}_{k-1}(\lambda_{\nu-1})]$; (13.2.5 b)

  for $j:=1$ **step** 1 **until** i **do** $NMGM(k,\tilde{u}_k(\lambda_\nu),0;\lambda_\nu)$ (13.2.5 c)

end;

. . . .

$NMGM(\ldots;\lambda)$ is the nonlinear multi-grid iteration for Eq. (1) at λ. Although this algorithm is as easy to execute as the usual nested iteration (9.3.14), the error of $\tilde{u}_k$ is not only proportional to h_k^κ but also to $h_k^\kappa \Delta\lambda$, where

$$\Delta\lambda := \lambda_\nu - \lambda_{\nu-1}. \tag{12.2.6}$$

The correction step (5 b) is used by Brandt [12, p. 306/7] in another connection (cf. § 13.2.4).

In § 9.5.1 we introduced spheres $\mathscr{U}_k(\varepsilon_k)$ for the single problem $\mathscr{L}(u) = 0$. The spheres $\mathscr{U}_k^\lambda(\varepsilon_k)$ with centre $u_k(\lambda)$ correspond to $L(u,\lambda) = 0$. The radius ε_k must be small enough so that $\mathscr{L}_k(\cdot,\lambda)$ is injective on $\mathscr{U}_k^\lambda(\varepsilon_k)$.

Proposition 13.2.1 (convergence of the modified nested iteration).

(i) For all $u_k^i \in \mathscr{U}_k^\lambda(\varepsilon_k)$, $0 \leqq k \leqq l$, $\lambda \geqq \lambda_0$, the iteration $NMGM$ is assumed to produce u_k^{i+1} with a uniform contraction number $\zeta < 1$:

$$\| u_k^{i+1} - u_k(\lambda) \|_{\mathscr{U}} \leqq \zeta \, \| u_k^i - u_k(\lambda) \|_{\mathscr{U}}.$$

(ii) Suppose

$$\| u_k(\lambda_{\nu-1}) - u_k(\lambda_\nu) - \tilde{p}\,[u_{k-1}(\lambda_{\nu-1}) - u_{k-1}(\lambda_\nu)] \|_{\mathscr{U}} \leqq C_1 \, h_k^\kappa \, \Delta\lambda. \tag{13.2.7}$$

(iii) Define C_2 by (9.5.13), with κ from (7).

(iv) Choose i and $\Delta\lambda$ such that

$$\zeta^i (1 + 2\,C_2) < 1. \tag{13.2.8 a}$$

$$C_1 \, \Delta\lambda \, h_k^\kappa / [1 - \zeta^i (1 + 2\,C_2)] \leqq \varepsilon_k, \quad 1 \leqq k \leqq l. \tag{13.2.8 b}$$

(v) Determine starting values $\tilde{u}_k(\lambda_0)$ at $\lambda = \lambda_0$ such that inequality (9) holds for $\nu = 0$.

(vi) Perform (5 a) so that inequality (9) holds for $k = 0$.

Then the modified nested iteration (5) is well-defined and produces approximations $\tilde{u}_k(\lambda_\nu)$, $0 \leqq k \leqq l$, at all $\lambda_\nu = \lambda_0 + \nu\,\Delta\lambda$, such that the errors are bounded by

$$\| \tilde{u}_k(\lambda_\nu) - u_k(\lambda_\nu) \|_{\mathscr{U}} \leqq \zeta^i \cdot C_1 \cdot \Delta\lambda \cdot h_k^\kappa / [1 - \zeta^i (1 + 2\,C_2)], \quad 0 \leqq k \leqq l. \tag{13.2.9}$$

Proof. We prove the inequality by induction over ν and, secondarily, over k. By (v) and (vi) we may presuppose inequality (9) for $\tilde{u}_{k-1}(\lambda_{\nu-1})$, $\tilde{u}_k(\lambda_{\nu-1})$, $\tilde{u}_{k-1}(\lambda_\nu)$ to prove (9) for $\tilde{u}_k(\lambda_\nu)$. The error in the starting value $u_k^o := u_k^0(\lambda_\nu) := \tilde{u}_k(\lambda_{\nu-1}(+ \tilde{p}\,[\tilde{u}_{k-1}(\lambda_{\nu-1}) - \tilde{u}_{k-1}(\lambda_\nu)]$ from (5 b) is

$$\begin{aligned}
&\| u_k^0 - u_k(\lambda_\nu) \|_{\mathscr{U}} \\
&\quad = \| \{u_k(\lambda_{\nu-1}) - u_k(\lambda_\nu) - \tilde{p}[u_{k-1}(\lambda_{\nu-1}) - u_{k-1}(\lambda_\nu)]\} + [\tilde{u}_k(\lambda_{\nu-1}) - u_k(\lambda_{\nu-1})] \\
&\qquad - \tilde{p}\,\{[\tilde{u}_{k-1}(\lambda_{\nu-1}) - u_{k-1}(\lambda_{\nu-1})] - [\tilde{u}_{k-1}(\lambda_\nu) - u_{k-1}(\lambda_\nu)]\} \|_{\mathscr{U}} \\
&\quad \leqq C_1 \, h_k^\kappa \, \Delta\lambda + \frac{\zeta^i C_1 \, \Delta\lambda}{1 - \zeta^i (1 + 2\,C_2)} h_k^\kappa + 2\,C_{20} \frac{\zeta^i C_1 \, \Delta\lambda}{1 - \zeta^i (1 + 2\,C_2)} h_{k-1}^\kappa \\
&\quad \leqq \left[C_1 + \frac{\zeta^i C_1}{1 - \zeta^i (1 + 2\,C_2)} (1 + 2\,C_2) \right] \Delta\lambda \, h_k^\kappa \\
&\quad = C_1 \, \Delta\lambda \, h_k^\kappa / [1 - \zeta^i (1 + 2\,C_2)].
\end{aligned}$$

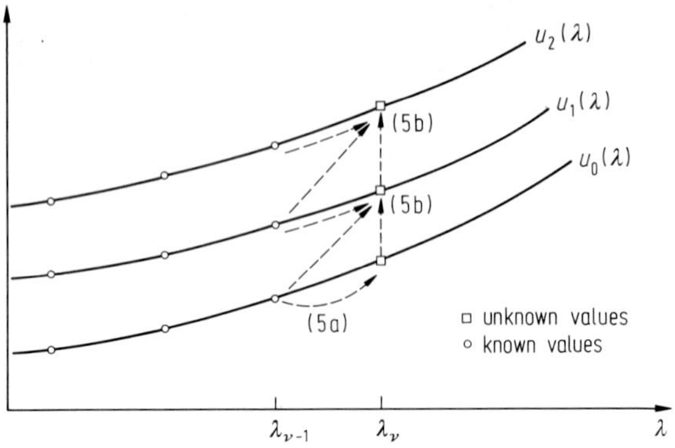

Fig. 13.2.2. Solution process of algorithm (5)

By (8 b), $u_k^0 \in \mathscr{U}_k^\lambda(\varepsilon_k)$ for $\lambda = \lambda_\nu$. Since iteration (5 c) is well-defined, Inequality (9) follows also for k, completing the proof. □

Note that Proposition 13.2.1 guarantees that all results $\tilde{u}_k(\lambda_\nu)$ belong to the same branch.

13.2.3 Modifications

The equidistant step size $\Delta\lambda$ has been chosen to simplify the analysis. The error estimate by $O(h_k|\lambda_\nu - \lambda_{\nu-1}|)$ can still be proved if the previous step size $\lambda_{\nu-1} - \lambda_{\nu-2}$ is not too large compared with $\lambda_\nu - \lambda_{\nu-1}$ (otherwise the approximations $\tilde{u}_k(\lambda_{\nu-1})$, which are involved in the computation of $u_k(\lambda_\nu)$, are too rough).

Remark 13.2.2. Suppose (i), (ii), (iii), and (8 a) from Proposition 13.2.1. Let $\beta \in (0, 1]$ satisfy

$$\beta > (1 + C_2)\zeta^i/(1 - C_2\zeta^i)$$

and define the maximum step width by

$$\Delta\lambda_{\max} := \max_{1 \leq k \leq l} \varepsilon_k[1 - C_2\zeta^i - (1 + C_2)\zeta^i/\beta]/[C_1 h_k^\kappa(1 - C_2\zeta^i)].$$

Choose the λ-steps such that

$$\beta(\lambda_{\nu-1} - \lambda_{\nu-2}) \leqq \lambda_\nu - \lambda_{\nu-1} \leqq \Delta\lambda_{\max} \quad \text{for } \nu = 2, 3, \ldots$$

Assume that the errors of $\tilde{u}_k(\lambda_0)$ at $\lambda = \lambda_0$ and that the errors of $\tilde{u}_0(\lambda_\nu)$ at the coarsest grid satisfy (10). Then algorithm (5) yields approximations with errors bounded by

$$\|\tilde{u}_k(\lambda_\nu) - u_k(\lambda_\nu)\|_{\mathscr{U}} \leqq C_1 h_k^\kappa(\lambda_\nu - \lambda_{\nu-1})\zeta^i/[1 - C_2\zeta^i - (1 + C_2)\zeta^i/\beta]. \tag{13.2.10}$$

Proof. Cf. Hackbusch [33]. □

If solutions at $\lambda_\nu = \lambda_0 + \nu\,\Delta\lambda'$ are desired, it might happen that $\Delta\lambda'$ is too large to satisfy condition (8 b). Then the choice $\Delta\lambda'$ could lead to divergence or to convergence to the wrong branch, etc. However, it is not necessary to use a smaller $\Delta\lambda < \Delta\lambda'$ at all levels.

Remark 13.2.3. Let $\Delta\lambda_0 \leqq \Delta\lambda_1 \leqq \ldots \leqq \Delta\lambda_l$ be step sizes with $\Delta\lambda_{k+1}/\Delta\lambda_k \in \mathbb{N}$ and

$$\Delta\lambda_k \leqq \varepsilon_k\, h_k^{-\kappa}\,[1 - \zeta^i(1 + 2\,C_2)]/C_1.$$

Algorithm (5) can be modified so that approximations $\tilde{u}_k(\lambda)$ are computed only at $\lambda = \lambda_0 + \nu\,\Delta\lambda_k$ ($\nu \in \mathbb{N}$). The step width $\Delta\lambda$ in (5 b) is to be replaced by $\Delta\lambda_k$ and the loop over k stops as soon as $\lambda_\nu - \lambda_0$ is not a multiple of $\Delta\lambda_k$. If, for example, $\Delta\lambda_k = 2^k\,\Delta\lambda_0$, the computational work is proportional to $1/\Delta\lambda_l$.

13.2.4 Frozen Truncation Error Technique

Let $\tilde{r}$ be the restriction from (9.3.9); e.g., $\tilde{r} = \tilde{r}_{k-1,k}: \mathscr{U}_k \to \mathscr{U}_{k-1}$ may be the trivial injection r_{inj}. The products of $\tilde{r}$ are denoted by $\tilde{r}_{kl} := \tilde{r}_{k,k+1}\,\tilde{r}_{k+1,k+2}\ldots\tilde{r}_{l-1,l}$. The truncation error at level k (relative to level l) is defined by

$$t_{kl}(\lambda) := \mathscr{L}_k(\tilde{r}_{kl}\,u_l(\lambda), \lambda) \tag{13.2.11}$$

(cf. § 14.1.2). By definition $\hat{u}_k(\lambda_\nu) := \tilde{r}_{kl}\,u_l(\lambda_\nu)$ is the solution of $\mathscr{L}_k(u_k, \lambda_\nu) = t_{kl}(\lambda_\nu)$. The so-called 'frozen-$\tau$ technique' of Brandt [6], [12] is based on the equation

$$\mathscr{L}_k(\hat{u}_k, \lambda_\nu) = t_{kl}(\lambda_{\nu-1}), \tag{13.2.12}$$

where the correct right-hand side $t_{kl}(\lambda_\nu)$ is replaced by the truncation error $t_{kl}(\lambda_{\nu-1})$, which can (approximately) be computed from the data at $\lambda_{\nu-1}$. Let $\hat{u}_k(\lambda_\nu)$ be the solution of Eq. (12). Under suitable smoothness conditions one can prove

$$\|\hat{u}_k(\lambda_\nu) - \tilde{r}_{kl}\,u_l(\lambda_\nu)\|_{\mathscr{U}} \leqq C_1'\,|\lambda_\nu - \lambda_{\nu-1}|\,h_k^\kappa. \tag{13.2.13}$$

Note that $\hat{u}_k(\lambda_\nu)$ approximates $u_l(\lambda_\nu)$ rather than $u_k(\lambda_\nu)$. If the right side in (13) is small enough, one might omit the computation of $\hat{u}_{k+1}(\lambda_\nu), \ldots, \hat{u}_l(\lambda_\nu)$. If $u_l(\lambda_\nu)$ is desired, the following modification of algorithm (5) can be applied:

> **comment** $\tilde{u}_l(\lambda_{\nu-1})$ is given; $t_{ll}(\lambda_{\nu-1}) := 0$;
> **for** $k := l-1$ **step** -1 **until** 0 **do**
> **begin** $\tilde{u}_k(\lambda_{\nu-1}) := \tilde{r}\,\tilde{u}_{k+1}(\lambda_{\nu-1})$; $t_{kl}(\lambda_{\nu-1}) := \mathscr{L}_k(\tilde{u}_k(\lambda_{\nu-1}), \lambda_{\nu-1})$ **end**;
> **comment** compute approximate solution $\tilde{u}_0(\lambda_\nu)$ of $\mathscr{L}_0(u_0(\lambda_\nu), \lambda_\nu) = t_{0l}(\lambda_{\nu-1})$;
> **for** $k := 1$ **step** 1 **until** l **do**
> **begin** $\tilde{u}_k(\lambda_\nu) := \tilde{u}_k(\lambda_{\nu-1}) + \tilde{p}\,(\tilde{u}_{k-1}(\lambda_\nu) - \tilde{u}_{k-1}(\lambda_{\nu-1}))$;
>   **for** $j := 1$ **step** 1 **until** i **do** $NMGM\,(k, \tilde{u}_k(\lambda_\nu), t_{kl}(\lambda_{\nu-1}); \lambda_\nu)$
> **end**;

13.3 Parabolic Initial-Boundary Value Problems

The simplest implicit time-discretisation of the parabolic problem

$$\partial u/\partial t + \mathscr{L}(u,t) = 0 \qquad \text{for } t \geqq 0, \quad \mathbf{x} \in \Omega, \tag{13.3.1 a}$$

$$u(\mathbf{x},0) = u_0(\mathbf{x}) \qquad \text{in } \Omega, \tag{13.3.1 b}$$

$$u(\cdot,t) = f_\Gamma(t) \qquad \text{on } \Gamma, \tag{13.3.1 c}$$

leads to the sequence

$$u_l(\cdot,t) + \Delta t_l\, \mathscr{L}_l(u_l(\cdot,t),t) = u_l(\cdot,t-\Delta t_l) \quad \text{for } t = \Delta t_l, 2\,\Delta t_l, \ldots \tag{13.3.2}$$

of elliptic problems. As in Eq. (1.1) the functions $u_l(\mathbf{x},t)$ depend smoothly on the parameter t, and again algorithm (2.5) applies. A more general approach is described by Hackbusch [35].

Another algorithm of Brandt [12] is based on the frozen truncation error technique and involves $u_k(\mathbf{x},t_\nu)$ only at grids as coarse as possible (i.e. $u_k(\cdot,t_\nu)$ is not necessarily computed at all levels $k = 0, \ldots, l$). However, this variant is restricted to problems with $u(\cdot,t) \to u_\infty$ $(t \to \infty)$, since, in contrast to problem (1.1), errors at previous times $t_0, \ldots, t_{\nu-1}$ generate an error at t_ν, which cannot be removed by further iterations at t_ν.

14. Extrapolation and Defect Correction Techniques

14.1 Extrapolation

14.1.1 Richardson's Extrapolation

In § 5.4 we have already described the use of Richardson's extrapolation from h_{k-1}, h_{k-2} to h_k, to provide sufficiently accurate starting iterates in the nested iteration (5.4.3).

The usual extrapolation from h_{l-1} and h_l to the limit (i.e. to $h = 0$) reads

$$\hat{u}_{l-1} := \tilde{r}\, u_l + [h_{l-1}^{\kappa}/h_l^{\kappa} - 1]^{-1}\,(\tilde{r}\, u_l - u_{l-1}), \tag{14.1.1}$$

where $\tilde{r}\colon \mathscr{U}_l \to \mathscr{U}_{l-1}$ is a suitable restriction (e.g. $\tilde{r} = r_{inj}$ if $\Omega_{l-1} \subset \Omega_l$). κ is the consistency order. The accuracy of the extrapolated value $\hat{u}_{l-1}$ is described in

Lemma 14.1.1 (Richardson's extrapolation). Let u be the exact solution and assume that $\tilde{R}_k\colon \mathscr{U} \to \mathscr{U}_k$ satisfies $\tilde{r}\,\tilde{R}_l = \tilde{R}_{l-1}$. The extrapolated value $\hat{u}_{l-1}$ from (1) is of order $\tau > \kappa$, i.e.

$$\hat{u}_{l-1} = \tilde{R}_{l-1}\, u + O\,(h_{l-1}^{\tau}), \tag{14.1.2 a}$$

if there exists an asymptotic expansion

$$u_k = \tilde{R}_k\, u + h_k^{\kappa}\, \tilde{R}_k\, e + O\,(h_k^{\tau}), \qquad \tau > \kappa, \tag{14.1.2 b}$$

of the global error for $k = l - 1, l$.

Note 14.1.2. To obtain order h_{l-1}^{τ} from approximations $\tilde{u}_{l-1}, \tilde{u}_l$, their iteration errors have to be at least of the same order. Such values can be obtained from the nested iterations of § 5.4.

Unfortunately, assumption (2 b) is often violated. It requires a regular discretisation method at both levels $l, l - 1$. If e.g. irregular difference schemes are used near the boundary, the corresponding error must be of order h_k^{τ} (cf. Pereyra-Proskurowski-Widlund [1], Marchuk-Shaidurov [1, § 4.2.1] and the references given therein). Nonetheless, even a finite element solution with irregular triangulation admits an extrapolation (1), if the triangulations $\mathscr{T}_{l-1}$ and

$\mathcal{T}_l$ are suitably constructed (cf. Lin Qun-Lu Tao-Shen Shumin [1]). Since expansions $u_k = \tilde{R}_k(u + h_k^{\kappa_1} e_1 + h_k^{\kappa_2} e_2) + O(h_k^\tau)$, $\kappa_1 < \kappa_2 < \tau$, are difficult to obtain, we shall not consider extrapolations of higher order.

14.1.2 Truncation Error Extrapolation

A very similar extrapolation technique is based on the truncation error

$$t_k := \mathcal{L}_k(\tilde{R}_k u) \quad \text{(where } u \text{ is the continuous solution)} \tag{14.1.3 a}$$

and on the relative truncation error

$$t_{kl} := \mathcal{L}_k(\tilde{r}_{kl} u_l), \qquad k < l \quad (\text{with } \tilde{r}_{kl}\tilde{R}_l = \tilde{R}_k) \tag{14.1.3 b}$$

(cf. Brandt [5, § 2.3]). Because $\mathcal{L}(u) = 0$ and $\mathcal{L}_l(u_l) = 0$, we may rewrite (3 a, b)

$$t_k = \mathcal{L}_k(\tilde{R}_k u) - R_k \mathcal{L}(u), \qquad t_{kl} = \mathcal{L}_k(\tilde{r}_{kl} u_l) - r_{kl}\mathcal{L}_l(u_l). \tag{14.1.3 c}$$

Assuming expansion (2 b) one expects from (3 c) that

$$t_k = h_k^\kappa R_k t + O(h_k^\tau), \qquad t_{kl} = (h_k^\kappa - h_l^\kappa) R_k t + O(h_k^\tau), \tag{14.1.3 d}$$

where $t = -Le$, $L := \mathcal{L}'(u)$. Using (3 d), one can approximate t_{l-1} from $t_{l-1,l}$:

$$\hat{t}_{l-1} := [1 - h_l^\kappa / h_{l-1}^\kappa]^{-1} t_{l-1,l} = t_{l-1} + O(h_{l-1}^\tau). \tag{14.1.3 e}$$

Lemma 14.1.3 (truncation error extrapolation). Let u_l and u_{l-1} be the discrete solutions of $\mathcal{L}_k(u_k) = 0$ $(k = l-1, l)$. Compute the relative truncation error $t_{l-1,l}$. Extrapolate $\hat{t}_{l-1}$ according to (3 e), and solve

$$\mathcal{L}_{l-1}(\hat{u}_{l-1}) = \hat{t}_{l-1}. \tag{14.1.3 f}$$

Then $\hat{u}_{l-1} = \tilde{R}_{l-1} u + O(h_{l-1}^\tau)$ is expected.

For a precise proof one has to specify a suitable norm $\|\cdot\|_{\mathcal{F}}$ (so that (3 d) becomes $\|t_k - h_k^\kappa R_k t\|_{\mathcal{F}} \leqq C h_k^\tau$) and formulate suitable assumptions on the inverse of $\mathcal{L}_k$.

Note 14.1.4. If $\mathcal{L}_k(u_k) = L_k u_k - f_k$ is affine, Richardson's extrapolation and the truncation error extrapolation yield the same results.

An obvious implication of Note 14.1.4 is

Note 14.1.5. The truncation error extrapolation requires the same assumptions on an asymptotic expansion (2 b) as Richardson's extrapolation.

14.1.3 τ-Extrapolation

The computation of $\hat{u}_{l-1}$ from (3 f) requires the (approximate) solution of three equations: $\mathcal{L}_{l-1}(u_{l-1}) = 0$, $\mathcal{L}_l(u_l) = 0$, $\mathcal{L}_{l-1}(\hat{u}_{l-1}) = \hat{t}_{l-1}$. Instead, one can apply the following iterative process which involves an extrapolation of the form (3 e) in each iteration step. This iteration is proposed by Brandt [12, p. 274] and called the τ-extrapolation.

τ-extrapolation process applied to $\mathscr{L}_l(u_l) = 0,\ l \geqq 1$ (14.1.4)

procedure *EXTRAPOLATED NMGM* (l, u); **integer** l; **array** u;
begin integer j; **array** $\tilde{u}, v, d$;

$$u := \mathscr{S}_l^{(\nu)}(u; 0); \tag{14.1.4 a}$$

$$\tilde{u} := \tilde{r} * u; \tag{14.1.4 b}$$

$$d := \mathscr{L}_{l-1}(\tilde{u}) - r * \mathscr{L}_l(u); \tag{14.1.4 c}$$

$$d := (1 - h_l^\kappa / h_{l-1}^\kappa)^{-1} * d; \tag{14.1.4 d}$$

$v := \tilde{u}$; **for** $j := 1$ **step** 1 **until** γ **do** *NMGM* $(l-1, v, d)$; (14.1.4 e)

$$u := u + p * (v - \tilde{u}) \tag{14.1.4 f}$$

end;

$\tilde{r}$ equals $\tilde{r}_{l-1,l}$ from (3 b).

Note 14.1.6. Without extrapolation (4 d), algorithm (4) would coincide with $NMGM(l, u, 0)$ from (9.3.13) for the special choice $\nu_2 = 0$ (i.e. no post-smoothing).

In general, $u_l = \mathscr{L}_l^{-1}(0)$ is not a fixed point of iteration (4). According to § 9.5, it suffices to study the two-grid version of (4), where line (4 e) is replaced by

$$v := \mathscr{L}_{l-1}^{-1}(d). \tag{14.1.4 e′}$$

Further, the nonlinear iteration may be replaced by the linearisation about the limit value $\hat{u}_l$, the existence of which is still to be proved. The linearisation coincides with algorithm (4 a–d, e′, f) applied to the linear equations $\mathscr{L}_k(\mathrm{u}_k) := L_k u_k - f_k = 0$. In this case the iteration becomes

two-grid τ-extrapolation process applied to the linear equations
$L_{l-1} u_{l-1} = f_{l-1},\ L_l u_l = f_l$ (14.1.5)

$$\bar{u}_l^i := \mathscr{S}_l^{(\nu)}(u_l^i, f_l); \tag{14.1.5 a}$$

$$d_{l-1} := (1 - h_l^\kappa / h_{l-1}^\kappa)^{-1} [(L_{l-1} \tilde{r} \bar{u}_l^i - f_{l-1}) - r(L_l \bar{u}_l^i - f_l)]; \tag{14.1.5 b}$$

$$u_l^{i+1} := \bar{u}_l^i + p[L_{l-1}^{-1}(f_{l-1} + d_{l-1}) - \tilde{r}\bar{u}_l^i]; \tag{14.1.5 c}$$

Lemma 14.1.7. Suppose that $\mathscr{S}_l^{(\nu)}(u_l, f_l) = S_l^{(\nu)} u_l + T_l^{(\nu)} f_l$. Iteration (5) can be written as

$$u_l^{i+1} = M_l' u_l^i + g_l, \tag{14.1.6 a}$$

where

$$M_l' = C_l' S_l^{(\nu)}, \quad C_l' = I - p\,[\tilde{r} - L_{l-1}^{-1}\,\sigma\,(L_{l-1}\,\tilde{r} - r\,L_l)] \tag{14.1.6 b}$$

$$\sigma = (1 - h_l^\kappa / h_{l-1}^\kappa)^{-1}, \tag{14.1.6 c}$$

$$g_l = C_l'\, T_l^{(\nu)} f_l + p\, L_{l-1}^{-1}\,[f_{l-1} + \sigma\,(r\,f_l - f_{l-1})]. \tag{14.1.6 d}$$

From the representation

$$M_l' = [\sigma\,(L_l^{-1} - p\,L_{l-1}^{-1}\,r) + (1-\sigma)\,(I - p\,\tilde{r})\,L_l^{-1}]\,L_l\,S_l^{(\nu)} \tag{14.1.6 e}$$

one obtains

Proposition 14.1.8. In addition to the smoothing property (6.1.7 a–c) and the approximation property (6.1.8), suppose

$$\|(I - p\,r)\,L_l^{-1}\|_{\mathscr{U} \leftarrow \mathscr{F}} \leqq C_I\, h_l^\alpha. \tag{14.1.7}$$

Then the two-grid convergence holds as in Theorem 6.1.7: There exist numbers $\underline{\nu}$ and $\bar{h} > 0$ such that

$$\|M_l'\|_{\mathscr{U} \leftarrow \mathscr{U}} \leqq (\sigma\, C_A + (\sigma - 1)\, C_I)\,\eta\,(\nu) < 1 \tag{14.1.8}$$

for all $\underline{\nu} \leqq \nu < \bar{\nu}(h_l)$, if $h_l \leqq \bar{h}$. σ is defined by (6 c).

Note that Inequality (7) already occured in (6.3.29 c). By (8) the matrix $I - M_l'$ is non-singular, and Lemma 14.1.7 implies

Remark 14.1.9. Assume $\|M_l'\|_{\mathscr{U} \leftarrow \mathscr{U}} < 1$. Iteration (5) converges to

$$\lim_{i \to \infty} u_l^i = \hat{u}_l = (I - M_l')^{-1}\, g_l \tag{14.1.9}$$

with M_l' and g_l from (6 b–d).

It remains to study the accuracy of $\hat{u}_l$. We introduce the errors

$$\hat{\delta}_l := \hat{u}_l - u_l^*, \qquad \bar{\delta}_l := \bar{u}_l - u_l^*, \qquad \text{where } u_l^* = \tilde{R}_l\, u, \qquad \bar{u}_l := \mathscr{S}_l^{(\nu)}(\hat{u}_l, f_l).$$

Note that $\lim_{i \to \infty} \bar{u}_l^i = \bar{u}_l$ differs from $\hat{u}_l$. Consider iteration (5) for $u_l^i = u_l^{i+1} = \hat{u}_l$ and $\bar{u}_l^i = \bar{u}_l$. (5 b) and (5 c) yield

$$\begin{aligned}\hat{u}_l &= \bar{u}_l - p\,[\tilde{r}\,\bar{u}_l - L_{l-1}^{-1}\,(f_{l-1} + d_{l-1})] \\ &= \bar{u}_l - p\,L_{l-1}^{-1}\,[L_{l-1}\,\tilde{r}\,\bar{u}_l - f_{l-1} - d_{l-1}] \\ &= \bar{u}_l - p\,L_{l-1}^{-1}\,[(1-\sigma)\,(L_{l-1}\,\tilde{r}\,\bar{u}_l - f_{l-1}) + \sigma\,r\,(L_l\,\bar{u}_l - f_l)].\end{aligned}$$

Since $L_l\,\bar{u}_l - f_l = L_l\,u_l^* - f_l + L_l\,\bar{\delta}_l = t_l + L_l\,\bar{\delta}_l$ and $L_{l-1}\,r\,\bar{u}_l - f_{l-1} = L_{l-1}\,u_{l-1}^* - f_{l-1} + L_{l-1}\,\tilde{r}\,\bar{\delta}_l = t_{l-1} + L_{l-1}\,\tilde{r}\,\bar{\delta}_l$ because of $\tilde{r}\,u_l^* = \tilde{r}\,\tilde{R}_l\,u = \tilde{R}_{l-1}\,u = u_{l-1}^*$, the error $\hat{\delta}_l = \hat{u}_l - u_l^*$ equals

$$\begin{aligned}\hat{\delta}_l &= \bar{u}_l - u_l^* - p\,L_{l-1}^{-1}\,[(1-\sigma)\,(t_{l-1} + L_{l-1}\,\tilde{r}\,\bar{\delta}_l) + \sigma\,r\,(t_l + L_l\,\bar{\delta}_l)] \\ &= \{I - p\,L_{l-1}^{-1}\,[(1-\sigma)\,L_{l-1}\,\tilde{r} + \sigma\,r\,L_l]\}\,\bar{\delta}_l - p\,L_{l-1}^{-1}\,[(1-\sigma)\,t_{l-1} + \sigma\,r\,t_l] \\ &= C_l'\,\bar{\delta}_l - p\,L_{l-1}^{-1}\,[(1-\sigma)\,t_{l-1} + \sigma\,r\,t_l]\end{aligned} \tag{14.1.10a}$$

with C_l' from (6 b). From

$$\begin{aligned}\bar{\delta}_l = \bar{u}_l - u_l^* &= \mathscr{S}_l^{(v)}(\hat{u}_l, f_l) - u_l^* = S_l^{(v)}\hat{u}_l + T_l^{(v)} f_l - u_l^* \\ &= S_l^{(v)}\hat{u}_l + (I - S_l^{(v)}) L_l^{-1} f_l - u_l^* = (I - S_l^{(v)})(u_l - u_l^*) + S_l^{(v)}(\hat{u}_l - u_l^*) \\ &= (I - S_l^{(v)})(u_l - u_l^*) + S_l^{(v)}\hat{\delta}_l \end{aligned} \tag{14.1.10 b}$$

(cf. (1.3.6)) we conclude that

$$\begin{aligned}\hat{\delta}_l &= C_l' S_l^{(v)}\hat{\delta}_l + C_l'(I - S_l^{(v)})[u_l - u_l^*] - p L_{l-1}^{-1}[(1-\sigma)t_{l-1} + \sigma r t_l] \\ &= M_l'\hat{\delta}_l + v_l + w_l, \end{aligned} \tag{14.1.10 c}$$

where

$$v_l = C_l'(I - S_l^{(v)})(u_l - u_l^*), \quad w_l = -p L_{l-1}^{-1}[(1-\sigma)t_{l-1} + \sigma r t_l]. \tag{14.1.10d}$$

It suffices to study v_l and w_l because of

Lemma 14.1.10. Let $\|M_l'\|_{\mathscr{U}\leftarrow\mathscr{U}} \leqq \zeta < 1$ (cf. (8)). Then (10 c) implies

$$\|\hat{\delta}_l\|_{\mathscr{U}} = \|\hat{u}_l - \tilde{R}_l u\|_{\mathscr{U}} \leqq (\|v_l\|_{\mathscr{U}} + \|w_l\|_{\mathscr{U}})/(1-\zeta). \tag{14.1.11}$$

If $t_k = h_k^\kappa R_k t + O(h_k^\tau)$ (cf. (3 d)), the following inequality should hold:

$$\|L_{l-1}^{-1}(r t_l - (h_l^\kappa/h_{l-1}^\kappa) t_{l-1})\|_{\mathscr{U}} \leqq C_t h_l^\tau, \quad \tau > \kappa. \tag{14.1.12}$$

Lemma 14.1.11. Suppose (12) and $\|p\|_{\mathscr{U}\leftarrow\mathscr{U}} \leqq \bar{C}_p$ (cf. (7.1.2)). Then

$$\|w_l\|_{\mathscr{U}} \leqq C_w h_l^\tau, \quad C_w := \bar{C}_p C_t \sigma \quad \text{with } \sigma \text{ from (6c)}. \tag{14.1.13}$$

Below we shall discuss the inequality $\|v_l\|_{\mathscr{U}} \leqq C_v h_l^\tau$, which may be written as

$$\|(I - p L_{l-1}^{-1} r L_l) T_l^{(v)} t_l\|_{\mathscr{U}} \leqq C_v h_l^\tau, \tag{14.1.14}$$

since $u_l - u_l^* = L_l^{-1}(f_l - L_l u_l^*) = -L_l^{-1} t_l$ and $T_l^{(v)} = (I - S_l^{(v)}) L_l^{-1}$. Lemmata 14.1.10–11 yield

Proposition 14.1.12. Assume $\tilde{r}\tilde{R}_l = \tilde{R}_{l-1}$, $\|M_l'\|_{\mathscr{U}\leftarrow\mathscr{U}} \leqq \zeta < 1$ according to (8). Suppose that the truncation errors t_l, t_{l-1} satisfy (12) and (14). Then the τ-extrapolation process (5) converges to a limit $\hat{u}_l$ with

$$\|\hat{u}_l - \tilde{R}_l u\|_{\mathscr{U}} \leqq h_l^\tau (C_v + C_w)/(1-\zeta). \tag{14.1.15}$$

It remains to study Inequality (14).

Note 14.1.13. Let $\|\cdot\|_{\mathscr{U}}$ be the Euclidean norm $|\cdot|_0$. Inequality (14) holds with $\tau = \kappa + 2m$ if

$$|I - p L_{l-1}^{-1} r L_l|_{0\leftarrow 0} \leqq C, \tag{14.1.16 a}$$

$$|T_l^{(v)}|_{0\leftarrow 0} \leqq C h_l^{2m}, \tag{14.1.16 b}$$

$$|t_l|_0 \leqq C h_l^\kappa. \tag{14.1.16 c}$$

Inequality (16 a) follows from the approximation property (6.1.8): $\|L_l^{-1} - p L_{l-1}^{-1} r\|_{\mathscr{U}\leftarrow\mathscr{F}} \leqq C_A h_l^\alpha$, and from $\|L_l\|_{\mathscr{F}\leftarrow\mathscr{U}} \leqq C_L h_l^{-\alpha}$ (smoothing statement we need scales $\mathscr{U}^s$, $\mathscr{F}^s$ of Banach spaces and their discrete analogues $\mathscr{U}_l^s$, $\mathscr{F}_l^s$. Let R_l, R_l', and $\tilde{R}_l$ be restrictions:

$$R_l: \mathscr{F}^\sigma \to \mathscr{F}_l^\sigma, \quad R_l': \mathscr{F}^\sigma \to \mathscr{F}_l^\sigma, \quad \tilde{R}_l: \mathscr{U}^\sigma \to \mathscr{U}_l^\sigma$$

such that the right-hand sides of Eq. (1 a, b) are given by

$$f_l = R_l f, \quad f_l' = R_l' f. \tag{14.2.5 a}$$

Proposition 14.2.3. Let $s \in \mathbb{R}$ and $i \geqq 0$ be fixed. Suppose (5 a) and (i) to (iii):
(i) consistency of L_l and L_l' with the respective orders κ and κ':

$$\| R_l L - L_l \tilde{R}_l \|_{\mathscr{F}_l^{\sigma-2} \leftarrow \mathscr{U}^{\sigma+\kappa/m}} \leqq C h_l^\kappa \quad \text{for } \sigma = s + i\kappa/m, \tag{14.2.5 b}$$

$$\| R_l' L - L_l' \tilde{R}_l \|_{\mathscr{F}_l^{\sigma-2} \leftarrow \mathscr{U}^{\tau}} \leqq C h_l^{\min(\kappa', (\tau-\sigma)m)}$$

$$\text{for } \sigma = s, s+\kappa/m, \ldots, s+(i-1)\kappa/m \text{ and } \tau = s+(i+1)\kappa/m, \tag{14.2.5 c}$$

where $2m$ is the order of the differential operator L_Ω;
(ii) relative consistency of L_l and L_l':

$$\| L_l' - L_l \|_{\mathscr{F}_l^{\sigma-2} \leftarrow \mathscr{U}_l^{\sigma+\kappa/m}} \leqq C h_l^\kappa \text{ for } \sigma = s, s+\kappa/m, \ldots, s+(i+1)\kappa/m \tag{14.2.5 d}$$

(iii) σ-regularity of L_l (cf. Definition 6.3.31):

$$\| L_l^{-1} \|_{\mathscr{U}_l^\sigma \leftarrow \mathscr{F}_l^{\sigma-2}} \leqq C \quad \text{for } \sigma = s, s+\kappa/m, \ldots, s+i\kappa/m. \tag{14.2.5 e}$$

Then the ith iterate u_l^i of the iterated defect correction (2) satisfies

$$\| u_l^i - u_l^* \|_{\mathscr{U}_l^s} \leqq C_i h_l^{\min\{\kappa', (i+1)\kappa\}} \| u \|_{\mathscr{U}^{s+(i+1)\kappa/m}}, \tag{14.2.6}$$

where $u_l^* = \tilde{R}_l u$ is the restriction of the solution u of $Lu = f$.

Proof. Use the representations

$$u_l^0 - u_l^* = L_l^{-1}(R_l L - L_l \tilde{R}_l) u,$$
$$u_l^i - u_l^* = u_l^{i-1} - u_l^* - L_l^{-1}(L_l' u_l^{i-1} - f_l')$$
$$= L_l^{-1}(L_l - L_l')(u_l^{i-1} - u_l^*) + L_l^{-1}(R_l' L - L_l' \tilde{R}_l) u \quad (i \geqq 1)$$

and prove the assertion by induction over i. □

As a consequence the optimal order of accuracy is obtained as soon as $i \geqq \kappa'/\kappa - 1$. Assumption (5 e) implies the stability of L_l; however, in contrast to Note 14.2.2, neither the stability nor the regularity of L_l' are required.

14.3 Combinations of the Multi-Grid Method and the Defect Correction Principle

14.3.1 The Multi-Grid Algorithm as a Secondary Iteration

Given a multi-grid program for solving $L_l u_l = f_l$ (or $\mathscr{L}_l(u_l) = f_l$ in the nonlinear case) the exact solution in (2.2 a, b) (or (2.4 a, b)) can be replaced by a certain number j of multi-grid iterations. To obtain sufficiently small iteration error, this number j has to satisfy

$$\zeta^j \leqq C h_l^\kappa,$$

where ζ is the contraction number of the multi-grid iteration.

Regarding iteration (2.2 b) or (2.4 b) as an infinite process we may replace L_l^{-1} or $\mathscr{L}_l^{-1}$ by one multi-grid step. In the general nonlinear case of (2.4 b) we obtain

$$\begin{aligned} &\tilde{u}_l := u_l^i; \\ &d_l := \mathscr{L}_l(\tilde{u}_l) - \mathscr{L}_l'(\tilde{u}_l); \\ &NMGM(l, \tilde{u}_l, d_l); \ \textbf{comment} \text{ multi-grid process for } \mathscr{L}_l(u_l) = d_l; \\ &u_l^{i+1} := \tilde{u}_l; \end{aligned} \tag{14.3.1}$$

Lemma 14.3.1. Let ζ be the contraction number of $NMGM$, while q is the contraction number of the iterated defect correction process (2.4b). The combined process (1) has the contraction number $\bar{q} \leqq \zeta + q + \zeta q$. If $\bar{q} < 1$, iteration (1) converges to the solution of $\mathscr{L}_l'(u_l') = 0$.

Proof (cf. Auzinger-Stetter [1]). Let $\bar{u}_l^{i+1} := \mathscr{L}_l^{-1}(d_l)$ be the exact result of (2.4 b) applied to u_l^i. The Lemma is proved by

$$\begin{aligned} \| u_l^{i+1} - u_l' \|_{\mathscr{U}} &\leqq \| \bar{u}_l^{i+1} - u_l^{i+1} \|_{\mathscr{U}} + \| \bar{u}_l^{i+1} - u_l' \|_{\mathscr{U}} \\ &\leqq \zeta \| u_l^i - \bar{u}_l^{i+1} \|_{\mathscr{U}} + q \| u_l^i - u_l' \|_{\mathscr{U}} \\ &\leqq (\zeta + q) \| u_l^i - u_l' \|_{\mathscr{U}} + \zeta \| \bar{u}_l^{i+1} - u_l' \|_{\mathscr{U}} \\ &\leqq (\zeta + q + \zeta q) \| u_l^i - u_l' \|_{\mathscr{U}}. \end{aligned}$$

□

14.3.2 The Defect Correction with Additional Smoothing

By Lemma 14.3.1, iteration (1) is fast only if both the multi-grid iteration and the iterated defect correction (2.2) are fast. Unfortunately, $\| I - L_l^{-1} L_l' \|_{\mathscr{U} \leftarrow \mathscr{U}} \leqq q \ll 1$ cannot be expected in general. The consistency implies $(I - L_l^{-1} L_l') v_l \approx 0$ only for smooth v_l. We are in the same situation as with the coarse-grid correction $(I - p L_{l-1}^{-1} r L_l) v_l$, which yields a small residual only if v_l is smooth. Therefore, each step of iteration (2.2 b) or (1) should be preceded by a smoothing step. Let $\mathscr{S}_l'$ be a smoothing iteration with respect to $\mathscr{L}_l'(u_l') = 0$. The resulting iteration reads

$$\tilde{u}_l := \mathscr{S}_l'^{(\mu)}(u_l^i, 0); \quad \text{(additional smoothing w.r.t. } \mathscr{L}_l'(u_l') = 0) \tag{14.3.2 a}$$

$$d_l := \mathscr{L}_l(\tilde{u}_l) - \mathscr{L}_l'(\tilde{u}_l); \tag{14.3.2 b}$$

$$NMGM(l, \tilde{u}_l, d_l); \quad \text{(multi-grid process w.r.t. } \mathscr{L}_l(u_l) = d_l) \tag{14.3.2 c}$$

$$u_l^{i+1} := \tilde{u}_l; \tag{14.3.2 d}$$

This iteration proposed by Auzinger-Stetter [1] requires no change in the multi-grid program. In the linear case of Eqs. (2.1 a, b) iteration (2 a–d) is equivalent to

$$\tilde{u}_l := \mathscr{S}_l'^{(\mu)}(u_l^i, f_l'); \quad v_l := 0; \quad MGM(l, v_l, L_l'\tilde{u}_l - f_l'); \quad u_l^{i+1} := \tilde{u}_l - v_l. \tag{14.3.3}$$

Remark 14.3.2. (i) $u_l' = L_l'^{-1} f_l'$ is a fixed point of iteration (3), provided that $\mathscr{S}_l'(u_l', f_l') = u_l'$.

(ii) The iteration matrix of iteration (3) is given by

$$(I - L_l^{-1} L_l') S_l'^{(\mu)} + M_l L_l^{-1} L_l' S_l'^{(\mu)}, \tag{14.3.4}$$

where $S_l'^{(\mu)}$ and M_l are the iteration matrices of $\mathscr{S}_l'^{(\mu)}$ and MGM, respectively.

Proof. (i) is obvious. (ii) The result of $MGM(\ldots)$ with starting value $v_l^{(0)} = 0$ is $v_l := M_l 0 + N_l(L_l'\tilde{u}_l - f_l')$. Hence, the iteration matrix is $(I - N_l L_l') S_l'^{(\mu)}$. Using $N_l = (I - M_l) L_l^{-1}$ (cf. (1.3.6)) we obtain the proof of the Remark. □

To prove convergence, the first term in (4) can be written as $(L_l'^{-1} - L_l^{-1}) L_l' S_l'^{(\mu)}$. The inequality

$$\| L_l'^{-1} - L_l^{-1} \|_{\mathscr{U} \leftarrow \mathscr{F}} \leqq C_1 h_l^{\alpha}$$

is similar to the approximation property, while

$$\| L_l' S_l'^{(\mu)} \|_{\mathscr{F} \leftarrow \mathscr{U}} \leqq \eta'(\mu) h_l^{-\alpha}$$

coincides with the smoothing property of $\mathscr{S}_l'^{(\mu)}$. The second term in (4) can be split into $\| M_l \|_{\mathscr{U} \leftarrow \mathscr{U}} \| L_l^{-1} L_l' S_l'^{(\mu)} \|_{\mathscr{U} \leftarrow \mathscr{U}}$. Assume $\| M_l \|_{\mathscr{U} \leftarrow \mathscr{U}} \leqq \zeta$ (ζ is the multi-grid contraction number) and $\| L_l^{-1} L_l' S_l'^{(\mu)} \|_{\mathscr{U} \leftarrow \mathscr{U}} \leqq C_2$. Then $C_1 \eta'(\mu) + C_2 \zeta$ is an upper bound of the contraction number of iteration (3).

A further iteration method proposed by Auzinger-Stetter [1, p. 338] is algorithm (2) with $\mathscr{S}_l'^{(\mu)}$ replaced by the smoother $\mathscr{S}_l^{(\mu)}$ with respect to the equation $\mathscr{L}_l(u_l) = 0$, which is also used inside the multi-grid iteration. In the linear case this version reads thus:

$$\tilde{u}_l := \mathscr{S}_l^{(\mu)}(u_l^i, f_l); \quad v_l := 0; \quad MGM(l, v_l, L_l'\tilde{u}_l - f_l'); \quad u_l^{i+1} := \tilde{u}_l - v_l. \tag{14.3.5}$$

The proof of the following result is left to the reader:

Remark 14.3.3. (i) The iteration matrix corresponding to (5) is given by $M_l' = [I - (I - M_l) L_l^{-1} L_l'] S_l^{(\mu)}$, which results from (4) by replacing S_l' by S_l.

(ii) Since u_l' is not a fixed point of $\mathscr{S}_l^{(\mu)}$, u_l' is also not a fixed point of iteration (5). The fixed point of (5) is represented by

$$\hat{u}_l = u_l' - (I - M_l')^{-1} [I - (I - M_l) L_l^{-1} L_l'](I - S_l^{(\mu)})(u_l' - u_l).$$

(iii) The error $\| \hat{u}_l - u_l' \|_{\mathscr{U}}$ is determined by $\| (I - S_l^{(\mu)})(u_l' - u_l) \|_{\mathscr{U}} = \| T_l^{(\mu)} L_l (u_l' - u_l) \|_{\mathscr{U}}$. For $T_l^{(\mu)}$ see (1.16 b).

14.3.3 The Defect Correction Inside the Multi-Grid Process

14.3.3.1 Algorithms

The algorithms from § 14.3.2 have performed the defect correction outside the multi-grid process. The obvious advantage of this is that standard multi-grid programs can be applied. On the other hand, the outer and inner smoothing iterations require more computational work than is necessary. The defect correction process can be combined with the coarse-grid correction $u_l \mapsto u_l - p L_{l-1}^{-1} r d_l$ by replacing the defect $d_l = L_l u_l - f_l$ by $d_l := L'_l u_l - f'_l$ (cf. Brandt [9], [12], Hackbusch [26], [30]). The resulting iteration can be written

$$\tilde{u}_l := \mathscr{S}_l^{(\nu)}(u_l^i, f_l); \tag{14.3.6 a}$$

$$d_{l-1} := r * (L'_l * \tilde{u}_l - f'_l); \tag{14.3.6 b}$$

$$v_{l-1} := 0; \quad \textbf{for } j := 1(1)\gamma \quad \textbf{do } MGM(l-1, v_{l-1}, d_{l-1}); \tag{14.3.6 c}$$

$$u_l^{i+1} := \tilde{u}_l - p * v_{l-1}; \tag{14.3.6 d}$$

where MGM is the usual multi-grid process (4.1.1). Specifying ν_1 and ν_2 in MGM by $\nu_1 = \nu$ and $\nu_2 = 0$, we can rewrite algorithm (6) as

multi-grid iteration for $L_l u_l = f_l$ with defect correction by $L'_l u'_l = f'_l$ (14.3.7)

procedure $MGMDC\,(k, u, f)$; **integer** k; **array** u, f;
comment k is restricted by $k \leqq l$,
$l > 0$ is the level number of the finest grid;
if $k = 0$ **then** $u := L_0^{-1} * f$ **else** (14.3.7 a)
begin integer j; **array** v, d;
$u := \mathscr{S}_l^{(\nu)}(u, f)$; (14.3.7 b)
$d := r *$ (**if** $k = l$ **then** $L'_l * u - f'_l$ **else** $L_k * u - f$); (14.3.7 c)
$v := 0$; **for** $j := 1(1)\gamma$ **do** $MGMDC(k-1, v, d)$; (14.3.7 d)
$u := u - p * v$ (14.3.7 e)
end;

Remark 14.3.4. The same modification can be built into the nonlinear multigrid iteration $NMGM$. For that purpose replace statement (9.3.12 d) in algorithm (9.3.12) by

$$d := r * (\textbf{if } l = l_{\max} \textbf{ then } \mathscr{L}'_{l_{\max}}(u) - f \textbf{ else } \mathscr{L}_l(u) - f). \tag{14.3.8}$$

The analogous replacement for $\mathscr{L}_l(u) - f$ can be used in (9.3.13c).

In general, neither $u_l = L_l^{-1} f_l$ nor $u'_l = L'^{-1}_l f'_l$ (provided L'_l is nonsingular) is a fixed point of iteration (7). However, we shall show that the iteration converges to a fixed point $\hat{u}_l$ with $\| \hat{u}_l - u_l^* \|_{\mathscr{U}} = O(h_l^{\min(\kappa', \kappa + 2m)})$, where κ and κ' are the consistency orders of the discretisations (2.1 a) and (2.1 b) respectively.

14.3.3.2 Convergence Analysis of the Multi-Grid Iteration with Defect Correction

Lemma 14.3.5. Let $u_l = L_l^{-1} f_l$ be a fixed point of $\mathscr{S}_l^{(\nu)}$ and denote the iteration matrix of the (usual) multi-grid iteration at level $l-1$ by M_{l-1}. Thus, iteration (7) at level l can be represented by

$$u_l^{i+1} = M_l' u_l^i + g_l, \tag{14.3.9 a}$$

where

$$M_l' = M_l'(\nu) = [I - p(I - M_{l-1}^{\gamma}) L_{l-1}^{-1} r L_l'] S_l^{(\nu)}, \tag{14.3.9 b}$$

$$g_l = g_l(f_l, f_l', \nu) \tag{14.3.9 c}$$

$$= [I - p(I - M_{l-1}^{\gamma}) L_{l-1}^{-1} r L_l'][I - S_l^{(\nu)}] L_l^{-1} f_l + p(I - M_{l-1}^{\gamma}) L_{l-1}^{-1} r f_l'.$$

From (2.5 d) [with $\kappa = \alpha$, $\sigma = s_{\mathscr{U}} m$] and (2.5 e) [with $\sigma = (s_{\mathscr{F}} + 2)m$ and $\sigma = s_{\mathscr{U}} m$] one concludes

$$\| L_l^{-1} (L_l - L_l') L_l^{-1} \|_{\mathscr{U} \leftarrow \mathscr{F}} \leqq C_1 h_l^{\alpha}, \tag{14.3.10}$$

where $\| \cdot \|_{\mathscr{U}}$ ($\| \cdot \|_{\mathscr{F}}$) is the norm of $\mathscr{U}_l^{s_{\mathscr{U}}}$ ($\mathscr{F}_l^{s_{\mathscr{F}}}$) and $\alpha = (s_{\mathscr{F}} - s_{\mathscr{U}} + 2)m$ according to Eq. (6.2.9). Note that α is identical with α from the smoothing and approximation properties.

Proposition 14.3.6. Suppose (10). Under the conditions of Theorem 7.1.2 with smoothing property (6.1.7 a) also holding for $\nu = 0$ (i.e. $S_l^{(\nu)} = I$), the multi-grid convergence stated in Theorem 7.1.2 remains valid for iteration (7):

$$\| M_l'(\nu) \|_{\mathscr{U} \leftarrow \mathscr{U}} \leqq C \eta(\nu) < 1, \tag{14.3.11}$$

if $\underline{\nu} \leqq \nu < \bar{\nu}(h_1)$, $h_1 \leqq \bar{h}$.

Proof. By (9 b),

$$M_l' = [I - p(I - M_{l-1}^{\gamma}) L_{l-1}^{-1} r L_l] S_l^{(\nu)}$$
$$+ p(I - M_{l-1}^{\gamma}) L_{l-1}^{-1} r L_l [L_l^{-1} (L_l - L_l') L_l^{-1}] L_l S_l^{(\nu)}.$$

The first term is the usual multi-grid iteration matrix M_l, while the second term can be estimated by

$$\| p(I - M_{l-1}^{\gamma}) L_{l-1}^{-1} r L_l \|_{\mathscr{U} \leftarrow \mathscr{U}} \| L_l^{-1} (L_l - L_l') L_l^{-1} \|_{\mathscr{U} \leftarrow \mathscr{F}} \| L_l S_l^{(\nu)} \|_{\mathscr{F} \leftarrow \mathscr{U}}$$
$$\leqq \| p \|_{\mathscr{U} \leftarrow \mathscr{U}} (1 + \| M_{l-1} \|_{\mathscr{U} \leftarrow \mathscr{U}}^{\gamma}) \| L_{l-1}^{-1} r L_l \|_{\mathscr{U} \leftarrow \mathscr{U}} C_1 \eta(\nu).$$

Using (7.1.2) and Lemma 7.1.5 [with $\nu = 0$] we obtain $\| M_l' \|_{\mathscr{U} \leftarrow \mathscr{U}} \leqq C \eta(\nu)$. □

Note 14.3.7. (i) In general, α from (10) cannot exceed the consistency order κ of L_l, provided that $\| \cdot \|_{\mathscr{U}} = \| \cdot \|_{\mathscr{F}}$.

(ii) The convergence result stated above involves the inverse of L_l but not of L'_l. Convergence holds even when L'_l is unstable or singular (cf. Note 14.2.2).

(iii) Thanks to (11) there exists a unique limit $\hat{u}_l = \hat{u}_l(v) := \lim_{j\to\infty} u_l^j$ of the iterates u_l^j of algorithm (7). $\hat{u}_l$ is the fixed point of iteration (7) and can also be characterised by means of (9 b, c):

$$\hat{u}_l = \hat{u}_l(v) = (I - M'_l)^{-1} g_l. \tag{14.3.12}$$

14.3.3.3 Error Estimates of the Limit $\hat{u}_l$

Let $u_l^* = \tilde{R}_l u$ be a restriction of the continuous solution $u = L^{-1} f$ onto the grid Ω_l. From (12), (9 c) and

$$[I - M'_l] u_l^* = [I - p(I - M_{l-1}^{\gamma}) L_{l-1}^{-1} r L'_l][I - S_l^{(v)}] u_l^* + p(I - M_{l-1}^{\gamma}) L_{l-1} r L'_l u_l^*$$

one obtains the following representation of the error $\hat{u}_l - u_l^*$.

Lemma 14.3.8. Suppose (11). The error $\hat{u}_l - u_l^* = \hat{u}_l - \tilde{R}_l u$ can be decomposed into

$$\hat{u}_l - u_l^* = [I - M'_l]^{-1} (v_l + w_l) \tag{14.3.13 a}$$

with

$$v_l = [I - p(I - M_{l-1}^{\gamma}) L_{l-1}^{-1} r L'_l][I - S_l^{(v)}] L_l^{-1} (f_l - L_l u_l^*), \tag{14.3.13 b}$$

$$w_l = p(I - M_{l-1}^{\gamma}) L_{l-1}^{-1} r (f'_l - L'_l u_l^*). \tag{14.3.13 c}$$

The equivalent of Lemma 14.1.10 is stated in

Lemma 14.3.9. Let $\| M'_l \|_{\mathcal{U}\leftarrow\mathcal{U}} \leqq \zeta < 1$ (cf. (11)). Then (13) implies

$$\| \hat{u}_l - u_l^* \|_{\mathcal{U}} \leqq (\| v_l \|_{\mathcal{U}} + \| w_l \|_{\mathcal{U}})/(1 - \zeta). \tag{14.3.14}$$

The decomposition into $v_l + w_l$ separates the influences of the defects $f_l - L_l u_l^*$ and $f'_l - L'_l u_l^*$. w_l can be expected to be of order $h_l^{\kappa'}$, where κ' is the consistency order of $L'_l u'_l = f'_l$:

Lemma 14.3.10. Suppose $\| \cdot \|_{\mathcal{U}} = \| \cdot \|_{\mathcal{U}_l^s}$ and let $s \leqq \tau \leqq s + \kappa'/m$. In addition to the conditions of Proposition 14.3.6 assume

$$\| L_l^{-1} \|_{\mathcal{U}_l^s \leftarrow \mathcal{F}_l^{s-2}} \leqq C, \qquad \| L'_l u_l^* - f'_l \|_{\mathcal{F}_l^{s-2}} \leqq C h_l^{(\tau - s)m} \| u \|_{\mathcal{U}^\tau}. \tag{14.3.15}$$

Then inequality (16) holds:

$$\| w_l \|_{\mathcal{U}} \leqq C h_l^{(\tau - s)m} \| u \|_{\mathcal{U}^\tau}. \tag{14.3.16}$$

Proof. The first factor in the right-hand side of

$$\| w_l \|_{\mathcal{U}} \leqq \| p(I - M_{l-1}^{\gamma}) L_{l-1}^{-1} r L_l \|_{\mathcal{U}\leftarrow\mathcal{U}} \| L_l^{-1} \|_{\mathcal{U}_l^s \leftarrow \mathcal{F}_l^{s-2}} \| L'_l u_l^* - f'_l \|_{\mathcal{F}_l^{s-2}}$$

is bounded by a constant as already used in the proof of Proposition 14.3.6. □

Finally, we state that all theoretical results and all algorithmical extensions also carry over to the nonlinear iteration of Remark 14.3.4.

14.3.4 An Application to the Convection Diffusion Equation

Consider Eq. (10.4.1): $-\varepsilon \Delta u + \mathbf{c} \operatorname{grad} u = f_\Omega$. Denote discretisation (10.4.5 a) by $L_l(\varepsilon_l)$. Define

$$L_l = L_l(\varepsilon_l), \qquad L_l' = L_l(\varepsilon), \tag{14.3.22}$$

where, e.g., ε_l is defined by

$$\varepsilon_l = \max\{\varepsilon, |c_1| h_l, |c_2| h_l\}$$

(cf. (10.4.9)). The consistency orders are $\kappa = 1$, $\kappa' = 2$. L_l is stable, whereas L_l' is unstable for sufficiently small ε/h_l (cf. § 10.4.2). As stated above the iterations (7) and (21) do not suffer from the instability of L_l'. The latter iteration has been introduced by Brandt [12] under the name 'double discretisation'.

Unfortunately, the analysis of § 14.3.3 does not apply so easily to (22). For instance, the defects $|L_l u_l^* - f_l|_0$ and $|L_l' u_l^* - f_l'|_0$ cannot be expected to be of order h_l^1 and h_l^2, respectively, since, in general, $u_l^* = \tilde{R}_l u$ is not sufficiently smooth but contains a boundary layer. Therefore, an analysis would be needed which yields $\hat{u}_l - u_l^* = O(h_l^2)$ outside the neighbourhood of the layer.

As mentioned in § 10.4.3.3, the convergence of the usual multi-grid iteration with L_l from (22) is relatively slow, because L_{l-1} is a poor approximation to L_l. In this situation it pays to apply iteration (21) instead of (7) (cf. Note 14.3.14ii), i.e. to use L_k' also on the coarser grids. A closely related statement is given in

Note 14.3.15. The usual coarse-grid correction is $u_l \mapsto u_l - p\, u_{l-1}$ with

$$u_{l-1} = L_{l-1}^{-1} d_{l-1}, \qquad d_{l-1} = r(L_l u_l - f_l),$$

where $L_{l-1} = L_{l-1}(\varepsilon_{l-1})$, $L_l = L_l(\varepsilon_l)$. Börgers [1, p. 119] considers a two-grid iteration, where the coarse-grid solution u_{l-1} is replaced by u_{l-1}^1, which is defined to be the first iterate of the defect correction process (2.2) with $L_{l-1} = L_{l-1}(\varepsilon_{l-1})$ instead of L_l and $L_{l-1}(\varepsilon_l)$ instead of L_l'. The explicit description is

$$u_l \mapsto u_l - p\, u_{l-1}^1,$$
$$u_{l-1}^1 := L_{l-1}(\varepsilon_{l-1})^{-1} [2 L_{l-1}(\varepsilon_{l-1}) - L_{l-1}(\varepsilon_l)] L_{l-1}(\varepsilon_{l-1})^{-1} d_{l-1}.$$

Under the conditions of Lemma 10.4.7, the contraction number and spectral radius of the iteration matrix M_l' satisfy

$$\| M_l' \|_{\mathscr{U} \leftarrow \mathscr{U}} \geqq \varrho(M_l') \geqq 0.25,$$

which has to be compared with (10.4.10).

Another so-called 'mixed defect correction process' is described and discussed by Hemker [6], [7], [12], Hemker-de Zeeuw [1].

15. Local Techniques

In this chapter we shall discuss techniques which improve the discrete solution by locally defined discretisations. The general framework is given by the '*local defect correction*' in § 15.1. A global discretisation of the continuous problem is completely avoided by the '*domain decomposition methods*' studied in § 15.3.

15.1 The Local Defect Correction

Let

$$L_l u_l = f_l \tag{15.1.1}$$

be a given approximation of the boundary value problem $Lu = f$ in Ω. For some reason (e.g., because of a re-entrant corner) this discretisation might be insufficient in a subregion

$$\Omega^1 \subset \Omega \tag{15.1.2}$$

with the boundary $\Gamma^1 \cup (\partial\Omega^1 \cap \Gamma)$, where $\Gamma^1 = \partial\Omega^1 \backslash \Gamma \subset \Omega$ and Γ: $\partial\Omega$.

One remedy is the use of a locally refined grid. In the finite element case a triangulation can be constructed with increasingly small elements as depicted in Fig. 3.8.6 (cf. also Schatz-Wahlbin [1]). Even difference equations can be formulated on grids Ω_l with locally refined grid widths (cf. Kaspar-Remke [1]).

A second remedy makes use of the known local behaviour of the solution. The finite element discretisation of boundary value problems with re-entrant corners is e.g. studied by Babuška-Rosenzweig [1] and Blum-Dobrowolski [1]. In the finite difference case one can use a special asymptotic expansion (cf. Zenger-Gietl [1]).

All approaches mentioned above yield one system of equations. Instead, we can consider two separated equations: firstly, discretisation (1) in a regular grid Ω_l, and secondly, a discretisation of the local boundary value problem

$$L_\Omega v = f_\Omega \quad \text{in } \Omega^1 \subset \Omega, \qquad L_\Gamma v = f_\Gamma \quad \text{on } \partial\Omega^1 \cap \Gamma, \tag{15.1.3 a}$$

$$v = u \quad \text{on } \Gamma^1, \tag{15.1.3 b}$$

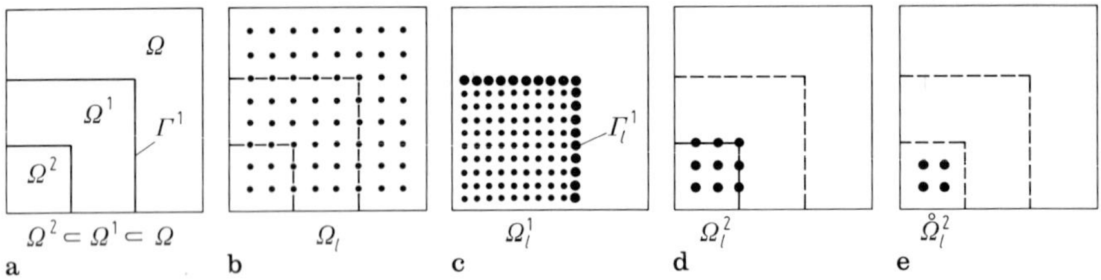

Fig. 15.1.1 a–e. Subregions and grids

where $u = L^{-1} f$ is the solution of $Lu = f$ in Ω. The second discretisation

$$L_l' v_l = f_l' \quad \text{in } \Omega_l^1, \qquad v_l = I_l u_l \quad \text{on } \Gamma_l^1 \tag{15.1.4}$$

is assumed to be adapted to the particular singularity of Eq. (3 a, b). In Eq. (4) the boundary conditions $v_l = I_l u_l$ on Γ_l^1 are formulated separately. I_l describes some interpolation of u_l in the Ω_l-grid at the boundary points of Γ_l^1. The grids Ω_l and Ω_l^1 may have different sizes (e.g. $h_l' < h_l$, where h_l' is the grid size of Ω_l^1, h_l the grid size of Ω_l) and Ω_l^1 may use other (local) coordinates (cf. Fig. 15.1.2).

The following iteration requires further subgrids

$$\Omega_l^2 \subset \Omega^1 \cap \Omega_l \quad \text{(cf. Fig. 15.1.1 d)}, \tag{15.1.5 a}$$

$$\mathring{\Omega}_l^2 := \{\mathbf{x} \in \Omega_l^2 \colon (L_l' v_l)(\mathbf{x}) \text{ involves only } v_l(\mathbf{y}) \text{ at } \mathbf{y} \in \Omega_l^2\} \quad \text{(cf. Fig. 15.1.1 e)}. \tag{15.1.5 b}$$

$\mathring{\Omega}_l^2$ consists of the 'interior points' of Ω_l^2. Possibly the inclusion $\Omega_l^2 \subset \Omega^1$ is strenghtened to

$$\text{distance } (\Omega_l^2, \Gamma^1) \geqq d > 0 \text{ independently of } l. \tag{15.1.5 c}$$

Since $\Omega_l^2 \subset \Omega^1$, it makes sense to interpolate v_l (defined on Ω_l^1) at $\mathbf{x} \in \Omega_l^2$. This interpolation will be denoted by J_l.

The 'local defect correction' combines the two discretisations (1), (4). Start with

$$f_l^0 := f_l \quad (f_l \text{ from Eq. (1)}). \tag{15.1.6 a}$$

The iterative process is defined by Step a) to Step g):

Step a) Compute the solution u^i of Eq. (6 b) on Ω_l (cf. Fig. 15.1.1 b):

$$L_l u_l^i = f_l^i; \tag{15.1.6 b}$$

Step b) Interpolate the boundary values on Γ_l^1 (cf. Fig. 15.1.1 c):

$$u_{l,\Gamma}^i := I_l u_l^i \quad (I_l \text{ from Eq. (4)}); \tag{15.1.6 c}$$

Step c) Solve the local problem (4) in Ω_l^1 (cf. Fig. 15.1.1 c):

$$L_l' v_l^i = f_l', \qquad v_l^i = u_{l,\Gamma}^i \quad \text{on } \Gamma_l^1; \tag{15.1.6 d}$$

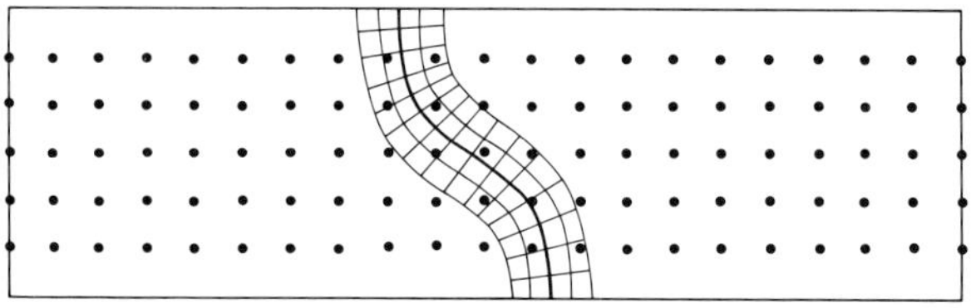

Fig. 15.1.2. Global grid Ω_l and local grid Ω_l^1 around an interface

Step d) Interpolate v_l^i on $\Omega_l^2 \subset \Omega_l$ (cf. Fig. 15.1.1 d):

$$\tilde{u}_l := J_l v_l^i; \tag{15.1.6 e}$$

Step e) Compute the defect in $\mathring{\Omega}_l^2$ (cf. Fig. 15.1.1 e):

$$d_l := L_l \tilde{u}_l - f_l \quad \text{in } \mathring{\Omega}_l^2 \subset \Omega_l; \tag{15.1.6 f}$$

Step f) Define the next right-hand side by

$$f_l^{i+1} := f_l + d_l \quad \text{on } \mathring{\Omega}_l^2, \qquad f_l^{i+1} := f_l \quad \text{on } \Omega_l \backslash \mathring{\Omega}_l^2. \tag{15.1.6 g}$$

Example 15.1.1. Consider the Poisson equation in the square grids of Fig. 15.1.3 with $h_l' < h_l$ (where, e.g., $h_l' = h_l/8$ is the grid size of Ω_l^1), and assume that the solution is $u(x, y) = \log(x^2 + y^2)$ (i.e. $f_\Omega = 0$). Both Eq. (1) and Eq. (4) are defined to be the standard five-point schemes with respect to the grid sizes h_l and h_l'. Because of the singularity at the lower left corner $x = y = 0$ the result of $L_l u_l = f_l$ is very poor, whereas the first iterate u_l^1 of (6 a–g) is already sufficiently accurate due to the finer grid size h_l'. Numerical results are reported by Hackbusch [37].

Equations (1) and (4) may be based on different types of discretisations. An extreme case is given in

Example 15.1.2. Sometimes local problems can be solved exactly without discretisation. Such an example is the Laplace equation $-\Delta u = 0$ in a sector Ω^1 (cf. Fig. 15.1.4), if $u = 0$ on the sides of the angle and $u = \sum_{\nu=1}^{N} \alpha_\nu \sin(\nu\pi\varphi/\varphi_0)$, $0 \leqq \varphi \leqq \varphi_0 = \frac{3}{2}\pi$, at the boundary point on Γ^1 corresponding to the angle φ. Therefore, we may choose Eq. (4) to be the (undiscretised) Laplace equation $-\Delta v_l = 0$ with the boundary condition $v_l = I_l u_l$, where $I_l u_l$ is a suitable representation $\sum_{\nu=1}^{N} \alpha_\nu \sin(\nu\pi\varphi/\varphi_0)$. Possible singularities emanating from the re-entrant corner are represented exactly and do not reduce the accuracy. For the complete definition of equations (1), (4) and for numerical results see Hackbusch [37].

The local defect correction process (6a–g) converges with a contraction number of order h_l^κ, where $\kappa > 0$ depends on the consistency order of the global and local discretisations and on the orders of the interpola-

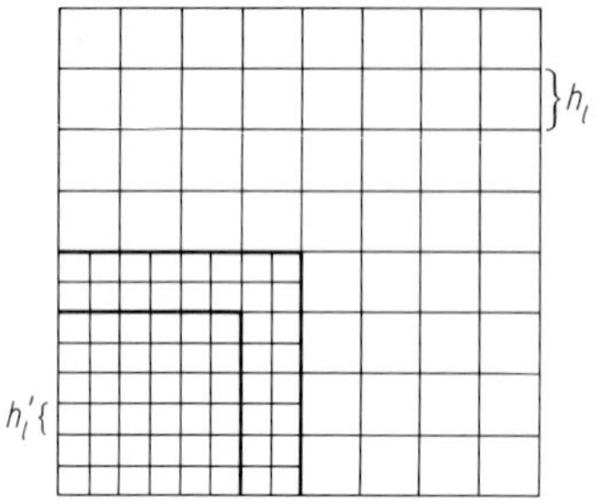

Fig. 15.1.3. Grids from Example 15.1.1

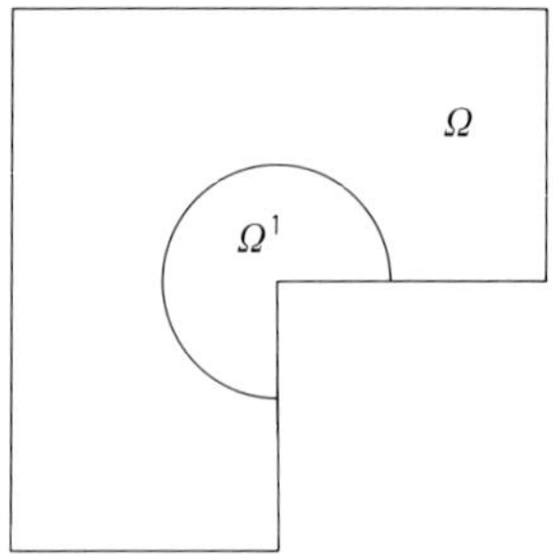

Fig. 15.1.4. Sector Ω^1 from Example 15.1.2

tions I_l, J_l:

$$u_l^i \to \hat{u}_l, \qquad \| u_l^{i+1} - \hat{u}_l \|_{\mathscr{U}} \leqq C h_l^{\kappa} \| u_l^i - \hat{u}_l \|_{\mathscr{U}}, \tag{15.1.7}$$

provided that h_l is sufficiently small and $d > 0$ in condition (5 c) (cf. Hackbusch [37]).

Remark 15.1.3. The limit value $\hat{u}_l$ in (7) is characterised by the system

$$\left[\begin{array}{c|c} L_l & A_l \\ \hline 0 & L_l' \\ \hdashline -I_l & 0 \quad I \end{array}\right] \begin{bmatrix} \hat{u}_l \\ \hline v_l \end{bmatrix} = \begin{bmatrix} f_{l,I} \\ \hline f_l' \\ \hdashline 0 \end{bmatrix} \tag{15.1.8 a}$$

where

$$f_{l,I} - A_l v_l := f_l + \chi (L_l J_l v_l - f_l),$$
$$(\chi d_l)(\mathbf{x}) = \begin{cases} d_l(\mathbf{x}) & \text{at } \mathbf{x} \in \mathring{\Omega}_l^2, \\ 0 & \text{elsewhere}. \end{cases} \tag{15.1.8 b}$$

Concerning the condition $d > 0$ in (5 c) we draw attention to

Note 15.1.4. In the case of $d = 0$ a counter-example can be given such that Eq. (8) is singular although both problems (1) and (4) are positive definite (cf. Hackbusch [37, Example 3.3.1]).

Let $u_l^* = \tilde{R}_l u$ be a restriction of the exact solution $u = L^{-1} f$ onto Ω_l, while $v_l^* = \tilde{R}_l^1 u$ is a restriction onto Ω_l^1. The error $\hat{u}_l - u_l^*$ is determined by the following quantities (cf. Hackbusch [37, § 3.3.2]):

(i) $L_l u_l^* - f_l$ on $\Omega_l \backslash \mathring{\Omega}_l^2$; it does not depend on $L_l u_l^* - f_l$ in $\mathring{\Omega}_l^2$, where the residual is supposed to be relatively large.
(ii) $L_l' v_l^* - f_l'$ on Ω_l^1; if Eq. (4) is chosen appropriately, this residual is assumed to be relatively small.
(iii) $J_l v_l^* - u_l^*$ on Ω_l^1 (interpolation error of J_l).
(iv) $I_l u_l^* - v_l^*$ on Γ_l^1 (interpolation error of I_l).

15.2 A Multi-Grid Iteration with a Local Grid Refinement

The local discretisation (1.4) can be regarded as a problem at level $l+1$ rather than l. Equation (1.8) becomes

$$L_l u_l - \chi L_l \tilde{r} u_{l+1} = (1-\chi) f_l, \tag{15.2.1 a}$$

$$L_{l+1} u_{l+1} = f_{l+1} \quad \text{on } \Omega_{l+1}, \qquad u_{l+1} = p_B u_l \quad \text{on } \Gamma_{l+1}, \tag{15.2.1 b}$$

when we rename: v_l as u_{l+1}, L_l' as L_{l+1}, f_l' as f_{l+1}, J_l as $\tilde{r}$, Γ_l^1 as Γ_{l+1}, I_l as p_B (p_B is a prolongation from the grid Ω_l only onto the boundary points Γ_{l+1} of the finer grid Ω_{l+1}). The grid $\Omega_{l+1} = \Omega_l^1 \backslash \Gamma_{l+1}$ covers only a part of Ω_l unlike the standard case (cf. Fig. 15.2.1).

The smoothing iteration $u_{l+1} \mapsto \mathscr{S}_{l+1}^{(v)}(u_{l+1}, f_{l+1})$ is performed on the grid Ω_{l+1}, while the boundary values of u_{l+1} on Γ_{l+1} are left fixed.

For the coarse-grid correction one has to compute the defects

$$d_l = L_l \tilde{u}_l - \chi L_l \tilde{r} \tilde{u}_{l+1} - (1-\chi) f_l \quad \text{in } \Omega_l, \tag{15.2.2 a}$$

$$d_{l+1} = L_{l+1} \tilde{u}_{l+1} - f_{l+1} \quad \text{in } \Omega_{l+1}, \tag{15.2.2 b}$$

$$d_{B,l+1} = \tilde{u}_{l+1} - p_B \tilde{u}_l \quad \text{on } \Gamma_{l+1}, \tag{15.2.2 c}$$

of the approximations $\tilde{u}_l$, $\tilde{u}_{l+1}$. The exact corrections $\delta u_l = \tilde{u}_l - \hat{u}_l$ and $\delta u_{l+1} = \tilde{u}_{l+1} - u_{l+1}$ are the solutions of

$$L_l \delta u_l - \chi L_l \tilde{r} \delta u_{l+1} = d_l \quad \text{in } \Omega_l, \tag{15.2.3 a}$$

$$L_{l+1} \delta u_{l+1} = d_{l+1} \quad \text{in } \Omega_{l+1}, \tag{15.2.3 b}$$

$$\delta u_{l+1} - p_B \delta u_l = d_{B,l+1} \quad \text{on } \Gamma_{l+1}. \tag{15.2.3 c}$$

A suitable coarse-grid analogue of system (3 a–c) is

$$L_l \delta u_l^0 - \chi L_l \delta u_l^1 = d_l \quad \text{in } \Omega_l, \tag{15.2.4 a}$$

$$L_l \delta u_l^1 = r d_{l+1} \quad \text{in } \Omega_l^1, \tag{15.2.4 b}$$

$$\delta u_l^1 - \delta u_l^0 = r_B d_{B,l+1} \quad \text{on } \Gamma_l^1, \tag{15.2.4 c}$$

where the pair $(\delta u_l^0, \delta u_l^1)$ is the coarse-grid approximation to $(\delta u_l, \delta u_{l+1})$. The grid Ω_l is unchanged, whereas Ω_{l+1} from Eq. (3 b) is replaced by Ω_l^1, which is a subset of Ω_l. For instance, one may choose

$$\Omega_l^1 = \Omega_l \cap \Omega^1 \quad (\Omega^1 \text{ from } (1.2)).$$

The equations $(L_l \delta u_l^1 - r d_{l+1})(\mathbf{x}) = 0$, $\mathbf{x} \in \Omega_l^1$, involve $\delta u_l^1(\mathbf{y})$ at the boundary points $\mathbf{y} \in \Gamma_l^1$. These values are prescribed by (4 c). r_B is a restriction from Γ_{l+1} onto Γ_l^1.

System (4 a–c) can be decoupled. Inserting (4 b) into (4 a), we can determine δu_l^0 by

$$L_l \delta u_l^0 = d_l + \chi r d_{l+1} \quad \text{in } \Omega_l. \tag{15.2.5 a}$$

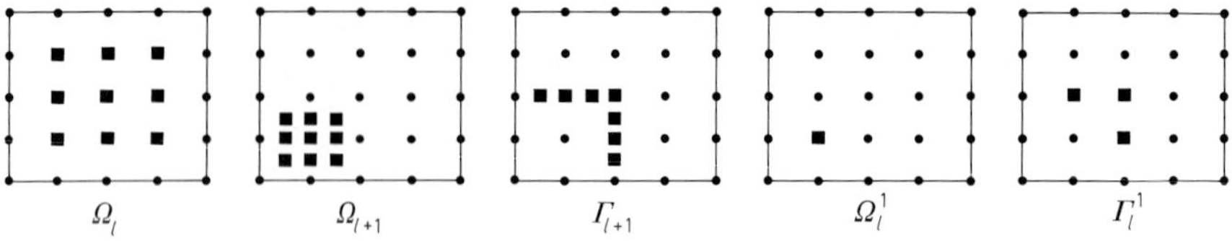

Fig. 15.2.1. Grids Ω_l, Ω_{l+1}, subgrid Ω_l^1, boundaries Γ_{l+1}, Γ_l^1

while δu_l^1 is the solution of

$$L_l \delta u_l^1 = r\, d_{l+1} \quad \text{in } \Omega_l^1, \qquad \delta u_l^1 = \delta u_l^0 + r_B\, d_{B,l+1} \quad \text{on } \Gamma_l^1. \tag{15.2.5 b}$$

The complete multi-grid iteration consists of the following steps:

(i) pre-smoothing: perform $\tilde{u}_{l+1} := \mathscr{S}_{l+1}^{(\nu_1)}(u_{l+1}^i, f_{l+1})$ on Ω_{l+1} with given boundary values $u_{l+1}^i(\mathbf{x})$, $\mathbf{x} \in \Gamma_{l+1}$. $\tilde{u}_l := u_l^i$ requires no smoothing.

(ii) coarse-grid correction:

(ii$_1$) Compute the defects d_l, d_{l+1}, $d_{B,l+1}$ by (2 a–c).

(ii$_2$) Solve the coarse-grid equation (5 a) by γ steps of the multi-grid iteration on Ω_l. Denote the approximate solution by $\widetilde{\delta u}_l^0$.

(ii$_3$) Solve the coarse-grid equation (5 b) by γ steps of the multi-grid iteration on Ω_l^1. Denote the approximate solution by $\widetilde{\delta u}_l^1$.

(ii$_4$) Correct $\tilde{u}_l$ and $\tilde{u}_{l+1}$ by

$$u_l^{i+1} = \tilde{u}_l - \widetilde{\delta u}_l^0, \qquad u_{l+1}^{i+1} := \tilde{u}_{l+1} - p\, \widetilde{\delta u}_l^1. \tag{15.2.6}$$

(iii) If post-smoothing is desired, perform $u_{l+1}^{i+1} \mapsto \mathscr{S}_l^{(\nu_2)}(u_{l+1}^{i+1}, f_{l+1})$ on Ω_{l+1} with renewed values $u_{l+1}^{i+1} = p_B u_l^{i+1}$ on Γ_{l+1}.

Lemma 15.2.1. The solution process becomes simpler if we assume:

a) $\mathring{\Omega}_l^2$, involved in the definition (1.8 b) of χ, coincides with Ω_l^1, i.e.

$$(\chi\, d_l)(\mathbf{x}) = \begin{cases} d_l(\mathbf{x}) & \mathbf{x} \in \Omega_l^1, \\ 0 & \text{elsewhere}. \end{cases} \tag{15.2.7 a}$$

b) $(L_l v_l)(\mathbf{x})$ at $\mathbf{x} \in \Omega_l^1$ involves only $v_l(\mathbf{y})$ at $\mathbf{y} \in \Omega_l^1 \cup \Gamma_l^1$.

c) Let w_l be a grid function on Ω_l, which coincides with v_l on $\Omega_l^1 \cup \Gamma_l^1$. Then

$$p_B w_l = p\, v_l \quad \text{on } \Gamma_{l+1}, \tag{15.2.7 b}$$

$$\tilde{r} p = I \tag{15.2.7 c}$$

must hold. Equation (7 b) can be regarded as a definition of p_B by means of p. Equations (7 c) requires $\tilde{r} = r_{inj}$ (cf. (3.5.1)).

Then the coarse-grid correction (ii) applied to $\tilde{u}_l$ and $\tilde{u}_{l+1}$ with

$$\tilde{u}_l = \tilde{r}\, \tilde{u}_{l+1} \quad \text{on } \Omega_l^1 \cup \Gamma_l^1, \qquad \tilde{u}_{l+1} = p_B \tilde{u}_l \quad \text{on } \Gamma_{l+1} \tag{15.2.7 d}$$

yields corrected values u_l^{i+1}, u_{l+1}^{i+1} also satisfying Eq. (7 d). In this case step (ii$_3$) can be omitted since

$$\delta u_l^1 = \delta u_l^0 . \tag{15.2.7 e}$$

Proof. (7 a) and (7 d) imply $d_l = 0$ on Ω_l^1 and $d_{B,l+1} = 0$. One concludes from assumption b) that the solution δu_l^0 of Eq. (5 a) also satisfies Eq. (5 b). Equation (5 c) holds for u_l^{i+1} and u_{l+1}^{i+1}, since

$$u_l^{i+1} = \tilde{u}_l - \delta u_l^0 = \tilde{r}\,\tilde{u}_{l+1} - \tilde{r}\,p\,\delta u_l^0 = \tilde{r}\,(\tilde{u}_{l+1} - p\,\delta u_{l+1}^1) = \tilde{r}\,u_{l+1}^{i+1} \quad \text{on } \Omega_l^1 \cup \Gamma_l^1,$$

$$u_{l+1}^{i+1} = \tilde{u}_{l+1} - p\,\delta u_l^1 = p_B\,\tilde{u}_l - p_B\,\delta u_l^0 = p_B\,(u_l - \delta u_l^0) = p_B\,u_l^{i+1} \quad \text{on } \Gamma_{l+1}. \ \square$$

Note 15.2.2. Suppose that $p_B\,\tilde{r}\,u_l$ depends only on $u_l(\mathbf{x})$ at $\mathbf{x} \in \Gamma_{l+1}$. Then condition (7 d) holds if $u_{l+1}^0 = p_B\,\tilde{r}\,u_{l+1}^0$ on Γ_{l+1} and if each smoothing step is followed by $\tilde{u}_l(\mathbf{x}) := \tilde{r}\,\tilde{u}_{l+1}(\mathbf{x})$ at all $\mathbf{x} \in \Omega_l^1 \cup \Gamma_l^1$.

The resulting iteration coincides with the algorithm briefly described by Brandt [3, p. 359]. The *nonlinear* version corresponding to algorithm (9.3.13) takes the following form.

Start: u_l^0, u_{l+1}^0 with $u_{l+1}^0 = p_B\,\tilde{r}\,u_{l+1}^0$ on Γ_{l+1}.

Iteration:

(i) pre-smoothing: perform $u_{l+1}^i \mapsto \tilde{u}_{l+1} = \mathscr{S}_{l+1}^{(\nu)}(u_{l+1}^i, f_{l+1})$ on Ω_{l+1};
(ii) set $\tilde{u}_l(\mathbf{x}) := \tilde{r}\,\tilde{u}_{l+1}(\mathbf{x})$ at all $\mathbf{x} \in \Omega_l^1 \cup \Gamma_l^1$;
(iii) coarse-grid correction:

$d_l := \mathscr{L}_l(\tilde{u}_l) - r(\mathscr{L}_{l+1}(\tilde{u}_{l+1}) - f_{l+1})$;

$v_l := \tilde{u}_l$; **for** $j := 1\,(1)\,\gamma$ **do** *NMGM* (l, v_l, d_l);

$u_{l+1}^{i+1} := \tilde{u}_{l+1} + p\,(v_l - \tilde{u}_l)$.

If $\Omega_l^1 = \Omega_l$, this algorithm coincides with iteration (9.3.13) at level $l + 1$.

Remark 15.2.3. Equation (7 e) and thus the simplified algorithm described above cannot be obtained if $\mathring{\Omega}_l^2$ is a proper subset of Ω_l^1. Note that $\mathring{\Omega}_l^2 = \Omega_l^1$ contradicts condition (1.5 c) with $d > 0$.

15.3 Domain Decomposition Methods

15.3.1 Domain Decomposition

The local defect correction methods described in §§ 15.1 – 2 enables us to introduce one or even more local grids, but a global grid is always required. On the other hand, there are domain decomposition methods, which involve only local grids. Two approaches are to be distinguished. The first approach is based upon a decomposition of Ω into two or more disjunct subdomains (cf.

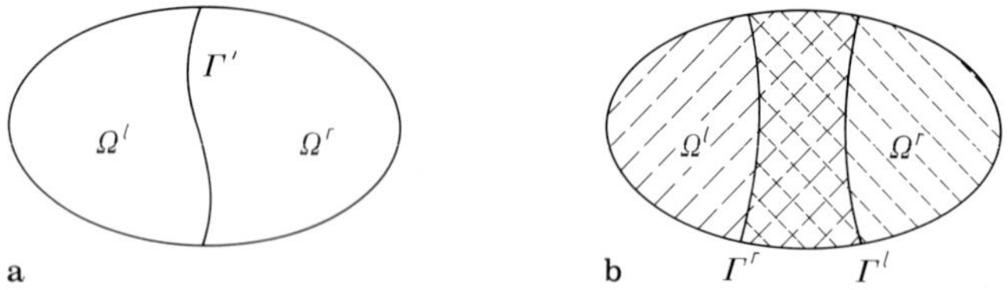

Fig. 15.3.1 a, b

Fig. 15.3.1 a). The second approach uses overlapping subdomains. In the situation of Fig. 15.3.1 b, there are two interior boundaries:

$$\Gamma^r = \partial\Omega^r \backslash \Gamma = \partial\Omega^r \cap \Omega, \qquad \Gamma^l = \partial\Omega^l \backslash \Gamma = \partial\Omega^l \cap \Omega.$$

Any solution of $Lu = f$ in Ω gives rise to solutions $u^l = u|_{\Omega^l}$, $u^r = u|_{\Omega^r}$ of the local problems

$$L_\Omega u^l = f_\Omega \quad \text{in } \Omega^l, \qquad L_\Gamma u^l = f_\Gamma \quad \text{on } \partial\Omega^l \backslash \Gamma^l, \tag{15.3.1 a}$$

$$L_\Omega u^r = f_\Omega \quad \text{in } \Omega^r, \qquad L_\Gamma u^r = f_\Gamma \quad \text{on } \partial\Omega^r \backslash \Gamma^r, \tag{15.3.1 b}$$

$$u^l = u^r \quad \text{on } \Gamma^l, \qquad u^r = u^l \qquad \text{on } \Gamma^r. \tag{15.3.1 c}$$

On the other hand, any pair of solutions u^l, u^r of Eq. (1) coincides with the restrictions of $u = L^{-1} f$ onto Ω^l and Ω^r, provided that the boundary value problem in $\Omega^l \cap \Omega^r$ is uniquely solvable.

The discrete analogue of Eq. (1) is

$$L_l^l u_l^l = f_l^l \qquad \text{in } \Omega_l^l, \qquad L_l^r u_l^r = f_l^r \qquad \text{in } \Omega_l^r, \tag{15.3.2 a}$$

$$u_l^l = \pi^l u_l^r \quad \text{on } \Gamma_l^l, \qquad u_l^r = \pi^r u_l^l \quad \text{on } \Gamma_l^r, \tag{15.3.2 b}$$

where $\pi^l u_l^r$ denotes the interpolation of the grid function u_l^r at points of the discrete interior boundary Γ_l^l. We permit different grids or different kinds of discretisations in $\Omega_l^l \cup \Omega^r$ and $\Omega_l^r \cap \Omega^l$. As a consequence there are two possibly different solutions in $\Omega^l \cap \Omega^r$.

It is also possible to discretise the original problem by Eq. (1.1) and to decompose the grid Ω_l into overlapping subgrids Ω_l^l and Ω_l^r as we did for the continuous problem. In this case the solutions u^l, u^r of (2 a, b) coincide in $\Omega_l^l \cap \Omega_l^r$ and the mapping $u_l^r \mapsto \pi^l u_l^r$ is the injection $\pi^l u_l^r(\mathbf{x}) = u_l^r(\mathbf{x})$ for all $\mathbf{x} \in \Gamma_l^l \subset \Omega_l^r$.

Domain decomposition methods with non-intersecting subdomains are presented for instance by Bjørstad-Widlund [1] and Dryja [3]. Overlapping subdomains are used, e.g., by Starius [2], Glowinski-Periaux-Dinh [1], and Hackbusch [14].

15.3.2 The Schwarz Iteration

A basic iterative method for solving the coupled local systems in Eqs. (1 a–c) is the Schwarz iteration:

$$u_l^{l,i+1} \text{ is the solution of } L_l^l u_l^l = f_l^l \quad \text{in } \Omega_l^l, \quad u_l^l = \pi^l u_l^{r,i} \quad \text{on } \Gamma_l^l, \tag{15.3.3 a}$$

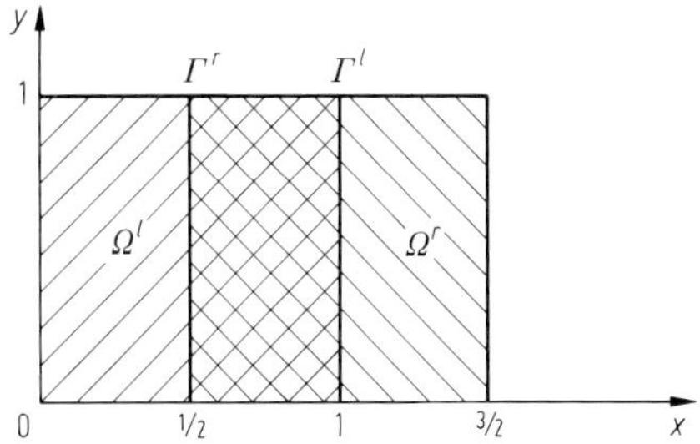

Fig. 15.3.2

$$u_l^{r,i+1} \text{ is the solution of } L_l^r u_l^r = f_l^r \quad \text{in } \Omega_l^r, \quad u_l^r = \pi^r u_l^{l,i} \quad \text{on } \Gamma_l^r, \tag{15.3.3 b}$$

which can be regarded as a particular block-Jacobi method.

Under suitable conditions one can prove the convergence of iteration (3). It turns out that the rate depends on the distance

$$d := \text{distance}\,(\Gamma^l, \Gamma^r) > 0 \tag{15.3.4}$$

(cf. Corollary 15.3.2).

For a precise analysis we consider Poisson's equation on Ω, which is decomposed into two overlapping squares Ω^l and Ω^r (cf. Fig. 15.3.2).

Lemma 15.3.1. Let L_l^l and L_l^r be the standard five-point schemes on the respective square grids Ω_l^l and Ω_l^r of Fig. 15.3.2 and define π^l and π^r by $(\pi^l u_l^r)(\mathbf{x}) = u_l^r(\mathbf{x})$ at $\mathbf{x} \in \Gamma_l^l$, $(\pi^r u_l^l)(\mathbf{x}) = u_l^l(\mathbf{x})$ at $\mathbf{x} \in \Gamma_l^r$. The convergence rate of Schwarz' iteration (3) is given by

$$0.5/\cosh\left(\frac{\tilde{\mu}\,\pi}{2}\right) = 0.199\ldots + O(h_l^2) \quad \text{with } \tilde{\mu} = \frac{2}{\pi\, h_l} \operatorname{Ar\,sinh}\left(\sin\frac{\pi\, h_l}{2}\right). \tag{15.3.5}$$

Proof. The error of $u_l^{r,0}$ on Γ_l^l can be represented by

$$u_l^{r,0}|_{\Gamma_l^l} = \sum_{\mu=1}^{N_l} \alpha_\mu \sin(\mu\,\pi\, y), \qquad N_l = h_l^{-1} - 1\,. \tag{15.3.6}$$

The boundary values $u_l^{l,1} = \sin(\mu\,\pi\, y)$ on Γ_l^l and $u_l^{l,1} = 0$ at the other boundary points yield the solution

$$u_l^{l,1} = \sin(\mu\,\pi\, y)\sinh((x-1)\,\tilde{\mu}\,\pi)/\sinh(\tilde{\mu}\,\pi),$$

where

$$\tilde{\mu} = \frac{2}{\pi\, h_l} \operatorname{Ar\,sinh}\left(\sin\frac{\mu\,\pi\, h_l}{2}\right).$$

Hence, the error of $u_l^{l,1}$ on Γ_l^r equals $\lambda_\mu \sin(\mu\,\pi\, y)$, where

$$\lambda_\mu = \sinh\left(\frac{\tilde{\mu}\,\pi}{2}\right)\Big/\sinh(\tilde{\mu}\,\pi) = 0.5/\cosh\left(\frac{\tilde{\mu}\,\pi}{2}\right). \tag{15.3.7}$$

given: $u_l = (u_l^l, u_l^r)$ and $f_l = (f_l^l, f_l^r, g_l^l, g_l^r)$;

a) set $\tilde{u}_l^l := \pi^l u_l^r + g_l^l$ on Γ_l^l, $\tilde{u}_l^l := u_l^l$ on Ω_l^r; (15.3.10 a)

b) apply one smoothing step $\mathscr{S}_l^l$ to $\tilde{u}_l^l$ on Ω_l^l;
denote the result again by $\tilde{u}_l^l$; (15.3.10 b)

c) set $\tilde{u}_l^r := \pi^r \tilde{u}_l^l + g_l^r$ on Γ_l^r, $\tilde{u}_l^r := u_l^r$ on Ω_l^r; (15.3.10 c)

d) apply one smoothing step $\mathscr{S}_l^r$ to $\tilde{u}_l^r$ on Ω_l^r;
denote the result again by $\tilde{u}_l^r$; (15.3.10 d)

define: $\mathscr{S}_l(u_l, f_l) := \tilde{u}_l = (\tilde{u}_l^l, \tilde{u}_l^r)$. (15.3.10 e)

The main part of the computational work of $\mathscr{S}_l^{(\nu)}$, either defined by (9) or (10), is the performance of ν local smoothing steps on Ω_l^l and Ω_l^r. The interpolations π^r and π^l have to be executed only once in the case of (9) and ν-times for (10). Under suitable conditions the smoothing iteration (10) converges to the solution of Eq. (2 a, b), unlike iteration (9). The multi-grid convergence improves, as can be seen from Table 15.3.1. However, as a consequence of Remark 15.3.4, the rates differ significantly for very coarse grid sizes only. Remark 15.3.3 (i) contrasts with

Remark 15.3.5. The steps (10 b) and (10 d) cannot be computed in parallel, since the boundary data needed in (10 d) are the result of (10 b, c).

16. The Multi-Grid Method of the Second Kind [1]

16.1 Fredholm's Integral Equation of the Second Kind

16.1.1 Historical Comments

Multi-grid methods were first developed for elliptic boundary value problems. However, a two-grid iteration closely related to a multi-grid algorithm for integral equations of the second kind (which we call a multi-grid iteration of the second kind) is already described in 1960 by Brakhage [1]. Abramov [1] and Šišov [1] also describe algorithms for integral eigenvalue problems that approach the two-grid idea. Atkinson [1, 2] took up the idea of Brakhage and developed automatic programs for linear integral equations.

Independently of the author, Hemker and Schipper formulated the multi-grid iteration of the second kind (Schippers [1], Hemker-Schippers [1] and Wolff [1]). The author's first report (1978) on this subject appeared in 1981 (Hackbusch [20]).

16.1.2 Fredholm's Integral Equation of the Second Kind

Fredholm's integral equation of the second kind is

$$u(\mathbf{x}) = \int_D k(\mathbf{x}, \mathbf{y})\, u(\mathbf{y})\, d\mathbf{y} + f(\mathbf{x}) \quad \text{for all } \mathbf{x} \in D \tag{16.1.1}$$

with given kernel function k defined on $D \times D$ and given f. A short notation for Eq. (1) is

$$u = Ku + f, \tag{16.1.2}$$

where $(Ku)(\mathbf{x}) = \int_D k(\mathbf{x}, \mathbf{y})\, u(\mathbf{y})\, d\mathbf{y}$ defines the integral operator K.

---

1 This chapter is based on the author's contribution to the proceedings volume of Holstein-Paddon [1]

Let $\mathscr{U}$ be the underlying Banach space involved in Eq. (2): for $f \in \mathscr{U}$ a solution of (2) is sought. Under the usual assumption that $K\colon \mathscr{U} \to \mathscr{U}$ is compact, the estimate

$$\|(I-K)^{-1}\|_{\mathscr{U} \leftarrow \mathscr{U}} \leqq C \tag{16.1.3}$$

holds, whenever $\lambda = 1$ is not an eigenvalue of K (cf. Yosida [1, § X.5]).

In the following, we assume that there exists some Banach subspace $\mathscr{V} \subset \mathscr{U}$ such that image$(K) \subset \mathscr{V}$, or

$$\|K\|_{\mathscr{V} \leftarrow \mathscr{U}} \leqq C. \tag{16.1.4}$$

The elements of $\mathscr{V}$ must have a higher order of differentiability than those of $\mathscr{U}$, as in the following examples. Inequality (4) expresses the smoothing effect of K.

Example 16.1.1. Suppose $k \in C^s(D \times D)$ with bounded D, where C^s is the Hölder space (cf. § 1.4.1). Then K satisfies (4) with $\mathscr{U} = C^0(D)$, $\mathscr{V} = C^s(D)$.

The assumption on k can be weakened. Weakly singular kernels are included, as shown:

Example 16.1.2. For $D = [0,1]$ and $k(x, y) = 1/|x-y|^\lambda$ with $0 \leqq \lambda \leqq 1$, inequality (4) is satisfied with $\mathscr{U} = C^0(D)$ and $\mathscr{V} = C^{1-\lambda}(D)$.

$\mathscr{U}$ and $\mathscr{V}$ may also be chosen as Sobolev spaces $\mathscr{U} = H^s(D)$ and $\mathscr{V} = H^t(D)$ with $t > s$ as in the following

Example 16.1.3. Let $D = [0,1]$ and $k(x, y) = 1/\sqrt{2-x-y}$. Inequality (4) holds for all pairs $\mathscr{U} = H^s(D)$, $\mathscr{V} = H^{s+1/2}(D)$ with constants $C = C(s)$ as long as $-1 < s < \frac{1}{2}$, where $H^{-t}(D)$ is the dual of $H^t(D)$ for $t > 0$.

In all the previous examples, $\mathscr{V}$ is compactly embedded in $\mathscr{U}$ and Note 16.1.4 applies.

Note 16.1.4. If (4) holds for $\mathscr{V}$ compactly embedded in $\mathscr{U}$, the operator $K\colon \mathscr{U} \to \mathscr{U}$ is compact.

16.1.3 Discretisation

There are various kinds of discretisation methods, e. g. kernel approximations, projection methods (in particular Galerkin and collocation methods), and Nyström's method (cf. CTH Baker [1]). Taking the trapezoidal rule as quadrature formula, we obtain the following very simple discretisation.

Example 16.1.5 (Nyström's method with repeated trapezoidal rule). Suppose $D = [0, 1]$, $k \in C^0(D \times D)$ and

$$h_l = 1/n_l, \quad n_l = 2^l n_0 \quad (l = 0, 1, \ldots) \tag{16.1.5 a}$$

with some $n_0 \in \mathbb{N}$. The desired solution u_l is a grid function defined on

$$D_l = \{0, h_l, 2h_l, \dots, 1 - h_l, 1\} \tag{16.1.5 b}$$

satisfying the system

$$u_l(x) = h_l \sum_{y \in D_l} w(y) k(x, y) u_l(y) + f(x), \qquad x \in D_l \tag{16.1.5 c}$$

of $n_l + 1$ equations. The weights are $w(0) = w(1) = \frac{1}{2}$ and $w(y) = 1$ elsewhere.

System (5 c) or any other discretisation with discretisation parameter h_l will be denoted by

$$u_l = K_l u_l + f_l. \tag{16.1.6}$$

A special collocation approach is described in

Example 16.1.6 ('panel method'). Let h_l as in (5 a) and define the jth element of the finite element basis by $\varphi_j(x) = 1$ for $jh_l \leqq x \leqq (j+1)h_l$ and $\varphi_j(x) = 0$ elsewhere. The entries of the matrix K_l are

$$K_{l,ij} = (K\varphi_j)((i + \tfrac{1}{2})h_l), \qquad 0 \leqq i, j \leqq n_l - 1. \tag{16.1.7 a}$$

The components $f_{l,i} = f((i + \frac{1}{2})h_l)$ form the grid function f_l. The ith component of u_l is the value of $u = \sum_j u_{l,j}\varphi_j$ at $x = (i + \frac{1}{2})h_l$. Therefore, the grid D_l is given by

$$D_l = \{h_l/2, 3h_l/2, \dots, 1 - h_l/2\}. \tag{16.1.7 b}$$

The grid functions u_l and f_l in Eq. (6) belong to $\mathscr{U}_l$, which is the discrete analogue of $\mathscr{U}$. Similarly, a discrete counterpart $\mathscr{V}_l$ of $\mathscr{V}$ is needed. For instance, the discrete space $\mathscr{V}_l = C^s(D_l)$ corresponding to $C^s(D)$ has the norm

$$\|v_l\|_{C^0(D_l)} = \max_{x \in D_l} |v_l(x)| \quad \text{if } s = 0,$$

$$\|v_l\|_{C^s(D_l)} = \max\{\|v_l\|_{C^0(D_l)}, \ |v_l(x) - v_l(x')|/|x - x'|^s : x, x' \in D_l, x \neq x'\}$$
$$\text{if } 0 < s \leqq 1,$$

etc. The respective norms of $\mathscr{U}_l$ and $\mathscr{V}_l$ are written as $\|\cdot\|_{\mathscr{U}}$ and $\|\cdot\|_{\mathscr{V}}$.

Under usual assumptions (cf. Anselone [1]) system (6) is soluble if h_l is sufficiently small. Further, $(I - K_l)^{-1}$ is uniformly bounded. The counterpart of inequality (3) is the *stability* assumption

$$\|(I - K_l)^{-1}\|_{\mathscr{U} \leftarrow \mathscr{U}} \leqq C_S \quad (l = 0, 1, \dots) \tag{16.1.8}$$

with C_S independent of l.

The analogue of inequality (4) is the *regularity* assumption

$$\|K_l\|_{\mathscr{V} \leftarrow \mathscr{U}} \leqq C_R \quad (l = 0, 1, \dots). \tag{16.1.9}$$

Example 16.1.7. Under the conditions of Example 16.1.1, the discretisation of Example 16.1.5 satisfies the regularity assumption (9) with

$$C_R = \max\{\|k(\cdot, y)\|_{C^s([0,1])} : 0 \leqq y \leqq 1\}.$$

Instead of Note 16.1.4 we have

Note 16.1.8. Assume (3) and (9), and let $\mathscr{V}$ be compactly embedded in $\mathscr{U}$. Suppose that there are interpolations $P_l: \mathscr{V}_l \to \mathscr{V}$ and restrictions $R_l: \mathscr{U} \to \mathscr{U}_l$ with $\|P_l\|_{\mathscr{V} \leftarrow \mathscr{V}} \leqq C$, $\|R_l\|_{\mathscr{U} \leftarrow \mathscr{U}} \leqq C$, satisfying $R_l P_l = I$. Under the consistency condition

$$\lim_{l \to \infty} \|(P_l K_l R_l - K) v\|_{\mathscr{U}} = 0 \quad \text{for all } v \in \mathscr{U},$$

the stability condition (8) follows for sufficiently large l.

Again, a restriction

$$r: \mathscr{U}_l \to \mathscr{U}_{l-1} \tag{16.1.10}$$

is needed in the multi-grid process. The restriction r may be weighted, but whenever $D_{l-1} \subset D_l$ and $\mathscr{U}_l = C^s(D_l)$, $\mathscr{V}_l = C^t(D_l)$, the trivial injection $r = r_{inj}$,

$$(r_{inj} u_l)(\mathbf{x}) = u_l(\mathbf{x}), \qquad \mathbf{x} \in D_{l-1}, \tag{16.1.11}$$

can be used (cf. (3.5.1)).

The *relative consistency* condition states that the expression $(r K_l - K_{l-1} r) v_l$ is small for smooth v_l; more precisely:

$$\|r K_l - K_{l-1} r\|_{\mathscr{U} \leftarrow \mathscr{V}} \leqq C_C h_l^\kappa \tag{16.1.12}$$

where κ is called the consistency order. κ depends on the discretisation (e.g. on the order of the quadrature formula) and on the choice of $\mathscr{U}$ and $\mathscr{V}$.

Example 16.1.9. Let K_l be the discretisation of Example 16.1.5, assume $\|k(\cdot, y)\|_{C^\alpha([0,1])} \leqq C$ and choose $r = r_{inj}$. Then for all $s \geqq 0$, $0 \leqq t \leqq \alpha$ the estimate

$$\|r K_l - K_{l-1} r\|_{C^t(D_{l-1}) \leftarrow C^s(D_l)} \leqq C(t, s)\, h_l^{\min(2, \alpha, s)}$$

holds. The appropriate choice of $\mathscr{U}_l$ and $\mathscr{V}_l$ will be $\mathscr{U}_l = C^0(D_l)$ and $\mathscr{V}_l = C^{\min(2,\alpha)}(D_l)$ corresponding to $t = 0$, $s = \kappa = \min(2, \alpha)$.

The stability of a prolongation $p: \mathscr{U}_{l-1} \to \mathscr{U}_l$ is characterised by

$$\|p\|_{\mathscr{U} \leftarrow \mathscr{U}} \leqq C_p. \tag{16.1.13}$$

The interpolation error of p should be small enough to satisfy

$$\|I - p r\|_{\mathscr{U} \leftarrow \mathscr{V}} \leqq C_I h_l^\kappa. \tag{16.1.14}$$

Example 16.1.10. Let $r = r_{inj}$, $h_{l-1} = 2 h_l$, p be a piecewise linear interpolation operator, $\mathscr{U}_l = C^0(D_l)$, $\mathscr{V}_l = C^\alpha(D_l)$. Then (13) holds with $C_p = 1$, while the interpolation error is $\|I - p r\|_{\mathscr{U} \leftarrow \mathscr{V}} \leqq C_I(\alpha)\, h_l^{\min(\alpha, 2)}$ with $C_I(\alpha) = 1$ for $0 \leqq \alpha \leqq 1$, $C_I(\alpha) = \frac{1}{2}$ for $\alpha > 1$, $\kappa = \min(2, \alpha)$.

Roughly speaking, the maximum κ satisfying (12) and (14) is the minimum of the interpolation order of p, the order of the quadrature method and the differentiability order of k.

Example 16.1.11. For the collocation method of Example 16.1.6 one should define r and p by

$$(r\,u_l)(x) = \tfrac{1}{2}u_l(x - h_l/2) + \tfrac{1}{2}u_l(x + h_l/2) \quad \text{for } x \in D_{l-1},$$
$$(p\,u_{l-1})(x \pm h_l/2) = u_{l-1}(x) \quad \text{for } x \in D_{l-1}.$$

Then, inequality (14) holds with $\kappa = 1$, $\mathscr{U}_l = C^0(D_l)$, $\mathscr{V}_l = C^1(D_l)$. Nonetheless, $\kappa = 2$ can be obtained by the choice of $\mathscr{U}_l$ equipped with

$$\|u_l\|_{\mathscr{U}} = \max\{|u_l(x)\,v_l(x) + u_l(x - h_l)\,v_l(x - h_l)| : x, x - h_l \in D_l, \|v_l\|_{C^1(D_l)} \leqq 1\}.$$

We conclude with estimates of the relative discretisation errors (error of u_{l-1} relative to u_l).

Proposition 16.1.12. Let $f_{l-1} = r f_l$. Regularity (9) and consistency (12) imply

$$\|r\,u_l - u_{l-1}\|_{\mathscr{U}} \leqq C_R\,C_C\,h_l^{\kappa}\,\|u_l\|_{\mathscr{V}}. \tag{16.1.15}$$

The similar statement

$$\|u_l - p\,u_{l-1}\|_{\mathscr{U}} \leqq (C_p\,C_R\,C_C + C_I)\,h_l^{\kappa}\,\|u_l\|_{\mathscr{V}} \tag{16.1.15'}$$

requires (13) and (14) (cf. (5.2.3)). We can estimate $\|u_l\|_{\mathscr{V}}$ from

$$\|u_l\|_{\mathscr{V}} \leqq C_R\,C_S\,\|f_l\|_{\mathscr{U}} + \|f_l\|_{\mathscr{V}} \leqq (C_R\,C_S\,\|I\|_{\mathscr{U}\leftarrow\mathscr{V}} + 1)\,\|f_l\|_{\mathscr{V}}. \tag{16.1.16}$$

Proof. Use $u_{l-1} - r\,u_l = (I - K_{l-1})^{-1}(r\,K_l - K_{l-1}\,r)\,u_l$ and $p\,u_{l-1} - u_l = p(u_{l-1} - r\,u_l) - (I - p\,r)\,u_l$. □

16.2 Multi-Grid Iteration

16.2.1 First Variant

16.2.1.1 Algorithm

Formally, the multi-grid iteration is the same as algorithm (4.1.1) with L_l replaced by $I - K_l$. However, the smoothing step is much simpler than in the elliptic case, where a careful choice of the smoothing iteration is essential. Here we use Picard's iteration $u_l \mapsto K_l\,u_l + f_l$ and perform only one ($\nu = 1$) iteration step. The choice $\nu \geqq 2$ is possible (cf. § 16.5.1), but has quite a different justification than in the elliptic case. The convergence of the Picard iteration (i.e. $\varrho(K_l) < 1$) is not required.

Two-grid iteration TGM for solving $u_l = K_l\,u_l + f_l$ (16.2.1)

$$u_l' := K_l\,u_l^j + f_l \quad \text{(smoothing step)} \tag{16.2.1 a}$$

$$u_l^{j+1} := u_l' - p(I - K_{l-1})^{-1}\,r(u_l' - K_l\,u_l' - f_l) \quad \text{(coarse-grid correction)} \tag{16.2.1b}$$

By (1 a), the error $u_l^j - u_l$ (u_l: solution of (1.6)) of the jth iterate is mapped into $u_l' - u_l = K_l(u_l^j - u_l)$. Because of the regularity assumption (1.9), this error is smoothed.

As in § 4, the recursive solution of $(I - K_{l-1})^{-1} d_{l-1}$, $d_{l-1} := r(u_l' - K_l u_l' - f_l)$ in (1 b) by two steps of the same iteration at level $l-1$, yields the first variant MGM' of the multi-grid iteration.

Multi-grid iteration MGM' for solving $u_l = K_l u_l + f_l$ (16.2.2)

procedure $MGM\,(l, u, f)$; **integer** l; **array** u, f;
if $l = 0$ **then** $u := (I - K_0)^{-1} f$ **else** (16.2.2 a)
begin integer i; **array** d, v;

$u := K_l * u + f$; (16.2.2 b)

$d := r(u - K_l * u - f)$; (16.2.2 c_1)

$v := 0$; **for** $i := 1, 2$ **do** $MGM\,(l-1, v, d)$; (16.2.2 c_2)

$u := u - p * v$ (16.2.2 c_3)

end;

Formally, algorithm (2) is the W-cycle variant of (4.1.1) (i.e. $\gamma = 2$) with L_l replaced by $I - K_l$ and $\mathscr{S}_l(u_l, f_l) := K_l u_l + f_l$. Three iterations in (2 c_2) yield no further improvement, while one iteration does not give the convergence results described in § 16.2.1.3.

16.2.1.2 Computational Work

The work of one multi-grid iteration is dominated by the multiplication $K_l * u_l$, since K_l is a full matrix. Therefore, the work required by r, p and vector additions can be neglected. Assume that there are numbers $n_l = \dim(\mathscr{U}_l) + O(1)$ (cf. Example 16.1.5) such that

$$u_l \mapsto K_l u_l \text{ requires} \leqq C_K n_l^2 + O(n_l) \text{ operations} \tag{16.2.3 a}$$

$$u_l \mapsto r K_l u_l \text{ requires} \leqq C_{rK} n_l^2 + O(n_l) \text{ operations} \tag{16.2.3 b}$$

while

$$n_{l-1} \leqq C_N n_l. \tag{16.2.4}$$

Note 16.2.1. For a usual matrix multiplication the constant in (3 a) is given by $C_K = 2$ (1 addition, 1 multiplication). In the case of a weighted restriction r the constant C_{rK} equals C_K. If r is the trivial injection r_{inj} (cf. (1.11)), $K_l u_l$ has to be evaluated only partially. One needs $2 n_{l-1} n_l \leqq 2 C_N n_l^2$ operations; hence $C_{rK} = C_K C_N$. The standard choice (1.5 a) of n_l satisfies (4) with $C_N = \frac{1}{2}$. Therefore, depending on r, (5 a) or (5 b) hold:

$$C_K = 2, \quad C_{rK} = 2, \quad C_N = \tfrac{1}{2}, \tag{16.2.5 a}$$

$$C_K = 2, \quad C_{rK} = 1, \quad C_N = \tfrac{1}{2}. \tag{16.2.5 b}$$

By the following observation one can save some matrix multiplications:

Note 16.2.2. The two multi-grid iterations in ($2\,c_2$) are started with $u_{l-1} = 0$. Whenever the starting iterate is zero, the smoothing step (2 b) requires no matrix multiplication.

Thanks to Note 16.2.2, two multi-grid iterations at level l starting with $u_l^0 = 0$ require $C_l^* n_l^2 + O(n_l)$ operations, where C_l^* satisfies the inequality $C_l^* \leqq C_K + 2C_{rK} + 2C_N^2 C_{l-1}^*$. Since the work of (2 a) can be neglected, we obtain $C_l^* \leqq (C_K + 2C_{rK})/(1 - 2C_N^2)$. One iteration (2) applied to $u_l^0 \neq 0$ requires $C'_{MGM} n_l^2 + O(n_l)$ operations, where $C'_{MGM} \leqq C_K + C_{rK} + C_N^2 C_l^*$. This proves

Proposition 16.2.3. Suppose $C_N < \sqrt{\frac{1}{2}}$. One step of the multi-grid iteration (2) requires $C'_{MGM} n_l^2 + O(n_l)$ operations, where

$$C'_{MGM} = [(1 - C_N^2)\, C_K + C_{rK}]/(1 - 2\,C_N^2)\,. \tag{16.2.6}$$

The constants from (5 a) and (5 b) yield

$$C'_{MGM} = 7 \quad \text{(case of (5 a))}, \qquad C'_{MGM} = 5 \quad \text{(case of (5 b)}\,. \tag{16.2.7}$$

The last number expresses that one multi-grid iteration costs as much as $2\frac{1}{2}$ Picard iterations. A more precise analysis is needed to answer the following question. Let a fixed finest step size $h = 1/n$, $n = n_{l_{\max}}$ be given. The choice of $l_{\max}$ determines $n_0 = n/2^{l_{\max}}$. What is the optimal n_0? Taking into account the work of the exact solution in (2 a), one finds $n_0 \approx \sqrt{3n/2}$. However, this choice gives only a very slight improvement: $C'_{MGM} = 5$ from (7) is replaced by $5 - \sqrt{6/n}$. The same calculation shows that the three-grid iteration ($l_{\max} = 2$) is cheaper than the two-grid iteration ($l_{\max} = 1$) as soon as $n > 10$.

16.2.1.3 Convergence Analysis

The two-grid iteration (1) can be written as

$$u_l^{j+1} = M_l^{TGM} u_l^j + N_l^{TGM} f_l$$

with the iteration matrix

$$M_l^{TGM} = [I - p(I - K_{l-1})^{-1} r(I - K_l)]\, K_l\,. \tag{16.2.8}$$

Theorem 16.2.4. Assume (1.8), (1.9), (1.12), (1.13), (1.14) with some $\kappa > 0$. Then the contraction number of the two-grid iteration (1) with respect to $\|\cdot\|_{\mathscr{U}}$ is given by

$$\| M_l^{TGM} \|_{\mathscr{U} \leftarrow \mathscr{U}} \leqq C_{TGM}\, h_l^{\kappa} \tag{16.2.9}$$

with $C_{TGM} \leqq (C_I + C_p C_S C_C)\, C_R$.

Proof. The splitting

$$\begin{aligned} M^{TGM} &= \{(I - pr) + p(I - K_{l-1})^{-1} [(I - K_{l-1})r - r(I - K_l)]\}\, K_l \\ &= [(I - pr) + p(I - K_{l-1})^{-1} (r K_l - K_{l-1} r)]\, K_l \end{aligned}$$

implies

$$\|M_l^{TGM}\|_{\mathscr{U}\leftarrow\mathscr{U}} \leqq [\|I - pr\|_{\mathscr{U}\leftarrow\mathscr{V}} + \|p\|_{\mathscr{U}\leftarrow\mathscr{U}}\|(I-K_{l-1})^{-1}\|_{\mathscr{U}\leftarrow\mathscr{U}}\|rK_l - K_{l-1}r\|_{\mathscr{U}\leftarrow\mathscr{V}}]\|K_l\|_{\mathscr{V}\leftarrow\mathscr{U}}$$
$$\leqq (C_I + C_p C_S C_C) C_R h_l^\kappa . \qquad \square$$

Inequality (9) implies convergence for sufficiently small h_l. Convergence cannot be guaranteed for all levels, since possibly $C_{TGM} h_1^\kappa \geqq 1$.

The multi-grid iteration (2) can be represented by

$$u_l^{j+1} = M_l' u_l^j + N_l' f_l .$$

As in Lemma 7.1.4 one proves

Lemma 16.2.5. The iteration matrix M_l' of the multi-grid algorithm (2) is recursively defined by $M_1' = M_1^{TGM}$ and

$$M_l' = [I - p(I - M_{l-1}'^2)(I-K_{l-1})^{-1} r(I-K_l)]K_l \tag{16.2.10}$$
$$= M_l^{TGM} + pM_{l-1}'^2(I-K_{l-1})^{-1} r(I-K_l)K_l \tag{16.2.10'}$$
$$= M_l^{TGM} + pM_{l-1}'^2[r - (I-K_{l-1})^{-1}(rK_l - K_{l-1}r)]K_l . \tag{16.2.10''}$$

Let r be bounded by

$$\|r\|_{\mathscr{U}\leftarrow\mathscr{U}} \leqq C_r . \tag{16.2.11}$$

In all previous examples $C_r = 1$ holds. Usually, the estimate $\|K_l\|_{\mathscr{U}\leftarrow\mathscr{U}} \leqq \|K_l\|_{\mathscr{V}\leftarrow\mathscr{U}} \leqq C_R$ is valid. We assume

$$\|K_l\|_{\mathscr{U}\leftarrow\mathscr{U}} \leqq C_K . \tag{16.2.12}$$

C_K may be small or large depending on the problem. Equations (10″) yields the inequality

$$\|M_l'\|_{\mathscr{U}\leftarrow\mathscr{U}} \leqq \|M_l^{TGM}\|_{\mathscr{U}\leftarrow\mathscr{U}} + C_p\|M_{l-1}'\|^2_{\mathscr{U}\leftarrow\mathscr{U}}[C_r C_K + C_S C_C C_R h_l^\kappa]. \tag{16.2.13}$$

If h_1 is small enough there exists C^* with

$$C_p C_{TGM}(1 + C^* h_{l-1}^\kappa)^2 [C_r C_K + C_S C_C C_R h_l^\kappa] \leqq C^* \quad \text{for } l \geqq 2. \tag{16.2.14}$$

This proves

Theorem 16.2.6 (convergence of the multi-grid iteration). Assume (1.8), (1.9), (1.12), (1.13), (1.14) with $\kappa > 0$, (11), and (12). Let h_1 be sufficiently small. Then C^* exists such that

$$\|M_l'\|_{\mathscr{U}\leftarrow\mathscr{U}} \leqq C_{TGM}(1 + C^* h_l^\kappa) h_l^\kappa . \tag{16.2.15}$$

Hence, asymptotically, two-grid and multi-grid iterations have the same contraction numbers. However, if h_1 is too large, there is no C^* satisfying (14), and $\|M_l'\|_{\mathscr{U}\leftarrow\mathscr{U}}$ diverges rapidly:

$$\|M_l'\|_{\mathscr{U}\leftarrow\mathscr{U}} = O([C_p C_r C_K]^{2^{l-1}}) \to \infty .$$

To gain a better understanding we shall suppose $rp = I$, which holds for $r = r_{inj}$ (cf. (1.11)) whenever p is an interpolation.

Lemma 16.2.7. If $rp = I$, a further representation of M_l' is

$$M_l' = M_l^{TGM} + p\,M_{l-1}'^{\,2}\,r(M_l^{TGM} - K_l). \tag{16.2.16}$$

For the usual values $C_p = C_r = 1$, Eq. (16) yields

$$\|M_l'\|_{\mathcal{U}\leftarrow\mathcal{U}} \leqq \|M_l^{TGM}\|_{\mathcal{U}\leftarrow\mathcal{U}} + \|M_{l-1}'\|^2_{\mathcal{U}\leftarrow\mathcal{U}}(\|M_l^{TGM}\|_{\mathcal{U}\leftarrow\mathcal{U}} + C_K) \tag{16.2.17}$$

with C_K from (12).

Note 16.2.8. Assume $rp = I$, $C_r = C_p = 1$, and define $\zeta = \sup\{\|M_l^{TGM}\|_{\mathcal{U}\leftarrow\mathcal{U}} : l \geqq 1\}$ (usually $\zeta = \|M_1^{TGM}\|_{\mathcal{U}\leftarrow\mathcal{U}}$). The inequalities (15) and $\|M_l'\|_{\mathcal{U}\leftarrow\mathcal{U}} < 2\zeta < 1$ implying multi-grid convergence are ensured by

$$4\zeta(\zeta + C_K) \leqq 1. \tag{16.2.18}$$

Obviously, it is hard to satisfy this condition if $C_K \gg 1$.

16.2.2 Second Variant

16.2.2.1 Algorithm

Hemker-Schippers [1] proposed a modification of the multi-grid iteration which is applicable also for large C_K. It is based on the equivalent formulation of the coarse-grid correction in the two-grid iteration (1) by

$$u_l^{j+1} = u_l' + pr(u_l^j - u_l') - p(I - K_{l-1})^{-1}[(I - K_{l-1})r(u_l^j - u_l') + rd_l], \tag{16.2.19}$$

where $d_l = u_l' - K_l u_l' - f_l$. The restriction r in front of $u_l^j - u_l'$ may also be replaced by another $\hat{r} \neq r$. The evaluation of the square bracket requires a further multiplication by K_{l-1}.

Although (19) and (1 b) are equivalent, the multi-grid versions differ.

Multi-grid iteration MGM'' for solving $u_l = K_l u_l + f_l$ (16.2.20)

procedure $MGM(l, u, f)$; **integer** l; **array** u, f;
if $l = 0$ **then** $u := (I - K_0)^{-1} f$ **else** (16.2.20 a)
begin integer i; **array** v, d;
  $d := u - K_l * u - f$; (16.2.20 b$_1$)
  $u := u - (I - pr)d$; (16.2.20 b$_2$)
  $d := ((I - K_{l-1}) * r - r * K_l) * d$; (16.2.20 b$_3$)
  $v := 0$; **for** $i := 1, 2$ **do** $MGM(l-1, v, d)$; (16.2.20 b$_4$)
  $u := u - p * v$ (16.2.20 b$_5$)
end;

The smoothing step is hidden in statement (20 b$_3$).

16.2.2.2 Computational Work

Note 16.2.2 remains valid. Instead of Proposition 16.2.3 we obtain

Proposition 16.2.9. Suppose $C_N < \sqrt{\frac{1}{2}}$. One step of the multi-grid iteration (20) requires $C''_{MGM} n_l^2 + O(n_l)$ operations, where

$$C''_{MGM} = (C_K + C_{rK})/(1 - 2C_N^2) \tag{16.2.21}$$

with C_K, C_{rK} from (3). The constants (5 a) or (5 b) yield

$$C''_{MGM} = 8 \text{ (case (5 a))}, \qquad C''_{MGM} = 6 \text{ (case (5 b))}. \tag{16.2.22}$$

Note 16.2.10. In special cases the work can be less than indicated by (22). Assume (5 b). If $r = r_{inj}$ and if K_l is obtained by the trapezoidal rule (1.5 c), the components of $r K_l v_l$ are the sum of $\frac{1}{2} K_{l-1} r v_l$ and $n_l/2$ further terms. Therefore, the evaluation of $(K_{l-1} r - r K_l) v_l$ in (20 b$_3$) requires only $2 n_{l-1} n_l \leqq 2 C_N n_l^2$ operations. One concludes that $C''_{MGM} = 5$, as in (7).

16.2.2.3 Convergence Analysis

Again, the two-grid iteration matrix M_l^{TGM} is given by (8), whereas the multi-grid iteration matrix M_l'' of algorithm (20) equals $M_1'' = M_1^{TGM}$ for $l = 1$ and

$$\begin{aligned} M_l'' &= I - \{I - pr + p(I - M''^2_{l-1})(I - K_{l-1})^{-1}[(I - K_{l-1})r - rK_l]\}(I - K_l) \\ &= M_l^{TGM} + p M''^2_{l-1}[r - (I - K_{l-1})^{-1}(rK_l - K_{l-1}r)K_l] \end{aligned} \tag{16.2.23}$$

for $l \geqq 2$. Note that (23) differs from (10″) only in that the initial term r in the square brackets of (23) has been replaced by $r K_l$ in (10″). Instead of (13) one obtains

$$\| M_l'' \|_{\mathcal{U} \leftarrow \mathcal{U}} \leqq \| M_l^{TGM} \|_{\mathcal{U} \leftarrow \mathcal{U}} + C_p \| M''_{l-1} \|^2_{\mathcal{U} \leftarrow \mathcal{U}} (C_r + C_S C_C C_R h_l^\kappa). \tag{16.2.24}$$

Theorem 16.2.11. Assume (1.8), (1.9), (1.12), (1.13), (1.14) with $\kappa > 0$, and (11). Let h_1 be sufficiently small. Then C^* exists such that (15) holds for M_l''. In contrast to Theorem 16.2.6, the upper bound of h_1 and the constant C^* do not depend on C_K (cf. (12)).

The situation becomes simpler in

Corollary 16.2.12. If $rp = I$, a further representation of M_l'' is

$$M_l'' = M_l^{TGM} + p M''^2_{l-1} r (M_l^{TGM} - I). \tag{16.2.25}$$

If in addition $C_p = C_r = 1$, the multi-grid iteration converges when $\| M_l^{TGM} \|_{\mathcal{U} \leftarrow \mathcal{U}} \leqq 1/[2(\sqrt{2} + l)] \approx 0.207$ for $l \geqq 1$.

Proof. M_l'' satisfies Inequality (17) with $C_K := 1$. Note 16.2.8 with $C_K = 1$ yields the desired result. □

16.2.3 Third Variant

16.2.3.1 Algorithm

The difficulty of the first variant was due to the presence of the factor K_l in the second term of (10″). This factor is not avoided in the second version but the dominant term $r\,K_l$ from (10″) is replaced by r in (23). In the case of the first variant (2) the multiplication by K_l yields a smoothing which has not been exploited hitherto. Since the defect d from (2 c$_1$) is already smooth, the first of the two iterations (2 c$_2$) can be performed without the smoothing step (2 b). Therefore, the resulting third variant is even cheaper than the first one. The resulting procedure MGM''' has a further parameter sm. If its value is **true** (**false**), the smoothing is (not) performed. One step $u_l^j \to u_l^{j+1}$ is defined by means of MGM (.,.,., **true**). The value sm = **false** is used only inside the algorithm at lower levels.

Multi-grid iteration MGM''' for solving $u_l = K_l u_l + f_l$ (16.2.26)

procedure $MGM(l, u, f, sm)$; **integer** l; **array** u, f; **Boolean** sm;
if $l = 0$ **then** $u := (I - K_0)^{-1} f$ **else** (16.2.26 a)
begin array v, d;
  if sm **then** $u := K_l * u + f$; (16.2.26 b)
  $d := v * (u - K_l * u - f)$; (16.2.26 c$_1$)
  $v := 0$; $MGM(l-1, v, d,$ **false**$)$; $MGM(l-1, v, d,$ **true**$)$; (16.2.26 c$_2$)
  $u := u - p * v$ (16.2.26 c$_3$)
end;

16.2.3.2 Computational Work

Proposition 16.2.13. Suppose $C_N < \sqrt{\frac{1}{2}}$. One step of the multi-grid iteration matrix (26) (with sm = **true**) requires $C'''_{MGM}\, n_l^2 + O(n_l)$ operations, where

$$C'''_{MGM} = (C_K + C_{rK})(1 - C_N^2)/(1 - 2\,C_N^2) \tag{16.2.27}$$

(cf. (3), (4)). The constants (5 a) or (5 b) yield

$$C'''_{MGM} = 6 \text{ (case (5 a))}, \qquad C'''_{MGM} = 4.5 \text{ (case (5 b))}. \tag{16.2.28}$$

16.2.3.3 Convergence Analysis

The two-grid iteration matrix (for all variants) can be written as $M_l^{TGM} = A_l^{TGM} K_l$ with

$$A_l^{TGM} = I - p\,r + p(I - K_{l-1})^{-1}(r\,K_l - K_{l-1}\,r)$$

(cf. (8)). The proof of Theorem 16.2.4 showed

$$\| A_l^{TGM} \|_{\mathscr{U} \leftarrow \mathscr{V}} \leqq (C_I + C_p C_S C_C) h_l^\kappa =: \alpha_l^{TGM}. \tag{16.2.29}$$

The iteration matrix of algorithm (26) is $M_l''' = A_l''' K_l$ with

$$\begin{aligned} A_l''' &= A_l^{TGM} + p A_{l-1}''' K_{l-1} A_{l-1}''' (I - K_{l-1})^{-1} r (I - K_l) \\ &= A_l^{TGM} + p A_{l-1}''' K_{l-1} A_{l-1}''' [r - (I - K_{l-1})^{-1} (r K_l - K_{l-1} r)] \end{aligned} \tag{16.2.30}$$

(cf. (10′) and (10″)). We need an estimate of $\|(I - K_{l-1})^{-1} r (I - K_l)\|_{\mathscr{V} \leftarrow \mathscr{V}}$. Inequality (1.16) proves

Lemma 16.2.14. The constant $C_S^* := C_R C_S \| I \|_{\mathscr{U} \leftarrow \mathscr{V}} + 1$ satisfies

$$\|(I - K_l)^{-1}\|_{\mathscr{V} \leftarrow \mathscr{V}} \leqq C_S^*. \tag{16.2.31}$$

Lemma 16.2.15. Inequalities (1.9), $\| I \|_{\mathscr{U} \leftarrow \mathscr{V}} \leqq$ const, (31), and

$$\| r \|_{\mathscr{V} \leftarrow \mathscr{V}} \leqq C_r^*, \tag{16.2.32}$$

imply $\|(I - K_{l-1})^{-1} r (I - K_l)\|_{\mathscr{V} \leftarrow \mathscr{V}} \leqq$ const.

$(I - K_{l-1})^{-1} r (I - K_l)$ equals $r - (I - K_{l-1})^{-1} (r K_l - K_{l-1} r)$. A natural estimate of the latter difference is

$$\| r K_l - K_{l-1} r \|_{\mathscr{V} \leftarrow \mathscr{V}} \leqq C_C^* h_l^\kappa \tag{16.2.33}$$

(cf. Example 16.1.9 with $\mathscr{V}_l = C^\kappa(D_l)$).

Lemma 16.2.16. (31), (32), and (33) imply

$$\|(I - K_l)^{-1} r (I - K_l)\|_{\mathscr{V} \leftarrow \mathscr{V}} \leqq C_r^* + C_S^* C_C^* h_l^\kappa. \tag{16.2.34}$$

Set $\beta_l :=$ const in the case of Lemma 16.2.15 or $\beta_l := C_r^* + C_S^* C_C^* h_l^\kappa$ in the case of Lemma 16.2.16. The norm $\alpha_l := \| A_l''' \|_{\mathscr{U} \leftarrow \mathscr{V}}$ can be estimated by

$$\alpha_l \leqq \alpha_l^{TGM} + C_p \alpha_{l-1}^2 C_R \beta_l.$$

Thus we obtain the inequality

$$\| M_l''' \|_{\mathscr{U} \leftarrow \mathscr{U}} \leqq \| A_l''' \|_{\mathscr{U} \leftarrow \mathscr{V}} \| K_l \|_{\mathscr{V} \leftarrow \mathscr{U}} \leqq \alpha_l C_R =: \zeta_l$$

with

$$\zeta_l \leqq \zeta_l^{TGM} + C_p \zeta_{l-1}^2 \beta_l, \quad \zeta_l^{TGM} := \alpha_l^{TGM} C_R, \quad \alpha_l^{TGM} \text{ from (29).}$$

This recursion formula is similar to (24), if β_l is defined by the right side of (34). Therefore, this third variant (26) as well as the second variant are both applicable for large values of $C_K = \sup\{\| K_l \|_{\mathscr{U} \leftarrow \mathscr{U}} : l \geqq 1\}$.

16.3 Numerical Results

16.3.1 Comparison of the Three Variants

Consider the integral equation

$$u(x) = \int_0^1 \lambda \cos(\pi x s) u(s) \, ds + f(x), \qquad x \in D = [0, 1], \tag{16.3.1}$$

with f such that $u(x) = e^x \cos(7x)$ is a solution (cf. Atkinson [1], [2], Hemker-Schippers [1]). The parameter λ will be chosen as 1, 10, and 100. We discretise Eq. (1) by the trapezoidal rule of Example 16.1.5 and use the step sizes $h_l = 1/n_l$ of (1.5 a). Since this quadrature formula is of second order, the appropriate prolongation p is the piecewise linear interpolation (2.3.9). The restriction can be chosen as $r = r_{inj}$ from (1.11). Then, conditions (1.8), (1.9), (1.12), (1.13), (1.14) are satisfied with $\kappa = 2$.

Table 16.3.1. Observed rate of convergence of the two-grid iteration ($n_1 = 2n_0$) for Eq. (1) with $\lambda = 1$

n_1	2	4	8	16	32	64	128
rate	$1.5_{10}-1$	$3.1_{10}-2$	$9.6_{10}-3$	$2.5_{10}-3$	$6.3_{10}-4$	$1.6_{10}-4$	$4.0_{10}-5$

The observed convergence rates of the two-grid iterations are listed in Table 16.3.1 for $\lambda = 1$. The 'observed convergence rate' is defined by $[\| u_l^5 - u_l \|_{\mathscr{U}} / \| u_l^0 - u_l \|_{\mathscr{U}}]^{1/5}$ (we do not claim that these numbers are always very close to the true rate). The corresponding convergence rates of the multi-grid iterations (2.2), (2.20) and (2.26) are almost independent of the number of levels. For instance, the two-grid iteration with $n_0 = 64$, $n_1 = 128$ has nearly the same rate (namely 0.00004) as any of the three multi-grid variants with $n_0 = 1, n_1 = 2, \ldots, n_7 = 128$. Hence, the numbers of Table 16.3.1 also hold for the multi-grid case (n_1 replaced by n_l).

Table 16.3.1 shows that the rates are proportional to $h_l^\kappa = h_l^2$ as predicted by Theorem 16.2.4. Theorem 16.2.6 as well as Theorem 16.2.11 and the corresponding statement in § 16.2.3.3 are also confirmed, since two-grid and multi-grid rates can hardly be distinguished. It can be checked that conditions (1.8), (1.9), (1.12), (1.14) are needed. If, e.g., the prolongation p is chosen as unsymmetric, piecewise constant interpolation (of order 1), inequality (1.14) is satisfied at most for $\kappa = 1$ and consequently the convergence rates are observed to be $O(h_l)$. Later (§ 16.11.2), numerical results will demonstrate that in the case of Example 16.1.3 ($\kappa = \frac{1}{2}$) the rates are proportional to $h_l^{1/2}$. In the present example, where the kernel is very smooth, one should employ higher order quadrature formulae. Simpson's rule is fourth order. To obtain a rate $O(h_l^4)$ one has to choose p as cubic interpolation ($\kappa = 4$).

As stated in Theorem 16.2.6, the success of the first variant (2.2) depends on the smallness of the bound C_K of $\| K_l \|_{\mathscr{U} \leftarrow \mathscr{U}}$. Obviously, this number is proportional to the factor λ in Eq. (1). Tables 16.3.2 and 16.3.3 contain the observed rates of the multi-grid variants for $\lambda = 10$ and $\lambda = 100$ respectively. The finest step size $1/n$ is determined by $n = n_l$. The coarsest grid width is $1/n_0$ on level 0 (cf. (1.5 a)). As predicted by Theorem 16.2.6, the first variant (2.2) diverges if the coarsest step size h_0 is not small enough ($h_0 \geqq \frac{1}{2}$ for $\lambda = 10$, $h_0 \geqq \frac{1}{8}$ for $\lambda = 100$).

Table 16.3.2. Observed multi-grid convergence rates for Eq. (1) with $\lambda = 10$

	n	$n_0 = 1$	$n_0 = 2$	$n_0 = 4$	$n_0 = 8$	$n_0 = 16$	$n_0 = 32$	$n_0 = 64$
1st variant	2	$7.2_{10}-1$	—	—	—	—	—	—
	4	$6.7_{10}+0$	$4.5_{10}-1$	—	—	—	—	—
	8	$6.8_{10}+1$	$1.2_{10}+0$	$1.2_{10}-1$	—	—	—	—
	16	$7.3_{10}+4$	$4.4_{10}+0$	$1.2_{10}-1$	$3.3_{10}-2$	—	—	—
	32	$4.7_{10}+10$	$1.6_{10}+2$	$2.5_{10}-2$	$1.4_{10}-2$	$8.3_{10}-3$	—	—
	64	$2.0_{10}+22$	$1.7_{10}+5$	$9.2_{10}-3$	$2.7_{10}-3$	$2.4_{10}-3$	$3.6_{10}-3$	—
	128	$2.5_{10}+45$	$4.0_{10}+11$	$5.1_{10}-4$	$5.7_{10}-4$	$5.5_{10}-4$	$5.5_{10}-4$	$5.6_{10}-4$
2nd variant	2	$7.2_{10}-1$	—	—	—	—	—	—
	4	$7.0_{10}-1$	$4.5_{10}-1$	—	—	—	—	—
	8	$5.8_{10}-1$	$1.7_{10}-1$	$1.2_{10}-1$	—	—	—	—
	16	$4.2_{10}-1$	$4.0_{10}-2$	$2.1_{10}-2$	$3.3_{10}-2$	—	—	—
	32	$2.2_{10}-1$	$7.9_{10}-3$	$8.0_{10}-3$	$7.5_{10}-3$	$8.3_{10}-3$	—	—
	64	$6.1_{10}-1$	$2.1_{10}-3$	$2.1_{10}-3$	$2.1_{10}-3$	$2.1_{10}-3$	$3.6_{10}-3$	—
	128	$4.7_{10}-3$	$5.3_{10}-4$	$5.3_{10}-4$	$5.3_{10}-4$	$5.3_{10}-4$	$5.3_{10}-4$	$5.6_{10}-4$
3rd variant	2	$7.2_{10}-1$	—	—	—	—	—	—
	4	$4.2_{10}-1$	$4.5_{10}-1$	—	—	—	—	—
	8	$1.9_{10}-1$	$1.5_{10}-1$	$1.2_{10}-1$	—	—	—	—
	16	$6.2_{10}-2$	$3.6_{10}-2$	$2.1_{10}-2$	$3.3_{10}-2$	—	—	—
	32	$5.8_{10}-3$	$8.8_{10}-3$	$7.9_{10}-3$	$7.5_{10}-3$	$8.3_{10}-3$	—	—
	64	$2.1_{10}-3$	$2.0_{10}-3$	$2.0_{10}-3$	$2.0_{10}-3$	$2.1_{10}-3$	$3.6_{10}-3$	—
	128	$5.3_{10}-4$	$5.3_{10}-4$	$5.3_{10}-4$	$5.3_{10}-4$	$5.3_{10}-4$	$5.3_{10}-4$	$5.6_{10}-4$

Table 16.3.3. Observed multi-grid convergence rates for Eq. (1) with $\lambda = 100$

	n	$n_0 = 1$	$n_0 = 2$	$n_0 = 4$	$n_0 = 8$	$n_0 = 16$	$n_0 = 32$	$n_0 = 64$
1st variant	2	$9.3_{10}+0$	—	—	—	—	—	—
	4	$7.7_{10}+3$	$5.0_{10}+0$	—	—	—	—	—
	8	$1.5_{10}+7$	$1.1_{10}+3$	$1.3_{10}+0$	—	—	—	—
	16	$1.7_{10}+17$	$1.0_{10}+7$	$8.5_{10}+1$	$4.1_{10}-1$	—	—	—
	32	$1.0_{10}+36$	$1.7_{10}+16$	$1.7_{10}+4$	$5.7_{10}+0$	$1.2_{10}-1$	—	—
	64	$1.0_{10}+74$	$8.2_{10}+33$	$4.6_{10}+10$	$2.5_{10}+2$	$3.8_{10}-1$	$3.2_{10}-2$	—
	128	—	$8.0_{10}+69$	$8.6_{10}+22$	$9.1_{10}+6$	$7.3_{10}-1$	$2.9_{10}-2$	$2.0_{10}-2$
2nd variant	2	$9.3_{10}+0$	—	—	—	—	—	—
	4	$3.7_{10}+2$	$5.0_{10}+0$	—	—	—	—	—
	8	$6.8_{10}+5$	$8.5_{10}+0$	$1.3_{10}+0$	—	—	—	—
	16	$5.7_{10}+11$	$3.5_{10}+1$	$1.3_{10}+0$	$4.1_{10}-1$	—	—	—
	32	$3.2_{10}+23$	$1.4_{10}+3$	$1.4_{10}+0$	$2.4_{10}-1$	$1.2_{10}-1$	—	—
	64	$1.0_{10}+47$	$1.8_{10}+6$	$1.8_{10}+0$	$6.7_{10}-2$	$4.3_{10}-2$	$3.2_{10}-2$	—
	128	—	$3.3_{10}+12$	$2.9_{10}+0$	$1.0_{10}-2$	$9.2_{10}-3$	$8.9_{10}-2$	$2.0_{10}-2$
3rd variant	2	$9.3_{10}+0$	—	—	—	—	—	—
	4	$1.2_{10}+2$	$5.0_{10}+0$	—	—	—	—	—
	8	$1.5_{10}+4$	$1.9_{10}+1$	$1.3_{10}+0$	—	—	—	—
	16	$2.1_{10}+8$	$3.1_{10}+2$	$1.1_{10}+0$	$4.1_{10}-1$	—	—	—
	32	$4.5_{10}+16$	$8.5_{10}+4$	$8.1_{10}-1$	$2.3_{10}-1$	$1.2_{10}-1$	—	—
	64	$1.9_{10}+33$	$6.2_{10}+9$	$8.0_{10}-1$	$7.9_{10}-2$	$4.3_{10}-2$	$3.2_{10}-2$	—
	128	$3.8_{10}+66$	$3.7_{10}+19$	$5.7_{10}-1$	$1.4_{10}-2$	$9.6_{10}-3$	$8.9_{10}-3$	$2.0_{10}-2$

The numbers of Tables 16.3.2 and 16.3.3 in the diagonal entries ($n = 2n_0$) correspond to the two-grid case. As mentioned above, all three variants are equivalent as two-grid iterations and their rates coincide. The two-grid rates of both Tables 16.3.2–3 show that Theorem 16.2.4 holds even in the case of divergence.

The numerical results illustrate that there may be two different reasons for divergence. The first variant diverges if $\| K_l \|_{\mathscr{U} \leftarrow \mathscr{U}}$ is too large and all three variants fail if the two-grid iteration with some coarsest grid size n_0 has a convergence rate of about 1.

The second and third variants yield very similar convergence rates; however, the latter variant is cheaper. It is remarkable that the third multi-grid version (2.26) converges for $\lambda = 100$, $n_0 = 4$, $n_l \geqq 32$, although the two- and three-grid iterations ($n \leqq 16$) diverge.

16.3.2 The Integral Equation Method for Elliptic Problems

Let $\Omega \subset \mathbb{R}^2$ be a connected domain with boundary Γ. Consider the Dirichlet problem

$$\Delta \Phi = 0 \quad \text{in } \Omega, \qquad \Phi = g \quad \text{on } \Gamma. \tag{16.3.2}$$

A harmonic function (i.e. a solution of $\Delta \Phi = 0$) in $\mathbb{R}^2 \backslash \Gamma$ is given by the 'double layer potential'

$$\Phi_d(\zeta) = \frac{1}{2\pi} \int_\Gamma \mu(z) \cos(n_z, z - \zeta) \, |z - \zeta|^{-1} \, d\Gamma_z, \qquad \zeta \notin \Gamma. \tag{16.3.3}$$

with arbitrary 'doublet distribution' μ. The term $\cos(n_z, z - \zeta)$ is the cosine of the vector $z - \zeta \in \mathbb{R}^2$ and the outward normal n_z at $z \in \Gamma$. The integral $\Phi_d(\zeta)$ can also be evaluated at $\zeta \in \Gamma$, provided that Γ is sufficiently smooth. The limits $\Phi_d^+(\zeta)$ and $\Phi_d^-(\zeta)$ of Φ_d at $\zeta \in \Gamma$ from the outer and inner side, respectively, differ by $\mu(\zeta)$ if μ is continuous at ζ:

$$\Phi_d^{\pm}(\zeta) = \Phi_d(\zeta) \mp \mu(\zeta)/2, \qquad \zeta \in \Gamma.$$

The solution of the Dirichlet problem (2) is Φ_d if $\Phi_d^-(\zeta) = g(\zeta)$ on Γ, i.e. $g(\zeta) = \Phi_d(\zeta) + \mu(\zeta)/2$. From this condition we obtain the integral equation

$$\mu(\zeta) = -\frac{1}{\pi} \int_\Gamma \mu(z) \cos(n_z, z - \zeta) \, |z - \zeta|^{-1} \, d\Gamma_z + 2g(\zeta) \tag{16.3.4}$$

for the unknown μ. Given a representation $\gamma(t)$, $0 \leqq t \leqq 1$, of the boundary Γ, the function $u(t) := \mu(\gamma(t))$ is the solution of

$$u(s) = -\frac{1}{\pi} \int_0^1 u(t) \cos(n_{\gamma(t)}, \gamma(t) - \gamma(s)) \, |\gamma(t) - \gamma(s)|^{-1} \, \| d\gamma/dt \| \, dt + 2g(\gamma(s)). \tag{16.3.4'}$$

The regularity (1.4) is proved by Schippers [5]:

Proposition 16.3.1. Assume $\Gamma \in C^{2+\alpha}$ with $0 < \alpha < 1$, i.e. there is a parametrisation of Γ by $\gamma(t)$, $0 \leqq t \leqq 1$, with $0 < \|d\gamma/dt\|$, such that the periodic extension of γ to $\mathbb{R}$ satisfies $\gamma \in C^{2+\alpha}(\mathbb{R})$. Then the integral operator associated with Eq. (4) is a bounded mapping from $L^\infty(\Gamma)$ (or $C^0(\Gamma)$) into $C^{1+\alpha}(\Gamma)$:

$$\|K\|_{C^{1+\alpha}(\Gamma) \leftarrow L^\infty(\Gamma)} \leqq C, \qquad \alpha \in (0,1).$$

Problem (4′) can be discretised by the collocation method of Example 16.1.6 (panel method).

An exterior boundary value problem arising from the non-circulatory potential flow around an aerofoil can be described by the same integral operator (Schippers [6]). The flow is represented by the potential Φ satisfying

$$\begin{aligned} &\Delta\Phi = 0 && \text{in} \quad \mathbb{R}^2 \backslash \bar{\Omega} \quad \text{(exterior of } \Omega\text{)}, \\ &\partial\Phi/\partial n = 0 && \text{on} \quad \Gamma \\ &\Phi(\zeta) \to \langle U, \zeta \rangle && \text{for} \quad \|\zeta\| \to \infty \end{aligned} \tag{16.3.5}$$

where $\langle \cdot, \cdot \rangle$ is the scalar product in $\mathbb{R}^2$. $U \in \mathbb{R}^2$ is the velocity of the undisturbed flow. The solution of Eq. (5) is given by the double layer potential (3) with μ satisfying the integral equation

$$\mu(\zeta) = -\frac{1}{\pi} \int_\Gamma \mu(z) \cos(n_z, z - \zeta) |z - \zeta|^{-1} \, d\Gamma_z - 2\langle U, \zeta \rangle.$$

The following numerical results are taken from Schippers [6]. He considers problem (5) for the Kármán-Trefftz aerofoil, which is defined in the complex plane by

$$z(s) = c(e^{is} - \sqrt{1 - \varrho^2} + i\varrho)^k / (e^{is} - \delta + i\varrho)^{k-1}, \qquad 0 \leqq s \leqq 2\pi, \tag{16.3.6}$$

with $c = 2L(\delta + \sqrt{1 - \varrho^2})^{k-1} / (2\sqrt{1 - \varrho^2})^k$, δ: thickness, ϱ: camber, k: trailing edge angle, L: length.

The conformal mapping

$$z \mapsto w = z(1 - \tilde{z}/z)^{1 - 1/k}, \qquad \tilde{z} \in \Omega \text{ fixed},$$

removes the corner ('trailing edge', cf. Fig. 16.3.1).

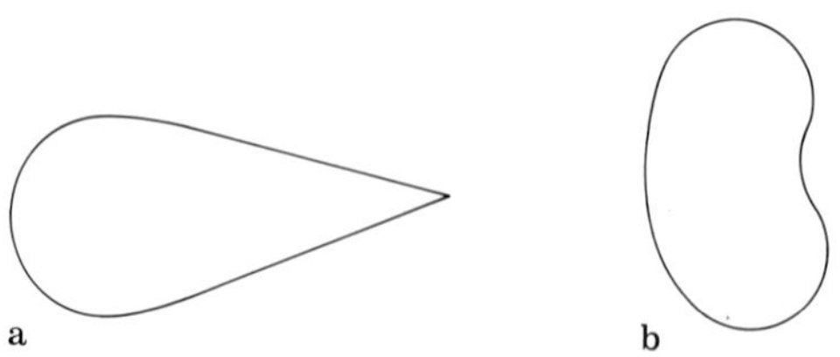

Fig. 16.3.1. Kármán-Trefftz aerofoil in (**a**) z- and (**b**) w-coordinates

The resulting boundary Γ is described by $w = w(s)$, $0 \leqq s \leqq 2\pi$. The parameters in Eq. (6) are $\varrho = 0$, $L = 1$, $\delta = 0.05$, $k = 1.9$ and $k = 1.99$.

Further data are $U = (1, 0)^T$, $\tilde{z} = 1.95$.

Following Schippers [6], the observed convergence rates of the multi-grid iteration (2.2) (first variant) with coarsest grid size $h_0 = \frac{1}{8}$ are listed in Table 16.3.4. The numbers confirm that the convergence rate is proportional to h_l^κ, $\kappa = 2$ (cf. Example 16.1.11).

Table 16.3.4. Multi-grid convergence rates for non-circulatory potential flow problem

k \ n	16	32	64	128	256
1.90	0.26	0.091	0.020	0.0045	0.0011
1.99	0.43	0.26	0.095	0.022	0.0054

The circulatory flow problem will be discussed in § 16.5.3.

16.4 Nested Iteration

16.4.1 Algorithm

The following nested iteration coincides with algorithm (5.1.3):

Standard nested iteration (16.4.1)

$\tilde{u}_0 := (I - K_0)^{-1} f_0;$ (16.4.1 a)

for $k := 1$ **step** 1 **until** l **do**

begin $\tilde{u}_k := \tilde{p}\,\tilde{u}_{k-1};$ (16.4.1 b)

for $j := 1$ **step** 1 **until** i **do** $MGM(k, \tilde{u}_k, f_k)$ (16.4.1 c)

end;

The work of the nested iteration with $i = 1$ is less than the work of two multi-grid iterations at level l.

Note 16.4.1. Assume (2.4): $n_{l-1} \leqq C_N n_l$ with $C_N < \sqrt{\frac{1}{2}}$. The work of (1 a) and (1 b) can be neglected. The total computational work of algorithm (1) is bounded by $i/(1 - C_N^2)$ times the work of one MGM-iteration at level l. For

$i = 1$ the nested iteration requires $Cn_l^2 + O(n_l)$ operations with C given in Table 16.4.1:

Table 16.4.1. The factor C in the formula for the computational work of the nested iteration (1)

multi-grid variant	first	second	third
case (2.5 a)	9.33	10,67	8.00
case (2.5 b)	6.67	8.00	6.00
case of Note 16.2.10	—	6.67	—

16.4.2 Accuracy

As in § 5.2 we can prove

Proposition 16.4.2. Assume

$$\| u_k - \tilde{p} u_{k-1} \|_{\mathcal{U}} \leqq C_D h_k^\kappa \quad \text{(relative discretisation error)} \tag{16.4.2 a}$$

$$\| \tilde{p} \|_{\mathcal{U} \leftarrow \mathcal{U}} h_{k-1}^\kappa / h_k^\kappa \leqq \text{const} \quad (\tilde{p} = \tilde{p}_{k \leftarrow k-1}) \tag{16.4.2 b}$$

$$\| M_k^{MGM} \|_{\mathcal{U} \leftarrow \mathcal{U}} \leqq C_{MGM} h_k^\kappa \tag{16.4.2 c}$$

for $1 \leqq k \leqq l$, where M_k^{MGM} is the respective iteration matrix $M_{k'}'$, M_k'' or M_k''' of the multi-grid algorithm (2.2), (2.20), (2.26). Then all results $\tilde{u}_k$ $(0 \leqq k \leqq l)$ of the nested iteration (1) satisfy

$$\| \tilde{u}_k - u_k \|_{\mathcal{U}} \leqq C_D h_k^\kappa C_{MGM} h_k^\kappa + O(h_k^{3\kappa}), \qquad 0 \leqq k \leqq l. \tag{16.4.3}$$

The relative discretisation error (2 a) is already mentioned in (1.15′). Proposition 16.4.2 shows that the iteration error of $\tilde{u}_k$ is much smaller than the (relative) discretisation error. Note that condition (2c) implies multi-grid convergence. In that case we know from Theorem 16.2.6 that $\| M_l^{MGM} \|_{\mathcal{U} \leftarrow \mathcal{U}} \leqq C_{TGM} h_k^\kappa + O(h_k^{2\kappa})$. Thereby, (3) may be replaced by

$$\| \tilde{u}_k - u_k \|_{\mathcal{U}} \leqq C_D h_k^\kappa C_{TGM} h_k^\kappa + O(h_k^{3\kappa}). \tag{16.4.3′}$$

16.4.3 Nested Iteration with Nyström's Interpolation

If a quadrature formula is used as discretisation method, one can replace the interpolation $\tilde{p}$ in (1 b) by Nyström's interpolation (cf. Nyström [1]):

$$\tilde{u}_k := K_{k,k-1} \tilde{u}_{k-1} + f_k, \tag{16.4.4 a}$$

where the rectangular matrix $K_{k,k-1}$ is defined by

$$(K_{k,k-1} u_{k-1})(x) = h_{k-1} \sum_{y \in D_{k-1}} w(y)\, k(x, y)\, u_{k-1}(y), \qquad x \in D_k. \tag{16.4.4 b}$$

$K_{l+1,l}u_l + f_{l+1}$ is the right-hand side of (1.5c) evaluated at $x \in D_{l+1}$ instead of $x \in D_l$. If $D_{k-1} \subset D_k$ and $f_k(x) = f_{k-1}(x)$ on D_{k-1}, the restriction of $K_{k,k-1}u_{k-1} + f_k$ to D_{k-1} equals $K_{k-1}u_{k-1} + f_{k-1}$.

The interpolation (4) is much more expensive. It requires

$$C_K n_k n_{k-1} \leqq C_K C_N n_k^2 \tag{16.4.5}$$

operations, whereas standard interpolations $\tilde{u}_k = \tilde{p}\tilde{u}_{k-1}$ take work of order $O(n_k)$. Nonetheless, it pays to apply Nyström's interpolation, since the resulting approximation $\tilde{u}_k$ is already smoothed. The first iteration of the following multi-grid algorithm (if $i = 1$ there is only one iteration) can be performed *without* a smoothing step. Combining the modified nested iteration ($i = 1$) with the third multi-grid variant (2.26), we obtain the following algorithm:

Nested iteration with Nyström's interpolation (16.4.6)

$\tilde{u}_0 := (I - K_0)^{-1} f_0;$ (16.4.6 a)

for $k := 1$ **step** 1 **until** l **do**

begin $\tilde{u}_k := K_{k,k-1}\tilde{u}_{k-1} + f_k;$ (16.4.6 b)

$MGM(k, \tilde{u}_k, f_k,$ **false**); **comment** MGM from (2.26) (16.4.6 c)

end;

Note 16.4.3. The nested iteration (6) requires Cn_l^2 operations with

$C = 6.67$, case (2.5 a),
$C = 4.67$, case (2.5 b).

The respective numbers are $C = 8$ and $C = 5.33$ if the first multi-grid variant (2.2) is used in (6c).

It is remarkable that $\frac{14}{3} n_l^2$ operations (equivalent to $2\frac{1}{3}$ Picard iterations) are sufficient for producing results $\tilde{u}_0, \tilde{u}_1, \ldots, \tilde{u}_l$ to an accuracy beyond the discretisation error.

We conclude with numerical results. Let $\tilde{u}_k$ be the results of the nested iteration (6) for the problem from § 16.3.1 with $\lambda = 10$. A comparison of the iteration error $\| u_k - \tilde{u}_k \|_\infty$ with the discretisation error $\| u_k - u \|_\infty$ is given in Table 16.4.2:

Table 16.4.2. Results of nested iteration (6) applied to problem (3.1)

k	0	1	2	3	4	5	6
h_k	$\frac{1}{2}$	$\frac{1}{4}$	$\frac{1}{8}$	$\frac{1}{16}$	$\frac{1}{32}$	$\frac{1}{64}$	$\frac{1}{128}$
iteration error	0.0	$4.8_{10}-1$	$8.2_{10}-2$	$6.7_{10}-3$	$1.4_{10}-4$	$6.3_{10}-6$	$6.0_{10}-7$
discretisation error	$1.8_{10}+0$	$4.9_{10}-2$	$1.5_{10}-2$	$4.9_{10}-3$	$1.7_{10}-3$	$5.9_{10}-4$	$2.1_{10}-4$

16.5 Comments and Modifications

16.5.1 More than One Picard Iteration

Multi-grid algorithms for elliptic equations use ν steps of some smoothing iteration, where ν must be sufficiently large (cf. Theorem 7.1.2 or Theorem 7.2.2). One can try to replace the smoothing step (2.2b) by

$$\textbf{for } i = 1 \textbf{ step } 1 \textbf{ until } \nu \textbf{ do } u := K_l * u + f; \tag{16.5.1}$$

As known from § 16.2.1, the choice of $\nu > 1$ is dangerous if $\| K_l \|_{\mathscr{U} \leftarrow \mathscr{U}} \gg 1$. Even if $\| K_l \|_{\mathscr{U} \leftarrow \mathscr{U}} \approx 1$ the analysis given above shows no advantage in taking $\nu > 1$. However, there are cases where $\nu > 1$ is reasonable.

In Example 16.1.3 the operator K does not map from $L^2(D)$ into $H^\gamma(D)$ for fixed $\gamma \in (\frac{1}{2}, 1)$, but K^2 does so. Another example follows.

Example 16.5.1. Assume $D = [0, 1]$ and let $k(x, y)$ be sufficiently smooth in the two triangles $0 \leqq x \leqq y \leqq 1$ and $0 \leqq y < x \leqq 1$, but discontinuous across the diagonal $x = y$. Then K maps from $C^s(D)$ into $C^{s+1}(D)$, $s \geqq 0$, but not into $C^t(D)$ with $t > s + 1$. As a consequence $K^2 \colon C^0(D) \to C^2(D)$ holds.

Assume that the multi-grid iteration (2.2) (with $\nu = 1$) is applied. In the case of Example 16.1.3 one has to choose $\mathscr{U}_l = L^2(D_l)$, $\mathscr{V}_l = H^{1/2}(D_l)$ (discrete analogues of $L^2(D)$, $H^{1/2}(D)$), which yields $\kappa = \frac{1}{2}$ as maximum exponent in the $O(h^\kappa)$ right-hand sides of (1.12), (1.14). For Example 16.5.1 we have $\mathscr{U}_l = C^0(D_l)$, $\mathscr{V}_l = C^1(D_l)$, and $\kappa = 1$. The possibly better orders of the interpolation p and of the quadrature formula cannot be exploited. The multi-grid convergence rate is only $O(h_l^\kappa)$ with $\kappa = \frac{1}{2}$ or $\kappa = 1$ respectively.

The situation improves if we perform $\nu \geqq 2$ smoothing iterations (cf. (1)). The regularity condition (1.9) can be replaced by

$$\| K_l^\nu \|_{\mathscr{V} \leftarrow \mathscr{U}} \leqq C_R.$$

Now the respective choices of $\mathscr{V}_l$ in the previous examples are $H^\gamma(D_l)$, $\gamma \in (\frac{1}{2}, 1)$, and $C^2(D_l)$. The multi-grid convergence rates become $O(h_l^\gamma)$ and $O(h_l^2)$ respectively, instead of $O(h_l^{1/2})$ and $O(h_l)$.

16.5.2 Other Smoothing Iterations

A general iteration can be written as

$$u_l \to A_l^{-1}(B_l u_l + f_l) \quad \text{with} \quad A_l - B_l = I - K_l. \tag{16.5.2}$$

We always used Picard's iteration: $A_l = I$, $B_l = K_l$. In the case of a weakly singular kernel $k(x, y) = k_1(x, y)\, f(x - y)$ with $f(s) = |s|^{-\lambda}$ or $f(s) = \log |s|$ etc. one might think of employing a Jacobi-like iteration, since the main diagonal and the neighbouring diagonals contain entries of larger size than outside the diagonal band. A natural choice of A_l and B_l would be

$$A_l = I - K_l^B, \quad B_l = K_l - K_l^B, \tag{16.5.3a}$$

where the entries of K_l^B are

$$K_{l,ij}^B = \begin{Bmatrix} K_{l,ij} & \text{if } |i-j| \leqq r_l, \\ 0 & \text{otherwise} \end{Bmatrix}, \tag{16.5.3b}$$

The number r_l determines the band width of A_l. For periodic functions the difference $i-j$ in $|i-j| \leqq r_l$ is to be understood modulo $\dim(\mathscr{U}_l)$. It has to be ensured that A_l is stable: $\| A_l^{-1} \|_{\mathscr{U} \leftarrow \mathscr{U}} \leqq \text{const}$.

The smoothing iteration (3) can be regarded as Picard's iteration applied to a modified integral equation. Set $k^\delta(x, y) = k(x, y)$ if $|x - y| \leqq \delta$, $k^\delta = 0$ elsewhere, and define the operator K^δ by means of the kernel k^δ. The integral equation $u = Ku + f$ can be written as

$$u = (I - K^\delta)^{-1} [(K - K^\delta) u + f]. \tag{16.5.4}$$

Iteration (2) with A_l, B_l from Eq. (3) is Picard's iteration applied to the discretisation of Eq. (4), if $\delta = r_l h_l$.

A somewhat modified splitting of $I - K_l$ may be better than (3). Define the integral operator $\tilde{K}$ by the kernel $\tilde{k}(x, y) = k(x, y)$ for $|x - y| \geqq \delta$, $\tilde{k}$ smoothly continued for $|x - y| \leqq \delta$. $K^\delta := K - \tilde{K}$ can be used in (4). The corresponding discrete version (3) involves a matrix B_l with improved smoothing property.

The previous iterations can also be applied to integral equations of the first kind as done by Oskam-Fray [1]. In this case Eq. (4) becomes $u = (K^\delta)^{-1} [(K - K^\delta) u + f]$. Another approach to equations of the first kind will be proposed in § 16.10.5.

Not only the diagonal entries of K_l can become large. The next subsection reports such an example and describes an efficient smoothing iteration.

16.5.3 Application of the Integral Equation Method to the Calculation of Circulatory Flow around an Aerofoil

Consider the exterior boundary value problem (3.5) in the case of an aerofoil with trailing edge ζ_t (cf. Figs. 16.3.1 a and 16.5.1). Again the double layer potential (3.3) is used. At both sides of the edge ζ_t the flow speed must coincide ('Kutta condition'). As a consequence

$$d\mu(\zeta)/d\tau \to 0 \quad \text{as} \quad \zeta \to \zeta_t, \quad \zeta \in \Gamma,$$

must hold for the doublet distribution μ, where $d/d\tau$ represents derivation in the tangential direction. The resulting integral equation for μ is

$$\mu = K\mu - 2\langle U, \zeta \rangle + \beta(\mu^- - \mu^+), \tag{16.5.5}$$

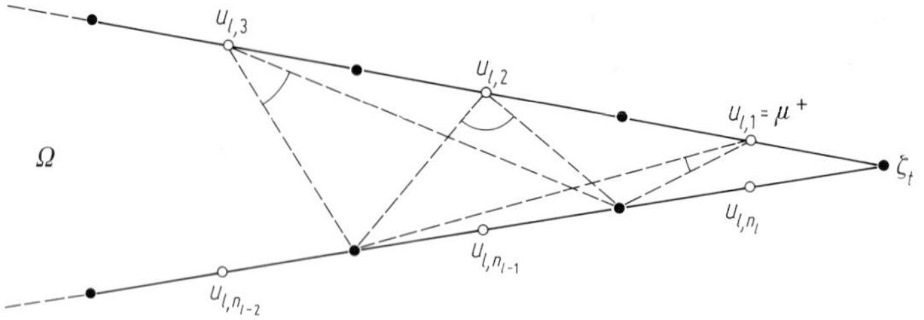

Fig. 16.5.1. Neighbourhood of the trailing edge

where $\beta(\zeta) = \frac{1}{\pi} \arg(\zeta_t - \zeta)$ and K is the integral operator associated with Eq. (3.4). The quantity μ^+ (μ^-) is the limit of $\mu(\zeta)$ along the upper (lower) side of the trailing edge (cf. Fig. 16.5.1). Schippers [4], [6] discretises the problem (5) by the panel method of Example 16.1.6. The parametrisation of Γ is chosen such that $\gamma(0) = \gamma(1) = \zeta_t$. For $s \in (0, \frac{1}{2})$ the boundary points $\gamma(s)$ and $\gamma(1-s)$ are lying on the upper and lower part of Γ having the same x-coordinates. Quantities μ^+ and μ^- are identified with the first and last components $u_{l,1}$ and u_{l,n_l} of the discrete grid function (cf. Fig. 16.5.1). The matrix K_l of the resulting discrete equation

$$u_l = K_l u_l + f_l \tag{16.5.6}$$

has large entries K_{l,i,n_l-i-1} in the cross-diagonal. The size of $K_{l,i,j}$ is determined by the angle at the ith grid point with sides intersecting the endpoints of the jth interval. Figure 16.5.1 depicts the respective angles involved by K_{l,i,n_l-1} for $i = 1, 2, 3$. Obviously the angle is extreme for $i = 2$ (in general for $i + j = n_l + 1$).

It turns out that the Picard iteration does not yield a fast converging multi-grid iteration. This is in accordance with the following result of Schippers [6].

Note 16.5.2. If a trailing edge is present, the integral operator K is not bounded as a mapping from $L^\infty(\Gamma)$ into $C^0(\Gamma)$.

Schippers [4], [6] proposed to smooth Eq. (6) either by block-Jacobi iteration (called 'paired Jacobi iteration'), where the ith block consists of $u_{l,i}$ and u_{l,n_l+1-i} $(1 \leqq i \leqq n_l/2)$, or by the corresponding 'paired Gauß-Seidel relaxation'. In the former case, the matrix A_l from (2) is given by

$$A_{l,ij} = \begin{cases} 1 - K_{l,ii} & \text{if } j = i, \\ -K_{l,i,n_l+1-i} & \text{if } j = n_l + 1 - i, \\ 0 & \text{otherwise.} \end{cases}$$

The multi-grid convergence is satisfactory if paired Jacobi iterations are used for smoothing. Even better results are obtained for paired Gauß-Seidel iteration (cf. Schippers [6], [7]).

16.5.4 Construction of the Coarse-Grid Discretisation

If one chooses a quadrature formula like the trapezoidal rule, one obtains a nested hierarchy of grids D_l. A different situation arises if one wants to apply a Gaussian quadrature formula. The first possibility would be to define each matrix K_l by a Gaussian quadrature rule with n_l grid points, where $n_0 < n_1 < \ldots < n_l$. Since $D_{l-1} \not\subset D_l$, the restriction r cannot be injection. For example, $r u_l$ may be the evaluation of the piecewise linearly interpolated grid function u_l at the points of D_{l-1}. Similarly, $p u_{l-1}$ can be defined.

A second possibility is explained below. For a fixed (finest) level l let K_l be a given discretisation using the grid D_l. Choose $D_0 \subset D_1 \subset \ldots \subset D_{l-1} \subset D_l$, and define r and p suitably, and set

$$K_{l-1} := r K_l p, \qquad K_{l-2} := r K_{l-1} p, \ldots. \tag{16.5.7}$$

For example, r may be the trivial injection (1.11), while p is a piecewise interpolation.

Proposition 16.5.3. Let $K_0, K_1, \ldots, K_{l-1}$ be constructed by (7). The two-grid iteration matrix (2.8) equals

$$M_k^{TGM} = [I + p(I - K_{k-1})^{-1} r K_k][I - pr] K_k, \qquad k = 1, 2, \ldots, l.$$

The multi-grid iteration matrices M_k' and M_k'' from (2.10) and (2.23), respectively, become

$$M_k' = M_k^{TGM} + p M_{k-1}'^{2} [r - (I - K_{k-1})^{-1} r K_k (I - pr)] K_k,$$
$$M_k'' = M_k^{TGM} + p M_{k-1}''^{2} [r - (I - K_{k-1})^{-1} r K_k (I - pr) K_k].$$

The relative consistency condition (1.12) follows from $r K_l - K_{l-1} r = r K_l [I - pr]$, while the regularity condition (1.9) at level l implies the same inequality at the coarse-grid levels $k < l$ if $\| p \|_{\mathscr{U} \leftarrow \mathscr{U}} \leqq 1$, $\| r \|_{\mathscr{V} \leftarrow \mathscr{V}} \leqq 1$.

16.5.5 Further Modifications

All multi-grid variants involved $\gamma = 2$ iterations of the coarse-grid equation (cf. (2.2 c_2), (2.20 b_4), (2.26 c_2)). It does not pay to perform $\gamma = 3$ iterations since the multi-grid rate is already sufficiently close to the two-grid rate (cf. Theorem 16.2.6 etc.).

In particular, if $\| K_l \|_{\mathscr{U} \leftarrow \mathscr{U}} < 1$, the choice $\gamma = 1$ is possible. However, if $\| K_l \|_{\mathscr{U} \leftarrow \mathscr{U}} \approx 1$, the contraction numbers $\| M_l' \|_{\mathscr{U} \leftarrow \mathscr{U}}$ and $\| M_l'' \|_{\mathscr{U} \leftarrow \mathscr{U}}$ will be of the size of $\| M_0^{TGM} \|_{\mathscr{U}\leftarrow\mathscr{U}}$ (instead of $\| M_l^{TGM} \|_{\mathscr{U}\leftarrow\mathscr{U}}$): A compromise is to choose

$$\gamma = 2 \quad \text{if } l \leqq l^*, \qquad \gamma = 1 \quad \text{if } l > l^*.$$

Then, $\| M_l'' \|_{\mathscr{U}\leftarrow\mathscr{U}} \approx \| M_{l^*}^{TGM} \|_{\mathscr{U}\leftarrow\mathscr{U}}$ can be expected for $l > l^*$. If $\| K_l \|_{\mathscr{U}\leftarrow\mathscr{U}} \gg 1$, the first variant cannot be used with $\gamma = 1$.

The convergence behaviour described in Theorems 16.2.6 and 16.2.11 is preserved by the choice

$$\gamma = 1 \quad \text{if } l \text{ is odd}, \qquad \gamma = 2 \quad \text{if } l \text{ is even}. \tag{16.5.8}$$

Note 16.5.4. Assume (2.5b). The nested iteration (4.1) with odd l and with MGM from (2.2) modified by (8) requires $Cn_l^2 + O(n_l)$ operations, where $C = \frac{82}{15} \approx 5.47$. The corresponding number for the analogous modification of the nested iteration (4.6) is $C = \frac{52}{15} \approx 3.47$.

For the sake of convenience only, the ratio $h_{l-1}/h_l = 2$ is proposed (cf. (1.5a)). One can, instead, use a sequence $\{h_l\}$ with increasing quotient h_{l-1}/h_l (cf. McCormick [1]).

16.6 The Nonlinear Equation of the Second Kind

The equation of the second kind is the fixed point equation

$$u = \mathscr{K}(u), \tag{16.6.1}$$

where $\mathscr{K}: \mathscr{U}^0 \subset \mathscr{U} \to \mathscr{U}$ is a (nonlinear) operator. Equations (1) is assumed to have at least one isolated solution $u^* \in \mathscr{U}^0$. If $\mathscr{K}$ is affine (i.e. $\mathscr{K}(u) = Ku + f$), we regain the linear problem (1.2).

Example 16.6.1. The nonlinear Fredholm integral equation of the second kind

$$u(\mathbf{x}) = \int_D k(\mathbf{x}, \mathbf{y}, u(\mathbf{y}))\, d\mathbf{y} + g(\mathbf{x}), \qquad \mathbf{x} \in D,$$

leads to Eq. (1) with $\mathscr{K}(u) = \int_D k(\,\cdot\,, \mathbf{y}, u(\mathbf{y}))\, d\mathbf{y} + g$. However, there are numerous other problems that can be formulated by Eq. (1). Examples will be given in §§ 16.8, 16.10, 16.11.

The discrete problems are denoted by

$$u_l = \mathscr{K}_l(u_l), \qquad l = 0, 1, \ldots, \tag{16.6.2}$$

where $\mathscr{K}_l: \mathscr{U}_l^0 \subset \mathscr{U}_l \to \mathscr{U}_l$ are nonlinear mappings. At least one isolated solution $u_l^* \in \mathscr{U}_l^0$ is supposed to exist. The derivative of $\mathscr{K}_l$ is denoted by

$$K_l(v_l) := \mathscr{K}_l'(v_l) \quad \text{for } v_l \in \mathscr{U}_l^0$$

and is assumed to be Lipschitz continuous:

$$\| K_l(v_l) - K_l(w_l) \|_{\mathscr{U} \leftarrow \mathscr{U}} \leqq C_K \| v_l - w_l \|_{\mathscr{U}} \quad \text{for all } v_l, w_l \in \mathscr{U}_l^0, \ l \geqq 0. \tag{16.6.3}$$

The derivative at $v_l = u_l^*$ (u_l^*: solution of Eq. (1)) is denoted by

$$K_l := K_l(u_l^*) = \mathscr{K}_l'(u_l^*). \tag{16.6.4}$$

We shall require that the matrix M_l^{TGM} from (2.8) defined with K_l and K_{l-1} from (4) satisfies

$$\| M_l^{TGM} \|_{\mathscr{U} \leftarrow \mathscr{U}} \leqq C_{TGM} h_l^\kappa, \quad M_l^{TGM} := [I - p(I - K_{l-1})^{-1} r(I - K_l)] K_l. \tag{16.6.5}$$

Consequently, the linear two-grid iteration applied to the linearised equation has a contraction number $\leqq C_{TGM} h_l^\kappa$. By Theorem 16.2.4, estimate (5) follows from (1.8), (1.9), (1.12), (1.13), (1.14).

Example 16.6.2. Discretise the nonlinear integral equation of Example 16.6.1 by the trapezoidal rule as in Example 16.1.5. $K_l(v_l)$ is given by Eq. (1.5c) with $k(x, y)$ replaced by $k_u(x, y, v_l(y))$. Relations $k_u \in C^1(D \times D \times \mathscr{U}^0)$ and $\mathscr{U}_l = C^0(D_l)$ imply Inequality (3). Concerning the choice of $\mathscr{V}_l$ and κ the examples of § 16.1.3 still apply.

The assumptions (3) and (5) are relatively weak. For example, Inequality (5) is to be satisfied only by $K_l = K_l(u_l^*)$, not by $K_l(v_l)$ for $v_l \neq u_l^*$. Note that (3) does not imply the regularity assumption (1.9) for $K_l(v_l)$.

For the multi-grid process it is necessary to consider the perturbed equations

$$u_l = \mathscr{K}_l(u_l) + f_l, \qquad l = 0, 1, \ldots \tag{16.6.6a}$$

with sufficiently small perturbations f_l. For example, we may require

$$\| f_l \|_{\mathscr{U}} \leqq \varepsilon_l. \tag{16.6.6b}$$

For simplicity, one may set

$$\mathscr{U}_l^0 := \{u_l\colon \ u_l \text{ is the solution of (6a) for } f_l \text{ with (6b)}\}.$$

Note 16.6.3. Let $u_l^* \in \mathscr{U}_l^0$ be a solution of (2). By the inverse mapping theorem, the assumptions (3) and (1.8) establish the unique solvability of (6a, b) in a neighbourhood of u_l^*, if $\varepsilon_l > 0$ from (6b) is sufficiently small.

The inverse mapping $f_l \mapsto u_l$ is denoted by Φ_l: $u_l = \Phi_l(f_l)$ is the solution of Eq. (6a). The mapping Φ_l is defined on $\{f_l \in \mathscr{U}_l\colon \| f_l \|_{\mathscr{U}} \leqq \varepsilon_l\}$. Its derivative is $(I - K_l(\Phi_l(f_l)))^{-1}$. In particular,

$$\Phi_l(0) = u_l^*, \quad \Phi_l'(0) = (I - K_l)^{-1} \tag{16.6.7}$$

holds.

16.7 Nonlinear Multi-Grid Method of the Second Kind

16.7.1 Two-Grid Iteration

The following algorithm corresponds to (9.3.7):

Nonlinear two-grid iteration for solving $u_l = \mathcal{K}_l(u_l) + f_l$ (16.7.1)

$$u_l' := \mathcal{K}_l(u_l^j) + f_l; \tag{16.7.1 a}$$

$$d_{l-1} := r * (u_l' - \mathcal{K}_l(u_l') - f_l); \tag{16.7.1 b$_1$}$$

$\tilde{u}_{l-1} \in \mathcal{U}_{l-1}^0$ must be given; possibly determined a posteriori by

$$\tilde{u}_{l-1} := r\, u_l'. \tag{16.7.1 b$_2$}$$

The defect of $\tilde{u}_{l-1}$ must be given also:

$$\tilde{f}_{l-1} := \tilde{u}_{l-1} - \mathcal{K}_{l-1}(\tilde{u}_{l-1}); \tag{16.7.1 b$_3$}$$

$$u_l^{j+1} := u_l' + p\,[\Phi_{l-1}(\tilde{f}_{l-1} - s\,d_{l-1}) - \tilde{u}_{l-1}]/s \tag{16.7.1 b$_4$}$$

Φ_k denotes the inverse of $I - \mathcal{K}_k$. $\Phi_k(f_k)$ is the solution of Eq. (6.6a): $u_k - \mathcal{K}_k(u_k) = f_k$. Φ_k replaces $\mathcal{L}_k^{-1}$ from § 9.

The equivalent of Remark 9.3.4 is

Note 16.7.1. If $\mathcal{K}_l$ and $\mathcal{K}_{l-1}$ are affine (i.e. $\mathcal{K}_k(u_k) = K_k u_k + g_k$), algorithm (1) with arbitrary $\tilde{u}_{l-1} \in \mathcal{U}_{l-1}$ and $s \in \mathbb{R}$ is algebraically equivalent to the linear iteration (2.1).

The function $\Phi_{l-1}(\tilde{f}_{l-1} - s\,d_{l-1})$ is only defined if

$$\| \tilde{f}_{l-1} - s\,d_{l-1} \|_{\mathcal{U}} \leqq \varepsilon_{l-1}. \tag{16.7.2}$$

Note 16.7.2. Let $\tilde{u}_{l-1}$ be given a priori with, e.g.,

$$\| \tilde{f}_{l-1} \|_{\mathcal{U}} \leqq \varepsilon_{l-1}/2. \tag{16.7.3}$$

Then the choice $s := \| \tilde{f}_{l-1} \|_{\mathcal{U}} / (2\, \| d_{l-1} \|_{\mathcal{U}})$ ensures (2). The fixed choice of $\tilde{u}_{l-1}$ has the advantage that $\tilde{f}_{l-1}$ (cf. (1 b$_3$)) has to be computed only once, whereas the definition of $\tilde{u}_{l-1}$ by (1 b$_2$) requires the computation of $\tilde{f}_{l-1}$ by (1 b$_3$) in every iteration.

Note 16.7.3. The analogue of the FAS scheme (9.3.10) is obtained by (1 b$_2$) and $s = 1$; $\tilde{f}_{l-1}$ has to be recomputed in every iteration. The inequality (2) in question becomes

$$\begin{aligned} &\| [r u_l' - \mathcal{K}_{l-1}(r u_l')] - r\,[u_l' - \mathcal{K}_l(u_l') - f_l] \|_{\mathcal{U}} \\ &\quad = \| r\, \mathcal{K}_l(u_l') - \mathcal{K}_{l-1}(r u_l') + r f_l \|_{\mathcal{U}} \leqq \varepsilon_{l-1}. \end{aligned}$$

The first term $\| r\mathscr{K}_l(u_l') - \mathscr{K}_{l-1}(r u_l')\|_{\mathscr{U}}$ may be expected to be of $O(h_l^\kappa)$, provided that u_l' is smooth. However, the smoothing step (1 a) guarantees the smoothness of the error $u_l' - \Phi_l(f_l)$, though not necessarily of u_l' itself.

16.7.2 Multi-Grid Iteration

As in (9.3.11) we assume that there exists an iterative method

$$(u_0^j, f_0) \mapsto u_0^{j+1} = \tilde{\Phi}_0(u_0^j, f_0) \tag{16.7.4}$$

for solving coarse-grid equations at level $l = 0$.

Note 16.7.4. If $\mathscr{K}_0$ is contractive (i.e. $\| \mathscr{K}_0(v_0) - \mathscr{K}_0(w_0)\|_{\mathscr{U}} \leqq L \| v_0 - w_0 \|_{\mathscr{U}}$, $L < 1$), a certain number of Picard iterations may serve as $\tilde{\Phi}_0$. Another possibility is a Newton iteration $\tilde{\Phi}_0$ requiring an approximation of the Jacobi matrix $K_0(u_0^j)$.

The multi-grid iteration needs arbitrary values

$$\tilde{u}_0 \in \mathscr{U}_0^0, \quad \tilde{u}_1 \in \mathscr{U}_1^0, \dots, \tilde{u}_{l-1} \in \mathscr{U}_{l-1}^0 \tag{16.7.5 a}$$

with defects

$$\tilde{f}_k = \tilde{u}_k - \mathscr{K}_k(\tilde{u}_k), \qquad 0 \leqq k \leqq l-1. \tag{16.7.5 b}$$

These values may be given a priori (cf. (8 b)) or a posteriori analogously to (1 b$_2$) by inserting statements

$$\tilde{u}_{l-1} := r u; \quad \tilde{f}_{l-1} := \tilde{u}_{l-1} - \mathscr{K}_{l-1}(\tilde{u}_{l-1})$$

between (6 b) and (6 c$_1$). The first variant (2.2) becomes

Nonlinear multi-grid iteration (first variant) for solving

$$u_l = \mathscr{K}_l(u_l) + f_l \tag{16.7.6}$$

procedure $NMGM(l, u, f)$; **integer** l; **array** u, f;
if $l = 0$ **then** $u := \tilde{\Phi}_0(u, f)$ **else** (16.7.6 a)
begin integer j; **real** s; **array** v, d;
  $u := \mathscr{K}_l(u) + f$; (16.7.6 b)
  $d := r * (u - \mathscr{K}_l(u) - f)$; (16.7.6 c$_1$)
  $s :=$ **if** $d = 0$ **then** l **else** $\sigma_{l-1} / \| d \|_{\mathscr{U}}$; (16.7.6 c$_2$)
  $d := \tilde{f}_{l-1} - s * d$; (16.7.6 c$_3$)

$v := \tilde{u}_{l-1};$ (16.7.6 c$_4$)

for $j := 1, 2$ **do** $NMGM(l-1, v, d);$ (16.7.6 c$_5$)

$u := u + p(v - \tilde{u}_{l-1})/s$ (16.7.6 c$_6$)

end;

. . . .

According to Note 16.7.2, one should choose $\sigma_{l-1} \leqq \varepsilon_{l-1}/2$. If $\tilde{f}_{l-1} \neq 0$ satisfies (3), the choice

$$\sigma_{l-1} = \| \tilde{f}_{l-1} \|_{\mathscr{U}}$$

is permitted. For the case of $d = 0$ see Note 9.3.3.

Note 16.7.5. Iteration (6) requires no derivative. Only the evaluation of $\mathscr{K}_l(u_l)$ is needed.

Also the second variant (2.20) can be generalised to nonlinear problems.

. . . .

Nonlinear multi-grid iteration (second variant) for solving

$$u_l = \mathscr{K}_l(u_l) + f_l \tag{16.7.7}$$

. . . .

procedure $NMGM(l, u, f)$; **integer** l; **array** u, f;

if $l = 0$ **then** $u := \tilde{\Phi}_0(u, f)$ **else** (16.7.7 a)

begin integer j; **real** s; **array** v, u', d;

$u' := \mathscr{K}_l(u) + f;$ (16.7.7 b)

$d := \mathscr{K}_{l-1}(ru') - \mathscr{K}_{l-1}(ru) + r * (u - \mathscr{K}_l(u') - f);$ (16.7.7 c$_1$)

$u := u - pr(u - u');$ $s :=$ **if** $d = 0$ **then** 1 **else** $\sigma_{l-1}/\| d \|_{\mathscr{U}};$ (16.7.7 c$_2$)

$d := \tilde{f}_{l-1} - s * d;$ (16.7.7 c$_3$)

$v := \tilde{u}_{l-1};$ **for** $j := 1, 2$ **do** $NMGM(l-1, v, d);$ (16.7.7 c$_4$)

$u := u + p * (v - \tilde{u}_{l-1})/s$ (16.7.7 c$_5$)

end;

. . . .

In (7 c$_1$) the difference $\mathscr{K}_{l-1}(ru') - \mathscr{K}_{l-1}(ru)$ may be replaced by $\mathscr{K}_{l-1}(\tilde{u}_{l-1} + r(u' - u)) - \tilde{u}_{l-1} + \tilde{f}_{l-1}$.

The nonlinear formulation of the third multi-grid variant (2.26) is obvious. Analogously to Note 16.7.1 one proves

Note 16.7.6. If the mappings $\mathscr{K}_k(u_k) = K_k u_k + g_k$, $0 \leqq k \leqq l$, are affine, the iterations (6) and (7) are algebraically equivalent to the respective linear multi-grid iterations (2.2) and (2.20), independently of the choice of σ_k and $\tilde{u}_{k-1}$.

16.7.3 Nested Iteration

The equivalent of algorithm (9.3.14) is the

nested iteration for solving $u_k = \mathscr{K}_k(u_k)$, $0 \leqq k \leqq l$ (16.7.8)

$\tilde{u}_0$: given approximation to u_0^*; (16.7.8 a)

for $k := 1$ **step** 1 **until** l **do**

begin $\tilde{f}_{k-1} := \tilde{u}_{k-1} - \mathscr{K}_{k-1}(\tilde{u}_{k-1})$; (16.7.8 b)

$\tilde{u}_k := \tilde{p}\,\tilde{u}_{k-1}$; (16.7.8 c)

for $j := 1$ **step** 1 **until** i **do** $NMGM(k, \tilde{u}_k, 0)$ (16.7.8 d)

end;

The nested iteration (8) is used not only to provide good starting guesses at the next level, but also to compute the auxiliary $\tilde{u}_k$ from (5 a) and their defects (5 b). When procedure $NMGM$ is called in (8 d) at level k, the values $\tilde{u}_0, \tilde{u}_1, \ldots, \tilde{u}_{k-1}$ and $\tilde{f}_0, \tilde{f}_1, \ldots, \tilde{f}_{k-1}$ are already computed. The statement (8 b) can be omitted if (1 $\mathrm{b}_{2,3}$) is used inside (6) or (7).

If $\mathscr{K}_l$ is defined by Example 16.6.2, also the nonlinear counterpart of the nested iteration (4.6) with Nyström's interpolation can be applied.

16.7.4 Convergence

In § 16.9.3 we shall analyse the convergence of the multi-grid iteration and the accuracy of the results $\tilde{u}_k$ of the nested iteration. A necessary condition for convergence $u_l^j \to u_l^*$ is stated in

Note 16.7.7. The exact solution u_l^* is a fixed point of the multi-grid iterations (6) and (7).

As in § 9.5 it will turn out that the nonlinear multi-grid iteration converges asymptotically as fast as the linear iteration applied to the linearised problem $u_l = K_l u_l + f_l$ with K_l from (6.4).

As with any nonlinear iteration, the multi-grid iteration may fail if the starting guess is not close enough to the solution. The starting values $\tilde{u}_k$ from (8 c) must be inside the 'convergence domain' $\mathscr{U}_k^* := \{v_k \in \mathscr{U}_k^0$: iteration with $u_k^0 = v_k$, $f_k = 0$ yields $u_k^i \to u_k^*\}$. However, $\tilde{p}\,\tilde{u}_{k-1} \in \mathscr{U}_k^*$ cannot be guaranteed in general even if $\tilde{u}_{k-1}$ is very acurate, since $\tilde{p}\,u_{k-1}^* \notin \mathscr{U}_k^*$ may hold. In that case, one should try to apply continuation techniques.

16.7.5 Computational Work

The work of vector additions and of multiplications by r, p or scalar factors can be neglected, since the dominating part of the computation is the evaluation of $\mathscr{K}_l(v_l)$. Assume

$$v_l \mapsto \mathscr{K}_l(v_l) \quad \text{requires} \leqq C_l \quad \text{operations,} \tag{16.7.9 a}$$

$$(v_0, f_0) \mapsto \tilde{\Phi}_0(v_0, f_0) \quad \text{requires} \leqq C_\Phi \quad \text{operations.} \tag{16.7.9 b}$$

Note 16.2.2 can be extended to the present case.

Note 16.7.8. The coarse-grid equations are solved in ($6c_5$) and ($7c_4$) by two iterations at level $l-1$ starting with $u_{l-1}^0 = \tilde{u}_{l-1}$. The evaluation of $\mathscr{K}_{l-1}(\tilde{u}_{l-1})$ can be omitted since $\mathscr{K}_{l-1}(\tilde{u}_{l-1}) = \tilde{u}_{l-1} - \tilde{f}_{l-1}$.

Taking into account Note 16.7.8 one proves

Proposition 16.7.9. Assume $\tilde{u}_k, \tilde{f}_k, 0 \leqq k \leqq l-1$, to be known. One multi-grid iteration at level l requires at most

$$C_l^{NMGM} = \begin{cases} 2C_l + 3\sum_{k=1}^{l-1} 2^{l-k-1} C_k + 2^l C_\Phi & \text{for 1st variant (6)} \\ \sum_{k=1}^{l} 2^{l-k+1} C_k + 2^{l-1} C_0 + 2^l C_\Phi & \text{for 2nd variant (7)} \\ 2C_l + \sum_{k=1}^{l-1} 2^{l-k} C_k + 2^l C_\Phi & \text{for 3rd variant} \end{cases} \tag{16.7.10}$$

operations. If $\tilde{u}_k$ and $\tilde{f}_k$ are computed in every iteration by means of ($1\,b_{2,3}$), the respective numbers are

$$C_l^{NMGM} = \begin{cases} \sum_{k=1}^{l} 2^{l-k+1} C_k + 2^{l-1} C_0 + 2^l C_\Phi & \text{for 1st variant,} \\ 2C_l + 5\sum_{k=1}^{l-1} 2^{l-k-1} C_k + 2^l (C_0 + C_\Phi) & \text{for 2nd variant,} \\ 2C_l + 3\sum_{k=1}^{l-1} 2^{l-k+1} C_k + 2^{l-1} C_0 + 2^l C_\Phi & \text{for 3rd variant.} \end{cases} \tag{16.7.11}$$

The nested iteration (8) requires at most

$$C_l^{\text{nested it.}} = \begin{cases} 2iC_l + \sum_{k=1}^{l-1} [1 + i(3 \cdot 2^{l-k} - 1)]\, C_k + C_0 \\ 2iC_l + \sum_{k=1}^{l-1} [1 + i(4 \cdot 2^{l-k} - 2)]\, C_k + [1 + i(2^l - 1)]\, C_0 \\ 2iC_l + \sum_{k=1}^{l-1} [1 + i2^{l-k+1}]\, C_k + C_0 \\ + i(2^{l+1} - 2)\, C_\Phi \end{cases} \tag{16.7.12}$$

operations. If $\tilde{u}_k$, $\tilde{f}_k$ are computed in each iteration and if (8 b) is omitted, the respective numbers are

$$C_l^{\text{nested it.}} = \begin{cases} \sum_{k=1}^{l} (2^{l-k+2} - 2)\, C_k + (2^l - 1)\,(C_0 + 2\,C_\Phi), \\ \sum_{k=1}^{l} (5 \cdot 2^{l-k} - 3)\, C_k + 2(2^{l-1} - 1)\,(C_0 + C_\Phi), \\ \sum_{k=1}^{l} (3 \cdot 2^{l-k} - 1)\, C_k + (2^l - 1)\,(C_0 + 2\,C_\Phi). \end{cases} \tag{16.7.13}$$

These formulae do not take into account that the evaluation of $r\,\mathcal{K}_l(u_l)$ may be cheaper than the evaluation of $\mathcal{K}_l(u_l)$.

16.8 Application to Nonlinear Elliptic Equations

Equation (6.1) of the second kind may be representative of equations other than integral equations. A prototype is studied in this section.

16.8.1 Reformulation of the Continuous Boundary Value Problem

Consider the elliptic boundary value problem

$$\begin{aligned} &-\Delta u(x, y) = g(x, y, u(x, y)) \quad \text{in} \quad \Omega = (0,1) \times (0,1), \\ &u = 0 \quad \text{on} \quad \Gamma = \partial\Omega \end{aligned} \tag{16.8.1}$$

with some (nonlinear) function $g \in C^0(\Omega \times \mathbb{R})$. Problem (1) is equivalent to

$$u = (-\Delta)^{-1}\, g(\cdot\,,\cdot\,, u), \tag{16.8.2}$$

where $w = (-\Delta)^{-1}\, v$ denotes the solution of $-\Delta w = v$ in Ω, $w = 0$ on Γ. Abbreviating the right side of (2) by

$$\mathcal{K}(u) := (-\Delta)^{-1}\, g(\cdot\,,\cdot\,, u), \tag{16.8.3}$$

we obtain the equation (6.1), $u = \mathcal{K}(u)$. The derivative of $\mathcal{K}$ is

$$K(v) := (-\Delta)^{-1}\, g_u(\cdot\,,\cdot\,, v),$$

if $g_u = \partial g/\partial u$ exists. The relation $w = (-\Delta)^{-1} v$ can be represented by means of Green's function $G(x, y;\, \xi, \eta)$: $w(x, y) = \int_\Omega G(x, y;\, \xi, \eta)\, v(\xi, \eta)\, d\xi\, d\eta$. Hence, formally, $\mathcal{K}(u)$ is the weakly singular, nonlinear integral operator

$$\mathcal{K}(u)(x, y) = \int_\Omega G(x, y;\, \xi, \eta)\, g(\xi, \eta, u(\xi, \eta))\, d\xi\, d\eta. \tag{16.8.4}$$

We have to check the regularity (1.4), $\| K \|_{\mathscr{V} \leftarrow \mathscr{U}} \leqq C$, of $K = K(u^*)$, since otherwise there is no hope of satisfying (1.9). We may choose

$$\mathscr{U} = C^0(\bar{\Omega}), \quad \mathscr{V} = C^{1+\alpha}(\bar{\Omega}) \quad \text{with fixed} \quad \alpha \in [0, 1). \tag{16.8.5}$$

The well-known estimate $\| (-\Delta)^{-1} w \|_{C^{1+\alpha}(\bar{\Omega})} \leqq C_\Omega \| w \|_{C^0(\bar{\Omega})}$ yields

$$\begin{aligned} \| Kw \|_{\mathscr{V}} &= \| (-\Delta)^{-1} g_u(\cdot, \cdot, u^*) w \|_{C^{1+\alpha}(\bar{\Omega})} \\ &\leqq C_\Omega \| g_u(\cdot, \cdot, u^*) \|_{C^0(\bar{\Omega})} \| w \|_{C^0(\bar{\Omega})}, \end{aligned}$$

hence

$$\| K \|_{\mathscr{V} \leftarrow \mathscr{U}} \leqq C_\Omega \max_{(x, y) \in \bar{\Omega}} | g_u(x, y, u^*(x, y)) |. \tag{16.8.6}$$

16.8.2 Discretisation

The boundary value problem (1) can be discretised by

$$L_l u_l = g_l(u_l), \tag{16.8.7}$$

where $g_l(u_l)$ is the grid function with values $g_l(u_l)(x, y) = g(x, y, u_l(x, y))$ on the grid

$$\begin{aligned} D_l &= \{(x, y) = (\nu h_l, \mu h_l) \in \mathbb{R}^2 : \nu, \mu \in \mathbb{Z}, 0 < \nu, \mu < 1/h_l\}, \\ h_l &= h_0/2^l. \end{aligned} \tag{16.8.8}$$

L_l may be the five-point scheme including the homogeneous boundary conditions. The reformulation (1)$\mapsto$(2) applied to (7) yields

$$u_l = \mathscr{K}_l(u_l) \quad \text{with} \quad \mathscr{K}_l(u_l) := L_l^{-1} g_l(u_l). \tag{16.8.9}$$

Note that the discretisation of $u = \mathscr{K}(u)$ by (9) is much easier than a discretisation of the integral operator (4). The performance of $u_l \mapsto \mathscr{K}_l(u_l)$ is relatively cheap:

Note 16.8.1. The evaluation of $\mathscr{K}_l(u_l)$ requires the evaluation of $g_l(u_l)$ $[O(h_l^{-2})$ operations$]$ and one call of a direct Poisson solver. For example Buneman's algorithm (cf. Buneman [1], Meis-Marcowitz [1]) can be used. The required work amounts to $O(h_l^{-2} |\log h_l|)$.

The derivative of $\mathscr{K}_l(u_l)$ at v_l equals

$$K_l(v_l) = L_l^{-1} g_l'(v_l), \quad \text{where} \quad (g_l'(v_l) w_l)(x, y) = [\partial g(x, y, v_l)/\partial u] w_l(x, y).$$

One can prove

Lemma 16.8.2. Assume $g, g_u \in C^0(\bar{\Omega} \times \mathbb{R})$, and suppose that there are solutions u_l^* of (7) bounded by $\| u_l^* \|_{C^0(D_l)} \leqq$ const for $l = 0, 1, 2 \ldots$. Then $K_l = \mathscr{K}_l'(u_l^*)$

satisfies the regularity condition (1.9) for the choice $\mathscr{U}_l = C^0(D_l)$, $\mathscr{V}_l = C^{1+\alpha}(D_l)$, $\alpha < 1$ (cf. § 16.1.3). Also, (1.12) and (1.14) can be shown to hold with $\kappa = 1 + \alpha$.

Although Eq. (9) is not a discrete integral equation, the multi-grid iterations (7.6) and (7.7) can be applied to Eq. (9), since according to Note 16.8.1 $\mathscr{K}_l(u_l)$ can easily be evaluated. The convergence depends on the conditions (1.8), (1.9), (1.12), (1.13), (1.14), not on the fact that $\mathscr{K}_l$ represents an integral equation.

Nevertheless, the abstract equation (9) of the second kind gives rise to some other properties of the multi-grid process. The reason is not the nonlinearity. The following comments hold even for a linear equation $u_l = \mathscr{K}_l(u_l)$, as obtained for $g(\cdot, \cdot, u) = g_1 + g_2 u$.

1) In the case of discrete integral equations (1.6), the entries $K_{l,ij}$ of the matrix K_l are explicitly known and they are used for the matrix multiplication. In the case of a linear Eq. (9) the entries of K_l are never computed and the mapping $u_l \mapsto K_l u_l = \mathscr{K}_l(u_l) - \mathscr{K}_l(0)$ is not performed as a matrix multiplication.
2) As a consequence, the modifications mentioned in § 16.5.2 cannot be generalised, since the matrix A_l involves the knowledge of special matrix components.
3) Nyström's interpolation (cf. § 16.4.3) cannot be extended to problem (9).
4) The computational work of the mapping $u_l \to \mathscr{K}_l(u_l)$ is not necessarily proportional to $(\dim \mathscr{U}_l)^2$. In fact, in the case of Eq. (9) the work is $O(n_l \log n_l)$, $n_l = \dim(\mathscr{U}_l)$ (cf. Note 16.8.1).
5) In the case of Eq. (9) the evaluation of $r_{inj} \mathscr{K}_l(v_l)$ is not cheaper than the evaluation of $\mathscr{K}_l(v_l)$.
6) Often, $\mathscr{K}_l(u_l)$ from (9) has to be approximated iteratively, as pointed out in § 16.9.

16.8.3 Computational Work

According to Note 16.8.1, the work C_l required in the evaluation of $\mathscr{K}_l(u_l)$ satisfies

$$C_{l-1} \leqq \tfrac{1}{4} C_l, \tag{16.8.10}$$

as for (one-dimensional) integral equations.

Note 16.8.3. Assuming (10) and neglecting C_Φ, one obtains the following operation counts for the nested iteration (7.8):

$$C_l^{\text{nested it.}} \leqq C_l * \begin{cases} \frac{1}{3} + 4\,\frac{2}{3} \cdot i & \text{(1st variant)} \\ \frac{1}{3} + 5\,\frac{1}{3} \cdot i & \text{(2nd variant)} \\ \frac{1}{3} + 4 \cdot i & \text{(3rd variant).} \end{cases} \tag{16.8.11}$$

16.8.4 Numerical Example

Consider the problem (1) with $g(x, y, u) = e^u$:

$$-\Delta u = e^u \quad \text{in} \quad \Omega = (0, 1) \times (0, 1), \qquad u = 0 \quad \text{on} \quad \Gamma \tag{16.8.12}$$

(cf. (9.2.2)). In (7) the five-point discretisation is chosen with $h_0 = \frac{1}{2}$, $h_1 = \frac{1}{4}, \ldots, h_5 = \frac{1}{64}$. At level $l = 0$ the discrete problem $u_0 = \mathscr{K}_0(u_0) + f_0$ is a scalar equation ($\dim \mathscr{U}_0 = 1$), which is solved approximately by

$$v_0 \mapsto \tilde{\Phi}_0(v_0, f_0) = 0.0668 - 1.07 f_0.$$

The nested iteration (7.8) is started with $\tilde{u}_0(\frac{1}{2}, \frac{1}{2}) = 0.066819$. The number of iterations in (7.8d) is $i = 1$. The same piecewise linear interpolation $\tilde{p} = p$ is used in (7.8c) as inside the multi-grid iteration (7.6). The resulting values $\tilde{u}_k$ at the midpoint $(\frac{1}{2}, \frac{1}{2})$ are listed in Table 16.8.1.

The iteration errors are much smaller than the discretisation errors. Richardson's extrapolation applied to the values $\tilde{u}_k(\frac{1}{2}, \frac{1}{2})$, $k = 4, 5, 6$, given in Table 16.8.1 yields the number 0.078101022605, which coincides with the accurate value of $u(\frac{1}{2}, \frac{1}{2})$ to nine digits. For more details, see Hackbusch [9].

Table 16.8.1. Results of the nested iteration applied to Eq. (12)

level k	0	1	2	3	4	5
grid size h_k	$\frac{1}{2}$	$\frac{1}{4}$	$\frac{1}{8}$	$\frac{1}{16}$	$\frac{1}{32}$	$\frac{1}{64}$
$\tilde{u}_k(\frac{1}{2}, \frac{1}{2})$	0.066819	0.074579	0.077190	0.077872	0.078043	0.078087
observed reduction factor	–	0.02	0.007	0.0026	0.001	0.0007

According to (11), the above computation requires a work $\leqq 5\,C_5$, equivalent to five calls of a direct Poisson solver. Were the problem (7) solved by Newton's iteration, direct Poisson solvers could not have been applied, since the resulting linear equations $-\Delta w = g_u w$ have variable coefficients $g_u(x, y)$.

16.9 Multi-Grid Methods with Iterative Computation of $\mathscr{K}_l(u_l)$

16.9.1 Notation

In the examples of § 16.8 the computation of $\mathscr{K}_l(u_l)$ involves the solution of a discrete Poisson equation. Applying a direct Poisson solver we are able to evaluate $\mathscr{K}_l(u_l)$ explicitly. However, if Eq. (8.7) involves a more complicated operator L_l, it cannot be inverted by a direct solver and one has to use an iterative method, e.g. a linear multi-grid iteration. Thus, $\mathscr{K}_l(u_l)$ will not be determined exactly. Instead, an approximation can be computed depending on the starting value and on the number of iterations. By

$$\mathscr{K}_l(v_l;\, w_l,\, \mu)$$

we denote the result of μ iterations with starting iterate $w_l =: \mathscr{K}_l(v_l;\, w_l,\, 0)$.

Example 16.9.1. Let $u_l^{j+1} = M_l u_l^j + N_l f_l$ be a (e.g. multi-grid) iteration solving $L_l u_l = f_l$ with L_l from (8.7). An iterative computation of $\mathscr{K}_l(v_l) = L_l^{-1} g_l(v_l)$ from (8.9) is given by

$$\begin{aligned} w_l =: \mathscr{K}_l(v_l;\, w_l,\, 0) &\mapsto \mathscr{K}_l(v_l;\, w_l,\, 1) := M_l w_l + N_l g_l(v_l) \mapsto \dots \\ &\mapsto \mathscr{K}_l(v_l;\, w_l,\, \mu) := M_l \mathscr{K}_l(v_l;\, w_l,\, \mu-1) + N_l g_l(v_l) \mapsto \dots \end{aligned}$$

converging to $\mathscr{K}_l(v_l)$ if $\varrho(M_l) < 1$. The explicit description of $\mathscr{K}_l(v_l;\, w_l,\, \mu)$ is

$$\mathscr{K}_l(v_l;\, w_l,\, \mu) = M_l^{\mu} w_l + \sum_{\chi=0}^{\mu-1} M_l^{\chi} N_l g_l(v_l), \tag{16.9.1 a}$$

$$\mathscr{K}_l(v_l;\, w_l,\, \mu) - \mathscr{K}_l(v_l) = M_l^{\mu}(w_l - \mathscr{K}_l(v_l)) \tag{16.9.1 b}$$

(cf. (1.3.3)). According to (1 b) we shall assume the error estimate

$$\| \mathscr{K}_l(v_l;\, w_l,\, \mu) - \mathscr{K}_l(v_l) \|_{\mathscr{U}} \leqq \varepsilon^{\mu} \, \| w_l - \mathscr{K}_l(v_l) \|_{\mathscr{U}}, \tag{16.9.2}$$

which is true for (elliptic) multi-grid iterations.

16.9.2 Algorithm

The multi-grid iteration (7.6) is to be reformulated with $\mathscr{K}_l(\,\cdot\,)$ replaced by $\mathscr{K}_l(\,\cdot\,;\,\cdot\,,\mu_l)$, μ_l depending on l. If some rough approximation $\tilde{u}_l$ of the solution $u_l = \mathscr{K}_l(u_l) + f_l$ is known, the obvious starting guess of $\mathscr{K}_l(\tilde{u}_l)$ is $\tilde{u}_l - f_l$. Hence, $\mathscr{K}_l(\tilde{u}_l)$ will be substituted by $\mathscr{K}_l(\tilde{u}_l;\, \tilde{u}_l - f_l,\, \mu_l)$.

Multi-grid iteration (first variant) for solving
$u_l = \mathscr{K}_l(u_l) + f_l$ with interior iterations (16.9.3)

procedure $NMGM(l, u, f)$; **integer** l; **array** u, f;
if $l = 0$ **then** $u := \tilde{\Phi}_0(u, f)$ **else** (16.9.3 a)
begin real s; **integer** j; **array** v, d;

$u := \mathscr{K}_l(u;\, u - f,\, \mu_l) + f$; (16.9.3 b)

$d := r * (u - \mathscr{K}_l(u;\, u - f,\, \mu_l) - f)$; (16.9.3 c_1)

if $d \neq 0$ **then** (16.9.3 c_2)

Then there exist constants C_u, C_f, $C_{\hat{f}}$, C_{NMGM} so that values $\tilde{u}_k$ and $\hat{f}_k$ produced by the nested iteration (5) satisfy

$$\|\tilde{u}_k - u_k^*\|_{\mathscr{U}} \leqq C_u h_k^{\kappa}(C_{NMGM} h_k^{\kappa})^i, \quad 0 \leqq k \leqq l, \tag{16.9.7a}$$

$$\|\hat{f}_k\|_{\mathscr{U}} \leqq C_{\hat{f}} h_k^{\kappa}(C_{NMGM} h_k^{\kappa})^i, \quad 0 \leqq k \leqq l-1, \tag{16.9.7b}$$

provided that $i \geqq 1$. $C_{NMGM} h_k^{\kappa}$ is the contraction number of the nonlinear multi-grid iteration (3). More precisely, if u_l^0, f_l, $\hat{f}_k$ satisfy

$$\|u_l^0 - \Phi_l(f_l)\|_{\mathscr{U}} \leqq C_u h_l^{\kappa}, \tag{16.9.8a}$$

$$\|f_l\|_{\mathscr{U}} \leqq \min(C_f h_l^{\kappa}, \varepsilon_l), \ \|\hat{f}_k\|_{\mathscr{U}} \leqq \min(C_{\hat{f}} h_k^{\kappa}, \varepsilon_k/2), \ 0 \leqq k \leqq l-1, \tag{16.9.8b}$$

iteration (3) with $f = f_l$ yields u_l^1 satisfying

$$\|u_l^1 - \Phi_l(f_l)\|_{\mathscr{U}} \leqq C_{NMGM} h_l^{\kappa} \, \|u_l^0 - \Phi_l(f_l)\|_{\mathscr{U}}. \tag{16.9.9}$$

The assumptions (6j), (6k) seem complicated. We shall propose a simple choice of σ_k in Corollary 16.9.5. Inequality (6l) excludes the case of rapidly decreasing ε_k. Usually, $\varepsilon_0 = \varepsilon_1 = \ldots$ will hold.

Proof. 1) First we prove the convergence statement (9).

1a) *Analysis of statement* (3b). The error of u_l^0 is $\delta_l^0 := u_l^0 - \Phi_l(f_l)$, $\Phi_l(f_l)$ solution of $u_l = \mathscr{K}_l(u_l) + f_l$. Set

$$u_l' := \mathscr{K}_l(u_l^0; u_l^0 - f_l, \mu_l) + f_l.$$

Then

$$\begin{aligned} \delta_l' := u_l' - u_l &= \mathscr{K}_l(u_l^0; u_l^0 - f_l, \mu_l) - \mathscr{K}_l(u_l) \\ &= \mathscr{K}_l(u_l^0) - \mathscr{K}_l(u_l) + \mathscr{K}_l(u_l^0; u_l^0 - f_l, \mu_l) - \mathscr{K}_l(u_l^0) \\ &= K_l(u_l)\,\delta_l^0 + O(\|\delta_l^0\|_{\mathscr{U}}^2) + \varepsilon^{\mu_l}\,\|(u_l^0 - f_l) - \mathscr{K}_l(u_l^0)\|_{\mathscr{U}} \end{aligned}$$

by (2). Since $u_l^0 - f_l - \mathscr{K}_l(u_l^0) = \delta_l^0 - (\mathscr{K}_l(u_l^0) - \mathscr{K}_l(u_l)) = O(\|\delta_l^0\|_{\mathscr{U}})$ and $K_l(u_l) = K_l + O(\|u_l - u_l^*\|_{\mathscr{U}}) = K_l + O(\|f_l\|_{\mathscr{U}})$, we conclude that

$$\delta_l' = K_l \delta_l^0 + O([\|\delta_l^0\|_{\mathscr{U}} + \varepsilon^{\mu_l} + \|f_l\|_{\mathscr{U}}]\,\|\delta_l^0\|_{\mathscr{U}}). \tag{16.9.10}$$

1b) *Analysis of statement* (3c$_1$). The defect

$$d_l := u_l' - \mathscr{K}_l(u_l'; u_l' - f_l, \mu_l) - f_l$$

can be written as

$$\begin{aligned} d_l &= u_l' - \mathscr{K}_l(u_l') - f_l + O(\varepsilon^{\mu_l}\,\|u_l' - \mathscr{K}_l(u_l') - f_l\|_{\mathscr{U}}) \\ &= [I - K_l(u_l)]\,\delta_l' + O(\|\delta_l'\|_{\mathscr{U}}^2 + \varepsilon^{\mu_l}\,\|\delta_l'\|_{\mathscr{U}}) \\ &= [I - K_l]\,\delta_l' + O([\|\delta_l'\|_{\mathscr{U}} + \varepsilon^{\mu_l} + \|f_l\|_{\mathscr{U}}]\,\|\delta_l'\|_{\mathscr{U}}). \end{aligned}$$

Equation (10) implies

$$d_l = [I - K_l]\,K_l \delta_l^0 + O([\|\delta_l^0\|_{\mathscr{U}} + \varepsilon^{\mu_l} + \|f_l\|_{\mathscr{U}}]\,\|\delta_l^0\|_{\mathscr{U}}). \tag{16.9.11}$$

1c) *Analysis of the coarse-grid equation.* Vector $d_{l-1} := \hat{f}_{l-1} - srd_l$ satisfies

$$\| d_{l-1} \|_{\mathscr{U}} \leqq \| \hat{f}_{l-1} \|_{\mathscr{U}} + |\sigma_{l-1}| \leqq (C_{\hat{f}} + \bar{C} C_{TGM}) h_{l-1}^{\kappa}$$

and $\| d_{l-1} \|_{\mathscr{U}} \leqq \varepsilon_{l-1}$. Choose C_f in (8b) large enough so that $C_f \geqq C_{\hat{f}} + \bar{C} C_{TGM}$. Let $v_{l-1} = \Phi_{l-1}(d_{l-1})$ be the exact solution of the coarse-grid equation. The error of the starting value $v_{l-1}^0 := \tilde{u}_{l-1}$ is

$$\begin{aligned} \| v_{l-1}^0 - v_{l-1} \|_{\mathscr{U}} &= \| \Phi_{l-1}(\tilde{f}_{l-1}) - \Phi_{l-1}(d_{l-1}) \|_{\mathscr{U}} \leqq C' \| \tilde{f}_{l-1} - d_{l-1} \|_{\mathscr{U}} \\ &= C' \| \tilde{f}_{l-1} - \hat{f}_{l-1} + srd_l \|_{\mathscr{U}} \leqq C' [\underline{C} C_{TGM} h_{l-1}^{\kappa} + 1] |\sigma_{l-1}| \\ &\leqq C' [\underline{C} C_{TGM} h_{l-1}^{\kappa} + 1] \bar{C} C_{TGM} h_{l-1}^{\kappa} \end{aligned}$$

with $C' := \sup \{ \| K_{l-1}(v_{l-1}) \|_{\mathscr{U} \leftarrow \mathscr{U}} : \; v_{l-1} = \Phi_{l-1}(f_{l-1}), \; \| f_{l-1} \|_{\mathscr{U}} \leqq \varepsilon_{l-1} \}$. Choose C_u so large that $C' [\underline{C} C_{TGM} h_{l-1}^{\kappa} + 1] \bar{C} C_{TGM} \leqq C_u$. Then, the data d_{l-1} and v_{l-1}^0 satisfy (8a, b). By induction, two iterations at level $l-1$ yield v_{l-1}^2 with

$$\begin{aligned} \| v_{l-1}^2 - v_{l-1} \|_{\mathscr{U}} &\leqq C_{NMGM}^2 h_{l-1}^{2\kappa} \| v_{l-1}^0 - v_{l-1} \|_{\mathscr{U}} \\ &\leqq C_{NMGM}^2 C' h_{l-1}^{2\kappa} (\| \delta f_{l-1} \|_{\mathscr{U}} + |\sigma_{l-1}|), \end{aligned}$$

where

$$\delta f_k := \tilde{f}_k - \hat{f}_k = \tilde{u}_k - \mathscr{K}_k(\tilde{u}_k) - \hat{f}_k. \tag{16.9.12}$$

1d) *Analysis of correction step* (3c$_6$). The new error δ_l^1 is

$$\begin{aligned} \delta_l^1 &:= u_l^1 - u_l = u_l' + p(v_{l-1}^2 - \tilde{u}_{l-1})/s - u_l \\ &= u_l' + p[\Phi_{l-1}(\hat{f}_{l-1} - srd_l) - \Phi_{l-1}(\hat{f}_{l-1})]/s - u_l \\ &\quad + p[(v_{l-1}^2 - v_{l-1}) + \Phi_{l-1}(\hat{f}_{l-1}) - \Phi_{l-1}(\tilde{f}_{l-1})]/s \\ &= \delta_l' - p[I - K_{l-1}(\Phi_{l-1}(\hat{f}_{l-1}))]^{-1} rd_l \\ &\quad + O(|s| \, \| rd_l \|_{\mathscr{U}}^2 + C_{NMGM}^2 h_l^{2\kappa} (\| \delta f_{l-1} \|_{\mathscr{U}} + |\sigma_{l-1}|)/s + \| \delta f_{l-1} \|_{\mathscr{U}}/s) \\ &= K_l \delta_l^0 - p[I - K_{l-1}]^{-1} r [I - K_l] K_l \delta_l^0 \\ &\quad + O(\| \hat{f}_{l-1} \|_{\mathscr{U}} \| rd_l \|_{\mathscr{U}} + |\sigma_{l-1}| \, \| rd_l \|_{\mathscr{U}} \\ &\quad + C_{NMGM}^2 h_l^{2\kappa} (\| \delta f_{l-1} \|_{\mathscr{U}} / |\sigma_{l-1}| + 1) \, \| rd_l \|_{\mathscr{U}} \\ &\quad + \| \delta f_{l-1} \|_{\mathscr{U}} \, \| rd_l \|_{\mathscr{U}} / |\sigma_{l-1}|) \end{aligned}$$

since $s = \sigma_{l-1} / \| rd_l \|_{\mathscr{U}}$. The first term equals $M_l \delta_l^0$ with M_l from (6a) and (2.9). Using (6a, e) and (11) one obtains

$$\begin{aligned} \delta_l^1 &= M_l \delta_l^0 + O([\| \hat{f}_{l-1} \|_{\mathscr{U}} + |\sigma_{l-1}| + \| \delta f_{l-1} \|_{\mathscr{U}} / |\sigma_{l-1}| \\ &\quad + C_{NMGM}^2 h_l^{2\kappa} + \| \delta_l^0 \|_{\mathscr{U}} + \varepsilon^{\mu_l} + \| f_l \|_{\mathscr{U}}] \, \| \delta_l^0 \|_{\mathscr{U}}), \\ \| \delta_1^1 \|_{\mathscr{U}} &\leqq C_{TGM} h_l^{\kappa} \\ &\quad \cdot \{ 1 + C[\bar{C} + \underline{C} + C_{\varepsilon} + (C_{\hat{f}} + 2C_c + C_f + C_{NMGM}^2 h_l^{\kappa}) / C_{TGM}] \} \, \| \delta_l^0 \|_{\mathscr{U}}. \end{aligned} \tag{16.9.13}$$

Make the ansatz $C_{NMGM} = \alpha C_{TGM}$. The expression in braces becomes $\{1 + C[\ldots + \alpha^2 C_{TGM} h_l^\kappa]\}$. By (6d), α can be chosen such that $\{\ldots\} \leqq \alpha$ yielding

$$\| \delta_l^1 \|_{\mathscr{U}} \leqq \alpha C_{TGM} h_l^\kappa \| \delta_l^0 \|_{\mathscr{U}} = C_{NMGM} h_l^\kappa \| \delta_l^0 \|_{\mathscr{U}}$$

and proving (9).

1e) *Start of the induction proof.* In step 1c) we used induction. At the lowest level $l = 0$ the claimed inequality (9) follows from (6c) if we choose $C_{NMGM} \geqq C_\Phi$.

2) Next, we prove (7a, b) by induction.

2a) $k = 0$. Equation (6g) implies (7a), if C_u is chosen such that $C_{u,0} \leqq C_u C_{NMGM}^i$. Equation (7b) follows from

2b) (7a) *implies* (7b). There is a constant C with

$$\begin{aligned} \| \tilde{f}_k \|_{\mathscr{U}} &= \| \tilde{u}_k - \mathscr{K}_k(\tilde{u}_k) \|_{\mathscr{U}} = \| \tilde{u}_k - \mathscr{K}_k(\tilde{u}_k) - (u_k^* - \mathscr{K}_k(u_k^*)) \|_{\mathscr{U}} \\ &\leqq C \| \tilde{u}_k - u_k^* \|_{\mathscr{U}} \end{aligned}$$

where $u_k^* = \Phi_k(0)$ is the solution of $u_k = \mathscr{K}_k(u_k)$. Equation (2) shows

$$\| \delta f_k \|_{\mathscr{U}} = \| \mathscr{K}_k(\tilde{u}_k; \tilde{u}_k, \mu_k) - \mathscr{K}_k(\tilde{u}_k) \|_{\mathscr{U}} \leqq \varepsilon^{\mu_l} \| \tilde{u}_k - \mathscr{K}_k(\tilde{u}_k) \|_{\mathscr{U}} = \varepsilon^{\mu_l} \| \tilde{f}_k \|_{\mathscr{U}} \tag{16.9.14}$$

(cf. (12)). The combination yields

$$\| \hat{f}_k \|_{\mathscr{U}} \leqq \| \tilde{f}_k \|_{\mathscr{U}} + \| \delta f_k \|_{\mathscr{U}} \leqq (1 + \varepsilon^{\mu_l}) \| \tilde{f}_k \|_{\mathscr{U}} \leqq C^* \| \tilde{u}_k - u_k^* \|_{\mathscr{U}}.$$

Set $C_{\hat{f}} = C_u C^*$; then (7b) is valid.

2c) It remains to prove (7a) for $k = l$, assuming that (7) holds for $k < l$. The error of the starting guess $u_l^0 := \tilde{p} \tilde{u}_{l-1}$ (cf. (5c)) is $\delta_l = u_l^0 - u_l^*$. It is bounded by

$$\begin{aligned} \| \delta_l^0 \|_{\mathscr{U}} &= \| \tilde{p} \tilde{u}_{l-1} - u_l^* \|_{\mathscr{U}} = \| \tilde{p}(\tilde{u}_{l-1} - u_{l-1}^*) + \tilde{p} u_{l-1}^* - u_l^* \|_{\mathscr{U}} \\ &\leqq C_c h_l^\kappa + C_{\tilde{p}} C_u h_{l-1}^\kappa (C_{NMGM} h_{l-1}^\kappa)^i. \end{aligned}$$

Choose C_u such that

$$C_u \geqq C_c / [1 - C_{\tilde{p}} C_H^\kappa (C_{NMGM} h_{l-1}^\kappa)^i],$$

which is possible by (6d), and $i > 0$. One obtains $\| \delta_l^0 \|_{\mathscr{U}} \leqq C_u h_l^\kappa$. Since (8a, b) are satisfied, (9) implies (7a). □

The choice of σ_{l-1} is still to be discussed. According to (13), the optimal value of σ_{l-1} is $O(\| \delta f_{l-1} \|_{\mathscr{U}}^{1/2})$. However, the magnitude $\| \delta f_{l-1} \|_{\mathscr{U}}^{1/2}$ is not known in practice. A very simple and practical choice is proposed in

Corollary 16.9.5. Set $\sigma_{l-1} := \| \hat{f}_{l-1} \|_{\mathscr{U}}$ with $\hat{f}_{l-1}$ from (5b). This value vanishes if and only if $\tilde{u}_{l-1} = u_{l-1}^*$ (i.e. $\tilde{f}_{l-1} = \hat{f}_{l-1} = 0$). The inequalities (6j, k) are satisfied with $\underline{C} = C_\varepsilon / (1 - C_\varepsilon C_{TGM} h_0^\kappa)$.

Proof. From $\| \delta f_k \|_{\mathscr{U}} \leqq \varepsilon^{\mu_k} \| \tilde{f}_k \|_{\mathscr{U}}$ (cf. (14)) and $\| \tilde{f}_k \|_{\mathscr{U}} \leqq \| \hat{f}_k \|_{\mathscr{U}} + \| \delta f_k \|_{\mathscr{U}}$ one concludes $\| \tilde{f}_k \|_{\mathscr{U}} \leqq \| \hat{f}_k \|_{\mathscr{U}} / (1 - \varepsilon^{\mu_k})$. Hence, $\sigma_k = 0$ implies $\hat{f}_k = \tilde{f}_k = 0$ and $\tilde{u}_k = u_k^*$. The term $\| \delta f_k \|_{\mathscr{U}} / |\sigma_k|$ is bounded by $\varepsilon^{\mu_k}/(1 - \varepsilon^{\mu_k}) \leqq C_\varepsilon C_{TGM} h_k^\kappa / (1 - C_\varepsilon C_{TGM} h_k^\kappa)$, proving the definition of $\underline{C}$. □

To avoid $\sigma_{l-1} = 0$ and the effects of rounding errors one should modify the choice σ_k by

$$\sigma_k := \| \hat{f}_k \|_{\mathscr{U}} + O(\sqrt{\| \tilde{u}_k \|_{\mathscr{U}} \cdot eps}), \tag{16.9.15}$$

where eps is the relative machine precision.

Corollary 16.9.6. If $\mathscr{K}_l(u_l)$ is evaluated exactly (i.e. $\mathscr{K}_l(u_l; \cdot, \cdot) = \mathscr{K}_l(u_l)$), the conditions (6b, e, j) can be omitted. Then $\hat{f}_k = \tilde{f}_k$.

There are several modifications of the nested iteration (5). Instead of computing $\hat{f}_{k-1}$ in (5b) after the multi-grid iterations at level $k-1$, one may define $\hat{f}_{k-1}$ together with the defect ($3c_1$) during the last multi-grid iteration at level $k-1$. Then $\| \hat{f}_{k-1} \|_{\mathscr{U}}$ and $\| \delta f_{k-1} \|_{\mathscr{U}}$ equal $O(\| \delta_{k-1}^{i-1} \|_{\mathscr{U}})$. Possible choices of σ_{k-1} are $\| \hat{f}_{k-1} \|_{\mathscr{U}}^{1/2}$ and $\| \hat{f}_{k-1} \|_{\mathscr{U}}$, or (15).

16.10 Applications to Elliptic Boundary Value and Eigenvalue Problems

16.10.1 Elliptic Boundary Value Problems

In § 16.8 we mentioned an application of the multi-grid iteration of the second kind to equations of the form $-\Delta u = g(u)$. The technique can be extended to

$$\mathscr{L}(u) := \mathscr{L}^I(u) - \mathscr{L}^{II}(u) = 0,$$

which can be reformulated as the fixed point equation

$$u = \Phi^I(\mathscr{L}^{II}(u)) \quad (\Phi^I \text{ is inverse of } \mathscr{L}^I),$$

provided that $\mathscr{L}^I$ contains the principal part of the (nonlinear) elliptic operator, whereas $\mathscr{L}^{II}$ is of lower order. Examples are given by see Hackbusch [9].

Interesting applications arise from systems of weakly coupled elliptic equations. A prototype of this form is

$$-\Delta v = g^I(v, w), \qquad -\Delta w = g^{II}(v, w) \text{ in } \Omega, \qquad v = w = 0 \text{ on } \Gamma,$$

which yields the system

$$v = (-\Delta)^{-1} g^I(v, w), \qquad w = (-\Delta)^{-1} g^{II}(v, w)$$

of two equations of the second kind. Set $u := (v, w)^T$. The evaluation of

$$\mathscr{K}(u) := ((-\Delta)^{-1} g^I(u), \quad (-\Delta)^{-1} g^{II}(u))^T$$

requires only the solution of two scalar Poisson equations. In the discrete case of $u_l = \mathscr{K}_l(u_l)$ one can make use of Poisson solvers.

Storage can be saved if the components of the elliptic system are coupled only by (linear or nonlinear) boundary conditions. Such systems arise, for instance, from *optimal control problems* (cf. Hackbusch [15]). We give a particular example.

Let $\Omega = (0, \pi) \times (0, \pi)$, $\Gamma = \partial\Omega$, $\Gamma_1 = \{(x_1, x_2) \in \Gamma\colon x_1 = \pi\} = \{\pi\} \times (0, \pi)$, $\Gamma_2 = \Gamma \backslash \Gamma_1$. For any $u \in L^2(\Gamma_1)$, $y(u)$ is the solution of

$$\begin{aligned} &\{-(1 + x_1^2)^{-1}(\partial^2/\partial x_1^2) - (\partial^2/\partial x_2^2)\}\, y(u) = 0 \quad \text{in} \quad \Omega, \\ &(1 + \pi^2)^{-1}(\partial y(u)/\partial x_1) = u \ \text{ on } \Gamma_1, \quad y(u) = 0 \ \text{ on } \Gamma_2. \end{aligned} \tag{16.10.1}$$

We seek the control u minimising the functional

$$J(v) = \| y(v)|_{\Gamma_1} - \sin x_2 \|^2_{L^2(\Gamma_1)} + (1 + \pi^2)^{-1} \| v \|^2_{L^2(\Gamma_1)}.$$

The solution u (optimal control) is determined by

$$u = -(1 + \pi^2)\, p(u)|_{\Gamma_1}, \tag{16.10.2}$$

where $p(u)$ is the solution of the adjoint problem

$$\begin{aligned} &-(\partial^2/\partial x_1^2)\,[p(u)/(1 + x_1^2)] - (\partial^2/\partial x_2^2)\, p(u) = 0 \quad \text{in} \quad \Omega, \\ &(\partial/\partial x_1)\,[p(u)/(1 + x_1^2)] = y(u)|_{\Gamma_1} - \sin x_2 \ \text{ on } \Gamma_1, \quad p(u) = 0 \ \text{ on } \Gamma_2. \end{aligned} \tag{16.10.3}$$

Equation (2) can be used to eliminate Eq. (1). Then we obtain a system of two elliptic equations for y and p, which are coupled by means of the boundary conditions. On the other hand, Eqs. (1), (3), (2) describe affine mappings $u \mapsto y(u) \mapsto p(u) \mapsto u$, which give rise to an equation $u = Ku + f$ of the second kind, where $Ku + f$ is the right-hand side of Eq. (2). It can be shown that K satisfies the regularity assumption (1.4) for $\mathscr{U} = L^2(\Gamma_1)$, $\mathscr{V} = H^2(\Gamma_1) \cap H_0^1(\Gamma_1)$.

Replacing the differential equations by difference schemes, we obtain $u_l = K_l u_l + f_l$ with K_l satisfying the regularity assumption (1.9) for the discrete analogues $\mathscr{U}_l$, $\mathscr{V}_l$. Conditions (1.12) and (1.14) hold with $\kappa = 2$. The evaluation of K_l requires the solution of the difference counterparts of Eqs. (1) and (3). The comparison of the rates of the multi-grid convergence shows that the rate is decreasing as $O(h_l^2)$, confirming $\kappa = 2$. The nested iteration (9.5) with $i = 1$ yields the following values $u_l(\frac{\pi}{2})$. For details see Hackbusch [15].

Table 16.10.1. Results of nested iteration (9.5) applied to Eqs. (1)–(3)

step size	$h_0 = \frac{\pi}{2}$	$h_1 = \frac{\pi}{4}$	$h_2 = \frac{\pi}{8}$	$h_3 = \frac{\pi}{16}$	$h_4 = \frac{\pi}{32}$	$h_5 = \frac{\pi}{64}$
$\tilde{u}(\frac{\pi}{2})$	0.0126	0.0919	0.1286	0.1456	0.1505	0.15128
iteration error $(\tilde{u}_l - u_l)(\frac{\pi}{2})$	0.0	$9.2_{10}-4$	$7.8_{10}-3$	$2.5_{10}-3$	$1.9_{10}-4$	$8.2_{10}-6$
discretisation error $(u_l - u)(\frac{\pi}{2})$	$1.4_{10}-1$	$5.9_{10}-2$	$1.5_{10}-2$	$3.3_{10}-3$	$7.8_{10}-4$	$1.9_{10}-4$

16.10.2 Eigenvalue Problems

Consider the eigenvalue problem

$$u = \lambda k u, \qquad u \neq 0, \tag{16.10.4}$$

where k may represent a linear integral operator. Let

$$u_l = \lambda_l k_l u_l, \qquad u_l \neq 0, \tag{16.10.5}$$

be the discrete analogues. According to § 12.2.1, the eigenvector can be normalised by $\varphi_l(u_l) = \lambda_l$ and Eq. (5) becomes

$$u_l = \varphi_l(u_l)\, k_l u_l =: \mathscr{K}_l(u_l). \tag{16.10.6}$$

The continuous analogue is

$$u = \varphi(u)\, k u =: \mathscr{K}(u). \tag{16.10.7}$$

Concerning the equivalence of Eqs. (5) and (6) see Note 12.2.2. Equation (7) is reasonable because of

Lemma 16.10.1. Let $0 \neq u \in \mathscr{U}$ be a solution of Eq. (7), $u = \mathscr{K}(u)$. Assume

$$\|k\|_{\mathscr{V} \leftarrow \mathscr{U}} \leqq C, \qquad \mathscr{V} \text{ compactly embedded in } \mathscr{U},$$

(cf. (1.4), Note (16.1.4)). Suppose that $\lambda = \varphi(u)$ is a single eigenvalue of k, i.e.

$$(I - \lambda k)\, w = \varrho u \text{ implies } \varrho = 0 \text{ and } w \in \operatorname{span}(u).$$

Further, let the derivative φ' of φ satisfy

$$\varphi'(u)\, u \neq 0, \qquad \varphi'(u)\colon \mathscr{U} \to \mathbb{C} \text{ bounded.}$$

Then the derivative $K = \mathscr{K}'(u)$ is compact and satisfies (1.3) and (1.4). Thus, the solution u of Eq. (7) is isolated.

Proof. Since $K v = \varphi(u)\, k v + (\varphi'(u)\, v)\, k u$, the estimate (1.4) follows and implies compactness of K. Therefore, it suffices to show that $v = K v$ implies $v = 0$. Now $v = K v$ is equivalent to

$$(I - \lambda k)\, v = \varrho u, \qquad \text{where } \lambda = \varphi(u), \quad \varrho = \varphi'(u)\, v / \varphi(u).$$

Since λ is a single eigenvalue, one concludes $v = \alpha u$, $\varrho = 0$. Hence, $\varrho = \alpha \varphi'(u)\, u / \varphi(u) = 0$, implying $\alpha = 0$ and $v = 0$. □

Similarly, one can prove the regularity assumption (1.9). The consistency condition (1.12) on K_l follows from the corresponding condition on k_l and simple assumptions on φ_l (cf. Hackbusch [18]).

Remark 16.10.2. The above reformulation of (4) as (6) can also be applied if the eigenvalue appears nonlinearly: $u = k(\lambda)\, u$. In that case, $\mathscr{K}_l$ becomes $\mathscr{K}_l(u_l) := k_l(\varphi_l(u_l))\, u_l$.

An example of $u = k(\lambda)\,u$ is given in §16.10.4. The operator k may arise from an elliptic boundary value problem. In §16.10.3 we show that even non-standard problems such as the Steklov eigenvalue problem can easily be solved.

16.10.3 First Example: Steklov Eigenvalue Problem

Consider the boundary value problem

$$-\Delta y = 0 \quad \text{in} \quad \Omega = (0,1) \times (0,1), \tag{16.10.8a}$$

$$\partial y/\partial n = \lambda y \quad \text{on} \quad \Gamma = \partial\Omega. \tag{16.10.8b}$$

This can be reformulated as (1), $u = \lambda k u$, with

$$k: \mathscr{U} \subset L^2(\Gamma) \to \mathscr{V} \subset H^1(\Gamma)$$

defined as follows. Let $\mathscr{U} = \{v \in L^2(\Gamma): \int_\Gamma v\,d\Gamma = 0\}$. For $u \in \mathscr{U}$ the Neumann problem

$$-\Delta y = 0 \quad \text{in} \quad \Omega, \qquad \partial y/\partial n = u \quad \text{on} \quad \Gamma,$$

has solutions y determined up to a constant. We make y unique if we require $y|_\Gamma \in \mathscr{U}$. Set $ku := y|_\Gamma$. It lies in $\mathscr{V} = \mathscr{U} \cap H^1(\Gamma)$. The eigenvalue problem $u = \lambda k u$ is equivalent to (8). According to the definition of $\mathscr{K}$ by (7), Eq. (8) is equivalent to $u = \mathscr{K}(u)$. The discrete problem can be written as Eq. (6): $u_l = \mathscr{K}_l(u_l)$, where u_l is defined on the boundary Γ_l.

In the following, the functional φ_l is defined by $\varphi_l(u_l) = u_l(0,0)$. In order to compute the fifth eigenvalue, one may start with the grid size $h_0 = \frac{1}{4}$. At level $l = 0$ the fifth eigenvalue and corresponding eigenvector can be computed to be

$$\tilde\lambda_0 = \tilde u_0(0,0) = 4.628, \;\; \tilde u_0(\tfrac{1}{4}, 0) = -0.726, \;\; \tilde u_0(\tfrac{1}{2}, 0) = -3.175.$$

The remaining values of $\tilde u_0$ on Γ_0 are defined by symmetries

$$u_0(x,y) = u_0(1-x, y) = u_0(x, 1-y). \tag{16.10.9}$$

Using this starting value $\tilde u_0$, the nested iteration (7.8) with $i = 1$ yields the numbers of Table 16.10.2.

Table 16.10.2. Fifth eigenvalue of the Steklov problem (8)

step size	$h_0 = \frac{1}{4}$	$h_1 = \frac{1}{8}$	$h_2 = \frac{1}{16}$	$h_4 = \frac{1}{32}$	$h_4 = \frac{1}{64}$
$\tilde\lambda = \tilde u_l(0,0)$	4.6277	4.6918	4.6562	4.651034	4.647942
exact λ_l	4.6277	4.6690	4.6552	4.649417	4.647822

Although the fifth eigenvalue is a double one, no extra computational difficulties arise since the symmetry (9) determines the eigenfunctions uniquely. For further details, see Hackbusch [18].

16.10.4 Second Example: Nonlinear Eigenvalue Problem

Consider the integrodifferential equation

$$-u''(x) = \int_0^1 e^{\lambda|x-y|} u(y)\,dy; \qquad u(0) = u(1) = 0; \qquad u \not\equiv 0. \tag{16.10.10}$$

This nonlinear eigenvalue problem can be written as

$$u = k(\lambda)\,u, \qquad u \neq 0,$$

where $k(\lambda)\,v$ is the solution w of

$$-w''(x) = \int_0^1 e^{\lambda|x-y|}\,v(y)\,dy, \qquad w(0) = w(1) = 0.$$

According to Remark 16.10.2, one obtains the nonlinear equation $u = \mathscr{K}(u)$ with $\mathscr{K}(u) := k(\varphi(u))\,u$. Here, we use $\varphi(v) = v(1/2)$.

The discrete nonlinear equation $u_l = \mathscr{K}_l(u_l)$ results by replacing $-u''$ by second differences and the integral by the trapezoidal rule. Starting with

$$\tilde{u}_0(0) = \tilde{u}_0(1) = 0,\ \tilde{u}_0(\tfrac{1}{4}) = \tilde{u}_0(\tfrac{3}{4}) = 6.925668,\ \tilde{u}_0(\tfrac{1}{2}) = 7.734301$$

for step size $h_0 = \frac{1}{4}$, we calculate the eigenvalues given in Table 16.10.3 without difficulty.

Table 16.10.3. Results of problem (10)

h_l	$\frac{1}{4}$	$\frac{1}{8}$	$\frac{1}{16}$	$\frac{1}{32}$	$\frac{1}{64}$
λ_l	7.734	6.535	6.3024	6.2975	6.2339

16.10.5 Elliptic Boundary Value Problems (Revisited)

In § 16.3.2 we formulated an elliptic problem as a Fredholm integral equation of the second kind. In this section we discuss integral equations of the first kind. Again, we consider Laplace's equation

$$\Delta u(\zeta) = 0 \qquad \text{in} \quad \mathbb{R}^2 \backslash \Gamma, \tag{16.10.11 a}$$

$$u(\zeta) = G(\zeta) \quad \text{on} \quad \zeta \in \Gamma, \tag{16.10.11 b}$$

with $u(\zeta) = B\log|\zeta| + O(1)$ as $|\zeta| \to \infty$. We represent $u(\zeta)$ by $\int_\Gamma v(z)\log|\zeta - z|\,d\Gamma_z$ with v satisfying the integral equation

$$g(\zeta) = \int_\Gamma v(z)\log|\zeta - z|\,d\Gamma_z \tag{16.10.12}$$

of the first kind. Let $z(t), 0 \leqq t \leqq 1$, be a parametrisation of Γ. Eq. (12) becomes

$$g(t) = \int_0^1 v(s) \log|z(t) - z(s)|\, ds, \qquad 0 \leqq t \leqq 1 \tag{16.10.13}$$

where $g(t) := G(z(t))$ and $v(s) := V(z(s))\,|dz(s)/ds|$. To avoid non-trivial solutions of the homogeneous problem, one should consider the system

$$g(t) = \int_0^1 v(s) \log|z(t) - z(s)|\, ds + \omega, \qquad \int_0^1 v(s)\, ds = B, \tag{16.10.14}$$

with given g, B for the unknowns v, ω. However, to simplify the situation, we shall discuss Eq. (13) instead of (14).

The multi-grid iteration (2.2) applies to integral equations of the second kind only. Equation (13) can be rewritten as

$$g(t) = \int_0^1 v(s) \log \delta(s-t)\, ds + \int_0^1 v(s) \log \frac{|z(s) - z(t)|}{\delta(s-t)}\, ds, \tag{16.10.15}$$

where $\delta(s-t) = \min\{|s-t|, |s-t-1|, \ |s-t+1|\}$ is the distance modulo 1. Equation (15) reads as

$$g = Av + kv, \tag{16.10.16}$$

where the first integral operator A is a convolution, whereas the second one has a smooth (non-singular) kernel, provided that Γ (hence $z(\cdot)$) is sufficiently smooth. It can be shown that A^{-1} is a pseudo-differential operator of first order. If $\Gamma \in C^r$ with $r > 2$, $k: L^2(0,1) \to H^{r-1}(0,1)$ holds and $A^{-1}: H^{r-1}(0,1) \to H^{r-2}(0,1)$ implies $K := -A^{-1}k: L^2(0,1) \to H^{r-2}(0,1)$. Therefore,

$$v = Kv + f \quad \text{with} \quad K = -A^{-1}k, \quad f = A^{-1}g, \tag{16.10.17}$$

is an equation of the second kind, to which the multi-grid process applies.

The discretisation (e.g. by the Galerkin method) leads us to the system

$$A_l v_l + k_l v_l = g_l, \tag{16.10.18}$$

where $A_l = (a_{jk}^{(l)})$ is a $n_l \times n_l$ Toeplitz matrix, i.e. $a_{jk}^{(l)} = a_{j'k'}^{(l)}$ for $j - k \equiv (j' - k') \bmod n_l$. The Galerkin discretisation (18) and its discretisation error are analysed by Hsiao-Kopp-Wendland [1].

According to Eq. (17), the discrete equation can be written as

$$v_l = K_l v_l + f_l, \tag{16.10.19}$$

with $K_l = -A_l^{-1} k_l$, $f_l = A_l^{-1} g_l$. Let T_l be the discrete Fourier transform $v_l \mapsto \hat{v}_l = T_l v_l$. Since for any Toeplitz matrix A_l the transformed matrix $D_l := T_l^{-1} A_l T_l$ is diagonal, A_l^{-1} equals $T_l D_l^{-1} T_l^{-1}$. The multiplication $K_l v_l$ requires the usual multiplication $k_l v_l$ ($O(n_l^2)$ operations), while T_l^{-1} and T_l can be performed by fast Fourier transforms ($O(n_l \log n_l)$ operations). One concludes that the computational work of the multi-grid iteration applied to (19) is almost the same as for a usual integral equation (1.6) of the second kind.

We conclude with numerical results for the system (14). Let Γ be an ellipse with axes of length 2 and 1. The discrete equations (18) are obtained by a Galerkin approach with piecewise constant finite elements. At the lowest level $l = 0$ the dimension of the finite element space is $n_0 = 4$. The observed multi-grid convergence rates are as follows:

n_l	16	32	64
Multi-grid rate	0.079	0.037	0.019

Obviously, the rates behave as $O(h_l^\kappa)$ with $\kappa = 1$.

16.11 Application to Parabolic Problems

We begin with the time-periodic problem, since this can more easily be formulated. The more interesting parabolic boundary control problem is discussed in § 16.11.2.

16.11.1 Time-Periodic Parabolic Problems

Let Ω be a domain in $\mathbb{R}^d$ with boundary Γ. $(0, T)$ is a fixed time interval. Set $Q = \Omega \times (0, T)$ and $\Sigma = \Gamma \times (0, T)$. We consider the parabolic equation

$$y_t + Ly = g_1 \quad \text{in } Q, \tag{16.11.1 a}$$

subject to some boundary condition

$$By = g_2 \quad \text{in } \Sigma. \tag{16.11.1 b}$$

The solution has to be T-periodic:

$$y(\cdot, T) = y(\cdot, 0) \quad \text{in } \Omega. \tag{16.11.1 c}$$

The usual initial value problem consists of (1 a, b) and $y(\cdot, 0) = u$. Denote its solution by $y(x, t; u)$ and define $\mathscr{K}(u) := y(\cdot, T; u)$. Therefore, Eq. (1 a–c) is equivalent to $u = \mathscr{K}(u)$. Here, $\mathscr{K}$ is affine: $\mathscr{K}(u) = Ku + f$ with f depending on g_1 and g_2.

Discretising (1 a, b) e. g. by a totally implicit difference scheme, we obtain a discrete analogue $\mathscr{K}_l(u_l)$. It involves the solution of one discrete initial-boundary value problem. The regularity assumption (1.9) is discussed in Hackbusch [25]. Therein, numerical examples are reported. In the case of $\Omega \subset \mathbb{R}^1$ and discretisation parameters $\Delta t_l = O(\Delta x_l^2)$, the nested iteration (4.1) requires less than 4.6 C_l operations, where C_l is the work required by one discrete initial-boundary value problem at level l.

The method described above has been applied by Schippers [2], [4], [6] to oscillating disk flow (rotating flow due to an infinite disk performing torsional oscillations). The governing nonlinear equations for the unknowns f, g, h are

$$\frac{\omega}{\Omega} f_t = \frac{\Omega}{2\omega} f_{zz} + 2h f_z - f^2 + g^2 - k,$$

$$\frac{\omega}{\Omega} g_t = \frac{\Omega}{2\omega} g_{zz} + 2h g_z - 2fg,$$

$$h_z = f.$$

16.11.2 Optimal Control Problems for Parabolic Equations

Let $\Omega, Q, \Sigma, (0, T)$ be defined as in § 16.11.1, and $\Sigma_0 \subset \Sigma$. For any $u \in L^2(\Sigma_0)$ the state $y(u)$ is defined as the solution of

$$y_t(u) + A\,y(u) = g_1 \quad \text{in } Q, \tag{16.11.2a}$$

$$By(u) = u \quad \text{on } \Sigma_0, \quad By(u) = g_2 \quad \text{on } \Sigma\backslash\Sigma_0, \tag{16.11.2b}$$

$$y(\cdot, 0; u) = y_0 \quad \text{on } \Omega, \tag{16.11.2c}$$

where A is an elliptic differential operator and B a boundary operator. We seek u minimising a cost function, e.g.

$$J(v) = \| y(\cdot, T; v) - z \|^2_{L^2(\Omega)} + \| v \|^2_{L^2(\Omega)}$$

for some given z. The optimal control u is characterised by

$$u = -\, p(u)|_{\Sigma_0}, \tag{16.11.3}$$

where $p(u)$ is the solution of the adjoint parabolic equation

$$-\, p_t(u) + A^*\, p(u) = 0 \quad \text{in } Q, \tag{16.11.4a}$$

$$Cp(u) = 0 \quad \text{on } \Sigma, \tag{16.11.4b}$$

$$p(\cdot, T; u) = y(\cdot, T; u) - z \quad \text{on } \Omega \tag{16.11.4c}$$

(cf. Hackbusch [12]). Eliminating u in (2b) by (3), we obtain a coupled system of two parabolic equations, where the second one (Eq. (4)) has reversed time direction. The discretised problem would be a large sparse system.

Instead, we consider another representation of the problem. Equation (2) describes a mapping $u \mapsto y(u)$, while Eq. (4) defines a mapping $y(u) \mapsto p(u)$. Set $\mathscr{K}(u) := p(u)|_{\Sigma_0}$. This is affine: $\mathscr{K}(u) = Ku + f$ with f depending on g_1, g_2, y_0, z. Equation (3) is equivalent to $u = \mathscr{K}(u)$.

Replacing Eqs. (2) and (4) by difference schemes, we can define analogous equations $u_l = \mathscr{K}_l(u_l)$. The evaluation of $\mathscr{K}_l(u_l)$ requires the solution of two discrete initial-boundary value problems, namely $u_l \mapsto y_l(u_l)$ and $y_l(\cdot, T; u_l) \mapsto p_l(u_l)$.

Consider the particular case of $\Omega = (0,1)$, $T = 1$, $\Sigma_0 = \{0\} \times (0,T)$, $A = A^* = -\partial^2/\partial x^2$, $B = C = \partial/\partial n$. Then the equations for $y(u), p(u)$ become

$$y_t = y_{xx} + g_1, \quad -p_t = p_{xx}, \quad y(x,0) = y_0, \quad p(x,T) = y(x,T) - z,$$
$$-y_x(0,t) = u(t), \quad y_x(1,t) = g_2, \quad p_x(0,t) = p_x(1,t) = 0.$$

It can be shown that K is a bounded mapping from $\mathscr{U} = L^2(\Sigma_0)$ into $\mathscr{V} = H^{1/2}(\Sigma_0)$, while $K_l = \mathscr{K}_l'$ satisfies the analogous discrete estimates. K can be represented as integral operator with the kernel function from Example 16.1.3. The inequalities (1.12) and (1.14) hold with $\kappa = \frac{1}{2}$ and $h_l = \Delta t_l = O(\Delta x_l^2)$.

The following numerical results are obtained for

$$g_1 = -\tfrac{\pi}{2}\cos(\tfrac{\pi}{2}(1-x)), \quad g_2 = 0, \quad y_0 = -\tfrac{\pi}{2}\cos(\tfrac{\pi}{2}(1-x)),$$
$$z = 1 - \tfrac{\pi}{2}\cos(\tfrac{\pi}{2}(1-x)).$$

Table 16.11.1 shows grid sizes Δx_l, Δt_l and the convergence rates ϱ_l of the multi-grid iteration (2.2). The rates decrease as $O(\Delta t_l^{1/2})$ confirming the above value of $\kappa = \frac{1}{2}$.

Table 16.11.1. Multi-grid convergence rates

l	0	1	2	3	4	5
Δt_l	$\frac{1}{2}$	$\frac{1}{4}$	$\frac{1}{8}$	$\frac{1}{16}$	$\frac{1}{32}$	$\frac{1}{64}$
Δx_l	$\frac{1}{4}$	$\frac{1}{8}$	$\frac{1}{8}$	$\frac{1}{16}$	$\frac{1}{16}$	$\frac{1}{32}$
ϱ_l	—	0.086	0.063	0.044	0.030	0.019

Extensions to control problems with bounded controls are discussed in Hackbusch [17]. Quite a different algorithm for parabolic bang-bang control problems is described by Hackbusch-Will [1].

Bibliography

Abramov AA: [1] O vydelenii glavnoi časti nekotorych algebraičeskich zadač. Ž vyčisl matem i matem fiz 2,1 (1962) 141–145 – engl transl: On the separation of the principal part of some algebraic problems. USSR Comput Math Phys 2,1 (1962) 147–151

Absi E, R Glowinski, P Lascaux, and H Veysseyre (eds): [1] Numerical methods in engineering, vol I. Proceedings, École Centrale de Paris, Dec 1980. Dunod, Paris 1980

Adams RA: [1] Sobolev spaces. Acad Press, New York 1975

Adamson TC and MF Platzer (eds): [1] Transonic flow problems in turbomachinery. Proceedings, Monterey, Febr. 1976. Hemisphere, Washington 1977

Agmon S, A Douglis, and L Nirenberg: [1] Estimates near the boundary for solutions of elliptic partial differential equations satisfying general boundary conditions II. Comm Pure Appl Math 17 (1964) 35–92

Albrecht J and L Collatz (eds): [1] Numerical treatment of integral equations. ISNM 53. Birkhäuser, Basel 1980

Alcouffe RE, A Brandt, JE Dendy Jr, and JW Painter: [1] The multi-grid method for the diffusion equation with strongly discontinuous coefficients. SIAM J Sci Statist Comput 2 (1981) 430–454

Anselone PhM: [1] Collectively compact operator approximation theory and applications to integral equations. Prentice-Hall, Englewood Cliffs 1971

Ansorge R, Th Meis, and W Törnig (eds): [1] Iterative solution of nonlinear systems. Proceedings, Oberwolfach, Febr 1982. Lecture Notes in Math 953. Springer, Berlin 1982

Asselt EJ van: see Van Asselt EJ

Astrachancev GP: [1] Ob odnom iteracionnom metode rešenija setočnych elliptičeskich zadač. Ž vyčisl matem i matem fiz 11,2 (1971) 439–448 – Engl transl: An iterative method of solving elliptic net problems. USSR Comput Math and Math Phys 11,2 (1971) 171–182

Astrachancev GP: [2] Iteracionnoe utočnenie sobstvennych značenij. Ž vyčisl matem i matem fiz 16,1 (1976) 131–139 – Engl transl: The iterative improvement of eigenvalues. USSR Comput Math and Math Phys 16,1 (1976) 123–132

Astrachancev GP and LA Ruchovec: [1] Metod RP Fedorenka dlja variacionno-raznostnych schem s ekstrapoljaciej (engl title*: Fedorenko's method for variational difference schemes with extrapolation). In: Lebedev [1] 20–26

Astrachancev GP and LA Ruchovec: [2] Metod relaksacii na posledovatel'nosti setok dlja elliptičeskich uravnenij s estestvennymi kraebymi uslovijami. Ž vyčisl matem i matem fiz 21,4 (1981) 926–944 – Engl transl: The method of relaxation on a sequence of meshes for elliptic equations with natural boundary conditions. USSR Comput Math and Math Phys 21,4 (1981) 111–130

Atkinson K: [1] Iterative variants of the Nyström method for the numerical solution of integral equations. Numer Math 22 (1973) 17–31

* for convenience of the reader translated by the author

Atkinson K: [2] An automatic program for linear Fredholm integral equations of the second kind. ACM Trans Math Software 2 (1976) 154–171

Auslender A, W Oettli, and J Stoer (eds): [1] Optimization and optimal control. Proceedings, Oberwolfach, March 1980. Lecture Notes in Control and Information Sci 30, Springer, Berlin 1981

Auzinger W and HJ Stetter: [1] Defect corrections and multigrid iterations. In: Hackbusch – Trottenberg [1] 327–351

Axelsson O: [1] Solution of linear systems of equations: iterative methods. In: Barker [1] 1–51

Axelsson O: [2] On multigrid methods of the two-level type. In: Hackbusch – Trottenberg [1] 352–367

Axelsson O, LS Frank, and A van der Sluis (eds): [1] Analytical and numerical approaches to asymptotic problems in analysis. Proceedings, Nijmegen, June 1980. North-Holland, Amsterdam 1981

Aziz AK: [1] The mathematical foundation of the finite element method with applications to partial differential equations. Acad Press, New York 1972

Aziz AK: see Babuška I et al

Babuška I: [1] The finite element method with penalty. Math Comp 27 (1973) 221–228

Babuška I and AK Aziz: [1] Survey lectures on the mathematical foundations of the finite element method. In: Aziz [1] 3–363

Babuška I, M Práger, and E Vitásek: [1] Numerical processes in differential equations. Interscience Publ, London 1966

Babuška I and WC Rheinboldt: [1] A-posteriori estimates for the finite element method. Int J Numer Meth Engrg 12 (1978) 1597–1615

Babuška I and MB Rosenzweig: [1] A finite element scheme for domains with corners. Numer Math 20 (1972) 1–21

Bachvalov NS: [1] O schodimosti odnogo relaksacionnogo metoda pri estestvennych ograničenijach na elliptičeskij operator. Ž vyčisl matem i matem fiz 6,5 (1966) 861–883 – Engl transl: On the convergence of a relaxation method with natural constraints on the elliptic operator. USSR Comput Math and Math Phys 6,5 (1966) 101–135

Bachvalov NS and JuA Kuznecov (eds): [1] Variacionno-raznostnye metody v matematiceskoj fizike (Variational difference methods in mathematical physics). Proceedings, May 1983, Moskow. AN SSSR 1984

Baker CTH: [1] The numerical treatment of integral equations. Oxford University Press, London 1977

Baker TJ: [1] The computation of transonic potential flow. In: Essers [1] 213–289

Balgovind R: see Ghil M et al

Bank RE: [1] A comparison of two multilevel iterative methods for nonsymmetric and indefinite elliptic finite element equations. SIAM J Numer Anal 18 (1981) 724–743

Bank RE: [2] A multi-level iterative method for nonlinear elliptic equations. In: Schultz [1] 1–16

Bank RE: [3] Analysis of a multilevel inverse iteration procedure for eigenvalue problems. SIAM J Numer Anal 19 (1982) 886–898

Bank RE: [4] PLTMG users' guide – June, 1981 version. Techn Report, University of San Diego at La Jolla, 1982

Bank RE and TF Chan: [1] PLTMGC: A multi-grid continuation program for parameterized nonlinear elliptic systems. SIAM J Sci Statist Comput (to appear)

Bank RE and CC Douglas: [1] Sharp estimates for multigrid rates of convergence with general smoothing and acceleration. SIAM J Numer Anal (to appear)

Bank RE and TF Dupont: [1] Analysis of a two-level scheme for solving finite element equations. Report CNA-159, University of Texas at Austin, 1980

Bank RE and TF Dupont: [2] An optimal order process for solving elliptic finite element equations. Math Comp 36 (1981) 35–51

Bank RE, JW Jerome, and DJ Rose: [1] Analytical and numerical aspects of semiconductor device modeling. In: Glowinski – Lions [1] 593–597

Bank RE and DJ Rose: [1] Global approximate Newton methods. Numer Math 37 (1981) 279–295

Bank RE and DJ Rose: [2] Analysis of a multilevel iterative method for nonlinear finite element equations. Math Comp 39 (1982) 453–465

Bank RE and AH Sherman: [1] Algorithmic aspects of the multi-level solution of finite element equations. In: Duff – Stewart [1] 62–89

Bank RE and AH Sherman: [2] The use of adaptive grid refinement for badly behaved elliptic partial differential equations. Math Comput Simulation 22 (1980) 18–24

Bank RE and AH Sherman: [3] An adaptive, multi-level method for elliptic boundary value problems. Computing 26 (1981) 91–105

Bank RE and AH Sherman: [4] A comparison of smoothing iterations for multilevel methods. In: Vichnevetsky – Stepleman [1] 143–147

Bank RE and A Weiser: [1] Some a posteriori error estimators for elliptic partial differential equations. Math Comp (to appear)

Bank RE, AH Sherman, and A Weiser: [1] Refinement algorithms and data structures for regular local mesh refinement. In: Scientific computing, IMACS, North-Holland, Amsterdam 1983

Barkai D and A Brandt: [1] Vectorized multigrid poisson solver for the CDC Cyber 205. Appl Math Comput 13 (1983) 215–227

Barker VA (ed): [1] Sparse matrix techniques. Proceedings, Copenhagen, Aug 1976. Lecture Notes in Math 572. Springer, Berlin 1977

Barnhill RE, JH Brown, and AR Mitchell: [1] A comparison of finite element error bounds for Poisson's equation. IMA J Numer Anal 1 (1981) 95–103

Bercovier M and O Pironneau: [1] Error estimates for finite element method solution of the Stokes problem in the primitive variables. Numer Math 33 (1979) 211–224

Bjørstad PE: [1] The direct solution of a generalized biharmonic equation on a disk. In: Hackbusch [36] 1–9

Bjørstad PE and OB Widlund: [1] Iterative methods for the solution of elliptic problems on regions partitioned into substructures. Report, New York University 1984

Blum H and M Dobrowolski: [1] On finite element methods for elliptic equations on domains with corners. Computing 28 (1982) 53–63

Boerstoel JW: [1] A fast-solver algorithm for steady transonic potential-flow computations with Newton iteration and multigrid relaxation. In: Viviand [1] 21–41

Boerstoel JW: [2] A multigrid algorithm for steady transonic flows around aerofoils using Newton iteration. J Comput Phys 48 (1982) 314–343

Böhmer K: [1] Discrete Newton methods and iterated defect corrections. Numer Math 37 (1981) 167–192

Börgers C: [1] Mehrgitterverfahren für eine Mehrstellendiskretisierung der Poissongleichung und für eine zweidimensionale singulär gestörte Aufgabe. Diplomarbeit (master thesis), Universität Bonn 1981

Braess D: [1] The contraction number of a multigrid method for solving the Poisson equation. Numer Math 37 (1981) 387–404

Braess D: [2] The convergence rate of a multigrid method with Gauß-Seidel relaxation for the Poisson equation. In: Hackbusch-Trottenberg [1] 368–386

Braess D: [3] The convergence rate of a multigrid method with Gauß-Seidel relaxation for the Poisson equation. Math Comp 42 (1984) 505–519

Braess D and W Hackbusch: [1] A new convergence proof for the multigrid method including the V-cycle. SIAM J Numer Anal 20 (1983) 967–975

Braess D, Hackbusch W, and U Trottenberg (eds): [1] Multigrid methods. Proceedings, Oberwolfach, Dec 1984. Notes on Numerical Fluid Mechanics 11. Vieweg, Wiesbaden (to appear)

Brakhage H: [1] Über die numerische Behandlung von Integralgleichungen nach der Quadraturformelmethode. Numer Math 2 (1960) 183–196

Bramble JH: [1] A second order finite difference analog of the first biharmonic boundary value problem. Numer Math 9 (1966) 236–249

Bramble JH, and BE Hubbard: [1] A theorem on error estimation for finite difference analogues of the Dirichlet problem for elliptic equations. Contribution to Diff Eq 2 (1963) 319–340

Bramble HJ and BE Hubbard: [2] A finite difference analog of the Neumann problem for Poisson's equation. SIAM J Numer Anal 2 (1964) 1–14

Bramble HJ and BE Hubbard: [3] Approximation of solutions of mixed boundary value problems for Poisson's equation by finite differences. J Assoc Comput Mach 12 (1965) 114–123

Branca H-W: see Meis Th et al

Brand K: [1] Multigrid bibliography. 3rd edition. Report, Arbeitspapiere der GMD Nr 59, St Augustin 1983

Brandt A: [1] Multi-level adaptive technique (MLAT) for fast numerical solution to boundary value problems. In: Cabannes – Temam [1] 82–89

Brandt A: [2] Multi-level adaptive techniques. Report RC 6026, IBM TJ Watson Research Center, Yorktown Heights 1976

Brandt A: [3] Multi-level adaptive solutions to boundary-value problems. Math Comp 31 (1977) 333–390

Brandt A: [4] Multi-level adaptive techniques (MLAT) for partial differential equations: ideas and software. In: Rice [1] 277–318

Brandt A: [5] Multi-level adaptive techniques (MLAT) for singular-perturbation problems. In: Hemker – Miller [1] 53–142

Brandt A: [6] Multi-level adaptive finite-element methods: I. variational problems. In: Frehse – Pallaschke – Trottenberg [1] 91–128

Brandt A: [7] Multi-level adaptive computations in fluid dynamics. AIAA J 18 (1980) 1165–1172

Brandt A: [8] Stages in developing multigrid solutions. In: Absi – Glowinski – Lascaux – Veysseyre [1] 23–44

Brandt A: [9] Numerical stability and fast solutions to boundary value problems. In: Miller [2] 29–49

Brandt A: [10] Multigrid solvers on parallel computers. In: Schultz [1] 39–83

Brandt A: [11] Multigrid solutions to steady-state compressible Navier-Stokes equations. In: Glowinski – Lions [1] 407–422

Brandt A: [12] Guide to multigrid development. In: Hackbusch – Trottenberg [1] 220–312

Brandt A: [13] Multigrid solvers for non-elliptic and singular-perturbation steady-state problems. To appear

Brandt A: [14] Multigrid techniques: 1984 guide with applications to fluid dynamics. GMD-Studien Nr 85, Bonn 1984

Brandt A: see Alcouffe RE et al

Brandt A: see Barkai D et al

Brandt A: see South Jr JC et al

Brandt A and CW Cryer: [1] Multigrid algorithms for the solution of linear complementary problems arising from free boundary problems. SIAM J Sci Statist Comput 4 (1983) 655–684

Brandt A, JE Dendy Jr, and H Ruppel: [1] The multigrid method for semi-implicit hydrodynamic codes. J Comput Phys 34 (1980) 348–370

Brandt A and N Dinar: [1] Multigrid solutions to elliptic flow problems. In: Parter [1] 53–147

Brandt A, S McCormick, and J Ruge: [1] Multigrid methods for differential eigenproblems. SIAM J Sci Statist Comput 4 (1983) 244–260

Brezzi F: [1] On the existence, uniqueness and approximation of saddle-point problems arising from Lagrangian multipliers. RAIRO Anal Numer 8 (1974) 129–151

Brezzi F and J Pitkäranta: [1] On the stabilization of finite element approximations of the Stokes equations. In: Hackbusch [36] 11–19

Brooks AN and TJR Hughes: [1] Streamline-upwind Petrov-Galerkin formulations for convection dominated flows with particular emphasis on the incompressible Navier-Stokes equations. Comp Meth Appl Mech Engng 32 (1982) 199–259

Brown JH: see Barnhill RE et al

Bulirsch R: see Stoer J et al

Bulirsch R, RD Griegorieff, and J Schröder (eds): [1] Numerical treatment of differential equations. Proceedings, Oberwolfach, July 1976. Lecture Notes in Math 631. Springer, Berlin 1978

Buneman O: [1] A compact non-iterative Poisson solver. SUIPR Report 294, Stanford 1969

Cabannes H and R Temam (eds): [1] Proceedings of the third international conference on numerical methods in fluid mechanics. Proceedings, Paris, July 1972. Lecture Notes in Phys 18. Springer, Berlin 1973

Chan TFC and HB Keller: [1] Arc-length continuation and multi-grid techniques for nonlinear eigenvalue problems. SIAM J Sci Statist Comput 3 (1982) 173–194

Chan TFC and R Schreiber: [1] Parallel networks for multi-grid algorithms: architecture and complexity. Report, Yale University 1983

Chan TF: see Bank RE et al

Chatelin F: [1] Spectral approximation of linear operators. Acad Press, New York 1983

Chatelin F and WL Miranker: [1] Acceleration by aggregation of successive approximation methods. Linear Algebra Appl 43 (1982) 17–47

Ciarlet PhG: [1] The finite element method for elliptic problems. North-Holland, Amsterdam 1978

Cimmino G: [1] La Ricerca Scientifica, ser II, vol 1. Included in: Pubblicazionio dell'Instituto per le Applicazioni del Calcolo 34 (1938)

Collatz L: [1] The numerical treatment of differential equations. Springer, Berlin 1966

Collatz L: see Albrecht J et al

Concus P, GB Golub, and DP O'Leary: [1] Numerical solution of nonlinear elliptic partial differential equations by a generalized conjugate gradient method. Computing 19 (1978) 321–339

Cryer CW: see Brandt A et al

Dahlquist G: see Kronsjö L et al

Daugavet IK: [1] Ob iterativnom rešenii urabnenij, voznikajuščich pri kompaktnoj approksimacii operatorov. Ž vyčisl matem i matem fiz 20,4 (1980) 1046–1049 – Engl transl: The iterative solution of equations arising in the compact approximation of linear operators. USSR Comput Math and Math Phys 20,4 (1980) 243–247

Deconinck H and Ch Hirsch: [1] A multigrid finite element method for the transonic potential equation. In: Hackbusch – Trottenberg [1] 387–409

Deconinck H and Ch Hirsch: [2] A multigrid method for the transonic full potential equation discretized with finite elements on an arbitrary mesh. J Comput Phys 48 (1982) 344–365

Dendy Jr JE: [1] Black box multigrid. J Comput Phys 48 (1982) 366–386

Dendy Jr JE: [2] Black box multigrid for nonsymmetric problems. Appl Math Comput 13 (1983) 261–283

Dendy Jr JE: see Alcouffe RE et al

Dendy Jr JE: see Brandt A et al

Dendy Jr JE and JM Hyman: [1] Multi-grid and ICCG for problems with interfaces. In: Schultz [1] 247–253

De Zeeuw PM and EJ van Asselt: [1] The convergence rate of multi-level algorithms applied to the convection-diffusion equation. SIAM J Sci Statist Comput 6 (1985) 492–503

De Zeeuw PM: see Hemker PW et al

De Zeeuw PM: see Sonneveld P et al

Dick E: [1] A multigrid method for Cauchy-Riemann and steady Euler equations based on flux-difference splitting. In: Hackbusch [36] 20–37
Dinar N: see Brandt A et al
Dinh GV: see Glowinski R et al
D'Jakonov EG: [1] O schodimosti odnoga iteracionnogo processa (On the convergence of an iterative process). Usp Mat Nauk 21 (1966) 179–182
Dobrowolski M: see Blum H et al
Douglas CC: see Bank RE et al
Douglis A: see Agmon S et al
Dryja M: [1] Apriornye ocenki v W_2^2 v vypukloj oblasti dlja sistem raznostnych elliptičeskich uravnenij. Ž vyčisl matem i matem fiz 12,6 (1972) 1595–1601 – Engl transl: Prior estimates in W_2^2 in a convex domain for systems of difference elliptic equations. USSR Comput Math and Math Phys 12,6 (1972) 291–300
Dryja M: [2] A capacitance matrix method for Dirichlet problem on polygon region. Numer Math 39 (1982) 51–64
Dryja M: [3] A decomposition method for solving finite element equations. IInf UW Report Nr 122, University of Warsaw, 1983
Duff IS: [1] Sparse matrix software for elliptic pde's. In: Hackbusch – Trottenberg [1] 412–426
Duff IS and GW Stewart (eds): [1] Sparse matrix proceedings 1978. Proceedings, Knoxville, Nov 1978. SIAM, Philadelphia 1979
Dupont TF: see Bank RE et al
Essers JA (ed): [1] Computational methods for turbulent, transonic, and viscous flows. Proceedings, Von Karman Institute for Fluid Dynamics, Rhode Saint Genese, March–April 1981. Hemisphere, Washington 1983
Fedorenko RP: [1] Relaksacionnyj metod rešenija raznostnych elliptičeskich uravenija. Ž vyčisl matem i matem fiz 1,5 (1961) 922–927 – Engl transl: A relaxation method for solving elliptic difference equations. USSR Comput Math and Math Phys 1,5 (1961) 1092–1096
Fedorenko RP: [2] O skorosti schodimosti odnogo iteracionnogo processa. Ž vyčisl matem i matem fiz 4,3 (1964) 559–564 – Engl transl: The speed of convergence of one iterative process. USSR Comput Math and Math Phys 4,3 (1964) 227–235
Fix GJ: see Strang et al
Fletcher CAJ: [1] Computational Galerkin methods. Springer, New York 1984
Foerster H, K Stüben, and U Trottenberg: [1] Non-standard multigrid techniques using checkered relaxation and intermediate grids. In: Schultz [1] 285–300
Foerster H and K Witsch: [1] On efficient multigrid software for elliptic problems on rectangular domains. Math Comput Simulation 23 (1981) 293–298
Foerster H and K Witsch: [2] Multi-grid software for the solution of elliptic problems on rectangular domains. MG00 (release 1). In: Hackbusch – Trottenberg [1] 427–460
Fox L (ed): [1] Numerical solution of ordinary and partial differential equations. Pergamon, New York 1962
Frank LS: see Axelsson O et al
Fray JMJ: see Oskam B et al
Frederickson PO: [1] Fast approximate inversion of large elliptic systems. Report # 7–74, Lakehead University, 1974
Frederickson PO: [2] Fast approximate inversion of large sparse linear systems. Report # 7–75, Lakehead University, 1975
Frederickson PO: [3] Fast solution of certain generalized eigenvalue problems. Report, Lakehead University
Frehse J, D Pallaschke, and U Trottenberg (eds): [1] Special topics of applied mathematics. Proceedings, Bonn, Oct 1979. North-Holland, Amsterdam 1980
Fuchs L: [1] A Newton-multi-grid method for the solution of nonlinear partial differential equations. In: Miller [2] 291–296

Fuchs L: [2] Transonic flow computation by a multi-grid method. In: Rizzi – Viviand [1] 58–65

Fuchs L: [3] Multi-grid solution of the Navier-Stokes equations on non-uniform grids. NASA conf publ 2202, 83–100, 1981

Fuchs L: [4] Multi-grid schemes for incompressible flows. In: Hackbusch [36] 38-51

Fuchs L: [5] Defect corrections and higher numerical accuracy. In: Hackbusch [36] 52–66

Gannon D and J van Rosendale: [1] Highly parallel multigrid solvers for elliptic PDEs: an experimental analysis. ICASE report No 82–36, 1982

Gary J: [1] The multigrid iteration applied to the collocation method. SIAM J Numer Anal 18 (1981) 211–224

Geller W: see Hirschel EH et al

George JA: [1] Solution of linear systems of equations: direct methods for finite element problems. In: Barker [1] 52–101

Ghil M and R Balgovind: [1] A fast Cauchy-Riemann solver. Math Comp 33 (1979) 385–635

Gietl H: see Zenger C et al

Girault V and PA Raviart: [1] Finite element approximation of the Navier-Stokes equations. Lecture Notes in Math 749. Springer, Berlin 1979

Glowinski R: [1] Numerical methods for nonlinear variational problems. 2nd ed. Springer, Berlin 1983

Glowinski R: see Absi E et al

Glowinski R and JL Lions (eds): [1] Computing methods in applied sciences and engineering V. Proceedings of the fifth international symposium, Versailles, Dec 1981. North-Holland, Amsterdam 1982

Glowinski R and JL Lions (eds): [2] Computing methods in applied sciences and engineering VI. Proceedings of the sixth international symposium, Versailles, Dec 1983. North-Holland, Amsterdam 1984

Glowinski R, J Periaux, and GV Dinh: [1] Domain decomposition methods for nonlinear problems in fluid dynamics. Report 147, INRIA, Le Chesnay 1982

Glowinski R, J Periaux, and O Pironneau: [1] An efficient preconditioning scheme for iterative numerical solutions of partial differential equations. Appl Math Modelling 4 (1980) 187–192

Golub G: [1] Direct methods for solving elliptic difference equations. In: Morris [1] 1–19

Golub GH: see Concus P et al

Greenspan D (ed): [1] Numerical solutions of nonlinear differential equations. Proceedings, Madison, May 1966. J Wiley & Sons, New York 1966

Grigorieff RD: see Bulirsch R et al

Griffiths DF (ed) [1] Numerical analysis. Proceedings, Dundee, June–July 1983. Lecture Notes in Math 1066. Springer, Berlin 1983

Hackbusch W: [1] A fast iterative method solving Poisson's equation in a general region. In: Bulirsch – Grigorieff – Schröder [1] 51–62

Hackbusch W: [2] Ein iteratives Verfahren zur schnellen Auflösung elliptischer Randwertprobleme. Report 76-12, Universität zu Köln 1976

Hackbusch W: [3] A fast iterative method for solving Helmholtz's equation in a general region. In: Schumann [1] 182–188

Hackbusch W: [4] A fast numerical method for elliptic boundary value problems with variable coefficients. In: Hirschel – Geller [1] 50–57

Hackbusch W: [5] On the convergence of a multi-grid iteration applied to finite element equations. Report 77-8, Universität zu Köln 1977

Hackbusch W: [6] A multi-grid method applied to a boundary value problem with variable coefficients in a rectangle. Report 77-17, Universität zu Köln 1977 – revised version: Bericht 2/83, Universität Kiel 1983

Hackbusch W: [7] A numerical method for solving parabolic equations with opposite orientations. Computing 20 (1978) 229–240

Hackbusch W: [8] On the multi-grid method applied to difference equations. Computing 20 (1978) 291–306

Hackbusch W: [9] On the fast solution of nonlinear elliptic equations. Numer Math 32 (1979) 83–95

Hackbusch W: [10] Survey of convergence proofs for multi-grid iterations. In: Frehse – Pallaschke – Trottenberg [1] 151–164

Hackbusch W: [11] On the computation of approximate eigenvalues and eigenfunctions of elliptic operators by means of a multi-grid method. SIAM J Numer Anal 16 (1979) 201–215

Hackbusch W: [12] On the fast solving of parabolic boundary control problems. SIAM J Control Optim 17 (1979) 231–244

Hackbusch W: [13] Convergence of multi-grid iterations applied to difference equations. Math Comp 34 (1980) 425–440

Hackbusch W: [14] The fast numerical solution of very large elliptic difference schemes. J Inst Maths Applics 26 (1980) 119–132

Hackbusch W: [15] Fast solution of elliptic control problems. J Optim Theory Appl 31 (1980) 565–581

Hackbusch W: [16] Numerical solution of nonlinear equations by a multi-grid iteration of the second kind. In: Taylor – Hinton – Owen [1] 1041–1050

Hackbusch W: [17] Numerical solution of linear and nonlinear parabolic control problems. In: Auslender – Oettli – Stoer [1] 179–185

Hackbusch W: [18] Multi-grid solutions to linear and nonlinear eigenvalue problems for integral and differential equations. Report 80-3, Universität zu Köln 1980 – Rostock Math Colloq 25 (1984) 79–98

Hackbusch W: [19] Analysis and multi-grid solutions of mixed finite element and mixed difference equations. Report, Ruhr-Universität Bochum 1980

Hackbusch W: [20] Die schnelle Auflösung der Fredholmschen Integralgleichung zweiter Art. Beiträge Numer Math 9 (1981) 47–62

Hackbusch W: [21] On the convergence of multi-grid iterations. Beiträge Numer Math 9 (1981) 213–239

Hackbusch W: [22] Error analysis of the nonlinear multi-grid method of the second kind. Apl Mat 26 (1981) 18–29

Hackbusch W: [23] On the regularity of difference schemes. Ark Mat 19 (1981) 71–95

Hackbusch W: [24] Optimal $H^{p,p/2}$ error estimates for a parabolic Galerkin method. SIAM J Numer Anal 18 (1981) 681–692

Hackbusch W: [25] Fast numerical solution of time-periodic parabolic problems by a multi-grid method. SIAM J Sci Statist Comput 2 (1981) 198–206

Hackbusch W: [26] Bemerkungen zur iterativen Defektkorrektur und zu ihrer Kombination mit Mehrgitterverfahren. Rev Roumaine Math Pures Appl 26 (1981) 1319–1329

Hackbusch W: [27] Introduction to multi-grid methods for the numerical solution of boundary value problems. In: Essers [1] 45–92

Hackbusch W: [28] Analysis of discretizations by the concept of discrete regularity. In: Whiteman [2] 369–376

Hackbusch W: [29] Multi-grid convergence theory. In: Hackbusch – Trottenberg [1] 177–219

Hackbusch W: [30] On multi-grid iterations with defect correction. In: Hackbusch – Trottenberg [1] 461–473

Hackbusch W: [31] A note on the penalty correction method. Zeitschrift für Analysis und ihre Anwendungen 2 (1982) 59–69

Hackbusch W: [32] On the regularity of difference schemes – part II: regularity estimates for linear and nonlinear problems. Ark Mat 21 (1983) 3–28

Hackbusch W: [33] Multi-grid solution of continuation problems. In: Ansorge – Meis - Törnig [1] 20–45

Hackbusch W: [34] Multi-grid convergence for a singular perturbation problem. Linear Algebra Appl 58 (1984) 125–145

Hackbusch W: [35] Parabolic multi-grid methods. In: Glowinski – Lions [2], 189–197
Hackbusch W (ed): [36] Efficient solvers for elliptic systems. Notes on numerical fluid mechanics 10. Vieweg, Braunschweig 1984
Hackbusch W: [37] Local defect correction method and domain decomposition techniques. Computing, Suppl 5 (1984) 89–113
Hackbusch W: [38] Simultaneous eigenvalue computation. In: Braess – Hackbusch – Trottenberg [1] 24–32
Hackbusch W: [39] Theorie und Numerik elliptischer Differentialgleichungen. To be published by Teubner, Stuttgart
Hackbusch W: see Braess D et al
Hackbusch W and G Hofmann: [1] Results of the eigenvalue problem for the plate equation. Z Angew Math Phys 31 (1980) 730–739
Hackbusch W and HD Mittelmann: [1] On multi-grid methods for variational inequalities. Numer Math 42 (1983) 65–76
Hackbusch W und U Trottenberg (eds): [1] Multi-grid methods. Proceedings, Köln-Porz, Nov 1981. Lecture Notes in Math 960. Springer, Berlin 1982
Hackbusch W and Th Will: [1] A numerical method for a parabolic bang-bang problem. Control Cybernet 12 (1983) 99–116
Hemker PW: [1] On the structure of an adaptive multi-level algorithm. BIT 20 (1980) 289–301
Hemker PW: [2] Fourier analysis of grid functions, prolongations and restrictions. NW 93/80, Mathematisch Centrum, Amsterdam 1980
Hemker PW: [3] The incomplete LU-decomposition as a relaxation method in multi-grid algorithms. In: Miller [2] 306–311
Hemker PW: [4] Introduction to multigrid methods. Nieuw Arch Wisk (3) 29 (1981) 71–101
Hemker PW: [5] A note on defect correction processes with an approximate inverse of deficient rank. J Comp Appl Math 8 (1982) 137–140
Hemker PW: [6] An accurate method without directional bias for the numerical solution of a 2-D elliptic singular perturbation problem. In: Eckhaus – de Jager [1]
Hemker PW: [7] Mixed defect correction iteration for the accurate solution of the convection diffusion equation. In: Hackbusch – Trottenberg [1] 485–501
Hemker PW: [8] Numerical aspects of singular perturbation problems. Preprint NW 133/82, Mathematisch Centrum, Amsterdam 1982
Hemker PW: [9] On the comparison of line-Gauss-Seidel and ILU relaxation in multigrid algorithms. In: Miller [3] 269–277
Hemker PW: [10] Extensions of the defect correction principle. In: Miller [4] 33–45
Hemker PW: [11] Multigrid methods for problems with a small parameter. In: Griffiths [1] 106–121
Hemker PW: [12] Mixed defect correction iteration for the solution of a singular perturbation problem. Computing, Suppl 5 (1984) 123–145
Hemker PW and PM de Zeeuw: [1] Defect correction for the solution of a singular perturbation problem. Preprint NW 132/82, Mathematisch Centrum, Amsterdam 1982 – In: Scientific Computing (RS Stepleman et al, eds) 113–118. North Holland, Amsterdam 1983
Hemker PW and PM de Zeeuw: [2] Some implementations of multigrid linear system solvers. In: Holstein – Paddon [1]
Hemker PW, R Kettler, P Wesseling, and PM de Zeeuw: [1] Multigrid methods: development of fast solvers. Appl Math Comput 13 (1983) 311–326
Hemker PW and JJH Miller (eds): [1] Numerical analysis of singular perturbation problems. Proceedings, Nijmegen, May–June 1978. Acad Press, London 1979
Hemker PW and H Schippers: [1] Multiple grid methods for the solution of Fredholm integral equations of the second kind. Math Comp 36 (1981) 215–232
Hemker PW and SP Spekreijse: [1] Multigrid solution of the steady Euler equations. In: Braess-Hackbusch-Trottenberg [1] 33–44

Hemker PW, P Wesseling, and PM de Zeeuw: [1] A portable vector-code for autonomous multigrid modules. Preprint NW 154/83, Mathematisch Centrum, Amsterdam 1983 – to appear in: PDE software: modules, interfaces and systems (B Engquist: ed), Procs IFIP WG 2.5 Working Conf. North Holland, Amsterdam 1984

Hestenes MR: [1] Conjugate direction methods in optimization. Springer, New York 1980

Hestenes MR and E Stiefel: [1] Methods of conjugate gradients for solving linear systems. J Res Nat Bur Standards 49 (1952) 409–436

Hinton E and DRJ Owen: [1] An introduction to finite element computations. Pineridge Press, Swansea 1979

Hinton E: see Taylor C et al

Hirsch Ch: see Deconinck H et al

Hirschel EH and W Geller (eds): [1] Second GAMM-conference on numerical methods in fluid mechanics. Proceedings, Köln-Porz, Oct 1977. DFVLR 1977

Hlaváček I: see Nečas J et al

Hockney RW: [1] A fast direct solution of Poisson's equation using Fourier analysis. J Assoc Comput Mach 12 (1965) 95–113

Hofmann G: [1] Analyse eines Mehrgitterverfahrens zur Berechnung von Eigenwerten elliptischer Differentialoperatoren. Doctoral thesis, Kiel 1985

Hofmann G: see Hackbusch W et al

Holstein H and DJ Paddon (eds): [1] Short course on multigrid methods. Proceedings, Bristol, Sept 1983. Oxford Press 1985

Hörmander L: [1] Linear partial differential operators. Springer, Berlin 1969

Hsiao GC, P Kopp, and WL Wendland: [1] A Galerkin collocation method for some integral equations of the first kind. Computing 25 (1980) 89–130

Hubbard DE: see Bramble JH et al

Hughes TJR: see Brooks AN et al

Hussaini MY: see Zang TA et al

Hyman JM: [1] Mesh refinement and local inversion of elliptic partial differential equations. J Comput Phys 23 (1977) 124–134

Hyman JM: see Dendy Jr JE et al

Il'in AM: [1] Differencing scheme for a differential equation with a small parameter affecting the highest derivative. Math Notes Acad Sc USSR 6 (1969) 596–602

Jacobi CGJ: [1] Über eine neue Auslösungsart der bei der Methode der kleinsten Quadrate vorkommenden linearen Gleichungen. Astronom Nachr 32 (1845) 297–306

Jameson A: [1] Acceleration of transonic potential flow calculations on arbitrary meshes by the multiple grid method. Paper AIAA-79-1458, AIAA fourth computational fluid dynamics conference. New York 1979

Jameson A: [2] Solution of the Euler equations for two dimensional transonic flow by a multigrid method. Appl Math Comput 13 (1983) 327–355

Jameson A: see Schmidt W et al

Jennings A: [1] Matrix computation for engineers and scientists. Wiley, London 1977

Jensen K and N Wirth: [1] PASCAL user manual and report. 2nd ed. Springer, New York 1978

Jerome JW: see Bank RE et al

Jespersen DC: [1] Design and implementation of a multigrid code for the Euler equations. Appl Math Comput 13 (1983) 357–374

Johnson C and U Nävert: [1] An analysis of some finite element methods for advection-diffusion problems. In: Axelsson – Frank – van der Sluis [1] 99–116

Johnson C and J Pitkäranta: [1] Analysis of some mixed finite element methods related to reduced integration. Math Comp 38 (1982) 375–400

Jones IP and CP Thompson (eds): [1] Numerical solution for a comparison problem on natural convection in an enclosed cavity. Report AERE-R.9955, Harwell 1981

Kaczmarz S: [1] Angenäherte Auflösung von Systemen linearer Gleichungen. Bulletin de l'Academie Polonaise des Sciences et Lettres A 35 (1937) 355–357

Kadlec J: [1] O reguljarnosti rešenija zadači Puassona na oblasti s granicej, lokol'no podobnoj granice vypukloj oblasti (On the regularity of the solution of the Poisson equation on a domain with boundary locally similar to the boundary of a convex domain). Czechoslovak Math J 14 (1964) 386–393

Kaspar W and R Remke: [1] Die numerische Behandlung der Poisson-Gleichung auf einem Gebiet mit einspringenden Ecken. Computing 22 (1979) 141–151

Kato T: [1] Perturbation theory for linear operators. Springer, Berlin 1966

Keller HB: [1] Numerical solution of bifurcation and nonlinear eigenvalue problems. In: Rabinowitz [1] 359–383

Keller HB: see Chan TFC et al

Kellogg RB and JE Osborn: [1] A regularity result for the Stokes problem in a convex polygon. J Funct Anal 21 (1976) 397–431

Kershaw DS: [1] The incomplete Cholesky-conjugate gradient method for the iterative solution of systems of linear equations. J Comput Phys 26 (1978) 43–65

Kettler R: [1] Analysis and comparison of relaxation schemes in robust multigrid and preconditioned conjugate gradient methods. In: Hackbusch – Trottenberg [1] 502–534

Kettler R: see Hemker PW et al

Kettler R and JA Meijerink: [1] A multigrid method and a combined multigrid-conjugate gradient method for elliptic problems with strongly discontinuous coefficients in general domains. Shell Publ 604, KSEPL, Rijswijk 1981

Kopp P: see Hsiao GC et al

Korneev VG: [1] Schemy metoda konečnych elementov vysokich porjadkov točnosti (High order finite element methods). Izdatel'stvo, Leningrad 1977

Korneev VG and SE Ponomarev: [1] Rešenie relakcionnym iteracionnym metodom schem mkė dlja elliptičeskich uravnenij porjadka 2n (Solution of elliptic equations of order 2n by the relaxation method). Čisl Metody Mech Splosn sredy 8,7 (1977) 72–86

Kreiss H-O: [1] Difference approximations for the initial-boundary value problem for hyperbolic differential equations. In: Greenspan [1] 141–166

Krejn SG, JuI Petunin, and EM Semenov: [1] Interpolation of linear operators. AMS, Providence, Rhode Island 1982

Křížek M: see Neittaanmäki P et al

Kronsjö L: [1] A note on the "nested iterations" method. BIT 15 (1975) 107–110

Kronsjö L and G Dahlquist: [1] On the design of nested iterations for elliptic difference equations. BIT 11 (1971) 63–71

Kuznecov JuA: see Bachvalov NS et al

Ladyženskaja OA and NN Ural'ceva: [1] Linear and quasilinear elliptic equations. Acad Press, New York 1968

Langer U: [1] O vybore iteracionnych parametrov v relaksacionnom metode na posledovatel'nosti setok. Ž vyčisl matem i matem fiz 22,5 (1982) 1118–1132 – Engl transl: On the choice of iterative parameters in the relaxation method on a sequence of meshes. USSR Comput Math and Math Phys 22,5 (1982) 18–114

Lapin AV: [1] Issledovanie schodimosti raznostnych schem v norme $W_2^{(2)}$ dlja kvazilinejnych elliptičeskich uravnenij. Ž vyčisl matem i matem fiz 14,6 (1974) 1516–1525 – Engl transl: Study of the $W_2^{(2)}$-convergence of difference schemes for quasilinear elliptic equations. USSR Comput Math and Math Phys 14,6 (1974) 140–149

Lascaux P: see Absi E et al

Lebedev VI (ed): [1] Variacionno-raznostnye metody v matematičeskoj fizike (Variational difference methods in mathematical physics). Proceedings, Oct 1980, Novosibirsk. AN SSSR 1981

Lehmann H: see Meis Th et al

Linde HJ van: see Van Linde HJ

Linden J: [1] Mehrgitterverfahren für die Poisson-Gleichung in Kreis und Ringgebiet unter Verwendung lokaler Koordinaten. Diplomarbeit (master thesis), Universität Bonn 1981

Lin Qun, Lu Tao, and Shen Shumin: [1] Asymptotic expansion for finite element approximations. Report IMS-11, Chengdu Branch of Academia Sinica, 1983

Lin Qun, Wang Junping: [1] Some expansions for the finite element approximations. Report IMS-15, Chengdu Branch of Academia Sinica, 1984

Lions JL: see Glowinski R et al

Lions JL and E Magenes: [1] Non-homogeneous boundary value problems and applications, vol I. Springer, Berlin 1972

Lions JL and E Magenes: [2] Non-homogeneous boundary value problems and applications, vol II. Springer, Berlin 1972

Lynch RE and JR Rice: [1] A high order difference method for differential equations. Math Comp 34 (1980) 333–372

Magenes E: see Lions JL et al

Maitre J-F and F Musy: [1] The contraction number of a class of two-level methods; an exact evaluation for some finite element subspaces and model problems. In: Hackbusch – Trottenberg [1] 535–544

Maitre J-F and F Musy: [2] Algebraic formalization of the multigrid method in the symmetric and positive definite case – A convergence estimation for the V-cycle. In: Holstein – Paddon [1]

Maitre J-F and F Musy: [3] Multigrid methods: convergence theory in a variational framework. SIAM J on Numer Anal 21 (1984) 657–671

Maitre J-F, F Musy, and P Nigon: [1] A fast solver for the Stokes equations using multigrid with a UZAWA smoother. In: Braess – Hackbusch – Trottenberg [1]

Mandel J: [1] Convergence of an iterative method for the system $Ax + y = x$ using aggregation. Ekonom-Mat Obzor 17 (1981) 287–291

Mandel J: [2] On multilevel iterative methods for integral equations of the second kind and related problems. Numer Math 46 (1985) 147–157

Mandel J: [3] A multilevel iterative method for symmetric, positive definite linear complementarity problems. Appl Math Optim 11 (1984) 77–95

Mandel J: [4] Algebraic study of a multigrid method for some free-boundary problem. Report, Prague 1984

Mandel J: [5] Algebraic study of multigrid methods for symmetric, definite problems. To appear

Mandel J and B Sekerka: [1] A local convergence proof for the iterative aggregation method. Preprint, Charles University, Prag 1982

Mandel J and B Sekerka: [2] Iterační agregace a minimalizace rezidua při řešení meziodvětvových vztahů (Iterative aggregation and residue minimisation for solving input-output relations). Ekonom-Mat Obzor 18 (1982) 428–441

Mansfield L: [1] On multigrid solution of finite element equations with isoparametric elements. Numer Math 37 (1981) 423–432

Manteuffel TA: [1] An incomplete factorization technique for positive definite linear systems. Math Comp 34 (1980) 473–497

Marcowitz U: see Meis Th et al

Marchuk GI: [1] Methods of numerical mathematics. Springer, New York 1975

Marchuk GI and VV Shaidurov: [1] Difference methods and their extrapolation. Springer, New York 1983

Marsal D: [1] Die numerische Lösung partieller Differentialgleichungen. Bibl Institut, Mannheim 1976

McCormick S: [1] Mesh refinement methods for integral equations. In: Albrecht – Collatz [1] 183–190

McCormick S: [2] A mesh refinement method for $Ax = \lambda Bx$. Math Comp 36 (1981) 485–498

McCormick S: [3] Multigrid methods for variational problems: further results. SIAM J Numer Anal 21 (1984) 255–263

McCormick S: see Brandt A et al

Meijerink JA and HA van der Vorst: [1] An iterative solution method for linear systems of which the coefficient matrix is a symmetric M-matrix. Math Comp 31 (1977) 148–162

Meijerink JA and HA van der Vorst: [2] Guidelines for the usage of incomplete decompositions in solving sets of linear equations as they occur in practical problems. J Comput Phys 44 (1981) 134–155

Meijerink JA: see Kettler R et al

Meis Th: see Ansorge R et al

Meis Th and H-W Branca: [1] Schnelle Lösung von Randwertaufgaben. Z Angew Math Mech 62 (1982) T263–T270

Meis Th, H Lehmann, and H Michael: [1] Application of the multigrid method to a nonlinear indefinite problem. In: Hackbusch – Trottenberg [1] 545–557

Meis Th and U Marcowitz: [1] Numerische Behandlung partieller Differentialgleichungen. Springer, Berlin 1978 – Engl transl: Numerical solution of partial differential equations. Springer, New York 1981

Menzel R and H Schwetlick: [1] Zur Lösung parameterabhängiger nichtlinearer Gleichungen mit singulären Jacobi-Matrizen. Numer Math 30 (1978) 65–79

Michael H: see Meis Th et al

Miller JJH (ed): [1] Topics in numerical analysis. Proceedings, Dublin, Aug 1972. Acad Press, London 1973

Miller JJH (ed): [2] Boundary and interior layers – computational and asymptotic methods. Proceedings, Dublin, June 1980. Boole Press, Dublin 1980

Miller JJH (ed): [3] Computational and asymptotic methods for boundary and interior layers. Proceedings, Dublin, June 1982. Boole Press, Dublin 1982

Miller JJH (ed): [4] An introduction to computational and asymptotic methods for boundary and interior layers. Short Course. Dublin, June 1982. Boole Press, Dublin 1982

Miller JJH: see Hemker PW et al

Miranda C: [1] Partial differential equations of elliptic type. Springer, Berlin 1970

Miranker WL: [1] Hierarchical relaxation. Computing 23 (1979) 267–285

Miranker WL: see Chatelin F et al

Mitchell AR: [1] Computational methods in partial differential equations. Wiley, London 1969

Mitchell AR: see Barnhill RE et al

Mitchell AR and R Wait: [1] The finite element method in partial differential equations. Wiley, London 1977

Mittelmann HD: [1] A fast solver for nonlinear eigenvalue problems. In: Ansorge – Meis – Törnig [12] 46–67

Mittelmann HD: [2] Multi-grid methods for simple bifurcation problems. In: Hackbusch – Trottenberg [1] 558–575

Mittelmann HD: see Hackbusch W et al

Mittelmann HD and H Weber: [1] Multi-grid solution of bifurcation problems. SIAM J Sci Statist Comput 6 (1985) 49–60

Mizohata S: [1] The theory of partial differential equations. University Press, Cambridge 1973

Mol WJA: [1] Numerical solution of the Navier-Stokes equations by means of multigrid methods and Newton-iteration. Preprint NW 92/80, Mathematisch Centrum, Amsterdam 1980

Mol WJA: [2] On the choice of suitable operators and parameters in multigrid methods. Report NW 107/81, Mathematisch Centrum, Amsterdam 1981

Mol WJA: [3] Smoothing and coarse grid approximation properties of multigrid methods. Report NW 110/81, Mathematisch Centrum, Amsterdam 1981

Mol WJA: [4] Computation of flows around a Karman-Trefftz profile. Preprint NW 114/81, Mathematisch Centrum, Amsterdam 1981

Morris JLl (ed): [1] Symposium on the theory of numerical analysis. Proceedings, Dundee, Sept 1970. Lecture Notes in Math 193. Springer, Berlin 1971

Morton KW: see Richtmyer RD et al

Munksgaard N: [1] Solving sparse symmetric sets of linear equations by preconditioned conjugate gradients. ACM Trans Math Software 6 (1980) 206–219

Musy F: [1] Sur les méthodes multigrilles: formalisation algébrique et démonstration de convergence. C R Acad Sc Paris, Ser I, 295 (1982) 471–474

Musy F: see Maitre J-F et al

Nävert U: see Johnson C et al

Nečas J: [1] Sur la coercivité des formes sesqui-linéaires elliptiques. Rev Roumaine Math Pures Appl 9 (1964) 47–69

Nečas J: [2] Les méthodes directes en théorie des équations elliptiques. Paris, Masson 1967

Nečas J and I Hlaváček: [1] Mathematical theory of elastic and elasto-plastic bodies: an introduction. Elsever, Amsterdam 1981

Neittaanmäki P and M Křížek: [1] Conforming FE-method for obtaining the gradient of a solution to the Poisson equation. In: Hackbusch [36] 74–86

Nicolaides RA: [1] On multiple grid and related techniques for solving discrete elliptic systems. J Comput Phys 19 (1975) 418–431

Nicolaides RA: [2] On the l^2 convergence of an algorithm for solving finite element equations. Math Comp 31 (1977) 892–906

Nicolaides RA: [3] On the observed rate of convergence of an iterative method applied to a model elliptic difference equation. Math Comp 32 (1978) 127–133

Nicolaides RA: [4] On multigrid convergence in the indefinite case. Math Comp 32 (1978) 1082–1086

Nicolaides RA: [5] On finite element multigrid algorithms and their use. In: Whiteman [1] 459–466

Nicolaides RA: [6] On some theoretical and practical aspects of multigrid methods. Math Comp 33 (1979) 933–952

Nirenberg L: see Agmon S et al

Nitsche J and JCC Nitsche: [1] Error estimates for the numerical solution of elliptic differential equations. Arch Rational Mech Anal 5 (1960) 293–306

Nitsche JCC: see Nitsche J et al

Noble B: [1] Error analysis of collocation methods for solving Fredholm integral equations. In: Miller [1] 211–232

Nowak ZP: [1] Use of the multigrid method for Laplacian problems in three dimensions. In: Hackbusch – Trottenberg [1] 576–598

Nowak ZP: [2] A quasi-Newton multigrid method for determining the transonic lifting flows around airfoils. Report, Delft 1984

Nowak ZP and P Wesseling: [1] Multigrid acceleration of an iterative method with application to transonic potential flow. In: Glowinski – Lions [2] 199–217

Nyström EJ: [1] Über die praktische Auflösung von linearen Integralgleichungen mit Anwendungen auf Randwertaufgaben der Potentialtheorie. Soc Sci Fenn Comment Phys-Math 4,15 (1928) 1–52

Oden JR and JN Reddy: [1] An introduction to the mathematical theory of finite elements. Wiley, New York 1976

Oettli W: see Auslender A et al

Oganesjan LA and LA Ruchovec: [1] Variacionno-raznostnye metody rešenija elliptičestich uravnenij (Variational difference methods for elliptic equations). Izdatel'stvo Akademii Nauk Armjanskoj SSR, Erevan 1979

O'Leary DP: see Concus P et al

O'Leary DP and O Widlund: [1] Capacitance matrix methods for the Helmholtz equation on general three-dimensional regions. Math Comp 33 (1979) 849–879

Ortega JM and WC Rheinboldt: [1] Iterative solution of nonlinear equations in several variables. Acad Press, New York 1970

Osborn JE: see Kellog RB et al

Oskam B and JMJ Fray: [1] General relaxation schemes in multigrid algorithms. J Comput Phys 48 (1982) 423–440
Owen DRJ: see Hinton E et al
Owen DRJ: see Taylor C et al
Paddon DJ: see Holstein H et al
Painter JW: see Alcouffe RE et al
Pallaschke D: see Frehse J et al
Parter SV (ed): [1] Numerical methods for partial differential equations. Proceedings, Madison (WI), Oct 1978. Acad Press, New York 1979
Pereyra V: [1] On improving an approximate solution of a functional equation by deferred corrections. Numer Math 8 (1966) 376–391
Pereyra V: [2] Highly accurate numerical solution of quasi-linear elliptic systems in n dimensions. Math Comp 24 (1970) 771–783
Pereyra V, W Proskurowski, and O Widlund: [1] High order fast Laplace solvers for the Dirichlet problem on general regions. Math Comp 31 (1977) 1–16
Periaux J: see Glowinski R et al
Petunin JuI: see Krejn SG et al
Pironneau O: see Bercovier M et al
Pironneau O: see Glowinski R et al
Pitkäranta J and T Saarinen: A multigrid version of a simple finite element method for the Stokes problem. Report-MAT-A 217, Espoo 1984
Pitkäranta J: see Brezzi F et al
Platzer MF: see Adamson TC et al
Ponomarev SE: see Korneev VG et al
Práger M: see Babuška I et al
Proskurowski W: see Pereyra V et al
Proskurowski W and O Widlund: [1] On the numerical solution of Helmholtz's equation by the capacitance matrix method. Math Comp 30 (1976) 433–468
Rabinowitz PH (ed): [1] Applications of bifurcation theory. Proceedings, Madison, Oct 1976. Academic Press, New York 1977
Rautmann R (ed): [1] Approximation methods for Navier-Stokes problems. Proceedings, Paderborn, Sept 1979. Lecture Notes in Math 771. Springer, Berlin 1980
Raviart PA: see Girault V et al
Reddy JN: see Oden JR et al
Reid JK: [1] On the method of conjugate gradients for the solution of large sparse systems of linear equations. In: Reid [2]
Reid JK (ed): [2] Large sparse sets of linear equations. Proceedings. Acad Press, New York 1971
Reid JK: [3] Solution of linear systems of equations: direct methods (general). In: Barker [1] 102–129
Remke A: see Kaspar W et al
Reutersberg H: see Schröder J et al
Rheinboldt WC: [1] Adaptive mesh refinement processes for finite element solutions. Int J Numer Meth Engrg 17 (1981) 649–662
Rheinboldt WC: see Babuška I et al
Rheinboldt WC: see Ortega JM et al
Rice JR (ed): [1] Mathematical software III. Proceedings, Madison (WI), March 1977. Acad Press, New York 1977
Rice JR: see Lynch RE et al
Richtmyer RD and KW Morton: [1] Difference methods for initial value problems. 2nd ed. Interscience Publ, New York 1967
Ries M, U Trottenberg, and G Winter: [1] A note on MGR methods. Linear Algebra Appl 49 (1983) 1–26

Ries M and U Trottenberg: [1] MGR – Ein blitzschneller elliptischer Löser. Preprint 227, Universität Bonn 1979

Ritz W: [1] Über eine neue Methode zur Lösung gewisser Variationsprobleme der mathematischen Physik. J f reine und angew Math 135 (1909) 1–61

Rivara MC: [1] Adaptive multigrid software for the finite element method. Doctoral thesis. Katholieke Universiteit Leuven 1984

Rivara MC: [2] Algorithms for refining triangular grids suitable for adaptive and multigrid techniques. Int J Numer Meth Engrg 20 (1984) 745–756

Rizzi A and H Viviand (eds): [1] Numerical methods for the computation of inviscid transonic flows with shock waves. Proceedings, Stockholm, Sept 1979. Notes on numerical fluid mechanics 3. Vieweg, Braunschweig 1981

Rose DJ: see Bank RE et al

Rosendale J van: see Van Rosendale J

Rosenzweig MB: see Babuška I et al

Ruchovec LA: see Astrachancev GP et al

Ruchovec LA: see Oganesjan LA et al

Ruge JW: [1] Multigrid methods for differential eigenvalue and variational problems and multigrid simulation. Doctoral thesis, Colorado State University, Fort Collins 1981

Ruge J: see Brandt A et al

Ruhe A: [1] Computations of eigenvalues and eigenvectors. In: Barker [1] 130–181

Ruppel H: see Brandt A et al

Saarinen T: see Pitkäranta J et al

Šajdurov V V: O rešenii spektral'noj variacionno-raznostnoj zadači na posledovatel'nosti setok (On the solution of eigenvalue problems of variational difference equations in a sequence of grids). In: Bachvalow – Kuznecov [1] 149–160

Schaffer S: [1] High order multi-grid methods to solve the Poisson equation. NASA conf publ 2202, 275–284, 1981

Schatz AH and LB Wahlbin: [1] Maximum norm estimates in the finite element method on plane polygonal domains. Part 2, refinements. Math Comp 33 (1979) 465–492

Schechter M: [1] Modern methods in partial differential equations. McGraw-Hill, New York 1977

Schippers H: [1] Multi-grid techniques for the solution of Fredholm integral equations of the second kind. In: te Riele [1] 29–46

Schippers H: [2] Multiple grid methods for oscillating disc flow. In: Miller [2] 410–414

Schippers H: [3] The automatic solution of Fredholm equations of the second kind. Report NW 99/80, Mathematisch Centrum, Amsterdam 1980

Schippers H: [4] Application of multigrid methods for integral equations to two problems from fluid dynamics. J Comput Phys 48 (1982) 441–461

Schippers H: [5] On the regularity of the principal value of the double-layer potential. J Engineering Math 16 (1982) 59–76

Schippers H: [6] Multiple grid methods for equations of the second kind with applications in fluid mechanics. Doctoral thesis, Mathematisch Centrum, Amsterdam 1982

Schippers H: [7] Multigrid methods for boundary integral equations. Numer Math (to appear)

Schippers H: see Hemker PW et al

Schmidt W and A Jameson: [1] Applications of multi-grid methods for transonic flow calculations. In: Hackbusch – Trottenberg [1] 599–613

Schreiber R: see Chan TF et al

Schröder J: see Bulirsch R et al

Schröder J and U Trottenberg: [1] Reduktionsverfahren für Differenzengleichungen bei Randwertaufgaben I. Numer Math 22 (1973) 37–68

Schröder J, U Trottenberg, and H Reutersberg: [1] Reduktionsverfahren für Differenzengleichungen bei Randwertaufgaben II. Numer Math 26 (1976) 429–459

Schröder J, U Trottenberg, and K Witsch: [1] On fast Poisson solvers and applications. In: Bulirsch – Grigorieff – Schröder [1] 153–187

Schultz MH (ed): [1] Elliptic problem solvers. Proceedings, Santa Fe, June–July 1980. Acad Press, New York 1981

Schumann U: [1] Fast elliptic solvers. Proceedings, Karlsruhe, March 1977. Advance Publ, London 1978

Schwarz HR: [1] Methode der finiten Elemente. Teubner, Stuttgart 1980

Schwetlick H: see Menzel R et al

Seidel PhL: [1] Über ein Verfahren, die Gleichungen, auf welche die Methode der kleinsten Quadrate führt, sowie lineare Gleichungen überhaupt, durch successive Annäherung aufzulösen. Münch Abh 2,3 (1874) 81–108

Sekerka B: [1] Agregace a desagregace ve strukturních modelech (Aggregation and disaggregation in structural models). Ekonom-Mat Obzor 14 (1978) 411–426

Sekerka B: see Mandel J et al

Sekerka B and J Typolt: [1] The method used in the overall reform of wholesale prizes in Czechoslovakia. Czech Economic Papers 10 (1968) 47–57

Semenov EM: see Krejn SG et al

Shaidurov VV: see Marchuk GI et al

Sherman AH: see Bank RE et al

Shortley GH and R Weller: [1] Numerical solution of Laplace's equation. J Appl Phys 9 (1938) 334–348

Šišov VS: [1] Primenenija gruppovogo isključenija neizvestnych v zadače ob otyskanii sobstvennych značenij i sobstvennych funkcij linejnogo integral'nogo operatora s simmetričeskim jadrom. Ž vyčisl matem i matem fiz 2,3 (1962) 398–410 – engl trans: The determination of eigenvalues and eigenfunctions of a linear integral operator with a symmetrical kernel by means of group elimination of unknowns. USSR Comput Math and Math Phys 2,3 (1962) 419–432

Sluis A van der: see Van der Sluis A

Sonneveld P, P Wesseling, and PM de Zeeuw: [1] Multigrid and conjugate gradient methods as convergence acceleration techniques. In: Holstein – Paddon [1]

Sonneveld P: see Wesseling P et al

South Jr JC and A Brandt: [1] Application of a multi-level grid method to transonic flow calculations. In: Adamson – Platzer [1] 180–206

Southwell RV: [1] Stress-calculation in frameworks by the method of "systematic relaxation of constraints" – parts I, II: Proc Roy Soc (A) 151 (1935) 56–95; part III: Proc Roy Soc (A) 153 (1935) 41–76

Southwell RV: [2] Relaxation methods in engineering science – a treatise on approximate computation. Oxford University Press, London 1940

Southwell RV: [3] Relaxation methods in theoretical physics. Clarendon Press, Oxford 1946

Starius G: [1] Constructing orthogonal curvilinear meshes by solving initial value problems. Numer Math 28 (1977) 25–48

Starius G: [2] Composite mesh difference methods for elliptic boundary value problems. Numer Math 28 (1977) 243–258

Stepleman RS: see Vichnevetsky R et al

Stetter HJ: [1] The defect correction principle and discretization methods. Numer Math 29 (1978) 425–443

Stetter HJ: see Auzinger W et al

Stewart GW: see Duff IS et al

Stiefel E: [1] Über einige Methoden der Relaxationsrechnung. Z Angew Math Phys 3 (1952) 1–33

Stiefel E: see Hestenes MR et al

Stoer J: see Auslender A et al

Stoer J and R Bulirsch: [1] Introduction to numerical analysis. Springer, Berlin 1980

Strachovskaja LG: [1] Iteracionnyj metod vyčislenija pervoj sobstvennoj funkcii elliptičeskogo operatora. Ž vyčisl matem i matem fiz 17,3 (1977) 649–663 – Engl transl: An iterative method for evaluating the first eigenvalue of an elliptic operator. USSR Comput Math and Math Phys 17,3 (1977) 88–101

Strang G and GJ Fix [1]: An analysis of the finite element method. Prentice-Hall, Englewood Cliffs 1973

Stüben K: [1] MG01: A multi-grid program to solve $\Delta u - c(x, y) u = f(x, y)$ (on Ω), $u = g(x, y)$ (on $\partial\Omega$), on nonrectangular bounded domains Ω. IMA-Report 82.02.02, GMD, St. Augustin 1982

Stüben K: [2] Algebraic multigrid (AMG): experiences and comparisons. Appl Math Comput 13 (1983) 419–451

Stüben K: see Foerster H et al

Stüben K: see Ruge J et al

Stüben K and U Trottenberg: [1] Multigrid methods: fundamental algorithms, model problem analysis and applications. In: Hackbusch – Trottenberg [1] 1–176

Stüben K and U Trottenberg: [2] On the construction of fast solvers for elliptic equations. VKI Lecture Series 1982–4, Von Karman Institute for Fluid Dynamics, Rhode-Saint-Genese 1982

Stüben K, U Trottenberg, and K Witsch: [1] Software development based on multigrid techniques. To appear in: Proceedings of the IFIP TC 2 Working Conference on "PDE software: modules, interfaces and systems", Söderköping, 1983 – Report, Arbeitspapiere der GMD Nr 84, St Augustin 1983

Stüben K, U Trottenberg, and K Witsch: [2] On the convergence of multigrid iterations: quantitative results for model problems. Manuscript

Ta'asan S: [1] Multi-grid methods for highly oscillatory problems. Research report, Weizmann Institute, Rehovot 1981

Taylor C, E Hinton, and DRJ Owen (eds): [1] Numerical methods for non-linear problems, vol 1. Proceedings, Swansea, Sept 1980. Pineridge Press, Swansea 1980

Temam R: [1] Navier-Stokes equations. North-Holland, Amsterdam 1977

Temam R: see Cabannes H et al

Te Riele HJJ (ed): [1] Colloquium numerical treatment of integral equations. Proceedings, Amsterdam 1979. MC Syllabus 41, Mathematisch Centrum, Amsterdam 1979

Thomasset F: [1] Implementation of finite element methods for Navier-Stokes equations. Springer, New York 1981

Thomée V: [1] Elliptic difference operators and Dirichlet's problem. Contributon to Diff Eq 3 (1964) 301–324

Thomée V and D Westergren: [1] Elliptic difference equation and interior regularity. Numer Math 11 (1968) 196–210

Thompson CP: see Jones IP et al

Törnig W: [1] Numerische Mathematik für Ingenieure und Physiker, vol 2. Springer, Berlin 1979

Törnig W: see Ansorge R et al

Trottenberg U: [1] Reduction method for solving discrete elliptic boundary value problems – an approach in matrix terminology. In: Schumann [1] 98–111

Trottenberg U: see Braess D et al

Trottenberg U: see Foerster H et al

Trottenberg U: see Frehse J et al

Trottenberg U: see Hackbusch W et al

Trottenberg U: see Ries M et al

Trottenberg U: see Schröder J et al

Trottenberg U: see Stüben K et al

Typolt J: see Sekerka B et al

Ural'ceva NN: see Ladyženskaja OA et al

Van Asselt EJ: [1] The multigrid method and artificial viscosity. In: Hackbusch – Trottenberg [1] 313–326

Van Asselt EJ: see De Zeeuw PM et al

Van der Sluis A: see Axelsson O et al

Van der Vorst HA: [1] Iterative solution methods for certain sparse linear systems with a non-symmetric matrix arising from pde-problems. J Comput Phys 44 (1981) 1–19

Van der Vorst HA: [2] A vectorizable variant of some ICCG methods. SIAM J Sci statist Comput 3 (1982) 350–356

Van der Vorst HA: [3] Preconditioning by incomplete decompositions. Doctoral thesis, Rijksuniversiteit, Utrecht 1982

Van der Vorst HA: see Meijerink JA et al

Van Linde HJ: [1] High-order finite-difference methods for Poisson's equation. Math Comp 28 (1974) 369–391

Van Rosendale J: [1] Algorithms and data structures for adaptive multigrid elliptic solvers. Appl Math Comput 13 (1983) 453–470

Van Rosendale J: see Gannon D et al

Varga RS: [1] Matrix iterative analysis. Prentice Hall, Englewood Cliffs 1962

Verfürth, R: [1] The contraction number of a multigrid method with mesh ratio 2 for solving Poisson's equation. Linear Algebra Appl 60 (1984) 332–348

Verfürth R: [2] A multilevel algorithm for mixed problems. SIAM J Numer Anal 21 (1984) 264–271

Verfürth R: [3] Numerical solution of mixed finite element problems. In: Hackbusch [36] 132–144

Verfürth R: [4] A combined conjugate gradient-multigrid algorithm for the numerical solution of the Stokes problem. IMA J Numer Anal 4 (1984) 441–455

Veysseyre H: see Absi E et al

Vichnevetsky R and RS Stepleman (eds): [1] Advances in computer methods for partial differential equations. Proceedings, Bethleham, June–July 1979. IMACS, New York, 1979

Vitásek E: see Babuška I et al

Viviand H (ed): [1] Proceedings of the fourth GAMM-conference on numerical methods in fluid mechanics. Vieweg, Braunschweig 1982

Viviand H: see Rizzi A et al

Vorst HA van der: see Van der Vorst

Wachspress EL: [1] Iterative solution of elliptic systems and applications to the neutron diffusion equations of reactor physics. Prentice Hall, Englewood Cliffs 1966

Wachspress EL: [2] A rational finite element basis. Acad Press, New York 1975

Wahlbin LB: see Schatz AH et al

Wait R: see Mitchell AR et al

Watson GA (ed): [1] Numerical analysis. Proceedings, Dundee, June 1979. Lecture Notes in Math 773. Springer, Berlin 1980

Weber H: [1] An efficient technique for the computation of stable bifurcation branches. SIAM J Sci Statist Comput 5 (1984) 332–348

Weber H: see Mittelmann HD et al

Weiser A: see Bank RE et al

Weller R: see Shortley GH et al

Wendland WL: see Hsiao GC et al

Wesseling P: [1] Numerical solution of the stationary Navier-Stodes equation by means of a multiple grid method and Newton iteration. Report NA-18, Technische Hogeschool Delft 1977

Wesseling P: [2] The rate of convergence of a multiple grid method. In: Watson [1] 164–184

Wesseling P: [3] Theoretical and practical aspects of a multigrid method. SIAM J Sci Statist Comput 3 (1982) 387–407

Wesseling P: [4] A robust and efficient multigrid method. In: Hackbusch – Trottenberg [1] 614–630

Wesseling P: [5] Multigrid solution of the Navier-Stokes equations in the vorticity-streamfunction formulation. In: Hackbusch [36] 145–154

Wesseling P and P Sonneveld: [1] Numerical experiments with a multiple grid and a preconditioned Lanczos type method. In: Rautmann [1] 543–562

Wesseling P: see Hemker PW et al

Wesseling P: see Nowak Z et al

Wesseling P: see Sonneveld P et al

Westergren B: see Thomée V et al

Whiteman JR (ed): [1] The mathematics of finite elements and applications III – MAFELAP 1978. Proceedings, Uxbridge, April 1978. Acad Press, London 1979

Whiteman JR (ed): [2] The mathematics of finite elements and applications IV – MAFELAP 1981. Proceedings, Uxbridge, April–May 1981. Acad Press, London 1981

Widlund O: [1] A Lanczos method for a class of nonsymmetric systems of linear equations. SIAM J Numer Anal 15 (1978) 801–812

Widlund O: see Bjørstadt PE et al

Widlund O: see O'Leary DP et al

Widlund O: see Pereyra V et al

Widlund O: see Proskurowski W et al

Will Th: see Hackbusch W et al

Winter G: [1] Fourieranalyse zur Konstruktion schneller MGR-Verfahren. Doctoral thesis, Bonn 1982

Winter G: see Ries M et al

Wirth N: see Jensen K et al

Witsch K: see Foerster H et al

Witsch K: see Schröder J et al

Witsch K: see Stüben K et al

Wloka J: [1] Partielle Differentialgleichungen. Sobolevräume und Randwertaufgaben. Teubner, Stuttgart 1982

Wolff H: [1] Multi-grid techniek voor het oplossen van Fredholm-integraalvergelijkingen van de tweede soort. Report NN 19/79, Mathematisch Centrum, Amsterdam 1979

Wong YS: see Zang TA et al

Yosida K: [1] Functional analysis. Springer, Berlin 1968

Young DM: [1] Iterative methods for solving partial differential equations of elliptic type. Doctoral thesis, Harvard University 1950

Young DM: [2] Iterative solution of large linear systems. Acad Press, New York 1971

Young DM: [3] On the accelerated SSOR method for solving large linear systems. Adv in Math 23 (1977) 215–271

Yserentant H: [1] Die Mehrstellenformeln für den Laplaceoperator. Numer Math 34 (1980) 171–187

Yserentant H: [2] On the convergence of multi-level methods for strongly nonuniform families of grid and any number of smoothing steps per level. Computing 30 (1983) 305–313

Zang TA, YS Wong, and MY Hussaini: [1] Spectral multi-grid methods for elliptic equations. J Comput Phys 48 (1982) 485–501

Zeeuw PM de: see De Zeeuw PM

Zenger C and H Gietl: [1] Improved difference schemes for the Dirichlet problem of Poisson's equation in the neighbourhood of corners. Numer Math 30 (1978) 315–332

Zienckiewicz OC: [1] The finite element method in engineering science. McGraw-Hill, New York 1977 – German translation: Methode der finiten Elemente. Carl Hanser Verlag, München 1984

Zlámal M: [1] Discretization and error estimates for elliptic boundary value problems of the fourth order. SIAM J Numer Anal 4 (1967) 626–639

Subject Index

Main references in *italics*